THEORY OF RELATIONS

Revised Edition

STUDIES IN LOGIC

AND

THE FOUNDATIONS OF MATHEMATICS

VOLUME 145

ELSEVIER

AMSTERDAM • LONDON • NEW YORK • OXFORD • PARIS • SHANNON • TOKYO

THEORY OF RELATIONS

Revised Edition

Roland FRAÏSSÉ
Institut de Mathématiques de Luminy
Marseille, France

with an Appendix by

Norbert SAUER
University of Calgary, Calgary, Canada

2000

ELSEVIER

AMSTERDAM • LONDON • NEW YORK • OXFORD • PARIS • SHANNON • TOKYO

ELSEVIER SCIENCE B.V.
Sara Burgerhartstraat 25
P.O. Box 211, 1000 AE Amsterdam, The Netherlands

First edition 2000

Library of Congress Cataloging in Publication Data
A catalog record from the Library of Congress has been applied for.

ISSN: 0049-237X
ISBN: 0-444-50542-3

⊗ The paper used in this publication meets the requirements of ANSI/NISO Z39.48-1992 (Permanence of Paper).

Printed in The Netherlands

Contents

Introduction

In this "Introduction", **we will mention in bold face each notion or statement which did not already appear in** "Theory of Relations 1986", denoted ToR-86.

Relation theory goes back to the 1940's. It originates in the theory of order types, which is mainly due to HAUSDORFF (Grundzuge der Mengenlehre 1914, reference [107] in our Bibliography), to SIERPINSKI (Leçons sur les nombres transfinis 1928, taken up again in Cardinal and ordinal numbers 1958, bibl. reference [230]) to [240] SZPILRAJN (Sur l'extension de l'ordre partiel 1930), to [45] DUSHNIK, MILLER (Concerning similarity transformations of linearly ordered sets 1940), to [95] GLEYZAL (Order types and structure of orders 1940), and to [109] HESSENBERG (Grundbegriffe der Mengenlehre 1906, introducing the negative and rational ordinals). At that time, relation theory only extended the elementary notions of order type and embeddability to arbitrary relations.

Relation theory intersects only weakly with graph theory, with which it is sometimes still confused. Firstly, techniques in relation theory only rarely distinguish between graphs, i.e. symmetric binary relations, and relations of arbitrary arity. In addition, as opposed to graph theory, in relation theory one generally considers equally the two truth values (+) and (-) taken on by a relation with base E for each element of E^2 (or of E^n for the arity n). Yet an important exception is the study of augmented binary relations, especially augmented posets: see 2.9; also 6.5 and 9.9.

Secondly, relation theory uses techniques especially from combinatorics. Anything concerning relations with finite bases, or counting isomorphism types of finite restrictions of a given relation, or again the study of permutations of the base which preserves a given relation (i.e. automorphisms of the relation), makes use of combinatorics. From a more technical viewpoint, see incidence matrix (section 3.4), combinatorial lemmas (3.5), profile of a relation (3.6).

As for mathematical logic, its intersection with relation theory is rather important. One can even say that the two principal sources for relation theory are the study of order types, already mentioned, and the study of universal formulas (prenex formulas only having universal quantifiers) and boolean connections thereof, with the particular case of quantifier-free formulas. From a semantic, or model-theoretic viewpoint, this is the study of TARSKI-VAUGHT universal classes (5.10) and of their boolean combinations.

If one presents mathematical logic from a relational theoretic viewpoint, the basic notion is that of local isomorphism, i.e. isomorphism of a restriction of the first relation onto a restriction of the second one: see 9.1. For example, the free interpretability of a relation S in another relation R with the same base, is algebraically defined by the condition that every local automorphism of R (local isomorphism from R into R) is also a local automorphism of S. Equivalently, free interpretability is logically defined by the existence of a quantifier-free formula which defines S in the structure of R. For example, if A is a chain, or total ordering, then the betweenness relation "z is between x and y" is defined by the quantifier-free formula $(Axz \wedge Azy) \vee (Ayz \wedge Azx)$.

More generally TARSKI's elementary equivalence between relations R and S, saying that R and S satisfy the same first-order formulas with equality, is equivalent to say that R, S are (k,p)-equivalent for all integers k, p, which is a purely algebraic notion: **see back-and-forth notions**, section 10.10.

Coming back to the universal case and its boolean combinations, as common notions and techniques in both mathematical logic and relation theory, we have those of 1-isomorphism, 1-extension, projection filter (a variant of ultraproduct): see 10.1. And for each ordinal α, the α-morphism (10.4), which is not exactly the translation of a logical notion, but is indispensable in relation theory for the study of embeddability.

From the 1960's, an important connection appears between relation theory and the theory of permutations. See the study of orbits (section 12.3), the theorem on the increasing number of orbits (LIVINGSTONE, WAGNER 12.4) and the theorem on set-transitive (alias homogeneous) groups (CAMERON, 13.8). **Recent aspects of this study is considerably developed by** SAUER in the Appendix.

Let us also mention, from the 1970's, some unexpected connections between relation theory and topology (sections 6.6 and 6.7); and even connections with linear algebra (n-adherence, 12.3).

We shall now briefly present the principal notions studied, by mentioning first that chapters 1 through 8 mainly concern the theory of partial and total orderings (posets and chains), while chapters 9 through 13 and the Appendix mainly concern the general study of relations.

In chapter 1, we review basic set theoretical results, in general without proofs, which enable the reader to know, for instance, in which precise sense we use the notion of finite set (TARSKI's sense rather than DEDEKIND's), or the notion of cardinality of a set (in SCOTT's sense). This allows us to precise, throughout the rest of the book, which axioms are used for each proof: ZF alone, the axiom of choice, dependent choice, the ultrafilter axiom, the continuum hypothesis, etc.

Moreover it seems that even among logicians, a few of them are aware that, while $\omega_1 > \omega$ is provable in ZF alone, yet the countable axiom of choice, for instance, is used to prove that ω_1 is regular. Or that KÖNIG's theorem (see 1.1.9), even in the very particular case of two ordered pairs of sets, is not provable in ZF alone. Or that the possible equivalence between the axiom of choice and the statement that the range of a function is subpotent with its domain, is still an open problem, already put forth by [218] RUBIN 1963. Thus this chapter could be useful as a memory brush-up for the axiomatic set theoretician.

In chapter 2, we introduce some notions which are no longer classical, yet which extend well-known concepts. For example the coherence lemma (see 2.4), a not well-known version of the ultrafilter axiom. Another example is the cofinality of a poset, as well as the related notion of cofinal height (2.12). Classically, the notion of cofinality is confined to the single case of chains, or total orderings, which is too much restrictive, though interesting.

In 2.14 we mention **the computation of finite posets (up to isomorphism) until cardinal 13** (CHAUNIER, LYGERŌS). For the cardinal 14, it presently remains only one unknown value: the number of posets with 47 ordered pairs (the total number of pairs being $14.13/2 = 91$).

In 2.15 a subtle exercise communicated by MILNER and POUZET concerns **posets and their linear augmentations (alias linear extensions)**.

In chapter 3, we present RAMSEY's theorem and some of its important refinements, due to GALVIN and to NASH-WILLIAMS. Furthermore, the "initial interval theorem" or GALVIN's theorem is presented twice, with very different proofs: POUZET's proof in section 3.2 and LOPEZ' proof using the Ramsey sequence of conditions, in 3.7. Then we are led to the partition theorems of DUSHNIK, MILLER and of ERDÖS, RADO.

We also present a combinatorial study of the incidence matrix, with the linear independence lemma due to KANTOR, combinatorial lemmas and colors (3.4 and 3.5) leading to the profile increase theorem (POUZET, 3.6). This was already in ToR-86, but not the counterexample of STANTON: **a non-unimodal profile of an equivalence relaton on 24 elements**.

Let us mention the curious notion of **polychromatic Ramsey number** which apparently goes back to GALVIN and to Denis DEVLIN in the case of denumerable order-types (see 3.8.4 and 5.12.2).

In chapter 4, we begin a serious study of posets, with the notion of good and bad sequence, the study of finitely free posets and well partial orderings. We present HIGMAN's characterization of a well partial ordering (the set of initial intervals is well-founded under inclusion, 4.4). Also RADO's well partial ordering. Then HIGMAN's theorem on words in a well partial ordering (see 4.5). Let us also mention the BONNET's countable set of ideals (4.7), the direct product of posets (4.8), dimension (4.9), DE JONGH-PARIKH's theorem and the extraction property (4.11). Finally in 4.13 comes POUZET's theorem "every directed well partial ordering has a cofinal restriction isomorphic with a direct product of regular alephs".

In chapter 5, we consider embeddability between posets and between relations, the existence of immediate extensions (HAGENDORF 5.1), Cantorian theorem for posets (DILWORTH, GLEASON 5.2), the well partial ordering of finite trees (KRUSKAL 5.4). Then the existence of strictly decreasing infinite sequences of chains of reals: the denumerable sequences due to DUSHNIK, MILLER and the continuum length sequence due to SIERPINSKI (5.5). We briefly study SUSLIN's chain and tree in connection with SUSLIN hypothesis (5.8); then ARONSZAJN tree and SPECKER chain (5.9). Finally we shortly study universal classes (TARSKI,

VAUGHT 5.10). After KUNEN and Arnold MILLER (5.11), **we construct an ω_1-strictly decreasing sequence of denumerable posets**, with a method going back to DU BOIS-REYMOND. Finally in 5.12.2 **Ramsey theory is extended to infinite order-types with respect to embeddability** (ERDÖS, RADO, GALVIN, DEVLIN).

In chapter 6, we introduce the scattered chain, which does not admit any embedding of the chain **Q** of the rationals, and the classical HAUSDORFF decomposition. We introduce the indecomposable, as well as the right and the left indecomposable chain (section 6.3). We present HAGENDORF's theorem of unique sum-decomposition of an indecomposable chain (6.3.4) and some connected results (JULLIEN, LARSON). We begin to study the covering of a chain by right or left indecomposable intervals, or by doublets of indecomposable intervals (6.4). In section 6.5, we present the augmentation of a scattered poset by a scattered chain (BONNET, POUZET).

In 6.6 comes a powerful and nice tool: the simple convergence topology, firstly on sets of integers, then on initial intervals of a given poset. In 6.7 comes the important notion of a topologically scattered poset. **An unsufficient proof (of a true statement), page 184 lines -15 to -5 in ToR-86 is replaced by the theorem 6.7.4: equivalence between a finitely free scattered poset and a topologically scattered poset. The theorem uses the poset of antichains $\mathcal{A}(X)$ associated with any poset X; also the thickness of any finitely free poset** (thickness of X equals the height of $\mathcal{A}(X)$).

Then we shortly study indivisible chains (6.8). **Several refinements: age-indivisible, S-indivisible, edge-indivisible relational systems**, are studied by SAUER in Appendix.

In 6.9 is introduced the rigid chain, introduced by BONNET **then studied by** BONNET, SHELAH 1985.

Exercise 6.10.4 says that every restriction of a topologically scattered poset is itself topologically scattered. Without the equivalence theorem 6.7.4, this needs an interesting result of PELCZYNSKI and SEMADENI 1959: **compact topologically scattered spaces are preserved under continuous images**. This is again a connection between relations and topology.

In chapter 7, we introduce the important notion of barrier due to NASH-WILLIAMS; the partition theorem (section 7.1), the existence of a minimal bad barrier sequence (7.2); the forerunner and successor barrier (7.3); the hereditarily indecomposable chain (7.4). These are the main tools in the proof of the fundamental theorem of LAVER: every set of scattered chains forms a well quasi-ordering under embeddability (7.5). In other words, there exists neither an infinite strictly decreasing sequence nor an infinite set of mutually incomparable scattered chains. LAVER proved even more, in extending his result to chains which are a countable union of scattered chains. However his proof has not yet been sufficiently simplified to be presented in a textbook of a reasonable size. In this chapter, we also study the better partial ordering (a notion due to NASH-WILLIAMS), both for its intrinsic interest and for its applications to chains (sections 7.6 and 7.7).

In chapter 8, we study faithful extensions between relations and between chains

(HAGENDORF, JULLIEN, 8.1 and 8.2). Then faithful infinite extensions with MALITZ' and LOPEZ' counterexamples (8.3).

Bivalent tableaux (8.4) were first studied by LOPEZ and by Claire RAUZY. Then following an idea of BONNET, POUZET and Claire RAUZY, **to each poset of height two is associated a finite set of equivalent tableaux giving a new look to the extensivity problem** (see 8.5, HAZIM-SHARIF).

The chapter ends in 8.6 with SZPILRAJN chains, studied by BONNET, JULLIEN, POUZET.

Curiously, the study of tableaux, born at the same time as theory of relations with apparently analogous difficulties, has not yet found its tools for becoming a substantial theory. It remains a wild space beside the cultivated gardens of relation theory. To our knowledge, nobody has solved the simple problem of extensive tableaux (8.4), neither positively nor by a counterexample, even in the particular case of two columns only. It seems that the official theory of bipartite graphs (which are exactly identical to bivalent tableaux) is sterile for our problem.

In chapter 9, we begin the general theory of relations, with the notion of local isomorphism, free interpretability and free operator (sections 9.1 to 9.3), which is the relationist version of a logical free formula, and links relation theory with logic. Then we study constant, chainable, monomorphic relations (9.4 to 9.6).

Then appear some connections between monomorphy and tournaments and JEAN's deep result (9.7.3): for a binary relation with cardinality $p \geq 5$, the $(p-2)$-monomorphy implies the $(p-1)$-monomorphy. A connection is mentioned with DEMBOWSKI's 2-homogeneous permutation groups, also called 2-set-transitive groups.

In 9.8 **we introduce the general notion of interval modulo a relation (= strong interval when applied to posets) and use it to expose in** 9.9 **an important Hausdorff-construction: any finitely free, scattered poset can be built from well partial orderings by (1) inversion, (2) lexicographic sum, (3) augmentation**: see ABRAHAM, BONNET, 1999.

In chapter 10, we define the age of a relation (set of its finite restrictions up to isomorphism). We construct a rich denumerable n-ary relation, which embeds every denumerable n-ary relation (10.3).

Then in 10.4 we define and study the α-morphism which allows useful classifications among relations (α-older or younger, non-embeddability rank, non-richness rank). From THOMASSÉ 1997 **we know the existence of ω_1 non-embeddability ranks** if we start from a countable relation with arity at least equal to 2.

We study relations which are rich for their age (10.5) or minimal for their age (10.7); also inexhaustible relations (10.6).

Then we study finitist and almost chainable relations (10.8 and 10.9). Finitist structures are used in exercise 10.10.1 to present a simple Fraenkel-Mostowski model of set theory with negation of the axiom of choice, and an example, due to HODGES, of a function f with a domain strictly subpotent with its range.

Finally in 10.10 we quickly expose **some back-and-forth notions:** (k,p)**isomorphism, equivalence, operator. The classical elementary extension**

is translated in terms of (k,p)-isomorphism. A comparison is made between the classical elementary formula and the corresponding (k,p)-operator (with interesting exceptions in the case of the empty base).

In chapter 11 we introduce relative restrictions and ages, maximalist (= existentially closed) relations (11.2), saturated relations (11.3) relations leading to the existence criterion of POUZET, VAUGHT (11.4).

Then we go to THOMASSÉ's notions of **solid and fragil family, relation and morphism (a subtle refinement of the α-morphism already exposed) giving a new, intuitive proof of the indivisibility theorem** (see 11.6.4).

Then in 11.7 and 11.8 we expose the notions of **interval-ultrafilter, interval-closure, the non-classical, and especially the normal ultraproduct and ultrapower; also the useful definition of an elementary extensive interval-closure** (due to ILLE 1990). These notions are used in 11.9 to expose **three theorems on interval-closures**, again due to ILLE.

Chapter 12 is concerned with correspondence between relation theory and permutations, the link between them being the homogeneous relations and relational systems, the notion of orbit, the amalgamation theorem, the adherence of a permutation group (12.1 to 12.3). Let us remember the (already mentioned) theorem of increasing number of orbits, due to LIVINGSTONE, WAGNER (12.4). The chapter ends with the study of pre-homogeneous relations (12.7) and the corresponding existence criterion (12.9).

In chapter 13, we introduce the bounds of a relation R: finite relations non-embeddable in R but whose proper restrictions are embeddable in R. We present the well relation (13.2), and the existence of finitely many bounds for every chainable relation (FRASNAY). The proofs are simplified and the results are extended by POUZET using the powerful tool of p-well multirelation in 13.2.1. Then we present the compatibility of chains modulo a permutation group, the important group-compatibility theorem and chainability theorem, both due to FRASNAY (13.3). Then the notions of dilated and contracted group (13.4 and 5), the indicative group and the indicator leading to FRASNAY's reduction theorem (13.6). The particular case of **Q**-indicative groups leads to the set-transitive group theorem of CAMERON (13.8). Note that these notions lead to a marginal study inside permutation group theory, with many open problems. In 13.10 are studied reduction, compatibility, and monomorphy thresholds.

Section 13.11 is added **in order to extend to any relation the usual notion of adjacent elements modulo a chain, to expose the adjacence lemma and finally get the exact calculation of the reduction threshold** (confirmed by FRASNAY's examples). All that goes back to HODGES, LACHLAN, SHELAH 1977, yet was unsufficiently known by the author of ToR-86.

The appendix written by Norbert SAUER is to a certain degree a continuation of chapter 12: indeed there was a need to expose recent results on amalgamable ages and homogeneous relations and relational systems. In A.1 the reader will find **the Rado graph and some characterizations of homogeneous systems due to** CHERLIN, GARDINER, LACHLAN, SCHMERL, WOODROW. In A.2, **various types of amalgamation, bounds, orbits, universal homo-**

geneous relations. In A.3, a study of **inexhaustible homogeneous systems and ages**. Section A.4 concerns mainly **age-indivisibility and orbits**. Finally A.5 concerns **coloring copies of relational systems, edge-indivisibility, partitioning pairs of rationals leading to** DEVLIN's **formula**.

Let us mention that several paragraphs of ToR-86 have not been maintained in the present book. Let us give some examples with a short explanation.

On page 27 of ToR-86 an exercise about Alexandroff-Fodor theorem seems unsufficiently connected with relation theory in its present state. Idem page 126 an exercise on Mac-Neille completion. Idem page 200 about Krasner's lemma on directed posets. Idem pages 243-244 where the notion of dimensional arity, although it seems nice, yet did not lead to any substantial study, to our knowledge. Idem pages 310-311 the construction of a rich relation in the uncountable case. Idem chapter 12 paragraph 2 (pages 357-359) about preservation of bounds when going from a relation towards some of its extensions.

In order to keep this book to a reasonable size, we suppressed planned chapter(s) concerning the celebrated problem of reconstruction, i.e. the problem to know in what cases a relation with base E is completely determined, up to isomorphism, by the isomorphism types of its restrictions to proper subsets of E. This problem goes back to [247] ULAM 1960. For the strict "ulamian" case which considers restrictions to the base minus one element, the reader may consult [237] STOCKMEYER 1977, [16] J.A. BONDY 1991. For other kinds of restrictions, see [198] POUZET 1979, LOPEZ, RAUZY 1992, [22] BOUDABBOUS, LOPEZ 1995, [121] ILLE, RAMPON 1998 (see Bibliography).

I would like to thank those among my colleagues - professors, researchers, ex-students - who solved or contributed to the solution of all the problems presented here; and those who, by simplifying the inordinately long or difficult proof of the original paper, have made these results accessible, hence suitable for presentation in this textbook. Their names are mentioned together with their contribution.

This book would not exist without an amiable welcome by Professors Jean-Paul BRASSELET and François BLANCHARD, successive Directors of the "Institut de Mathématiques de Luminy" (Centre National de la Recherche Scientifique at Marseille) and Professor Jean-Yves GIRARD, Research Director. Thank also to my colleague Pierre ILLE who kindly shared his office with me, and to my colleagues Julien CASSAIGNE and Laurent REGNIER who enabled me to learn and to utilize the now indispensable LaTeX-code.

Chapter 1

Review of axiomatic set theory, ordinal, cardinal, aleph, relation, poset, chain

The purpose of this chapter is to situate precisely "theory of relations" within the framework of axiomatic set theory, which initially will be that of ZERMELO-FRAENKEL. The axioms of ZF are introduced below in sections 1.1 and 1.2. Our initial notations will be introduced there. In referring to the first and sometimes second chapter, we will indicate throughout the book which statements require only the axioms of ZF and those which require, to our knowledge, the axiom of choice, or rather the weaker ultrafilter axiom (boolean prime ideal axiom), or the axiom of dependent choice, etc. Most of the proofs, as well as classical definitions from the first and second chapter, are left to the reader.

1.1 First group of axioms for ZF; finite set, axiom of choice, König's theorem

We begin with the axioms of **extensionality, pair, union, power set** and the **scheme of separation**, all supposed known to the reader and going back to [252] ZERMELO 1908 (see Bibliography). **We denote the empty set by** 0, **inclusion** $\subseteq$, **strict inclusion** $\subset$. **We denote the union of the set** a **by** $\cup a$ (set of all elements of elements of a) and **the power set by** $\mathcal{P}(a)$ (set of all subsets of a). If $b \subseteq a$, we designate the **difference** by $a - b$. Singletons, unordered pairs (simply called **pairs**) are denoted by $\{a\}$, $\{a, b\}$, etc. Note that $\cup\{a\} = a$, that $\cup\ \{a, b\}$ is usually denoted $a \cup b$, etc.

The successor set $a \cup \{a\}$ **of** a **is denoted by** $a+1$. So that $1 = 0+1 = \{0\}$ is the successor of the empty set; $2 = 1 + 1 = \{0, 1\}$ is the successor of 1, etc. This notation coincides with the notation for ordinal addition, introduced in 1.3 below.

1.1.1 Finite set: Tarski's definition

Following [241] TARSKI 1924, we define a set a to be **finite** iff every non-empty set b of subsets of a contains an element which is minimal with respect to inclusion, i.e. an element $c \in b$ such that no $x \in b$ satisfies $x \subset c$. Taking complements, it is equivalent to say that a is finite exactly when every non-empty set of subsets of a contains a maximal element. A non-finite set is said to be infinite.

The empty set, a singleton, a pair are finite sets. Every subset of a finite set is finite. If a is finite, then so is the set composed of a together with an additional element. In particular, the successor $a + 1$ of a is finite. If a and b are both finite, so is their union.

If a set a and all its elements are finite, then the union $\cup a$ is finite. Indeed there exists at least a subset b of a such that $\cup b$ is finite (singletons for example); take a maximal b and then prove that $b = a$. This is often expressed in the following form called **pigeonhole principle**: if we partition an infinite set into finitely many subsets, then at least one of these subsets is infinite.

Scheme of induction for finite sets. If a condition C is true for the empty set, and if for every set a satisfying C and every set u, the set $a \cup \{u\}$ satisfies C, then C is true for every finite set.

1.1.2 Couple (= ordered pair), cartesian product, function

Given two sets a, b , the **couple** or ordered pair (a, b) is the set $\{\{a\}, \{a, b\}\}$ formed of the singleton $\{a\}$ and the (unordered) pair $\{a, b\}$. This definition goes back to KURATOWSKI 1921 (see also AJDUKIEWICZ). The set a is said to be the **first term** and b the **second term of the couple**. Clearly two couples are equal iff they have the same first and the same second terms.

The **cartesian product** $a \times b$ is the set of couples (x, y) where x belongs to a and y belongs to b.

A **function** or **mapping** from a onto b is a subset f of $a \times b$ such that every element x of a appears as first term in exactly one couple (x, y) belonging to f and every element y of b appears as a second term in at least one couple belonging to f. The set $a = \operatorname{Dom} f$ is called the **domain**, the set $b = \operatorname{Rng} f$ is the **range** of f. For each element x of a, the second term y of the unique couple (x, y) having first term x is denoted $y = f(x)$ or $y = fx$ and is called the **value** of f on x, or the **image** of x under f. For every superset $c \supset \operatorname{Rng} f$ we say that f is a function from a into c.

The transformation $\overline{f}$ and its inverse. If $u \subseteq \operatorname{Dom} f$, we denote by $\overline{f}(u)$ the set of elements fx where $x \in u$. The function thus denoted $\overline{f}$ is a function on the set of subsets of Dom f and is called the **transformation associated with f**. This transformation preserves inclusion, in the sense that $u \subseteq v$ implies $\overline{f}(u) \subseteq \overline{f}(v)$. However strict inclusion is not preserved.

If $v \subseteq \operatorname{Rng} f$, then the inverse image of v by f, denoted $\overline{f^{-1}}(v)$, is the set of elements x such that fx belongs to v. So we define the inverse transformation associated with f, denoted $\overline{f^{-1}}$. It preserves strict inclusion as well as inclusion.

The function f is said to be an **injection** or **injective function** iff $x \neq x'$ implies $fx \neq fx'$ for all x, x' in Dom f. If a is the domain, b the range, then an injection is said to be a **bijection** from a onto b. **The inverse of an injection** f is denoted by f^{-1}, so that in the case of f injective, the transformation associated with f^{-1} coincides with $\overline{f^{-1}}$ (the latter exists for every function f).

Given a function f, injective or not, if Dom f is finite, then Rng f is finite. For f injective, the converse is true.

A **permutation** of a is a bijection from a onto a. Given two elements x, y of a, the **transposition** $(x, y)_a$ is the permutation of a which interchanges x and y and is the identity on every other element of a.

1.1.3 Fixed point lemma

Let a be a set and h a function which takes each subset x of a to a subset hx of a. **Suppose that** h **is increasing under inclusion**: $x \subseteq y$ implies that $hx \subseteq hy$ for every $x, y \subseteq a$; then:

(1) **there exist sets** $x \subseteq a$ **majorized by** h , **in the sense that** $x \subseteq hx$; for example x can be taken as the empty set;

(2) **if** x **is majorized by** h , **then** hx **is majorized by** h ;

(3) **the union** u **of all majorized subsets satisfies** $hu = u$.

Bibliography: [136] KNASTER 1928, generalized by [244] TARSKI 1955.

1.1.4 Restriction, extension, composition, equipotence and Dedekind-finite set

Given a function f with domain a and a subset b of a , we call the **restriction** of f to b , denoted f/b , the set of ordered pairs belonging to f of which the first term belongs to b . Putting $g = f/b$, we say that f is an **extension** of g to the domain a .

We leave it to the reader to define the **composition** $g \circ f$ of the functions f and g , with $\mathrm{Dom}(g \circ f) = \overline{f^{-1}}(\mathrm{Dom}\, g \cap \mathrm{Rng}\, f)$ (where $\cap$ denotes the intersection).

A set b is said to be **equipotent** with a iff there exists a bijection of a onto b.

A set b is said to be **subpotent** with a iff there exists a subset of a equipotent with b . A set b is **strictly subpotent** with a iff b is subpotent with a but a is not subpotent with b. By BERNSTEIN-SCHRÖDER theorem 1.1.5 below, this is equivalent to saying that b is subpotent but not equipotent with a.

Every set equipotent with a finite set is itself finite. Every finite set is strictly subpotent with every infinite set. Two finite sets are always comparable, one being subpotent with the other.

If a and b are finite, then the cartesian product $a \times b$ is finite. If a is finite, then so is the power set $\mathcal{P}(a)$.

A finite set is not equipotent with any of its proper subsets.

Equivalently: if a **is finite then every injection of** a **into** a **is a permutation of** a.

• Suppose that f is an injection satisfying $\overline{f}(a) \subset a$. Take a subset m of a which is minimal among all subsets x of a satisfying $\overline{f}(x) \subset x$. Then $\overline{f}(\overline{f}(m)) \subset \overline{f}(m)$ by the injectivity of f: this contradicts the minimality. •

A set a is said to be **Dedekind-finite** iff a is not equipotent with any proper subset of itself (see [38] DEDEKIND 1888); it is **Dedekind-infinite** in the opposite case. Every finite set is Dedekind-finite. The converse will be proved in 1.2.6 by using the denumerable subset axiom (weaker than the axiom of choice).

1.1.5 Bernstein-Schröder theorem

Given sets a and b, if a is subpotent with b and b subpotent with a, then a is equipotent with b. The following proof is in [68] FRAENKEL 1953 and attributed to WHITAKER. It does not use the notion of integer, which is used in the classical "mirror proof"; see also [238] SUPPES 1960.

• Let f be an injection from a into b, and g be an injection from b into a. It suffices to find a subset u of a such that $b - \overline{f}(u)$ is sent to $a - u$ by the function $\overline{g}$, or equivalently $u = a - \overline{g}(b - \overline{f}(u))$. To do this, consider the function which takes each subset x of a into $a - \overline{g}(b - \overline{f}(x))$. This function is increasing under inclusion. By the fixed-point lemma 1.1.3, the union u of all x such that $x \subseteq a - \overline{g}(b - \overline{f}(x))$ satisfies the above. •

1.1.6 Cantor's lemma and theorem

Lemma. Let a be a set. There is no function, injective or otherwise, with domain a and range $\mathcal{P}(a)$ (set of subsets of a). Use the diagonal argument: let f be a mapping from a onto $\mathcal{P}(a)$; take the set of elements $x \in a$ such that $x \notin fx$.

Theorem. (1) Every set a is strictly subpotent with $\mathcal{P}(a)$.

(2) **If a is non-empty, then every set of mutually disjoint subsets of a is strictly subpotent with $\mathcal{P}(a)$.**

1.1.7 Exponential

Given sets a and b, the **exponential** or **power** ${}^{a}b$ is the set of functions from a into b. Thus ${}^{0}b = \{0\} = 1$ for each b. In particular ${}^{0}0 = 1$. However ${}^{a}0 = 0$ for each non-empty set a. For each set a, the set $\mathcal{P}(a)$ of subsets of a is equipotent with ${}^{a}2$, where $2 = \{0, 1\}$.

We have the following equipotences. For b and c disjoint, ${}^{(b \cup c)}a$ is equipotent with the cartesian product $({}^{b}a) \times ({}^{c}a)$. The set ${}^{c}(a \times b)$ is equipotent with the product $({}^{c}a) \times ({}^{c}b)$. Finally ${}^{c}({}^{b}a)$ is equipotent with ${}^{(b \times c)}a$.

1.1.8 Choice set and choice function

Let a be a set of non-empty mutually disjoint sets x. A **choice set** for a is a set whose intersection with each element x of a is a singleton. If a is finite, then there is a choice set for a (proof by induction).

Let a be a set of non-empty sets x. A **choice function** for a is a function f which to every element x of a associates an element fx of x . If a is finite, then there is a choice function for a .

Axiom of choice. Every set, even infinite, of non-empty mutually disjoint sets admits a choice set. Equivalently every set of non-empty sets admits a choice function. ([251] ZERMELO 1904)

An immediate consequence of the axiom of choice is the following. Given a function f, injective or otherwise, Rng f **is subpotent with** Dom f. In other words, given a non-empty set a, every set of mutually disjoints subsets of a is subpotent with a.

Problem. Does the preceding statement imply the axiom of choice (problem mentioned in [218] RUBIN 1963 p. 5 note 1).

A seemingly weaker consequence of the axiom of choice is the assertion that Dom f **is never strictly subpotent with** Rng f. This does not follow from ZF alone, i.e. from the axioms mentioned in 1.2.4 below. See 10.11.1, where a FRAENKEL-MOSTOWSKI model is constructed with Dom f strictly subpotent with Rng f, a result which is transferable to ZF via the theorem of JECH-SOCHOR (observation due to HODGES).

1.1.9 Generalized cartesian product, König's theorem

Let a be a non-empty set whose elements are non-empty. The (generalized) **cartesian product of** a is the set of choice functions which, to each element x of a associate an element of x.

If a reduces to the pair $\{b, c\}$, we have again the cartesian product $b \times c$ of 1.1.2.

If a is infinite, it follows from the axiom of choice that the cartesian product of a is non-empty.

König's theorem. Let I be a non-empty set of elements i (called indices), to each of which is associated a couple of sets a_i, b_i with a_i strictly subpotent with b_i. Then the union of the $a_i (i \in I)$ is strictly subpotent with the cartesian product of the b_i (axiom of choice is used).

• Suppose there exists a bijection h from the union A of a_i's onto the product B of the b_i's. For each i and each $x \in a_i$, take the function $hx \in B$ and take its value $(hx)(i)$ which belongs to b_i. Thus we define a function from a_i into b_i. By the axiom of choice, the range of this function is subpotent with a_i, thus strictly subpotent with b_i. Hence there is an element $u_i \in b_i$ which is not the value $hx(i)$ for any $x \in a_i$. The choice function which to each i associates u_i is not in the range B of h: contradiction. We leave it to the reader to see that the union of the a_i's is subpotent with the product of the b_i's. •

Problem. Can the above theorem be proved from only the axioms of ZF (defined in 1.2.4 below) in the case where the set I of indices is finite with cardinality greater than or equal to 2. Note that if, in addition to I being finite, we have for each index i that $\mathcal{P}(a_i)$ is subpotent with b_i, then by CANTOR's lemma 1.1.6, the axioms of ZF suffice for the proof.

For Card $I = 2$, KÖNIG's theorem is a consequence of ZF plus the axiom which asserts that Dom f is never strictly subpotent with Rng f, or of ZF plus the apparently weaker axiom which asserts that if a (resp. a') is strictly subpotent with b (resp. b') and b, b' disjoint, then $a \cup a'$ is strictly subpotent with $b \cup b'$.

1.2 Second group of axioms: foundation, infinity, substitution; ordinal, integer, denumerable set

Axiom of foundation. Every non-empty set a admits an element disjoint from a.

It follows that $x \notin x$ for any x . Moreover for any x, y it is impossible that $x \in y$ and $y \in x$, etc. The axiom of foundation was introduced by [253] ZERMELO 1930, inspired by a statement of [180] von NEUMANN 1929. As to its consistency, supposing that all other axioms of ZF are consistent, see below 1.9.1.

Given a set a, the successor $a + 1 = a \cup \{a\}$ is distinct from a (since $a \in a$ is impossible).

Moreover if $a + 1 = b + 1$ then $a = b$; otherwise we would have $a \in b + 1$ with $a \neq b$, so $a \in b$ and similarly $b \in a$, contradicting the axiom of foundation.

Given a set c, the set whose successor is c (which is unique if it exists) is called the **predecessor** of c, denoted by $c - 1$.

Given a set a and its successor $a + 1$, there is no intermediary set x such that $a \in x \in a + 1$.

1.2.1 Transitive set, ordinal

A set a is **transitive** iff, for every x, y, conditions $y \in x \in a$ imply $y \in a$. If a is transitive and non-empty, then every element of a is a proper subset of a. Also 0 $\in a$; note that 0 is the only element of a which is disjoint from a.

Every union and intersection of transitive sets is transitive.

If a is transitive, then so is $a + 1$.

A set a is **totally ordered** (by membership relation) iff, for every $x, y \in a$, either $x \in y$ or $y \in x$ or $x = y$. For example all singletons are totally ordered. However the singleton of 1, i.e. $\{1\} = \{\{0\}\}$ is not transitive. The set $\{0, 1, \{1\}\}$ is transitive but not totally ordered. The set $\{0, \{1\}\}$ is neither transitive nor totally ordered.

Every intersection of totally ordered sets is totally ordered.

A union of such sets is not necessarily totally ordered; however if the set of totally ordered sets is directed under inclusion (i.e. any two such sets are included in a third such set), then the union is totally ordered. Finally if a is totally ordered by membership, then so is $a + 1$.

An **ordinal** is a transitive set which is totally ordered by membership. For example 0, $1 = \{0\}$, $2 = \{0,1\}$.

Every element of an ordinal is an ordinal.

The successor set of an ordinal is an ordinal. The predecessor (if it exists) is an ordinal.

The intersection of any set of ordinals is an ordinal.

An ordinal a is said to be **less than or equal** to an ordinal b, denoted $a \leq b$, iff $a \in b$ or $a = b$; an ordinal a is **strictly less** than b, denoted $a < b$, iff $a \in b$. Hence $<$ is synonymous with $\in$ between ordinals.

If $a \leq b+1$, then $a \leq b$ or $a = b+1$.

Similarly $\geq$ (greater than or equal to) and $>$ (strictly greater than) are defined.

Given two ordinals a and b, the condition $a \in b$ (or $a < b$) is equivalent to strict inclusion $a \subset b$. Hence $a \leq b$ is equivalent to $a \subseteq b$.

• By transitivity $a \in b$ implies $a \subset b$.

Conversely suppose that $a \subset b$. Let $d \in b - a$ be an element disjoint from $b - a$. As $d \in b$, this d is an ordinal and $d \subset b$. Also $d \subseteq a$ since d is disjoint from $b - a$. So either $d = a$ (yielding $a \in b$), or $d \subset a$. If the latter occurs, let $u \in a - d$ so that $u \in a \subset b$. As b is an ordinal and $u \in b$ and $d \in b$, we have either $u \in d$ or $d \in u$ or $u = d$. If $u \in d$, this contradicts $u \in a - d$. If $d \in u$, then since $u \in a$ we have $d \in a$ which contradicts $d \in b - a$. If $u = d$, then $d \in a - d$ so $d \in a$, again contradicting $d \in b - a$. •

Trichotomy lemma. Given any two ordinals a, b, either $a \in b$ or $b \in a$ or $a = b$.

• As we know, the intersection $a \cap b$ is an ordinal. Either $a \cap b = a$ or $a \cap b = b$ or $a \cap b$ is strictly included in both a and b. In the first case $a \subseteq b$ so $a = b$ or $a \subset b$ and thus $a \in b$. A similar conclusion is reached in the second case. In the third case we have $a \cap b \in a$ and $a \cap b \in b$, so that $a \cap b$ belongs to itself, contradicting the axiom of foundation. •

We leave it to the reader to define the **maximum** and the **minimum** ordinal of a set of ordinals, denoted Max, Min.

Every non-empty set u of ordinals admits a minimum m: take m belonging to u and disjoint from u.

Scheme of the minimum ordinal. More generally, given a condition C which is satisfied by at least one ordinal, there is a minimum ordinal satisfying C.

Every transitive set of ordinals, every union of a set of ordinals is an ordinal.

We define an **upper bound** α of a set u of ordinals as any ordinal α greater than or equal to each element of u. The **supremum** denoted $\text{Sup}(u)$ is the least upper bound of u, or smallest ordinal $\geq$ each element of u.

It is easily seen that $\text{Sup}(u)$ **equals the union** $\cup u$.

If α is an ordinal and u is a set of ordinals such that $\beta \in u$ implies $\beta \leq a$, then $\text{Sup}\ u \leq \alpha$. In other words if $\alpha < \text{Sup}\ u$, then there exists an ordinal β in u with $\beta > \alpha$.

1.2.2 Ordinal-indexed sequence, α-sequence; extracted sequence

Given an ordinal α, an **α-sequence** or **ordinal-indexed sequence** is a function with domain α. In this case α is called the **length** of the sequence.

Given a sequence u, the elements or **terms** of u are all couples (i, u_i) for which the first term i is any ordinal strictly less than α. The i's are called **indices** of u, or u is **indexed** by $i < \alpha$. The second terms of the couples (which are arbitrary sets) are called the **values** of u and denoted u_i or $u(i)$.

In the particular case of an α-sequence with ordinal values, we leave it to the reader to define **increasing, decreasing, strictly increasing and strictly decreasing** sequences.

Given an ordinal α and an α-sequence u , we define an **extracted sequence** from u to be a sequence v with length $\beta \leq \alpha$, obtained by composition of u with h , where h is a strictly increasing β-sequence with values in α. So $v = u \circ h$ and $v_i = u_{h(i)}$ for each $i < \beta$.

The notion of extracted sequence is reflexive and transitive, but not antisymmetric. For instance, by the axiom of infinity introduced in 1.2.4 below, given two distinct elements a, b, we define the ω-sequences $a, b, a, b, ..$ and $b, a, b, a, ..$, each extracted from the other.

1.2.3 Integer, n-element set, word, n-tuple

By **non-negative integer**, or **integer**, or **natural number**, we mean a finite ordinal. Every element of an integer is an integer. Every non-zero (i.e. non-empty) integer has an integer predecessor.

If a is an integer and $b \in a$ (or $b < a$), then b is strictly included in a. As a is finite, b is strictly subpotent with a . Thus equipotent integers are identical.

Scheme of induction. If a condition C holds for 0 , and if for each integer a the condition $C(a)$ implies $C(a+1)$, then C holds for every integer.

Every finite set a is equipotent with an integer, which is called the **cardinal** or **cardinality** of a and denoted Card a.

A set equipotent with the integer n is called an **n-element set**.

A finite sequence or **word** is an **n-sequence**, where n is an integer; such a word of length n is called an **n-tuple**.

When restricted to words, the notion of extracted sequence becomes antisymmetric; i.e. two words each of which is extracted from the other are identical.

1.2.4 Axiom of infinity, set of integers; successor or limit ordinal; axioms of ZF

The simplest version of the **axiom of infinity** should assert the existence of an infinite set. **A more useful and stronger version is generally admitted: it asserts the existence of a Dedekind-infinite set; more precisely the**

existence of a set a which contains the element 0 and such that if x belongs to a, then the successor $x + 1 = x \cup \{x\}$ belongs to a.

Using the separation scheme, the axiom of infinity yields the existence of the **set of integers**, denoted by ω. The set ω is an infinite ordinal, the smallest non-zero ordinal without a predecessor.

The ordinal α is called a **successor ordinal** iff there exists a predecessor β with $\alpha = \beta + 1$. Otherwise it is called a **limit ordinal**. Examples: 0 and ω are limit ordinals.

Substitution scheme. At this point we replace the separation scheme by the more general substitution scheme, due to [67] FRAENKEL 1925; the reader is assumed to be familiar of it.

Now we can define, for example, $\omega + \omega = \omega.2$: beginning with the set ω of integers, associate to each integer i the ordinal $\omega + i$ defined below in section 1.3.1. Then using the substitution scheme define the set of $(\omega + i)$'s as i runs through ω.

Another example: denote by $\mathbf{N}_0 = \omega$ the set of integers, and for each integer i let $\mathbf{N}_{i+1} = \mathcal{P}(\mathbf{N}_i)$ (the set of subsets of $\mathbf{N}_i$). The substitution scheme allows one to define the set of $\mathbf{N}_i$ for i running through ω.

Axioms of ZF. The axioms previously introduced in the first lines of 1.1, where the separation scheme is replaced by the stronger substitution scheme, plus the axiom of foundation (first lines of 1.2) plus the axiom of infinity just described, are called **the axioms of ZF** (ZERMELO completed by FRAENKEL).

If no special assumption is explicitely mentioned in a theorem, lemma or proposition, then this indicates that our proof needs ZF alone. If the axiom of choice or other supplementary axioms (for the most part weakened versions of the axiom of choice, stated below) are used, then we clearly indicate such.

Recall that the axiom of choice has been proved consistent with ZF (if ZF itself is consistent) by GÖDEL in 1938 (see reference [98] of 1940). The negation of the axiom of choice has been proved consistent with ZF by COHEN in 1963 (reference [35] of 1966).

1.2.5 Denumerable set, countable set, countable axiom of choice

A set is said to be **denumerable** (resp. **countable**) if it is equipotent (resp. subpotent) with ω, the set of integers. ZF alone suffices to show that the union of two denumerable sets, the cartesian product of two denumerable sets, and the set of all finite subsets of ω are all denumerable. Following 1.2.2, we call an ω-**sequence** a sequence of length ω, hence indexed by the set of integers.

The **countable axiom of choice** is a particular case of the axiom of choice. It states that **for every countable set of non-empty disjoint sets, there is a choice set**. This axiom is strictly weaker than the axiom of choice; i.e. if ZF is consistent, then there is a model of ZF and countable choice which satisfies the negation of the general axiom of choice ([125] JECH 1973).

The countable axiom of choice implies that every denumerable union of denumerable sets is denumerable. Indeed this axiom allows one to choose, for each of the denumerable sets in the union, a bijection from that set onto the integers. On the other hand, the above statement is not provable from ZF alone: there is a model of ZF in which the continuum is a denumerable union of denumerable sets (Azriel LEVY, unpublished).

1.2.6 Denumerable subset axiom

This axiom states that every infinite set has a denumerable subset. It follows from the countable axiom of choice.

• Let a be an infinite set. For each integer i, associate the set of i-tuples of elements from a. By countable choice, we can associate to each integer i one of these i-tuples. It remains to take the ω-sequence formed from the terms of the chosen 1-tuple, 2-tuple, ... •

The denumerable subset axiom is strictly weaker than the countable choice ([125] JECH 1973).

Let a to be an infinite set; the following three conditions are equivalent:

(1) there exists a denumerable subset of a;

(2) there exists a bijection of a onto a proper subset of a; in other words, a is Dedekind-infinite (see 1.1.4);

(3) there exists a choice function f which to each finite subset x of a associates an element fx in the complement $a - x$.

Consequently, the denumerable subset axiom is equivalent to saying that finiteness coincides with Dedekind-finiteness. However with ZF alone, there can exist an infinite set having for each integer i a subset equipotent with i, yet having no denumerable subset.

Decreasing membership sequence. Having defined the integers, we can now complete our initial remarks about the axiom of foundation, by saying that there are no decreasing ω-sequences u with $u_{i+1} \in u_i$ for each integer i. In particular for every integer r there is no cycle $u_1 \in u_2 \in \ldots \in u_r \in u_1$.

1.2.7 Axiom scheme of foundation

The axiom of foundation is equivalent to the following axiom scheme.

Let C be a condition which holds for 0 and such that, if C holds for each element of a given set a, then C holds for a. Under these hypotheses, C holds for every set. Note that we can eliminate the hypothesis " C holds for 0 "; this being a particular case of the second hypothesis, made precise as follows: either there exists an element of a satisfying "not C", or a satisfies C.

• Let a be a set which falsifies the axiom of foundation: then a is non-empty and $0 \notin a$. Let C be the condition holding for every set which does not belong to a: in particular $C(0)$. Then C satisfies our hypotheses, but C does not hold for every element of a.

Conversely, let C be a condition satisfying our hypotheses, and let a be a set satisfying "not C". Thus a is non-empty. Let $a_1 \subseteq a$ be the (non-empty) set of elements of a satisfying "not C ". Let $a_2 \subseteq \cup a$ be the (non-empty) set of elements of $\cup a_1$ satisfying "not C ". Let $a_3 \subseteq \cup \cup a$ be the (non-empty) set of elements of $\cup a_2$ satisfying "not C ", etc. Then the union of the a_i's (i integer) falsifies the axiom of foundation. •

1.2.8 Hereditarily transitive set

A set a is said to be **hereditarily transitive** iff for every finite sequence $a_0, a_1, ..., a_n$ with $a_0 = a$ and $a_{i+1} \in a_i$ for $i < n$, then a_n is transitive.

A set is hereditarily transitive iff it is an ordinal. Use 1.2.1: every transitive set of ordinals is an ordinal. Equivalently a is an ordinal iff a and all elements of a are transitiive (see for instance [209] POWELL 1975 p. 223). Analogously , we leave it to the reader to prove that a set a is an integer iff either a is empty, or a is a successor set, and every element of a is either empty or a successor set (communicated by HATCHER in 1977).

1.2.9 Axiom of choice for finite sets

This is an important weakening of the axiom of choice, which asserts **the existence of a choice set for every non-empty, finite, mutually disjoint sets**. This weakened form is not implied by and does not imply the countable axiom of choice (1.2.5), nor the denumerable subset axiom (1.2.6). Even the axiom of dependent choice, which is stronger than countable choice, does not imply choice among finite sets: see 1.8.

1.2.10 Induction

We shall call **induction**, or **transfinite induction**, the following reasoning. Suppose that if a condition C holds for every ordinal strictly less than α then C holds for α; under this hypothesis, C holds for every ordinal. This is a form of the scheme stated in 1.2.1: if "not C" is satisfied by at least one ordinal, there is a least ordinal satisfying "not C ".

Often, induction is broken up into a proof for 0, a proof for the transition between an arbitrary ordinal α and its successor, and a proof for α a limit ordinal.

A definition by recursion is made by introducing a statement which uniquely associates a set a to each ordinal α. This statement will usually be of the following form:

there exists one and only one function f with domain $\alpha + 1$ (the successor of α), such that the initial couple $(0,u)$ belongs to f (where u is arbitrarily given), the final couple (α, a) belongs to f , and such that for each $\beta \leq \alpha$ the couple (β, b) belongs to f, provided that b has been obtained in a certain (suitably defined) manner from the set of couples belonging to f with first term $< \beta$.

Because of the uniqueness of f, when $\alpha' > \alpha$, the function f' corresponding to α' will be an extension of f to the domain $\alpha' + 1$.

Some examples of definition by recursion: sum, product, exponentiation for ordinals; aleph rank in 1.6.6. The definition of fundamental rank in 1.4.2 is also by recursion, if one begins by associating to each ordinal α the set of all sets with fundamental rank α.

Note that definition by recursion using the axiom of ZF is easier to justify than definition by simple recursion in first-order Peano arithmetic, such as is generally presented today (however, the original text of [182] PEANO 1894 is written in second-order logic). In order to justify definition by recursion in first-order arithmetic, one is led to use the "chinese remainder theorem". For instance, one defines $b = a!$ as an abreviation for the following: "there exists two integers u, v such that the remainder after division of u by $v + 1$ is 1; the remainder after division of u by $(a + 1)v + 1$ is b; and for each $i(1 \leq i \leq a)$ one obtains the remainder after division of u by $(i + 1)v + 1$ from the remainder after division of u by $iv + 1$ by multiplying the latter by i ". This procedure goes back to [97] GÖDEL 1931; the reader can consult the english translation by [37] DAVIS 1965, p.31 lemma 1.

1.3 Review of ordinal algebra, Cantor normal form, indecomposable ordinal

1.3.1 Sum

We say that $\alpha+\beta = \gamma$ (where α, β, γ are ordinals) iff there exists a function f with domain $\beta + 1$ (hence for each $u \leq \beta$ there is one and only one couple (= ordered pair) belonging to f with first term u), such that the initial couple $(0, \alpha)$ and final pair (β, γ) belong to f; if (u, v) belongs to f where $u < \beta$, then $(u + 1, v + 1)$ belongs to f; and finally such that if f contains couples (x, y) for which the first term x admit a supremum $(\text{Sup } x) \leq \beta$, then the couple $(\text{Sup}\, x, \text{Sup}\, y)$ belongs to f .

Given α and β , the reader can prove by induction on β the existence and uniqueness of the preceding function, hence the existence and uniqueness of the ordinal $\gamma = \alpha + \beta$. In the same manner, one proves for every α, β the equalities $\alpha+0 = 0+\alpha = \alpha$ and $\alpha+(\beta+1) = (\alpha+\beta)+1$, and for every α and every set of ordinals u the supremum equality $\alpha + (\text{Sup } u) = \text{Sup}(\alpha + u)$.

For all α, $\beta \neq 0$ we have $\alpha+\beta > \alpha$. For all α, β we have $\alpha + \beta \geq \beta$ where equality is possible with non-zero α : for instance $1 + \omega = \omega$. The supremum equality does not hold on the left: if i is an arbitrary integer then $\text{Sup}(i+\omega) = \omega \neq (\text{Sup } i) + \omega = \omega + \omega$.

Ordinal addition is associative. Commutativity holds for integers, or finite ordinals; however $1+\omega = \omega \neq \omega + 1$. For all α and $\beta \geq \alpha$, there exists one and only one γ satisfying $\alpha + \gamma = \beta$; this γ is called the difference $\beta - \alpha$.

The inequality $\beta \leq \gamma$ implies $\alpha + \beta \leq \alpha + \gamma$ and conversely. Also the same result for strict inequality $<$ Hence addition is left cancellable, i.e. $\alpha + \beta = \alpha + \gamma$

implies $\beta = \gamma$.

The inequality $\alpha \leq \beta$ implies $\alpha + \gamma \leq \beta + \gamma$. This does not hold in general for $<$, as $0 + \omega = 1 + \omega$. Hence addition is not right cancellable. Finally the ordinal 1 and consequently every finite ordinal is **absorbed** by every infinite ordinal, in the sense that $1+\alpha = \alpha$ for α infinite.

1.3.2 Product

We say that $\alpha.\beta = \gamma$ iff there exists a function f with domain $\beta+1$, such that the initial couple (0,0) and the final couple (β, γ) belong to f , and such that if $(u, v) \in f$ where $u < \beta$ then $(u + 1, v + \alpha) \in f$, and such that if $(x, y) \in f$ for all x belonging to a set which admits a supremum (Sup x) $\leq \beta$ then (Sup x, Sup y) $\in f$.

For all α, β we have $\alpha.0 = 0.\alpha = 0$ and $\alpha.(\beta + 1) = \alpha.\beta + \alpha$. For every α and every set of ordinals u we have the supremum equality $\alpha.(\text{Sup } u) = \text{Sup}(\alpha.u)$. Moreover $\alpha.\beta = 0$ is equivalent to $\alpha = 0$ or $\beta = 0$.

Ordinals of the form $\alpha.u$, with α fixed and u an arbitrary ordinal, are called the **multiples** of α. For example 0 is a multiple of every ordinal; α is a multiple of itself.

Every multiple of α, augmented by α, yields a multiple of α. The supremum of a set of multiples of α is a multiple of α. Finally every multiple of α is obtained from 0 by these two indicated processes. More rigorously if a condition is true for 0 and is preserved in the passage from an ordinal u to $u + \alpha$ as well as in the passage to supremum, then the condition is true for every multiple of α.

The supremum equality on the right, given above, does not hold on the left: if i designates an arbitrary integer, then $\text{Sup}(i.2) = \omega \neq (\text{Sup}\, i).2 = \omega.2$.

Multiplication is associative and distributive on the right: $\gamma.(\alpha+\beta) = \gamma.\alpha+\gamma.\beta$. Distributivity on the left and commutativity hold for integers; however $(\omega+1).\omega = \omega.\omega \neq \omega.\omega + 1.\omega$ and $2.\omega = \omega \neq \omega.2$. It can happen that $\alpha.\beta$ is not a multiple of β, e.g. $(\omega + 1).2 = \omega.2 + 1$: it is not a multiple of 2.

For $\alpha \neq 0$ the inequality $\beta \leq \gamma$ implies $\alpha.\beta \leq \alpha.\gamma$ and conversely. The same result holds for strict inequality. Thus multiplication is cancellable on the left except for 0; i.e. for α non-zero $\alpha.\beta = \alpha.\gamma$ implies $\beta = \gamma$.

The inequality $\alpha \leq \beta$ implies $\alpha.\gamma \leq \beta.\gamma$. This does not subsist for $<$ since $1.\omega = 2.\omega = \omega$. Multiplication is thus not cancellable on the right.

Given two ordinals α and $\beta \neq 0$, there is a unique γ called the **quotient**, and a unique ordinal ϵ called the **remainder** in the **division** of α by β , with $\alpha = \beta\gamma + \varepsilon$ and $\varepsilon < \beta$. Consequence of the supremum equality: there exists a maximum ordinal u such that $\beta.u \leq \alpha$.

1.3.3 Power or exponentiation

We say that $\alpha^\beta = \gamma$ iff there exists a function f with domain $\beta + 1$, such that the initial couple (0,1) and the final couple (β, γ) belong to f, and such that if

$(u, v) \in f$ where $u < \beta$ then $(u + 1, v.\alpha) \in f$, and such that if $(x, y) \in f$ for x belonging to a set which admits a supremum $\operatorname{Sup} x \leq \beta$, then $(\operatorname{Sup} x, \operatorname{Sup} y) \in f$.

For all α, β we have $\alpha^0 = 1$ and $\alpha^{(\beta+1)} = \alpha^\beta.\alpha$. For all α and every set of ordinals u, we have the supremum equality $\alpha^{(\operatorname{Sup} u)} = \operatorname{Sup}(\alpha^u)$.

Moreover $\alpha^\beta = 0$ is equivalent with $\alpha = 0$ and $\beta \neq 0$. Also $\alpha^\beta = 1$ is equivalent to $\alpha = 1$ or $\beta = 0$.

Ordinals of the form α^u, with α fixed and u an arbitrary ordinal, are called **powers** of α. For example 1 is a power of every α . If v is a power of α then so is $v.\alpha$. The supremum of a set of powers of α is a power of α. Finally, every power of α is obtained from 1 by these two processes.

The supremum equality on the right given above does not subsist on the left: if i designates an arbitrary integer, then $\operatorname{Sup}(i^2) = \omega \neq (\operatorname{Sup}\ i)^2 = \omega^2$.

We have $\alpha^{(\beta+\gamma)} = \alpha^\beta.\alpha^\gamma$ for all α, β, γ . It is impossible in general to interchange terms in the exponent; for example $2^{(\omega+1)} = \omega.2 \neq 2^1.2^\omega = \omega$.

We have $(\alpha^\beta)^\gamma = \alpha^{(\beta.\gamma)}$ for all α, β, γ. It is impossible in general to interchange terms of the product in the exponent: $(2^\omega)^2 = \omega^2 \neq 2^{(2.\omega)} = \omega$.

The equality $(a^c).(b^c) = (a.b)^c$ which holds for integers, does not subsist in general, even for finite exponents: $(\omega^2).2^2 = \omega^2.4 \neq (\omega.2)^2 = \omega^2.2$. The equality does not subsist for a, b finite and an infinite exponent: $(2^\omega).(2^\omega) = \omega^2 \neq 4^\omega = \omega$.

For $\alpha \geq 2$, the inequality $\beta \leq \gamma$ implies $\alpha.\beta \leq \alpha.\gamma$ and conversely; same result with $<$. Thus we have cancellation: $\alpha.\beta = \alpha.\gamma$ implies $\beta = \gamma$.

The functions $2^\alpha, \omega^\alpha$ etc. are strictly increasing. Moreover $\alpha \leq 2^\alpha \leq \omega^\alpha$ (proof by induction using the supremum equality). The inequality $\alpha \leq \beta$ implies $\alpha^\gamma \leq \beta^\gamma$; this does not subsist for $<$ since $2^\omega = 3^\omega = \omega$.

Finally **for $\alpha < \beta$ the ordinal ω^α is absorbed by** ω^β, i.e. $\omega^\alpha + \omega^\beta = \omega^\beta$.

From the precedings we see that 2^α, ω^α etc. are strictly increasing functions, and that $\alpha \leq 2^\alpha \leq \omega^\alpha$ (proof by induction using the supremum equality).

We can have $\alpha = \omega^\alpha$ already with a denumerable α called **the epsilon number**: consider the sequence $a(0) = \omega$, $a(1) = \omega^{a(0)} = \omega^\omega$, ... , $a(i+1) = \omega^{a(i)}$ then take $\operatorname{Sup} a(i)$ for all integers i.

1.3.4 Maximum exponent

Given two ordinals α and $\beta \geq 2$, there is a unique ordinal u which is the maximum exponent satisfying $\beta^u \leq \alpha$: consequence of the supremum equality.

Given α and $\beta \geq 2$ and the maximum exponent γ such that $\beta^\gamma \leq \alpha$, there exists a maximum ordinal δ such that $(\beta^\gamma.\delta \leq \alpha$. Moreover $\delta < \beta$. Note that δ is the quotient in the division of α by β^γ. Given β, γ and a strictly decreasing, thus finite sequence of ordinals $\gamma > \gamma(1) > \gamma(2) > ...$ and a corresponding sequence of ordinals $\delta(1), \delta(2)$,... each strictly less than β, we have $\beta^\gamma > \beta^{\gamma(1)}.\delta(1) + \beta^{\gamma(2)}.\delta(2) + ...$ (proof by induction on γ).

1.3.5 Cantor normal form

Given α and $\beta \geq 2$, there exists a decomposition of α into a finite sum of terms $\beta^{\gamma}.\delta$, with coefficients $\delta < \beta$ and exponents γ strictly decreasing. Furthermore this decomposition is unique. It is called the **Cantor decomposition** of α into powers of β or **Cantor normal form** of α in **base** β.

In the most usual case of the base ω, the coefficients δ are integers.

1.3.6 Decomposable or indecomposable ordinal

An ordinal α is called **decomposable** iff there exist $\beta < \alpha$ and $\gamma < \alpha$ with $\alpha = \beta + \gamma$; otherwise α is called **indecomposable**. If α is indecomposable, then every sum of two non-zero ordinals which is equal to α has second term equal to α, and conversely.

A non-zero ordinal α is indecomposable iff α is a power of ω. This follows from the existence and uniqueness of the Cantor decomposition into powers of ω, together with the absorption statement (end of 1.3.3).

1.4 Transitive closure, hereditarily finite set, fundamental rank, cardinal

1.4.1 Transitive closure, hereditarily finite set

For every set a, there exist transitive supersets of a, and among these there exists one which is included in all the others. This set is formed from the values of all finite sequences $x_1, ..., x_h (h$ integer) such that $x_1 \in a$ and $x_{i+1} \in x_i$ for each $i (1 \leq i < h)$. We shall call this set the **transitive closure** of a.

For each non-empty set a , the transitive closure of a is the union of a together with the transitive closures of the elements of a.

If $a \subseteq b$ then (Closure of a) $\subseteq$ (Closure of b).

Hereditarily finite set. This is a set whose transitive closure is finite. Examples: every finite transitive set; every integer (i.e. every finite ordinal). The singleton $\{1\} = \{\{0\}\}$ is non-transitive yet hereditarily finite.

Every hereditarily finite set is finite. Every element and every subset of a hereditarily finite set is hereditarily finite. Every finite set of hereditarily finite sets is hereditarily finite. Similarly for finite unions, finite cartesian products, and the power set of hereditarily finite sets.

A necessary and sufficient condition for a set a to be hereditarily finite is that, for every finite sequence $x_0, ..., x_h (h$ integer) with $x_0 = a$ and $x_{i+1} \in x_i$ for each $i < h$, the terms x_i are finite.

1.4.2 Fundamental rank

Let a be a set and c be the transitive closure of the singleton $\{a\}$. We say that the ordinal α is the **fundamental rank** of a, iff there exists a function f with domain c , taking ordinal values $\leq \alpha$, such that the initial ordered pair (0,0) and the final ordered pair (a, α) belong to f: so that $f(0) = 0$ and $f(a) = \alpha$; and such that if $u \in c$ then the value $f(u)$ is the smallest ordinal strictly greater than $f(x)$ for all x belonging to u.

It follows from the axiom of foundation that **every set has a unique fundamental rank**. Indeed, the empty set 0 has rank 0. Suppose that a is non-empty and that every element of a has a rank. Then by the preceding definition, a has rank equal to the smallest ordinal which is strictly greater than the ranks of all its elements. The existence of rank results from the axiom scheme of foundation 1.2.7.

For every ordinal α the fundamental rank is α itself.

1.4.3 Set of all sets with a given rank

For every ordinal α, there is a set $\mathbf{V}_\alpha$ of all sets with ranks $< \alpha$. Moreover $\mathbf{V}_\alpha$ has fundamental rank α.

• This is obvious for 0 since $\mathbf{V}_0$ is empty. If this is true for α, then it is true for $\alpha + 1$ with $\mathbf{V}_{\alpha+1}$ = set of elements and subsets of $\mathbf{V}_\alpha$. Finally for α a limit ordinal, $\mathbf{V}_\alpha$ is the union of the $\mathbf{V}_i$ for i strictly less than α. •

Note that for each ordinal α, the set $\mathbf{V}_{\alpha+1} - \mathbf{V}_\alpha$ of sets of rank α is non-empty, since $\mathbf{V}_\alpha$ and α belong to this set.

For i an integer, or finite ordinal, the set of sets of rank i is finite. It follows that every infinite set has rank at least equal to ω.

Note that a set is hereditarily finite iff its fundamental rank is finite. The set of hereditarily finite sets is the intersection of all sets which contain 0 and which, if they contain x and y , also contain $x \cup \{y\}$ as an element.

1.4.4 Cardinal (or cardinality)

Given a set a, consider sets equipotent with a and among these, those of minimum fundamental rank. By the preceding, these form a non-empty set which has been used by [225] SCOTT 1955 in order to define cardinals. For the **cardinal** or **cardinality** of a given set a, denoted Card a, we adopt the following artificial but general and rigorous definition:

If a is equipotent with an ordinal, then the smallest such is Card a; otherwise Card a is the set of all sets of minimum fundamental rank which are equipotent with a.

Thus **every set has a cardinal, and two sets are equipotent iff they have the same cardinal**.

Note that if there is no ordinal equipotent with a, then a is equipotent, not to (Card a), but to an arbitrary element of (Card a). This is only a minor inconvenience.

Given two cardinals a and b, the ordering of less than or equal to, or greater than or equal to, denoted $a \leq b$, means that every set of cardinal a is subpotent with every set of cardinal b. Obvious definition of strict ordering; notation $a < b$.

1.4.5 Cardinal sum, cardinal product and exponentiation

Let a and b be cardinals; the **cardinal sum** $a \uplus b$ is defined as the cardinal of the union of two disjoint sets of cardinal a, respectively b. This notation allows a clear distinction between ordinal sum $+$ and cardinal sum $\uplus$ composed of the $+$ and the symbol "union", remembering that we take the union of two disjoint sets with given cardinals.

The cardinal $a \uplus b$ does not depend upon the choice of disjoint sets of cardinal a and cardinal b.

Cardinal addition is commutative and associative. We have $a \uplus 0 = a$. Finally $a \leq a'$ and $b \leq b'$ imply $a \uplus b \leq a' \uplus b'$.

The **cardinal product** $a \times b$ is defined as the cardinal of the cartesian product of a set of cardinal a with a set of cardinal b. There will be no inconvenience in using the same symbol for cardinal multiplication and for the cartesian product of two sets. The cardinal product does not depend upon the choice of the sets of cardinal a, resp. b. Example $\omega \times \omega = \omega$.

Cardinal multiplication is commutative, associative, and distributive over cardinal addition: $(a \uplus b) \times c = (a \times c) \uplus (b \times c)$. We have $a \times 0 = 0$ and $a \times 1 = a$. Finally $a \leq a'$ and $b \leq b'$ imply $a \times b \leq a' \times b'$.

The **cardinal power or exponentiation** ${}^a b$ is defined as the cardinal of the power between sets of cardinal a, resp. b (notation from 1.1.7). Cardinal exponentiation does not depend upon the choice of the sets of cardinal a, resp. b.

We have ${}^0 a = 1$, and ${}^1 a = a$, and ${}^a 1 = 1$ for all a; also ${}^a 0 = 0$ for $a \neq 0$. For $b \neq 0$, conditions $a \leq a'$ and $b \leq b'$ imply ${}^a b \leq {}^{(a')} b'$.

Finally the equipotences indicated in 1.1.7 become cardinal equalities: ${}^{(b \uplus c)} a = ({}^b a) \times ({}^c a)$ and ${}^c (a \times b) = ({}^c a) \times ({}^c b)$ and ${}^c ({}^b a) = {}^{(b \times c)} a$.

Since cardinal operations are denoted by $a \uplus b$ and $a \times b$ and ${}^a b$, there will be no confusion with ordinal operations $a+b$ and $a.b$ and b^a. In particular the traditional cardinal notation $\aleph_\alpha$ becomes useless: we shall replace it by ω_α. Nevertheless the word "aleph" will be saved and currently used: it means the cardinal of any well-orderable set; see 1.6.3.

Following our definition of cardinal (preceding subsection), we identify, for instance, $\mathrm{Card}\, \omega$ with ω itself, and write the following cardinal equality $\omega \uplus 1 = \omega \uplus \omega = \omega$, as well as the usual ordinal inequality $\omega + 1 > \omega$.

Note that we can use the same symbol $(=)$ for cardinal and ordinal equality, and the same symbols $(<, \leq$ for cardinal and ordinal comparisons. Indeed the signification of such a symbol is clearly decided by the presence, on one or both sides, either of an ordinal operator $+$, $.$, right exponent, or of a cardinal operator $\uplus$, $\times$, left exponent, or an expression under the symbol Card.

1.5 Cardinality of the continuum; the axiom called continuum hypothesis

1.5.1 Cardinality of the continuum

A set is said to be **equipotent with the continuum** or equivalently to have the **cardinality of the continuum** iff it is equipotent with $\mathcal{P}(\omega)$, the power set of the set of integers, or equivalently with ${}^{\omega}2$, the set of functions on ω taking the values 0 or 1.

By Cantor's theorem 1.1.6, every countable set is strictly subpotent with $\mathcal{P}(\omega)$. By the preceding section, we have that ${}^{\omega}2 \times {}^{\omega}2 = {}^{(\omega \uplus \omega)}2 = {}^{\omega}2$ since $\omega \uplus \omega = \omega$ (cardinal sum). Hence the cartesian product of two sets each equipotent with the continuum is itself equipotent with the continuum. The same result holds for the cartesian product of a countable set with the continuum.

Similarly ${}^{\omega}({}^{\omega}2) = {}^{(\omega \times \omega)}2 = {}^{\omega}2$, since $\omega \times \omega = \omega$. Hence if a has the cardinality of the continuum, then the same is true for the set of ω-sequences with values in a.

1.5.2 Substraction of a countable set

If we substract an arbitrary denumerable subset a from a set c having the cardinality of the continuum, then the difference $c - a$ has the cardinality of the continuum. This is a special case of the following proposition.

Let a be an infinite set which is equipotent with the cartesian product $2{\times}a$, and let $c = \mathcal{P}(a)$. Then the difference set, obtained by removing from c an arbitrary subset which is equipotent with a , is equipotent with c.

• Since a is equipotent with $2{\times}a$, the set c , which is equipotent with ${}^{a}2$, is also equipotent with $c \times c$ by 1.1.7. Hence the difference of c and a subset which is equipotent with a is equipotent with the difference of $c \times c$ and the range of a bijection f with domain a. Each element x of a is associated to a couple $fx = (y, z)$ of elements y, z of c . Let us associate to each x the first term y of this pair. The function thus obtained has domain a and cannot have range $c = \mathcal{P}(a)$, by CANTOR's lemma 1.1.6.

Thus there exists an element u of c for which (u, z) is not the value by f of an element of a, for any z belonging to c. Hence the difference of $c \times c$ and $\overline{f}(a)$ includes a subset which is equipotent with c. Then use BERNSTEIN-SCHRŐDER theorem. •

1.5.3 Denumerable partition of the continuum

Let a be a set equipotent with the continuum. For every partition of a into denumerably many subsets, one of the subsets is equipotent with the continuum (uses the axiom of choice).

• Suppose on the contrary that there is a partition of a into disjoint subsets a_i (i integer), and that every a_i is strictly subpotent with a. Then by KŐNIG's theorem 1.1.9 (axiom of choice), the union a of the a_i's is strictly subpotent with the cartesian product of an ω-sequence of sets, each equipotent with the continuum. Yet this cartesian product is equipotent with the continuum: contradiction. •

1.5.4 Continuum hypothesis, generalized continuum hypothesis

The axiom called continuum hypothesis asserts the non-existenceof a set which is strictly intermediate, with respect to subpotence, between ω and $\mathcal{P}(\omega)$. This axiom is logically independent of ZF, and even of ZF plus the axiom of choice ([35] COHEN 1963, bibl. reference of 1966).

The axiom called generalized continuum hypothesis asserts the non-existence of a set strictly intermediate, with respect to subpotence, between a and $\mathcal{P}(a)$, for every infinite set a. When added to the axioms of ZF, this implies the axiom of choice (see 1.9.3 below).

1.6 Binary relation, quasi-ordering, poset, chain, well-ordering, aleph, Hartogs, aleph rank

1.6.1 Generalities, partial ordering = poset, chain, isomorphism

Let a be a set; a **binary relation** with **base** a, or based on a, is a set A of couples (x, y) with $x, y \in a$. The base is denoted $|A| = a$.

A binary relation A is said to be a **quasi-ordering** iff A is reflexive and transitive (these notions are assumed to be known). Instead of $(x, y) \in A$ we say that x is less than y, denoted by $x \leq y \bmod A$, or that y is greater than x , denoted $y \geq x \bmod A$.

An **equivalence relation** A is a symmetric quasi-ordering. Instead of $(x, y) \in A$ we say that x is equivalent to y , denoted $x \simeq y$ modulo A. As the reader already knows, the equivalence relation defines a partition of its base into disjoint equivalence classes.

Given a quasi-ordering A, the equivalence relation generated by A is the relation with same base, defined by $x \simeq y$ iff $x \leq y$ and $y \leq x(\bmod A)$. Thus for each element x the equivalence class of $x(\bmod A)$ is the set of all y such that $x \leq y \leq x(\bmod A)$. For example, subpotence defines a quasi-ordering among sets, and the generated equivalence relation is equipotence.

Classically we say that x is strictly less than y, denoted $x < y(\bmod A)$ iff $x \leq y$ but not $\geq y$; equivalently $y > x(\bmod A)$.

Given two elements x, y in the base, either $x \simeq y$ or $x < y$ or $x > y$ or these three conditions are falsified, in which case x and y are said to be **incomparable** modulo the quasi-ordering: denoted by $x|y(\bmod A)$.

Partial ordering or poset. This is an antisymmetrical quasi-ordering: for every x, y in the base of A, if $x \leq y$ and $y \leq x \bmod A$ then $x = y$. In other words, the equivalence generated by A reduces to the identity relation. Consequently the strict inequality $x < y$ means that $x \leq y$ and $\neq y$. Another consequence: given x, y four possibilities occur: either $x = y$ or $x < y$ or $x > y$ or finally $x|y$.

Example of a poset. Given a set a, any subset of the power set $\mathcal{P}(a)$ is partially ordered by inclusion.

Let A be a poset, D a subset of the base; we assume that the notion of **maximum** of $D(\bmod A)$, denoted $\operatorname{Max} D$, is known; similarly for the **minimum**, denoted $\operatorname{Min} D$. Recall that an element is said to be **maximal in** $D(\bmod A)$, iff it belongs to D and there is no element of D which is strictly greater. Analogous notion of a **minimal element**.

The maximum, if it exists, is maximal, but the converse is false; similarly for the minimum.

Example. Divisibility defines a partial ordering on positive integers. If the integer 1 is in the base, every integer is divisible by 1, so that 1 is the minimum. Now if 1 is excluded, there is no minimum, yet each prime integer is minimal modulo divisibility.

Start with a quasi-ordering A and take as a new base the set of equivalence classes, generated by A; then write (Equiv. class of x) $\leq$ (Equiv. class of y) iff $x \leq y(\bmod A)$. We thus obtain the partial ordering generated by the quasi-ordering A. For example, starting with subpotence, we generate the partial ordering of cardinals (which becomes a total ordering modulo the axiom of choice).

Chain or total ordering. A chain based on a is a poset with all elements mutually comparable: for any $x, y \in a$, either $x = y$ or $x < y$ or $x > y$ modulo the chain. Examples. The chain $\mathbf{N}$ based on integers (or natural numbers) with $x \leq y(\bmod \mathbf{N})$ iff $x \leq y$ as defined in 1.2.1, i.e. if $x \in y$ or $x = y$. The chain $\mathbf{Z}$ of the positive and negative integers with the usual comparison; also the chain $\mathbf{Q}$ of the rationals and the chain $\mathbf{R}$ of reals, introduced in 2.1 below.

Isomorphism, automorphism. Let A a binary relation based on a and A' based on a', and f a bijection from a onto a'. We say that f transforms or takes A into A', denoted $A' = \overline{f}(A)$ or that f is an isomorphism of A onto A', iff for any $x, y \in a$ the condition $(x, y) \in A$ is equivalent to $(fx, fy) \in A'$. An automorphism of A is an isomorphism from A onto A.

A binary relation A' is said to be **isomorphic** with A iff there exists an isomorphism from A onto A'. This condition is reflexive, symmetric and transitive, yielding an equivalence relation on every set of binary relations.

1.6.2 Well-ordering, well-orderable set, well-ordering axiom

We say that a chain A is a **well-ordering** iff every non-empty subset of the base has a minimum element $(\bmod A)$.

Examples. Every finite chain; the chain $\mathbf{N}$ of integers. Although one commonly confuses both notions, it is sometimes useful to precise that, given an ordinal α, the well-ordering canonically associated to α is the set of those couples (x, y) such that $x \in y \in \alpha$ or $x = y \in \alpha$. In particular, the chain **N** of integers is canonically associated with ω.

Given a well-ordering A, identity is the only automorphism of A. In other words two isomorphic ordinals are identical.

• Let f be an automorphism of A, with $f(i) \neq i$ for some i in the base $|A|$. We may assume $f(i) < i (\operatorname{mod} A)$ by eventually changing f into f^{-1}. Take the minimum element satisfying this inequality; then $f(f(i)) < f(i) (\operatorname{mod} A)$: contradiction. •

Given a well-ordering A, there exists one and only one ordinal which is isomorphic with A.

• Associate the ordinal 0 to the minimum element a_0 in A. Let i be a non-zero ordinal; assume that each ordinal $j < i$ has been associated to an element of the base $|A|$, but that there still remain elements to which no ordinal $< i$ has been associated. Then associate i to the minimum element among these, and call it a_i. We so define an isomorphism from the ordinal $\alpha = \operatorname{Sup}(i)$ onto the well-ordering A. Moreover α is unique, since identity is the only automorphism of an ordinal. •

The ordinal i so associated with a_i is called the **ordinal rank** of a_i (mod A). Equivalently, given an element x in a well-ordering A, the ordinal rank of x is the ordinal isomorphic with the initial interval of elements strictly less than $x \operatorname{mod} A$.

We say that a set a is **well-orderable** iff there exists a well-ordering on a.

A set a is well-orderable iff there exists a choice function on the set of non-empty subsets of a.

• Let f be a choice function on non-empty subsets of a. Let $a_0 = f(a)$. Let u be a non-zero ordinal, D_u the set of all $a_i (i < u)$. Let $a_u = f(a - D_u)$, as long as possible. Now the subsets D of a form a set, yet the ordinals do not; so we necessary reach a final ordinal u for which $D_u = E$.

Conversely, if we denote by A a well-ordering based on a, then to each non-empty subset b of a we associate $f(b) =$ the minimum element of $b \operatorname{mod} A$. •

By the above statement, the axiom of choice is equivalent to the **well-ordering axiom** which states that **every set is well-orderable**.

1.6.3 Aleph

In the case of a well-orderable set a, by 1.4.4 the cardinal of a is the smallest ordinal equipotent with a. Such a cardinal is called an **aleph**. In other words an aleph is an ordinal a which is equipotent with no ordinal $< a$ (less than with respect to the ordering of the ordinals). In particular, the finite alephs are the integers, the first infinite aleph is ω.

Note that if a and b are two equipotent ordinals, then every intermediate ordinal is equipotent to them (use BERNSTEIN-SCHRÖDER). Moreover, for any infinite ordinal a, the successor $a + 1$ is equipotent with a.

It follows that **every infinite aleph is a limit ordinal**.

1.6.4 Hartogs, trichotomy axiom for cardinals

Let a be an infinite set. we say that an ordinal u is **injectable** in a iff there exists an injection of u into a, or equivalently iff there exists a subset of a equipotent with u.

If u is injectable in a, then every ordinal $\leq u$ and every ordinal equipotent with u is injectable in a. Since a is infinite, every integer is injectable in a. However in order that ω be injectable in a, it is necessary that a be Dedekind-infinite: see 1.2.6.

Given a set a, the ordinals injectable in a form a set.

• For each ordinal u and each injection f of u into a, we associate to the couple (u, f) the set of couples (fx, fy) for which $x \leq y < u$. Such a set is a well-ordering. Each pair in a belongs to $\mathcal{P}(a)$; each couple belongs to $\mathcal{PP}(a)$; each relation, in particular each well-ordering on a belongs to $\mathcal{PPP}(a)$: by the axioms of ZF all those well-orderings form a set. Finally to each well-ordering corresponds one and only one ordinal u as a possible domain of f, since two isomorphic ordinals are identical. So that those ordinals which are injectable in a form a set, by using the substitution scheme: see 1.2.4. •

Since every ordinal equipotent with an injectable ordinal is itself injectable, the set of all ordinals injectable in a is an aleph, which we shall call the Hartogs aleph or simply the **Hartogs** of a. This is also the smallest ordinal which is not injectable in a ([106] HARTOGS 1915).

Trichotomy axiom for cardinals. The axiom of choice is equivalent to the trichotomy axiom which states that, given two cardinals a, b, either $a < b$ or $a = b$ or $a > b$.

• If every cardinal is an aleph, then trichotomy holds as an immediate consequence of ordinal trichotomy (1.2.1). Conversely, given a set a and the hartogs α of a, if trichotomy holds then necessarily a is subpotent to α, hence a is well-orderable. •

1.6.5 Successor aleph, limit aleph

Successor aleph (a_+). If a is itself an aleph, then the hartogs of a is the unique aleph immediately greater than a, in the sense that it is $\geq a$ and there is no strictly intermediate ordinal (with respect to subpotence) between a and its hartogs. We shall denote it by a_+ and call it the successor aleph of a .

For example $\omega_1 = \omega_+$ denotes **the successor aleph of** ω , and is the set of all countable ordinals; or again the least uncountable ordinal. Letting $\omega_0 = \omega$, for each integer i we let $\omega_{i+1} = (\omega_i)_+$.

Modulo the axiom of choice, every cardinal is an aleph, since every set is well-orderable. Hence for each cardinal a there exists a unique successor a_+ (immediately greater than a in the preceding sense: no strictly intermediate cardinal). Moreover every cardinal $> a$ is $\geq a_+$.

The axiom of choice is equivalent to the existence, for each cardinal a , of a successor in the stronger sense: a cardinal $a_+ > a$ such that every cardinal $> a$ is

$\geq a_+$; see [243] TARSKI 1954.

Limit aleph. The union or supremum ordinal of an arbitrary set of alephs is again an aleph. For example, from the preceding ω_i(i integer) we get $\omega_\omega = \text{Sup}(\omega_i)$, which is an aleph. We call a non-successor aleph, such as ω or ω_ω, a limit aleph.

1.6.6 Aleph rank

We generalize the preceding notation. Given an arbitrary ordinal u, for an infinite aleph a we write $a = \omega_u$ iff there exists a function f with domain $u + 1$ (the successor ordinal of u), such that the initial couple (0,ω) and the final couple (u, a) belong to f; and such that, if the couple (x, y) belongs to f with x an ordinal $< u$ and y an aleph, then $(x + 1, y_+)$ belongs to f; and finally such that if f contains a set of couples of the form (x, y) with Sup $x \leq u$ then the couple (Sup x, Sup y) belongs to f.

Thus for each ordinal u, there exists a unique aleph ω_u. Conversely for every infinite aleph a there exists a unique ordinal u such that $a = \omega_u$. We call u the **aleph rank** of a.

The ordinal inequality $u < v$ implies the cardinal inequality $\omega_u < \omega_v$.

We have $u \leq \omega_u$; **equality is possible**: consider the sequence $a(0) = \omega$, $a(1) = \omega_{a(0)} = \omega_\omega$, ... , $a(i + 1) = \omega_{a(i)}$, .. then take Sup $a(i)$ for all integers i.

For every ordinal u we have $\omega_{u+1} = (\omega_u)_+$, the successor aleph of ω_u. Moreover, for every set of ordinals x, we have the supremum equality: $\text{Sup}\,\omega_x = \omega_{(\text{Sup}\,x)}$.

Hence **an infinite aleph is a successor or a limit aleph, according to whether its aleph rank is a successor ordinal or limit ordinal** (including 0, since $\omega_0 = \omega$).

1.6.7 Concerning continuum hypothesis

In the presence of the axiom of choice, the continuum hypothesis is equivalent to the equality $^\omega 2 = \omega_1$.

However, **with the axioms of ZF alone in the absence of the axiom of choice, the preceding equality is, a priori, a stronger assertion than the continuum hypothesis**. Indeed, there may exist a model of ZF without choice, where there is no strictly intermediate set (with respect to subpotence) between ω and $^\omega 2$, yet where ω_1 is incomparable with $^\omega 2$. In other words, in such a model ω has two incomparable successors in the weak sense.

The situation is different with the generalized continuum hypothesis, which implies the axiom of choice (see 1.9.3). Thus the generalized continuum hypothesis implies the equality $^a 2 = a_+$ for each infinite aleph a. However, it seems possible to construct a model of ZF satisfying $^a 2 = a_+$ for each infinite aleph a and yet having non-aleph cardinals which are mutually incomparable: thus negating the axiom of choice.

1.6.8 Cardinal sum, product and exponentiation between alephs

(1) **Let a be an infinite aleph; then $a \uplus a = a \times a = a$.**

(2) **Let a, b be two alephs, at least one of which is infinite; then $a \uplus b = a \times b =$ Max(a, b) (for the product we assume that $a, b \neq 0$).**

It suffices to establish $a \times a = a$. Indeed $a \uplus a = a$ follows by BERNSTEIN-SCHRŐDER; by the same argument (2) follows from (1).

• To each couple of ordinals (α, β) we associate as follows the ordinal rank $C(\alpha, \beta)$. First we classify couples by increasing values of their maximum Max(α, β). Then inside each equivalence class, we order them lexicographically, by first difference. We get a well-ordering and denote by $C(\alpha, \beta)$ the ordinal rank of the couple (α, β).

For instance the unique couple (0,0) whose maximum is 0 has rank $C(0,0) = 0$. Then the three couples whose maximum is 1 have ranks $C(0,1) = 1, C(1,0) = 2, C(1,1) = 3$. Then the five couples whose maximum is 2 have ranks $C(0,2) = 4, ..., C(2,2) = 8$. In general if a couple (α', β') comes immediately after (α, β) then let $C(\alpha', \beta') = C(\alpha, \beta) +1$; if it comes immediately after the infinite sequence of the (α_i, β_i) then let $C(\alpha', \beta') = \operatorname{Sup} C(\alpha_i, \beta_i)$.

Note that Card $C(0,1) = 1{\times}1$; that Card $C(0,2) = 2{\times}2$; in general since $(0,\alpha)$ comes immediately after all couples (β, γ) with $\beta, \gamma < \alpha$, the cardinal of $C(0, \alpha)$ equals Card$(\alpha) \times$Card(α). Moreover we see by induction that $\alpha \leq C(0, \alpha)$; intuitively, it suffices to consider the increasing α-sequence $C(0,0) < C(0,1) < ... < C(0,i) < ..$ for all $i < \alpha$.

Everything comes down to proving that, for α infinite, the value $C(0, \alpha)$ is equipotent with α. This is true for ω since $C(0, \omega) = \omega$. Denote by α the smallest infinite ordinal which is strictly subpotent to $C(0, \alpha)$, thus to Card$(\alpha) \times$ Card(α). This α does not have any ordinal $\alpha' < \alpha$ which is equipotent with α. Indeed we would have α' equipotent with Card$(\alpha') \times$ Card(α'), hence α equipotent with Card$(\alpha) \times$ Card(α).

Thus α is necessarily an aleph. Denote by α_i the ordinals $< \alpha$. By hypothesis $C(0, \alpha_i)$ is equipotent with α_i, and so strictly subpotent to α. Thus $\operatorname{Sup} C(0, \alpha_i) \leq \alpha$. Consequently $C(0, \alpha) = \operatorname{Sup} C(0, \alpha_i) \leq \alpha$ strictly subpotent to $C(0, \alpha)$: contradiction. •

Let a be an infinite aleph and let $2 \leq b \leq a$; then ${}^a b = {}^a 2$. Indeed $a = a \times b$, so ${}^a 2 = {}^{(a \times b)} 2 = {}^a ({}^b 2) \geq {}^a b$.

Modulo the axiom of choice, this identity extends to arbitrary cardinals.

1.6.9 A classification of cardinals by their Hartogs

Given an infinite cardinal a, let α be the Hartogs of a.

Either $\alpha = \omega$; then a is infinite but not Dedekind-infinite (see 1.1.4). Then the cardinal sum $a + 1$ is immediately greater than a , in the weak sense: it is strictly greater (as a cardinal) and there is no set strictly intermediate with respect to subpotence.

Or $\alpha \geq \omega_1$. Then the cardinal sum $a + \alpha$ is immediately greater than a in the preceding weak sense (see [243] TARSKI 1954). Indeed a restriction of α either is isomorphic with α or with an ordinal β strictly less than α so that β is injectable into a . Then $a+$ Card$\beta = a$.

1.7 Relation, multirelation, arity, restriction, extension, isomorphism type

Let E be a set and n an integer. In 1.2.3 we defined the notion of n-**tuple** with values in E. We set aside two elements called **values**, which are denoted + and - (for instance, these can be defined by 0 and 1).

An n-**ary relation with base** E, or **based on** E, is a function R which associates the value $R(x_0, ..., x_{n-1}) = +$ or - to each n-tuple $(x_0, ..., x_{n-1})$ in E. By traditionalism, we often denote the n-tuple by its indices 1 to n instead of from 0 to $n - 1$.

The set E, the base of R, will be denoted $|R|$.

The integer n will be called the **arity** of R. For $n = 1, 2, 3$ we will say a **unary, binary, ternary relation**. In 1.6.1 we already introduced the binary relation with $(x_1, x_2) \in R$ or $\notin R$ instead of $R(x_1, x_2) = +$ or = - for every x_1, x_2 in the base. Both presentations are clearly equivalent.

For $n = 0$, we adopt the convention that there exist two 0-**ary relations** based on E, which we denote by $(E, +)$ and $(E, -)$: the 0-ary relations with value (+) and value (-).

We adopt the convention that, for each positive n, there exists a unique n-ary relation with empty base. However, there exist two 0-ary relations with empty base: (0,+) and (0,-). These conventions agree with the calculation of the number of n-tuples with values taken from a base of finite cardinal p; i.e. the number p^n

The number of n-ary relations based on p elements is (2 to the power p^n): for n, p finite, ordinal (p^n) and cardinal exponentiation ($^n p$) coincide.

Examples of relations. Several examples of binary relations are given in 1.6.1. A group is a ternary relation taking the value (+) when $x_1 \circ x_2 = x_3$ and the value (-) when $x_1 \circ x_2 \neq x_3$ where ($\circ$) is the composition law of the group. Instead of x_1, x_2, x_3 we shall often use x, y, z.

A **multirelation with base** E is a finite sequence R of relations $R_1, ..., R_h (h$ integer), each with base E. Each $R_i (i = 1, ..., h)$ is called a **component** of the multirelation R.

We call the **arity** of R the sequence $(n_1, ..., n_h)$ of arities of the components $R_1, ..., R_h$. We say then that the multirelation R is $(n_1, ..., n_h)$-**ary**. The length h of the sequence of indices can be zero: in this case the multirelation is reduced to its base E.

Instead of R_1, R_2, R_3, we shall often use R, S, T. In the case where $h = 2, 3, 4$, we will say resp. **a birelation, a trirelation, a quadrirelation**. Finally, the base of a multirelation R shall be denoted $|R|$.

Example. An ordered group is a (3,2)-ary birelation which is formed of the ternary group relation and the binary ordering relation.

A relation or multirelation will be called **finite, infinite, countable, denumerable or continuum-equipotent**, according to whether its base is finite, infinite, countable, denumerable or continuum-equipotent.The **cardinal of the multirelation** R is the cardinal of its base $|R|$.

Given two multirelations R, S with common base E, we call the **concatenation** of R and S, denoted (R, S), the sequence of components of R followed by the components of S, in which case for the latter the indices are increased by the number of terms in R.

1.7.1 Restriction, extension

Let R be an n-ary relation with base E, and let F be a subset of E. We call the **restriction** of R to F, denoted by R/F, the n-ary relation taking the same value for each n-tuple with values in F. For the arity 0, the restriction to F of the 0-ary relation $(E, +)$ will be $(F, +)$; similarly with (-) ; this remains valid for empty F.

The notion of restriction of a function in 1.1.4 was more general: in the simple case of a unary function, which corresponds to a binary relation R, we had the possibility of restrict R to all couples having their first term in F, the second term running in the whole E.

Given a relation R with base E and a superset E^+ of E, we call an **extension** of R to E^+ any relation with base E^+ whose restriction to E is R.

Let R, R' be two n-ary relations with common base E. If for every subset X of E with cardinal $\leq n$ we have $R/X = R'/X$, then $R = R'$. Indeed to each n-sequence $(x_1, ..., x_n)$ with values in E, it suffices to associate the set $X = \{x_1, ..., x_n\}$.

Given a multirelation $R = (R_1, ..., R_h)$ with base E and a subset F of E, we define the **restriction** of R to F, denoted by R/F, to be the multirelation $(R_1/F, ..., R_h/F)$. Given a multirelation R with base E and a superset E^+ of E, we call an **extension** of R to E^+ any multirelation with base E^+ whose restriction to E is R. Equivalently, any sequence $(R_1^+, ..., R_h^+)$ where each R_i^+ is an extension of R_i to $E^+ (i = 1, ..., h)$.

Let R, R' be two multirelations with common arity $(n_1, ..., n_h)$ and common base E. If for each subset X of E with cardinal $\leq \text{Max}(n_1, ..., n_h)$, we have $R/X = R'/X$, then $R = R'$.

1.7.2 Compatible relations

Two relations (or multirelations) with the same arity are said to be **compatible** iff they have the same restriction to the intersection of their bases.

Let $\mathcal{R}$ be a set of mutually compatible relations (or multirelations):

(1) there exists a common extension of the relations in $\mathcal{R}$, based on the union of their bases;

(2) let us denote by E the union of the bases and by n the common arity, or the maximum of the common arity (for multirelations); if each n-element subset of E is covered by one of the bases, then the common extension is unique.

1.7.3 Amalgamation lemma

Let A, B be two compatible posets (i.e. having the same restriction to the intersection of the bases). **Then there exists a poset which is an extension of both A and B, based on the union of the bases.**

• Write $x \leq y$ when $x, y \in |A|$ and $x \leq y (\mathrm{mod}\, A)$, or when we have the same condition for B, or when x belongs to $|A|$ and y belongs to $|B|$ and there exists an element t in the intersection with $x \leq t (\mathrm{mod}\, A)$ and $t \leq y (\mathrm{mod}\, B)$, or when we have the same condition when interchanging A and B. Finally write $x|y$ in the other cases. •

Note that this lemma extends to the case of two chains, the common extension itself being a chain. It does not extend to trees: see 2.11.6 below.

1.7.4 Negation, conjunction, disjunction

Given a relation R, its **negation** $\neg R$ is the relation with same base and arity, always taking the opposite value. Given R, S with the same base and arity, the **conjunction** $R \wedge S$ takes the value (+) iff R and S take the value (+). The **disjunction** $R \vee S$ takes the value (+) iff either R or S takes the value (+).

1.7.5 Converse of a binary relation, retro-ordinal

Given a binary relation R, its **converse**, denoted by R^-, is the relation with the same base, such that $R^-(x, y) = R(y, x)$ for every x, y.

In particular we consider an ordinal α as the binary relation based on the set already denoted by α (the set of ordinals $< \alpha$), and taking the value (+) iff $x \in y$ or $x = y$ (denoted already by $x \leq y$). The converse relation α^- will be called a **retro-ordinal**.

1.7.6 Isomorphism, automorphism, empty function

Let n be a non-negative integer, R an n-ary relation with base E and R' an n-ary relation with base E'; let f be a bijection from E onto E'. We immediately extend the isomorphism and automorphism already introduced for binary relations (see 1.6.1), keeping the same notation $R' = \overline{f}(R)$ and specifying that now $R'(fx_1, ..., fx_n) = R(x_1, ..., x_n) = +$ or -. For $n = 0$, either $R = (E, +)$ and we set $\overline{f}(R) = (E', +)$ or $R = (E, -)$ and then $\overline{f}(R) = (E', -)$.

Isomorphism defines an equivalence relation on every set of relations of a given arity. As in previous section, we say that R and R' are **isomorphic**.

We adopt the convention that the empty function, which is a bijection of the empty set onto itself, is also an automorphism of each

relation with empty base. In particular it is an automorphism of the 0-ary relation with empty base and value (+), denoted (0,+), and also of (0,-). However (0,+) and (0,-) **are not isomorphic.**

These definitions and conventions extend to multirelations. Given a multirelation $R = (R_1, ..., R_h)$ with base E and $R' = (R'_1, ..., R'_h)$ with base E', a bijection f from E onto E' transforms R into R', denoted $R' = \overline{f}(R)$, iff for each $i = 1, ..., h$ the function f is an isomorphism of the component R_i onto the component R'_i. In other words $\overline{f}(R_1, ..., R_h) = (\overline{f}(R_1), ..., \overline{f}(R_h))$.

1.7.7 Isomorphism type, order type

Modeled after the definition of "cardinal" in 1.4.4 (excluding the case of isomorphy with an ordinal), we consider the relations isomorphic with a given R, and among such, those whose base has minimum fundamental rank. These form a set, called the **isomorphism type** of R (the order type if R is a chain). Thus two relations are isomorphic iff they have the same type.

1.8 Axiom of dependent choice

Let R be a binary relation with base E, such that for each x of E there exists at least one y of E satisfying $(x, y) \in R$. **The axiom of dependent choice asserts that, given such an R, there exists an ω-sequence of elements a_i of E satisfying $(a_i, a_{i+1}) \in R$ for each integer i** ([176] MOSTOWSKI 1948).

The axiom of dependent choice follows from the axiom of choice: take a choice function which to each x in E associates an element of the (non-empty) set of those y which satisfy $(x, y) \in R$.

It is proved that the dependent choice is strictly weaker than the axiom of choice: see for instance [125] JECH 1973 p.122 and following.

The countable axiom of choice, stated in 1.2.5, **follows from the axiom of dependent choice.**

• Start from an ω-sequence of non-empty mutually disjoint sets a_i(i integer), take R to be the binary relation based on the union of the a_i , defined by $(x, y) \in R$ iff there exists an i with $x \in a_i$ and $y \in a_{i+1}$. •

The countable axiom of choice is strictly weaker than the axiom of dependent choice: see [125] JECH 1973 p.119 and following. Finally, from the axiom of dependent choice, assumed to be consistent, one cannot deduce the axiom of choice for finite sets, stated in 1.2.10. The proof is due to [176] MOSTOWSKI 1948 without the axiom of foundation, and to [62] FEFERMAN 1965 with foundation.

1.9 Exercises

1.9.1 Consistency of the axiom of foundation

To see that the addition of the axiom of foundation does not imply a contradiction, if the other axioms of ZF are consistent, we define an ordinal to be a set not only transitive and totally ordered by the membership relation, but also satisfying the following foundation condition. For every non-empty subset x of a, there exists an element y of x which is disjoint from x: see [12] BERNAYS 1968 p. 80.

Next, one must redefine an integer as a finite ordinal. Then anticipating 2.2.2, we say that a set a is **well-founded** iff every sequence x_i indexed by integers,with $x_0 = a$ and x_{i+1} belonging to x_i for each i, is finite. Finally, one verifies that the well-founded sets with the membership relation satisfy all the axioms of ZF, including the axiom of foundation.

Hint. It is not sufficient to define directly a well-founded set (other than an ordinal) by the foundation condition.

1.9.2 A classical interpretation of ordinal exponentiation

Let α, β be two ordinals. Consider functions f with domain β, taking values in α. We say that such an f is **almost zero** iff $f(i) \neq 0$ for at most finitely many elements i of β. Given two almost zero functions f and g let $f < g$ iff there exists an i in β with $f(i) < g(i)$ in the usual ordering for ordinals, and $f(j) = g(j)$ for all j such that $i < j < \beta$. Let $f \leq g$ iff $f < g$ or $f = g$. In other words, the set of such almost zero functions is ordered by last difference.

1 - Show that $\leq$ is awell-ordering on the set of almost zero functions.

2 - Show that thiswell-ordering is isomorphicwith the ordinal exponential α^β (induction on β).

1.9.3 The generalized continuum hypothesis implies the axiom of choice (Sierpinski)

The reader may consult [228] SIERPINSKI 1947; also [35] COHEN 1966.

1 - Let A, B be two sets. If $A \cup B$ is subpotent to the set $\mathcal{P}(2 \times A)$, then $\mathcal{P}(A)$ is subpotent to B.

• Let f be a bijection from $A \cup B$ onto $\mathcal{P}(2 \times A) = \mathcal{P}(A) \times \mathcal{P}(A)$ (here $=$ means "equipotent"). Each element x of A is transformed into the couple $f(x) = (y, z)$ of elements of $\mathcal{P}(A)$. To each x associate the first term y of this couple. By Cantor's lemma 1.1.6, there exists an element u of $\mathcal{P}(A)$ for which none of the couples (u, z) is the image of any x of A. Hence there exists a subset of B which is bijectively transformed by f into the set $\mathcal{P}(A)$ of the second terms z of the couples (u, z). •

2 - For each set A, let $P_0(A) = A$; let $P_1(A) = \mathcal{P}(A)$, the set of subsets of A; for each integer i, let $P_{i+1}(A) = \mathcal{P}(P_i(A))$. If A has a denumerable subset, then $P_i(A) = 2 \times P_i(A)$ for each i.

• By hypothesis A is equipotent with A augmented by an element. Hence we have $\mathcal{P}(A) = {}^{A}2$ equipotent with $2\times({}^{A}2)$. It follows that $\mathcal{P}(A)$ is equipotent with itself augmented by an element: then we iterate. •

3 - Let A be an infinite set, assumed to be equipotent with $2\times A$. If for $i =$ 0,1,2,3 the sets $P_i(A)$ and $P_{i+1}(A)$ do not have any set which is strictly intermediate (with respect to subpotence), then there exists a well-ordering of A.

• The set A is equipotent with a proper subset of itself, hence has a denumerable subset. Hence $P_i(A)$ is equipotentwith $2\times P_i(A)$ for each integer i: see statement (2). Consider the well-orderings based on subsets of A, and the isomorphism classes of thesewell-orderings. From this point on, denote $P_i(A)$ by P_i. The pairs of elements of A belong to P_1. The couples belong to P_2. The well-orderings based on subsets of A belong to P_3. Finally the isomorphism classes belong to P_4. The set H of isomorphism classes is equipotent with the hartogs of A (see 1.6.4) and is included in P_4. The union $H \cup P_3$ is intermediate, under subpotence, between P_3 and $P_3 \cup P_4$, the latter equipotentwith P_4, since P_3 is equipotentwith a subset of P_4 as well as with $2\times P_3$. By hypothesis $H \cup P_3$ is equipotent eitherwith P_4 or with P_3. In the first case, it is equipotentwith $\mathcal{P}(2 \times P_3)$, and by the previous statement (1) the set P_4 is equipotentwith H, hence well-orderable, so A is well-orderable as well.

Consider the second case: $H \cup P_3$ is equipotent with P_3. Then H and P_2 are both subpotent to P_3, hence $H \cup P_2$ subpotent to $2\times P_3$ and hence to P_3. By hypothesis $H \cup P_2$ is equipotent either to P_3 or to P_2.

In the first case, by the previous (1) the set P_3 is subpotent to H, hence well-orderable, so that A is well-orderable. In the second case $H \cup P_1$ is subpotent to P_2. By iterating, we obtain that $H \cup A$ (since $A = P_0$) is subpotent to P_1, hence $H \cup A$ is equipotent with A or with P_1. The first case is excluded, since the hartogs of A is never subpotent to A. The previous (1) shows that P_1 is subpotent to H, hence well-orderable, and consequently A itself is well-orderable. •

4 - Let A be an arbitrary set (finite or infinite). The union $A \cup \omega$ is equipotent with itself augmented by an element. Hence $B = \mathcal{P}(A \cup \omega)$ is equipotent with $2\times B$. Using the generalized continuum hypothesis, the statement (3) shows that B is well-orderable, so $A \cup \omega$ and hence A as well.

Chapter 2

Real, well-founded poset, coherence lemma, cofinality, regular or singular aleph, tree, net or ideal

2.1 Rational, real (chains Q and R)

2.1.1 Real

We leave it to the reader to redefine positive and negative integer, and then **real**, as a couple formed from an integer which is called the **integer part**, and an infinite set of non-negative integers. The latter set will be identified with an ω-sequence of terms u_i(i non-negative integer) with $u_i = 0$ or 1 according to whether i belongs to the infinite set of integers or not. This sequence is called the **binary expansion** of the real, which always contains infinitely many occurrences of zero.

The notions of **rational real** and **dyadic real**, i.e. rational whose denominator is a power of 2, are assumed to be familiar, as well as the denumerability of the set of rationals. We denote by **Q the chain of rationals**. The set of reals is equipotent with the continuum: remove from the set of all sets of integers, the denumerable set of finite sets of integers.

The reader is assumed to be familiar with the construction of **the chain of reals**, denoted by **R**, by cuts in the chain **Q** of rationals. Anticipating on the general definitions in 2.6.4 below, we define each **cut** in **Q** as a couple (A, A') where A is an initial interval of **Q** and A' is the complementary final interval. We say that a cut is **trivial** iff either A has a maximum (rational) element or if A' has a minimum. Otherwise the cut defines in Dedekind's manner an irrational number.

We shall use indifferently the notation $\mathbf{R}(x, y) = +$ or the notation less than or equal to: $x \leq y \bmod \mathbf{R}$, greater than or equal to $(y \geq x)$, and the related strict inequalities modulo $\mathbf{R}$. Also the reader can define the notion of **dense, cofinal, coinitial set of reals** (an example being the rationals or the dyadic reals). The reader can define a **closed, open, half-open interval of reals, an initial, final interval, an upper bound and lower bound of a set of reals, the maximum, the minimum, a real valued sequence which is strictly (or otherwise) increasing, decreasing**.

Every set of mutually disjoint intervals of reals which are not reduced to singletons is countable: enumerate the rationals and associate to each interval the first rational which belongs to it. Consequently, every strictly increasing (or strictly decreasing) ordinal-indexed sequence of reals is countable.

2.1.2 Dedekind's theorem

For historical information, see [38] DEDEKIND 1888.

If we partition the reals into an initial interval a and its complement the final interval b, both non-empty, then either a has a maximum element or b has a minimum element.

Consequently, for any set a of reals, if there exists an upper bound, then there exists a least upper bound called the **supremum** of a and denoted $\operatorname{Sup} a$. Analogous definition of the **infimum** which is denoted $\operatorname{Inf} a$.

In other words, for every set a of reals, there exists a smallest interval (with respect to inclusion) including a: either the interval $(\operatorname{Inf} a, \operatorname{Sup} a)$ which is closed, open or half-open; or the initial interval $(-\infty, \operatorname{Sup} a)$ or final interval $(\operatorname{Inf} a, +\infty)$ or $(-\infty, +\infty)$ containing all the reals.

When useful, we will use the sum and product of reals, which the reader is presumed to know.

2.1.3 Completion of real functions

The reader is familiar with the notion of **real function, increasing, decreasing** (strictly or otherwise) real function.

Let E be the set of reals and $\mathbf{R}$ their chain, F a subset of E and F^+ the smallest interval including F: with endpoints $\operatorname{Inf} F$ and $\operatorname{Sup} F \,(\bmod \mathbf{R})$.

For each increasing $(\bmod \mathbf{R})$ function f from F into E, there exists at least one increasing function f^+ which is an extension of f to the domain F^+.

• For each element u of $F^+ - F$, let $f^+(u)$ be the infimum of values $f(x)$ for all x in F such that $x > u$. •

If F is dense for the chain $\mathbf{R}/F^+$, then the extension f^+ of f is unique. Moreover if f is strictly increasing and F dense for $\mathbf{R}/F^+$, then f^+ is strictly increasing and hence is an isomorphism from $\mathbf{R}/F^+$ into $\mathbf{R}$.

2.1.4 Set of increasing real functions

(1) **Let E be the set of reals, F a non-empty subset of E; then the set of increasing functions from F into E is equipotent with the continuum.**

• Assume first of all that F is an interval. An increasing function from F into E has countably many points of discontinuity, since to these points correspond non-singleton mutually disjoint intervals. Hence an increasing function is defined by its values on the rationals for example, plus its values on the points of discontinuity: altogether making a countable sequence of real values.

In the case of an arbitrary F, let F^+ be the smallest interval including F. To each increasing function f from F into E, associate the function f^+ from F^+ into E which is increasing and an extension of f, obtained by the previous proposition. The set of f^+ is equipotent with the continuum, hence so is the set of restrictions f of f^+ to the domain F. •

(2) **Let F be a set of reals. If for each automorphism h of $\mathbf{R}$ which is different from the identity, there exists an element $x \in F$ with $h(x) \neq x$, then F is dense in $\mathbf{R}$.**

• Let $a < b$ be reals. There exists at least one automorphism, or strictly increasing function f from $\mathbf{R}$ onto $\mathbf{R}$, satisfying $f(x) = x$ for all $x \leq a$ and all $x \geq b$, and $f(x) \neq x$ for all $x(a < x < b)$. Hence F must have an element between a and b. •

2.1.5 Concerning the axiom of choice

To see some initial difficulties provided by the axiom of choice, which indicate that this axiom is not "obvious", note that it is impossible in ZF plus the axiom of choice, to define and prove uniqueness of a choice function which associates to each non-empty set of reals one of its elements. Similarly it is impossible to uniquely define a choice set picking one function from each pair of real functions $h, -h$ where for each real x, the value of $-h$ on x is the additive inverse of $h(x)$.

To obtain a proof of uniqueness, completing the existence (which is guaranteed by the axiom of choice), it is necessary for example to add to ZF **the axiom of constructibility of** [98] GŐDEL 1940.

2.2 Well-founded poset, maximal chain, Hausdorff-Zorn axiom

2.2.1 Interval, initial and final interval

Given a poset A, the reader is assumed to be familiar with the notion of an element z **between** x **and** y (mod A), or z **intermediate between** x **and** y, as well as that of an element **strictly intermediate**. An **interval** of A is a subset of the base which is closed with respect to the notion of "intermediate modulo A". An **initial interval** or of A is a subset closed with respect to "less than". A **final interval** is a subset closed with respect to "greater than".

Let A be a poset. Then every subset of the base $|A|$ which has no minimal element is infinite. Similarly for a subset without a maximal element.

• To each element x of the subset D under consideration, associate the set D_x of elements of D which are less than or equal to x (mod A). None of the D_x is minimal under inclusion (see 1.1.1, definition of a finite set). •

2.2.2 Well-founded poset, well-founded quasi-ordering

We say that a poset or a quasi-ordering is **well-founded** iff every non-empty subset of its base has at least one minimal element. Note that a well-founded chain is a well-ordering. Every finite partial ordering is well-founded. Every restriction of a well-founded poset is well-founded.

Given a poset A, the reader is assumed to know the notion of a sequence with values in A which is **increasing, decreasing, strictly or otherwise**.

Every well-founded poset A satisfies the following conditions:

(1) **there is no strictly decreasing ω-sequence in A**;

(2) **every totally ordered restriction of A is a well-ordering; equivalently every non-empty totally ordered restriction of A has a minimum**.

Conversely, each of the conditions (1), (2) implies, hence is equivalent to saying that A is well-founded. This uses the axiom of dependent choice, yet ZF suffices if A is countable, or if its base is well-orderable. In the general case, apply dependent choice to the relation $y < x \bmod A$.

2.2.3 Maximal chain

Let A be a poset and U a chain or totally ordered restriction of A. This chain U is said to be **maximal** (under inclusion, modulo A) iff every totally ordered restriction of A extending U is identical to U.

Maximal chain lemma. Let A be a poset. Denote by U any chain which is a restriction of A. **Let f be a function which to each chain U associates a chain $f(U)$ extension of U; then**:

(1) **to each U is associated a chain V which is an extension of U and verifies the "fixed point condition" $f(V) = V$**;

(2) **assume that for each U the chain $f(U)$ is a strict extension of U, unless U be maximal (in which case $f(U) = U$); then each preceding "fixed point" V is maximal** (mod A).

• Index by ordinals a sequence of chains U_i starting with $U_0 = U$. Set $U_{i+1} = f(U_i)$ and, for i a limit ordinal, let U_i be the common extension of $U_j (j < i)$ to the union of their bases. Now the restrictions of A form a set, yet the ordinals do not; hence we necessary reach an ordinal v such that $f(U_v) = U_v$ denoted V. The conclusion (1) follows; conclusion (2) is trivial. •

2.2.4 Maximal chain axiom (Hausdorff-Zorn axiom)

This axiom, going back to [107] HAUSDORFF 1914, then taken up by KURATOWSKI, MOORE and then ZORN, is stated as follows.

Given a poset A and a chain U which is a restriction of A, there exists a chain which is an extension of U and maximal mod A.

Modulo the axiom of choice, given A there exists a choice function f which verifies the strong hypothesis of the preceding subsection, i.e. $f(U)$ strict extension of U unless U be maximal. The maximal chain axiom follows.

2.2.5 The well-founded interval ordering

Let E be a set; denote by X any well-ordering based on a subset of E. Write $X \leq X'$ iff X is an initial interval of X'. The well-founded poset thus defined on the set of X's will be called **the interval-ordering on** E.

A set E is **well-orderable** iff there exists a maximal chain which is a restriction of the interval-ordering on E. Consequently, the maximal chain axiom implies the well-ordering axiom.

Now by 1.6.2 **the well-ordering axiom is equivalent to the axiom of choice. Finally the maximal chain axiom is equivalent to the axiom of choice.**

2.2.6 Free subset, antichain, maximal antichain

Given a poset A, a subset of its base is called **free** (mod A) iff its elements are mutually incomparable (mod A). The restriction A/D to such a free subset D is called an **antichain** (mod A). It reduces to the identity relation based on D.

A free subset and the corresponding antichain are called **maximal** (under inclusion) iff there is no proper superset which is free.

Given a poset A and a free subset D, there exists a maximal free subset including D (uses axiom of choice; ZF suffices if A is countable): apply the maximal chain axiom to the inclusion among free subsets.

2.2.7 A poset with cardinality ω_1 and only countable chains and antichains

Let us call a **Sierpinski poset** the following one. To every countable ordinal u injectively associate a real $r(u)$ (axiom of choice). Define the poset on the countable ordinals $u, v, ..$ by setting $u < v$ iff simultaneously $u < v$ modulo the usual ordering of ordinals and $r(u) < r(v)$ modulo the usual ordering of reals.

To obtain a totally ordered restriction of this poset, one must take a strictly increasing sequence of reals: such a sequence is countable, by 2.1.1. To obtain a free set, one must take a strictly decreasing sequence of reals.

A connection with Ramsey infinite numbers will appear in 3.3.1.

Note that the existence of a tree (as defined in 2.11 below) of cardinal ω_1 in which every chain and every antichain is countable, is not provable in ZF and even not provable using the axiom of choice and the continuum hypothesis. This problem is related to SUSLIN's axiom or hypothesis: see 5.8.

2.3 Filter, ultrafilter axiom

2.3.1 Filter

Given a set a, recall that a **filter** on a is a non-empty set F of non-empty subsets of a, such that

(1) if $x \in F$ and $x \subseteq y \subseteq a$, then $y \in F$;

(2) if $x, y \in F$, then the intersection $x \cap y \in F$; thus every finite intersection of elements of F is an element of F.

For example, the set of complements of finite subsets of ω is a filter on ω.

Every intersection of filters on a is a filter on a. Let F be a filter on a, and $b \subseteq a$ such that $b \cap x$ is non-empty for every element x of F. Then the set of intersections $b \cap x$ constitutes a filter on b, called the filter **induced** by F on b.

Let A be a set of subsets of a for which every finite intersection of its elements is non-empty. Then the set of supersets of these intersections constitutes a filter on a, called the filter **generated** by A.

Let F, G be two filters on the same set. G is said to be **finer** than F if it includes F; **strictly finer** if it strictly includes F.

Given a filter F on a and $b \subseteq a$, either $b \in F$ or $a - b \in F$ or every intersection of any $x \in F$ with b and with $a - b$ is non-empty. Hence there exists a filter finer than F which contains b or a filter finer than F which contains $a - b$ (both cases may coexist).

Let a be a set. If a set of filters on a is totally ordered by the comparison relation "finer than", or more generally if this comparison relation is a directed partial ordering (i.e. given two filters, there is a third filter which is finer than both), then the union of the filters is a filter on a.

2.3.2 Ultrafilter, ultrafilter axiom

Given a set a, an **ultrafilter** on a is a filter for which there is no strictly finer filter on a. For example if $u \in a$, then the set of subsets of a containing u is an ultrafilter, said to be **trivial**.

Already for the set ω, the axioms of ZF alone are not sufficient to prove the existence of a non-trivial ultrafilter. We have to add, for instance, the **ultrafilter axiom** (also called boolean prime ideal axiom: see 2.13 below), which asserts that **for every set a and every filter F on a, there exists an ultrafilter on a which is finer than F.**

For example this implies the existence of an ultrafilter on ω which contains as elements all complements of finite subsets of ω.

A necessary and sufficient condition that a filter F on a be an ultrafilter, is that for every subset x of a, either $x \in F$ or $a - x \in F$.

Let F be an ultrafilter and $x \in F$. Then for every partition of x into a finite number of disjoint subsets, one and only one of these subsets belongs to F.

Every filter F on a is the intersection of all ultrafilters on a which are finer than F (uses the ultrafilter axiom).

To see the impossibility of a countably set of subsets of ω generating an ultrafilter, see following 2.3.4 below.

The number of filters on a set. Let a be an infinite set; then the cardinal of the set of filters, as well as the cardinal of the set of ultrafilters, equals (2 to the power ${}^{(\mathrm{Card}\, a)}2$) (TARSKI; see [8] BELL, SLOMSON 1969, p.108 theorem 1.5; uses the axiom of choice).

Since each filter is a set of subsets of a, then the set of filters has at most the announced cardinality. Hence it suffices to construct a set of ultrafilters having this cardinal. See below 2.15.2.

2.3.3 Separation lemma

Let E be an infinite set, I a set of subsets $E_i (i \in I)$ of E where I is subpotent to E and each E_i is equipotent with E.

(1) **There exist two disjoint subsests C, D of E such that for each i the intersections $C \cap E_i$ and $D \cap E_i$ are non-empty.**

(2) More strongly, **there exist disjoint C, D such that $C \cap E_i$ and $D \cap E_i$ are equipotent with E** (uses axiom of choice; ZF suffices when E is countable); see [13] BERNSTEIN 1908 or [143] KURATOWSKI 1966 p. 514.

• (1) Well-order E according to its cardinal α, and I according to its cardinal $\beta \leq \alpha$. Put the first element x_0 of E_0 into C, the second y_0 into D. In general for each ordinal $i < \beta$, put the first x_i of E_i which is distinct from all x_j and $y_j (j < i)$ into C, and the second y_i distinct from all x_j and y_j into D. •

• (2) Replace the set I by a sequence of the terms E_i with repetitions, as follows. If $\beta = \alpha$ so that I is equipotent with E, then take the first term E_0, then E_0 followed by E_1, and in general for each i, take the sequence of $E_j (0 \leq j \leq i)$. If $\beta < \alpha$, then keep the E_i, but keep repeating so as to obtain a sequence with length $\alpha = \mathrm{Card}\, E$. In either case, repeat the preceding proof: for each set E_i the intersections $C \cap E_i$ and $D \cap E_i$ have cardinal α. •

2.3.4 Application to ultrafilters

If E is denumerable and F is a non-trivial ultrafilter on E, then there is no countable set of subsets E_i of E which generates F, in the sense that the elements of F are obtained by taking supersets of finite intersections of the E_i.

• Denote by E_i the E_i from the statement and also their finite intersections. Then the sets C and D, which are disjoint, would both belong to the ultrafilter: contradiction. •

More generally, **let E be infinite and F be an ultrafilter on E whose elements are equipotent with E. Then F cannot be generated by a set,**

equipotent with E, of elements of F.

2.3.5 The ultrafilter axiom follows from the axiom of choice

• Consider the set of filters on a given set, with the comparison ordering "finer filter than". Using 2.2.4, take a maximal chain extending the chain reduced to the singleton of a given filter F. The ultrafilter generated by the union of the filters belonging to the maximal chain is finer than F. •

The ultrafilter axiom is strictly weaker than the axiom of choice: [105] HALPERN, LEVY 1971 p. 83-134.

For a model of ZF having no ultrafilter other than trivial ultrafilters, see [15] BLASS 1977.

2.4 Coherence lemma, ordering axiom

2.4.1 Coherence lemma

Consider a set $\mathcal{J}$ of sets F for each of which we have a finite non-empty set U_F of multirelations based on F (all of the same arity) with the following hypotheses:

(1) **$\mathcal{J}$ is a directed system: if F, F' belong to $\mathcal{J}$, then there exists an $F"$ in $\mathcal{J}$ with $F" \supseteq F \cup F'$.**

(2) **if F, F' belong to $\mathcal{J}$ and $F' \subseteq F$, then every multirelation belonging to U_F, when restricted to F', yields an element of $U_{F'}$;**

in this case, there exists a multirelation R based on the union of the sets F in $\mathcal{J}$ such that for each F the restriction R/F belongs to U_F (uses the ultrafilter axiom; ZF suffices if the F are finite and their union countable).

• Denote by E the union of the F in $\mathcal{J}$. To each F associate the set V_F of extensions to E of multirelations belonging to U_F. The supersets of the V_F constitute a filter on the set of multirelations based on E with the given arity. Indeed if F, F' belong to $\mathcal{J}$, then there exists in $\mathcal{J}$ an $F" \supseteq F \cup F'$; hence $V_F \cap V_{F'}$ is a superset of $V_{F"}$.

Take an ultrafilter extending this filter. For each F of $\mathcal{J}$, partition V_F into a finite number of classes, each class defined by the restriction to F of the multirelations belonging to V_F. One and only one of these classes belongs to our ultrafilter: denote by R_F the corresponding restriction, so that R_F belongs to U_F. Hence the R_F are mutually compatible (common restriction to the intersection of their bases): the existence of the mutirelation R follows. If E is countable and the F are finite subsets of E, then the ultrafilter becomes superfluous. •

2.4.2 The coherence lemma is equivalent to the ultrafilter axiom

• It suffices to prove that the coherence lemma implies the ultrafilter axiom.

Let e be a set, $\mathcal{P}(e)$ be the set of subsets of e, and H a filter on e. Let F be a finite set of subsets of e which is closed with respect to union, intersection and taking complements (in e). To each F associate the set U_F of unary relations X with base F which satisfy the following conditions:

for each $a \in F$, if $a \in H$ then the value $X(a) = +$; if $e - a \in H$ then $X(a) = -$;
for each $a \in F$, we have opposite values $X(e - a) \neq X(a)$;
if $a, b \in F$ (so $a \cap b \in F$) and $X(a) = X(b) = +$, then $X(a \cap b) = +$;
if $a, b \in F$ and $a \subseteq b$ and $X(a) = +$, then $X(b) = +$.

The set U_F is non-empty for each F. The set of the F's forms a directed system, so we can apply the coherence lemma. Consequently there exists a unary relation based on $\mathcal{P}(e)$ whose restrictions to each F belongs to U_F. The subsets of e which give the value (+) to this unary relation constitute an ultrafilter on e which is finer than H. •

2.4.3 Variant of the coherence lemma

Following [210] RADO 1949, consider a set of finite mutually disjoint non-empty sets a, and for each finite set I of sets a, consider a choice function f_I which associates to each a of I an element $f_I(a) \in a$. Then there exists a choice function f whose domain is the set of all the a, and for each finite set I of the a, there exists a finite superset J of I with the restriction f/I equal to f_J/I. The preceding RADO's lemma plus the axiom of choice for finite sets is equivalent to the coherence lemma ([9] BENEJAM 1970).

2.4.4 Orderable set, ordering axiom

We say that a set E is **orderable** iff there exists a chain based on E. Using only the axioms of ZF, every finite set, every denumerable set, every set equipotent with the continuum is orderable. The ordering axiom asserts that every set is orderable.

The ordering axiom follows from the ultrafilter axiom, or equivalently from the coherence lemma.

• Let E be a set; to each finite subset F of E associate the set U_F of chains based on F. By the coherence lemma, there exists a relation R based on E every of whose finite restriction is a chain; thus R is a chain. •

The ordering axiom is strictly weaker than the ultrafilter axiom ([125] JECH 1973 p.100).

The ordering axiom implies the axiom of choice for finite sets (see 1.2.9).

• Given a set of mutually disjoint finite sets, it suffices to take a chain A based on their union: to each finite set we associate its minimum modulo A. •

The axiom of choice for finite sets is strictly weaker than the ordering axiom: see [150] LAUCHLI 1964 for ZF without foundation, completed for ZF by [186] PINCUS 1972. The axiom of choice for finite sets does not follow from the axiom of dependent choice (see 1.8). Hence the ordering axiom does not follow from dependent choice.

2.5 Set of initial intervals

2.5.1 Characterization of the set of initial intervals of a poset

Let E be a set, $\mathbf{A}$ a set of subsets of E. Two distinct elements of E are said to be separable by $\mathbf{A}$ iff there exists an element of $\mathbf{A}$ which contains one and not the other.

Given a poset A, denote by $\mathcal{J}(A)$ **the partial ordering of the initial intervals of** A (with respect to inclusion). Obviously $\mathcal{J}(A)$ is closed under union and intersection (finite or infinite). Moreover any two elements x, y of the base $|A|$ are separable by $\mathcal{J}(A)$: indeed we can assume that $x < y$ or $x|y$ (mod A), and it suffices to take the initial interval $\leq x(\text{mod}\, A)$ to separate x and y.

Conversely, **given a set A of subsets of E which is closed under union and intersection and such that any two elements of E be separable by A, then there exists a poset A based on E with $\mathbf{A} = \mathcal{J}(A)$**

Moreover A **is uniquely determined by A**. In other words $\mathcal{J}$ is an injection whose domain is the set of posets on E, and whose range is formed by sets of subsets of E which are closed under union and intersection and have the separation property.

• Define the binary relation A based on E by the condition that, when given two elements x, y of E, we put $x \leq y(\text{mod}\, A)$ iff every element of $\mathbf{A}$ which contains y as an element also contains x as an element. Then A is obviously reflexive and transitive. To see that A is antisymmetric, consider two distinct elements x and y belonging to E. If every element of $\mathbf{A}$ which contains x also contains y and conversely, then the separability hypothesis is falsified.

Finally, the uniqueness of A follows from the fact that two distinct partial orderings A and A' based on E yield two distinct $\mathcal{J}(A)$ and $\mathcal{J}(A')$. •

2.5.2 Case of a chain

(1) Let $\mathbf{C}$ be a set of subsets of E, which we assume to be totally ordered under inclusion. **For C to be a maximal chain with respect to inclusion, it is necessary and sufficient that C be closed under union and intersection (finite or otherwise), and that any two distinct elements of E be separable by C.**

(2) More generally, let $\mathbf{A}$ be a set of subsets of E, closed under union and intersection. Let $\mathbf{C}$ be a subset of $\mathbf{A}$ which is totally ordered by inclusion.

For C to be maximal among totally ordered restrictions of A, it is necessary and sufficient that C be closed under union and intersection, and that any two distinct elements of E which are separable by A are also separable by C.

• Assume that $\mathbf{C}$ is a chain which is maximal among the restrictions of $\mathbf{A}$. Then $\mathbf{C}$ is closed under union and intersection, for otherwise we could add the unions and intersections of elements of $\mathbf{C}$. Moreover, if x, y are two distinct elements of

E which are separated by an element X of $\mathbf{A}$ with x belonging to X yet not y, then denote by U the union of the elements of $\mathbf{C}$ which contain neither x nor y, and by V the intersection of the elements of $\mathbf{C}$ which contain both x and y . By the preceding discussion, U and V are elements of $\mathbf{C}$.

Since $\mathbf{C}$ is totally ordered by inclusion, then U is strictly included in V. Either $\mathbf{C}$ contains an element separating x and y, in which case our conclusion holds. Or every element of $\mathbf{C}$ is a subset of U or a superset of V. Hence U and V are consecutive with respect to inclusion in the chain $\mathbf{C}$. So $(U \cup X) \cap V$ is an element of $\mathcal{A}$, situated between U and V with respect to inclusion, and distinct from U and V as it contains x but not y. This contradicts the maximality of $\mathbf{C}$.

Conversely, assume that the chain $\mathbf{C}$ is not maximal among the totally ordered restrictions of $\mathbf{A}$. We can thus add to $\mathbf{C}$ a subset W of E which is an element of $\mathbf{A}$, and either including or included in every set which belongs to $\mathbf{C}$. Denote by U the union of those elements of $\mathbf{C}$ which are included in W, and by V the intersection of those elements of $\mathbf{C}$ which include W. Then either it is the case that U or V does not belong to $\mathbf{C}$: so that $\mathbf{C}$ is either not closed under union or under intersection. Or it is the case that U and V belong to $\mathbf{C}$: so that W is distinct from U and V, and hence is properly situated between U and V. Let u be an element of $W - U$ and v an element of $V - W$: then u and v are separated by the element W of $\mathbf{A}$, and yet are not separated by any element of $\mathbf{C}$. •

Going back to the general case of a partial ordering A, it is obvious that $\mathbf{A} = \mathcal{J}(A)$ is totally ordered (with respect to inclusion) iff A is a chain (or total ordering). This leads to the following consequence (3) of the preceding (1):

(3) Given a totally ordered (with respect to inclusion) set $\mathbf{C}$ of subsets of E, then $\mathbf{C}$ is maximal iff there exists a chain C such that $\mathbf{C} = \mathcal{J}(C)$.

2.5.3 Axiom of the maximal chain of inclusion

This axiom asserts that, for every set E, **the partial ordering of inclusion among subsets of E admits a maximal total ordering among its restrictions**.

It is considerably weaker than the general maximal chain axiom, as stated in 2.2.4: the HAUSDORFF-ZORN axiom, equivalent to the axiom of choice. By the preceding, the axiom of maximal chain of inclusion is equivalent to the ordering axiom (2.4.4). Consequently it follows from the ultrafilter axiom.

2.6 Ordinal sum and product of chains, homomorphic image, cut, Dedekind's generalized statement

2.6.1 Ordinal sum

Let A, B be two chains with disjoint bases. We call the **ordinal sum**, or simply the **sum** $A + B$ the chain based on the union of the bases, which is the common extension of A and B for which each element in $|A|$ precedes each element in $|B|$. This generalizes the notion of sum among ordinals.

This sum agrees with isomorphism, in the sense that if A' is isomorphic with A and B' with B , then $A' + B'$ is isomorphic with $A + B$. It is associative but not commutative, even up to isomorphism: $\omega + 1$ and $1 + \omega$ are not isomorphic.

2.6.2 Homomorphic image of a chain

Let A be a chain. Consider a partition of its base into mutually disjoint non-empty intervals A_i, again denoted i. Take the new base to be the set of these intervals, and define the chain I by putting $i < j$ iff each element of $|A_i|$ is strictly less than each element of $|A_j|$. This chain I is called a **homomorphic image** of A. The set of intervals A_i is called a decomposition of A. The chain A is called the I-**sum** or the **sum along** I **of the** A_i: notation $A = \Sigma A_i(i \in I)$.

Each homomorphic image of a chain A is isomorphic to a restriction of A (uses axiom of choice): take a choice set having one element in each interval of the decomposition. The converse is false. For example, the chain ω of the integers is a restriction of $\omega + 1$, but each decomposition of $\omega + 1$ into intervals has a least interval. Hence ω is not a homomorphic image of $\omega + 1$.

Another example. The chain $\mathbf{Q}$ of rationals is a restriction of the chain $\mathbf{R}$ of reals, but not a homomorphic image. Indeed Dedekind's theorem would be violated by a decomposition along $\mathbf{Q}$ of the chain $\mathbf{R}$, and a partition of $\mathbf{Q}$, thus a partition of $\mathbf{R}$ into an initial interval without a maximum and the complementary final interval without a minimum.

Note that a homomorphic image of a homomorphic image of A is again an image of A.

2.6.3 Ordinal product of chains

Given two chains A, B the **ordinal product** $A.B$ is the chain based on the cartesian product of the bases, defined by putting $(y, x) \leq (y', x')$ iff $y < y' (\operatorname{mod} B)$ or $y = y'$ and $x \leq x' (\operatorname{mod} A)$. In other words, by associating to each y of $|B|$ the chain A_y obtained by replacing each x of A by the couple (y, x), then taking the B-sum of the A_y. This product generalizes the notion of product among ordinals.

This product agrees with isomorphism: if A' is isomorphic with A and B' with B, then $A'.B'$ is isomorphic with $A.B$. It is associative but not commutative, even up to isomorphism: $\omega.2$ is not isomorphic with $2.\omega = \omega$. It is distributive on the right: $C.(A + B) = C.A + C.B$ but not on the left, as already seen with ordinals.

2.6.4 Cut; Dedekind's generalized statement

Start with a chain A: initial intervals of A are totally ordered by inclusion. To each initial interval X of A associate the final interval X', the complement of X relative to the base $|A|$. Then the couple (X, X') is said to be an A-**cut**. The comparison ordering between cuts is defined by inclusion of their first terms, which are initial intervals.

Separate the set of cuts of A into two complementary subsets C and D, such that every cut in C is less than every cut in D; we have DEDEKIND's generalized result: **either C has a maximum cut, or D has a minimum cut**.

• Take the union U of initial intervals of all cuts in C; then take the final interval U' complement of U. Then (U, U') is the supremum of C. If this cut belongs to C then it is the maximum; if it belongs to D then it is the minimum. •

2.6.5 Right and left bounds; trivial cut

Given an A-cut (U, U'), the maximum element of $U(\operatorname{mod} A)$, when it exists, is called the **left bound** of the cut; analogously the minimum of U' is called the **right bound**. If either the left bound or the right bound exists (obviously both can exist), then the cut (U, U') is said to be **trivial**.

Restriction of a cut; element inside a cut.

Let A be a chain, D be a restriction of A. If $(A', A")$ is an A-cut, then the common restriction D' of D and A' to the intersection of their bases, and the common restriction $D"$ of D and $A"$ constitute a D-cut $(D', D")$, called the restriction of $(A', A")$ to D.

An element x of the base $|A|$ is said to be inside $(D', D")$ iff $x >(\operatorname{mod} A)$ every element of D' and $x <(\operatorname{mod} A)$ every element of $D"$.

2.6.6 Dense set in a chain

Given a chain A, a subset D of its base $|A|$ is said to be **dense in A** iff between any two distinct elements $x, y \in |A|$ with $y > x(\operatorname{mod} A)$, there exists at least one element $z \in D$ with $x \leq z \leq y(\operatorname{mod} A)$. If D is dense in A, then inside any (A/D)-cut there exists at most one element of the base $|A|$.

2.7 Height, cofinal subset, cofinality

2.7.1 Height of an element, height of a well-founded poset

Let A be a well-founded poset (definition in 2.2.2). To each element x of the base $|A|$, associate as follows an ordinal called the **height of $x(\operatorname{mod} A)$** and denoted $\operatorname{Ht} x$.

If x is a minimal element, let $\operatorname{Ht} x = 0$. Let α be a non-zero ordinal; assume that each ordinal $< \alpha$ has been associated to at least one element, but that there still remain elements in the base to which no height $< \alpha$ has been associated. Then associate the height α to minimal elements among these.

Given a well-founded poset A, there is a unique height associated to each element of the base $|A|$.

Moreover, **for each element x of height α and every ordinal $\beta < \alpha$, there exists at least one element $< x$ $(\operatorname{mod} A)$ with height β.**

However, **given x of height α and an ordinal $\beta > \alpha$, even if there exist elements $> x$ of heights $> \beta$, it is possible that no element $> x$ exists with height β.**

• Let $a < b < c < d$ and $a < e < d$ with $e|b$ and $e|c$. Then e has height 1 and its only strict upper bound is d with height 3. •

If $x < y(\bmod A)$, then $\mathrm{Ht}\, x < \mathrm{Ht}\, y$.

Two distinct elements of the same height are incomparable. But two incomparable elements may have different heights.

The heights of the elements of a well-founded poset A constitute an ordinal, called the **height of** A and denoted $\mathrm{Ht}\, A$. Equivalently we have $\mathrm{Ht}\, A = \mathrm{Sup}((\mathrm{Ht}\, x) + 1)$ for all $x \in |A|$. The height of an element $x(\bmod A)$ is also the height of the restriction of A to elements $< x$.

In particular, the fundamental rank of a set a (1.4.2) is the height of the well-founded poset based on the transitive closure of a, and defined by $x < y$ iff there exists a finite sequence $t_0 = x, ..., t_h = y$ with h integer and $t_i \in t_{i+1}$ for each $i < h$.

Every well-ordering A is isomorphic with the ordinal $\mathrm{Ht}\, A$. The isomorphism transforms each element x into $\mathrm{Ht}\, x$, already called the ordinal rank of x (1.6.2).

(1) **Let A be a well-founded poset. Then each restriction B of A is a well-founded poset with** $\mathrm{Ht}\, B \leq \mathrm{Ht}\, A$. Indeed for each element $x \in |B|$ we have $\mathrm{Ht}\, x(\bmod B) \leq \mathrm{Ht}\, x(\bmod A)$.

(2) **If B is an initial interval of A containing x, then** $\mathrm{Ht}\, x \ (\bmod B) = \mathrm{Ht}\, x \ (\bmod A)$ (induction on $\mathrm{Ht}\, x$).

Given a well-founded poset A, it can happen that a maximal chain restriction of A does not reach $\mathrm{Ht}\, A$. Even the supremum of the heights of the maximal chains can be strictly less than $\mathrm{Ht}\, A$.

• For each integer i, take a chain isomorphic to i; these chains are assumed to be mutually incomparable. We obtain a well-founded poset with height ω, but in which every maximal chain is finite. •

Another example. Take denumerably many copies isomorphic to the previous poset. Above the j-th copy (j integer), put a chain isomorphic to j. Let these different posets with disjoint bases be mutually incomparable. We obtain a well-founded poset with height $\omega.2$, in which every maximal chain is finite.

2.7.2 Cofinal, co-initial subset and restriction

Let A be a poset. A subset D of $|A|$ is said to be **cofinal** (modulo A), and A/D is said to be a **cofinal restriction** of A iff for each x in $|A|$ there exists a y in D with $y \geq x \ (\bmod A)$. Analogous definition for a **co-initial subset** and a **co-initial restriction**.

If the base is non-empty, then every cofinal or co-initial subset is non-empty. Each superset of a cofinal set is cofinal; similarly with co-initial. Each cofinal restriction of a cofinal restriction is a cofinal restriction; similarly with co-initial.

Theorem. Let A be a poset, B a well-founded poset with the same base E. Remove those y such that there exists at least an x which is

both $> y \pmod A$ and $< y \pmod B$. Then the set D obtained after all such removals is cofinal $(\bmod A)$, and any two distinct elements of D are never ordered in opposite senses modulo A and modulo B.

Moreover **if B is a well-ordering, then A/D is a well-founded poset which is a cofinal restriction of A** ([197] POUZET 1979).

• First we see that D is cofinal $(\bmod A)$. Indeed for each y which is removed, there exists an x of least height $(\bmod B)$ among those satisfying $x > y \pmod A$ and $x < y \pmod B$. This x belongs to D: if it were removed, then there would exist an $x' > x > y \pmod A$ and $x' < x < y \pmod B$, contradicting the minimality of the height of x. Now consider the case that B is a well-ordering. For each non-empty subset X of D, let x be the minimum of B/X: then x is a minimal element $(\bmod A/X)$. For otherwise there would exist a $y \in X$ with $y < x \pmod A$. Since $x, y \in D$, we do not have $y > x (\bmod B)$. Hence $y < x (\bmod B)$, contradicting the minimality of x modulo B/X. •

From an intuitive point of view, note that when B is a well-ordering, if we denote by b_i the element with height $i \pmod B$, then D is the set of elements c defined as follows among the b. Let $c_0 = b_0$. Then $c_1 = b_{i(1)}$ = the b_i of least height $i(1) \neq 0 \pmod B$, among the elements $>$ or $|c_0 \bmod A$. Then $c_2 = b_{i(2)}$ = the b_i of least height $i(2) > i(1) (\bmod B)$, among the elements $>$ or $|c_0$ and $>$ or $|c_1 \bmod A$. In general for each ordinal α, assume that the $c_u = b_{i(u)} (u < \alpha)$ are defined. Then $c_\alpha = b_{i(\alpha)}$ = the b_i of least height $i(\alpha) > i(u)$ for all $u < \alpha$, among the elements which are simultaneously $>$ or $|c_0$, and $>$ or $|c_1$ and ... and $>$ or $|c_u (\bmod A)$ for all $u < \alpha$.

Corollary. For each poset A, there exists a cofinal set D such that A/D is a well-founded poset (uses axiom of choice; ZF suffices if A is countable or has well-orderable base; ZF clearly suffices also if A is well-founded). Take a well-ordering of $|A|$ and use the preceding proposition.

2.7.3 Cofinality, co-initiality

Let A be a poset. If, among the cofinal sets $(\bmod A)$, there exists one of least cardinal, then this cardinal is called the **cofinality** of A, denoted by $\operatorname{Cof} A$. Analogous definition of **co-initiality**.

With the axiom of choice, every cardinal is an aleph, hence the cofinality and co-initiality exist for each poset. With only the axioms of ZF, these only exist in particular cases, for example when the base is well-orderable. Their study is very different in the case of a chain (total ordering), the classical case considered here, from the general case of a poset, such a case introduced and studied by POUZET: see 2.12 below.

Let A be a chain with well-orderable base. Then **there exists a cofinal subset U of the base, with A/U a well-ordering isomorphic with** $\operatorname{Cof} A$; same result with co-initiality.

• Take a cofinal subset D with least cardinal, hence of cardinal equal to Cof A. Totally order D according to its cardinal. Then apply 2.7.2, replacing E by D and A by A/D, and B by a well-ordering of D isomorphic with Card D (i.e. the least ordinal based on D), which is Cof A. •

Corollary. Let α be an ordinal; Cof α is the least ordinal u for which there exists a u-sequence of successor ordinals whose union is α (equivalently: a strictly increasing u-sequence of successor ordinals whose union is α).

2.7.4 Cofinal subset of an indecomposable ordinal

Let α be an indecomposable ordinal, C a cofinal subset of α, with α/C of least order type, hence equal to Cof α. To each element i of C we associate an arbitrary ordinal α_i such that $i \leq \alpha_i < a$.

Then we have $\Sigma\alpha_i(i \in C) = (\mathbf{Sup}\,\alpha_i) = \alpha$, where Σ designates the ordinal sum of the α_i along the increasing i's.

• Obviously $\Sigma\alpha_i(i \in \mathrm{C}) \geq (\mathrm{Sup}\ \alpha_i) = \alpha$. It remains to eliminate the case of strict inequality.

Suppose that $\Sigma\alpha_i > \alpha$ and let u be the least element of C for which $\Sigma\alpha_i(i < u) \geq \alpha$. Either u has a predecessor $v (u = v + 1)$, in which case α is equal to $\Sigma\alpha_i(i < v)$ plus a non-zero ordinal which is $\leq \alpha_v$ hence $< \alpha$. In this case α is not indecomposable: contradiction.

Or u is a limit ordinal, hence $\Sigma\alpha_i(i < u) = \alpha$. Then the set of minimums of the disjoint intervals $\alpha_i(i < u)$ yields an ordinal isomorphic with u, and is cofinal for α. Hence $u \geq$ Cof α, and so the ordinal of α /C is strictly greater than Cof α: contradiction. •

Note that **it is necessary to assume that α is indecomposable**: take the counterexample $\alpha = \omega.2$ with i integer and $\alpha_i = \omega + i$.

It is also necessary to take α/C isomorphic with Cof α, and not only of cardinal Cof α: take the counterexample $\alpha = \omega^2$ with $i < \omega.2$ and the following distinction of two cases: $\alpha_i = \omega + i$ for finite i, and $\alpha_i = \omega.2 + \omega.(i - \omega)$ for $\omega \leq i < \omega.2$.

2.8 Regular or singular aleph, accessibility

2.8.1 Regular or singular aleph

An aleph (i.e. the cardinal of a well-orderable set) is said to be **regular** iff, considered as an ordinal, every cofinal subset is equipotent with it. In other words, its cofinality is equal to it. An aleph is said to be **singular** in the opposite case where its cofinality is strictly smaller.

For example 1 and ω are regular. Each integer ≥ 2 has cofinality 1, hence is singular. The cardinal ω_ω has cofinality ω, hence is singular.

Modulo the countable axiom of choice, ω_1 is regular.

• Given a countable subset D of ω_1, take the countable union of those countable ordinals which are elements of D. By 1.2.5 using the countable axiom of choice, this union is countable, hence cannot be the entire set ω_1. •

Note that the cardinal inequality $\omega_1 > \omega$ was already obtained in 1.6.5 using only ZF (indeed ω_1 is the Hartogs of ω).

Let A be a chain with well-orderable base; then $\operatorname{Cof} A$ is a regular aleph. Consequence of the fact that a cofinal restriction of a cofinal restriction is itself cofinal (see 2.7.2).

2.8.2 Every successor aleph is regular

• We use axiom of choice. Our aleph is of the form ω_{a+1} where a is an ordinal. Let u be its cofinality. Take a u-sequence of successor ordinals whose union is ω_{a+1}: see corollary in 2.7.3. From some point on, these ordinals are equipotent with ω_a. Suppose u strictly less than ω_{a+1}, hence $\operatorname{Card} u \leq w_a$. Then the union of the ordinals in our u-sequence has cardinal at most equal to $\omega_a \times \omega_a$ hence at most equal to ω_a (axiom of choice giving a bijection of each ordinal onto ω_a). Contradiction proving that $u = \omega_{a+1}$. •

2.8.3 Condition for singular alephs

Let a be an infinite aleph. **For a to be singular, it is necessary and sufficient that there exists a set u, strictly subpotent to a, of elements strictly subpotent to a, whose union yields a** (uses axiom of choice).

• If a is singular, then our conclusion is obvious. Conversely, if a is regular, then let u be a set of subsets of a whose union is a. Either one of the subsets is cofinal, hence of cardinal a. Or each subset is bounded above, and the set of supremums (= least upper bounds) is cofinal in a, hence of cardinal a. Replace each supremum by one of the corresponding subsets (axiom of choice): the set u has at least cardinal a. •

2.8.4 Regular or singular cardinal

The preceding proposition suggests the following generalization. A cardinal a (not necessarily an aleph) is said to be **singular** iff it is the union of a set strictly subpotent with a, whose elements are strictly subpotent with a; it is said to be regular otherwise.

In the presence of the axiom of choice, every cardinal is an aleph, and we have the classical definition 2.8.1. In the absence of the axiom of choice, we do not know whether this generalized definition of regular and singular cardinal yields interesting results.

With the axiom of choice and the continuum hypothesis, we know that the cardinal of the continuum equals ω_1 and so is regular. With only the axiom of choice, there exist models where the continuum is a regular aleph, and others where the continuum is a singular aleph. It can have any infinite cofinality except ω. For example it cannot equal ω_ω. This restriction on the cofinality results from the fact that, for any partition of the continuum into a countable number of subsets, there is at least one which is equipotent with the continuum: see 1.5.3 (KÖNIG's theorem using the axiom of choice).

2.8.5 Equality and inequality between alephs

(1) **Let a be a regular aleph; for every b $(1 \leq b < a)$ we have $^b a = a$** ([242] TARSKI 1938; uses generalized continuum hypothesis; ZF suffices for $a = \omega$; for $a = \omega_1$, ZF plus choice plus continuum hypothesis).

(2) **Let a be a limit aleph; for every $c, d < a$ we have $^c d < a$** (Ibid. prop.9; generalized continuum hypothesis is used).

• (1) The statement is true for $a = \omega$. Suppose that a is an infinite successor aleph, hence of the form $a = {}^c 2$ (generalized continuum hypothesis), and b satisfies $1 \leq b \leq c$. Then we have $^b a = {}^b({}^c 2) = {}^{(b \times c)}2 = {}^c 2 = a$ since $b \times c = c$ by 1.6.8.

Now suppose that a is a limit aleph which is still regular and strictly greater than ω. There exists an increasing a-sequence of successor cardinals $a_i < a (i$ runs through $a)$ with $a = \operatorname{Sup} a_i$. From a certain point on, we have $a_i > b$, hence $^b(a_i) = a_i$ by the previous discussion. Since a is regular, no b-sequence is cofinal in a. Hence the set of all$b$-sequences with values in a is the union, as i varies, of the sets of b-sequences with values in each a_i. Hence $^b a = \operatorname{Sup}({}^b(a_i)) = \operatorname{Sup} a_i = a$. •

• (2) Let u be a successor aleph satisfying $c < u < a$ and $d < u < a$. Then u is regular, hence $^c d \leq {}^c u = u$ by (1); hence $^c d < a$. •

Statement (1) does not hold for singular alephs.

• Let $a = \omega_\omega$ and $b = \omega$. Decompose a into the union of the $\omega_i (i$ integer) and associate to each ω_i its successor ω_{i+1}. By KÖNIG's theorem the union of the ω_i is strictly subpotent to the cartesian product of the ω_{i+1}. Hence we have $(\omega_\omega) < {}^\omega(\omega_\omega)$. •

Even in ZF (KÖNIG's theorem uses the axiom of choice), we have the following counterexample.

• Let $a_0 = \omega$, and for each integer i let us define $a_{i+1} =$ (2 to the power a_i), and then $a = \operatorname{Sup} a_i$. Then we have $^a 2 \leq {}^\omega a$, hence $a < {}^\omega a$: to get the first inequality, associate to each subset u of a the sequence of the intersections $u_i = a_i \cap u$, and note that $u_i \in a_{i+1}$. •

2.8.6 Regular limit aleph

Let a be a regular limit aleph; then the aleph rank of a is itself; in other words $\omega_a = a$.

• Assume the contrary, that $a < \omega_a$, and let b be the aleph rank of a, so $a = \omega_b (b < a)$. Then either b is a successor ordinal, so a is a successor aleph. Or the sequence of the $\omega_i (i < b)$ is cofinal in a, hence a is not regular. •

On the other hand, we can have $\omega_a = a$ where a is singular, as already seen in 1.6.6 by taking the limit of $a(0) = \omega$, $a(1) = \omega_{a(0)} = \omega_\omega, \ldots, a(i+1) = \omega_{a(i)}, \ldots$ for each integer i: the aleph obtained has cofinality ω, hence is singular.

2.8.7 An ordinal equality

Let a **be an aleph** $\geq \omega_1$; **then** $\omega^a = a$ (ordinal equality; uses the axiom of choice).

Consequently by 1.3.6, every infinite aleph is an indecomposable ordinal.

• For every ordinal a we have $\omega^a \geq a$ by 1.3.3. It remains to prove that $\omega^a \leq a$ when a is an aleph $\geq \omega_1$.

Firstly prove the case $a = \omega_1$. Note that for any countable ordinal u we have ω^u denumerable (induction on u using the regularity of ω_1, hence the countable axiom of choice: see 2.8.1). Then take a sequence of countable ordinals $u(i)$ which is cofinal to ω_1 and apply the supremum equality $\omega^{\omega_1} = \omega^{\mathrm{Sup}\, u(i)} = \mathrm{Sup}(\omega^{u(i)}) \leq \omega_1$.

Secondly use the preceding argument in the recurrence step from an aleph $a \geq \omega_1$ to its successor aleph $b = a_+$. We prove by induction that for any ordinal u equipotent with a then ω_u is itself equipotent with a ; here we use the regularity of b, the successor aleph (see 2.8.2). Then as precedently we take ordinals $u(i)$ each equipotent with a and whose sequence is cofinal to b.

Finally if a is a limit aleph, take an increasing sequence of alephs $b(i)$ which is cofinal to a; then by hypothesis $\omega^{b(i)} = b(i)$ for each i so that $\omega^a = \mathrm{Sup}\, \omega^{b(i)} = \mathrm{Sup}\, b(i) \leq a$.

Note that we need 1.6.8 proposition (2), for instance in the inductive argument to prove that $\omega^{u+1} = \omega^u.\omega$ is equipotent with $\mathrm{Card}\, \omega^u \times \omega = \mathrm{Card}\, \omega^u$. •

2.8.8 A property of infinite alephs

(1) **Let** a **be an infinite aleph; then** $a = \Sigma i (i < a)$. In other words $a =$ the ordinal sum of all its strict initial intervals (uses axiom of choice).

• Either a is a regular aleph; then a is indecomposable, so that our statement is a particular case of 2.7.4 by taking for C the entire base $|a|$.

Or a is a singular infinite aleph, thus $a = \mathrm{Sup}\, a_i (i < \mathrm{Cof}\, a)$ where each a_i is an infinite successor aleph, hence a regular aleph: see 2.8.2 (axiom of choice). Then replacing each element x in a by the corresponding strict initial interval $< x$, we reach a sum equal to a_i in place of each a_i, so that finally we exactly reach a. •

(2) **Let** a **be an infinite aleph; then** $a = \Sigma \alpha_i (i < a)$ **where** $1 \leq \alpha_i < a$, **provided that either** a **be regular, or** α_i **be an increasing function defined on the entire domain** a.

• If a is regular, use 2.7.4 and 2.8.7 (every infinite aleph is indecomposable).

For α_i increasing function, consider again a sequence of regular alephs a_m whose limit is a, then construct an increasing function $\beta_m < a$ such that $\beta_m = \Sigma \alpha_i (i < a_m)$, each $\beta_m < a$.

If a is singular and α_i not increasing, we have the following counterexample: $a = \omega_\omega, \alpha_i = \omega_i$ for i integer and 1 for $i \geq \omega$: then $\Sigma \alpha_i = \omega_\omega.2$. •

Problem. The statement (1) is obviously false for $a = \omega^2$ and more generally for $a = \omega^n (n$ integer); is it true for ω^ω; is there an easy criterion for those powers of ω which satisfy it.

2.8.9 Accessible cardinal; axiom of accessibility

An infinite cardinal a is said to be **accessible** iff either $a = \omega$, or there exists a set b strictly subpotent to a and whose elements are strictly subpotent to a, which satisfies $\cup b = a$; or finally there exists a set c strictly subpotent with a, and where a is subpotent to $\mathcal{P}(c)$. A cardinal is said to be **inaccessible** otherwise.

The axiom of accessibility asserts that every set has accessible cardinality. Under the assumption that ZF is consistent, the theory ZF plus the axiom of accessibility is consistent: see [226] SHEPHERDSON 1952.

Every accessible limit aleph different from ω is singular (uses generalized continuum hypothesis). Consequently in view of 2.8.2, for an accessible aleph $a \neq \omega$, this a is regular iff it is a successor aleph.

• Let ω_α be our aleph, where α is a non-zero limit ordinal. Every strictly smaller cardinal is an $\omega_i (i < \alpha)$. There is no set strictly subpotent with ω_α, hence of cardinality ω_i, with ω_α subpotent to $\mathcal{P}(\omega_i)$: indeed $\mathcal{P}(\omega_i) = \omega_{i+1}$ (generalized continuum hypothesis) and $i + 1 < \alpha$. Finally as our aleph is assumed to be accessible and distinct from ω, there exists a set strictly subpotent to ω_α, whose elements are as well strictly subpotent to ω_α, and whose union is ω_α: our aleph is thus singular. •

2.8.10 Weakly inaccessible aleph

An aleph is said to be **weakly inaccessible** iff it is different from ω, regular and a limit aleph. The preceding proposition is equivalent to saying that, with the axiom of choice and the generalized continuum hypothesis, every weakly inaccessible aleph is inaccessible.

Using only the axiom of choice, every inaccessible aleph is weakly inaccessible. Indeed, it is different from ω, regular, and there is no strictly smaller c for which it is subpotent to $\mathcal{P}(c)$: hence it is not a successor aleph.

Problem. Assume that ZF plus choice plus the existence of a weakly inaccessible aleph less than the continuum is consistent. Then does this theory remain consistent if we require that the continuum itself be weakly inaccessible. Classically we have an affirmative answer if we assume the consistency of ZF plus choice plus the existence of a weakly inaccessible aleph greater than the continuum.

Some results of BLASS, in a letter to Hodges, 1982. Consider the following statements:

(1) every countable union of countable sets is countable;

(2) if a set a and its elements have cardinals $<$ or $\mid \omega_1$, then the union of a has cardinal $<$ or $\mid \omega_1$;

(3) given a function f , if $\mathrm{Rng}\, f = \omega_1$, then ω_1 is subpotent to $\mathrm{Dom}\, f$.

Then it is proved that neither of (1) and (2) implies the other, and that (2) is equivalent to (3) plus the regularity of ω_1.

2.9 Augmentation of a relation, linear augmentation of a poset

2.9.1 Augmentation, weakening

Let A be an n-ary relation with base E. An n-ary relation B with the same base is said to be an **augmentation** of A, and A is a **weakening** of B, iff every n-tuple in E having the value $(+)$ in A still has the value $(+)$ in B.

Augmentation defines a partial ordering on the set of n-ary relations with a given base E.

A **strict** or **proper augmentation** of A is an augmentation distinct from A. Similarly we speak of a strict or proper weakening.

Let A be a poset which is not a chain. We obtain an **augmented poset** of A by taking two elements u, v which are incomparable $(\mathrm{mod}\, A)$, then defining, for any two distinct elements x, y of the base, $x < y$ iff either $x < y$ $(\mathrm{mod}\, A)$ or $x \leq u$ and $v \leq y$ $(\mathrm{mod}\, A)$.

Generalize this procedure as follows. **Let A be a poset with base E. Let $D \subseteq E$ and B be a poset with base D which is an augmentation of the restriction A/D. Then there exists a poset C which is an augmentation of A and an extension of B to E.**

Moreover, **there exists a poset C_0 which is minimal among these augmentations** C, i.e. every C which verifies the preceding is an augmentation of C_0.

• Let x, y be two elements of E. Put $x \leq y$ iff $x \leq y(\mathrm{mod}\, A)$ or if there exist two elements x', y' of D with $x \leq x'(\mathrm{mod}\, A)$ and $x' \leq y'(\mathrm{mod}\, B)$ and $y' \leq y(\mathrm{mod}\, A)$. The poset thus obtained is the minimal augmentation C_0; the C's are augmentations of C_0. •

2.9.2 Case of well-founded posets

(1) **A weakened well-founded poset is well-founded.**

On the other hand, an augmented well-founded poset is not necessarily well-founded: augment the identity on a denumerable base into the converse ω^- of the ordinal ω.

(2) **For each well-founded poset A, there exists a well-ordered augmentation of A** (uses axiom of choice; ZF suffices if A is countable or has well-orderable base).

• For each ordinal i strictly less than the height $\mathrm{Ht}\, A$, take a well-ordering C_i baset on the free set of elements with height i (axiom of choice). Then take the sum of the C_i according to increasing i. •

(3) **Let A, B be two well-founded posets with the same base. If B is an augmented poset of A, then for each x in the base** $\mathrm{Ht}\, x(\mathrm{mod}\, B) \geq \mathrm{Ht}\, x(\mathrm{mod}\, A)$. Consequently $\mathrm{Ht}\, B \geq \mathrm{Ht}\, A$. Proof by induction.

2.9.3 Augmentation axiom

This axiom asserts that **each poset has a totally ordered, or linear augmentation**. The augmentation axiom follows from the ultrafilter axiom ([240] SZPILRAJN 1930); ZF suffices in the case of a countable poset.

• Let A be a poset with base E. For each finite subset F of E, we let U_F denote the set of all chains with base F each of which is an augmentation of A/F. The U_F are non-empty and verify the hypotheses of the coherence lemma 2.4.1. Hence there exists a relation C based on E which satisfies $C/F \in U_F$ for each F (uses ultrafilter axiom). This C is a chain, since its finite restrictions are chains; finally C is an augmentation of A. •

The augmentation axiom can be stated equivalently: **given a poset A and a chain C which is an augmentation of a restriction of A, there exists a chain which is simultaneously an augmentation of A and an extension of C.** It suffices to use 2.9.1.

The augmentation axiom implies the ordering axiom of 2.4.4. Given a set E, it suffices to augment into a chain the poset which reduces to the identity on E.

Note that the augmentation axiom is strictly weaker than the ultrafilter axiom: see [63] FELGNER, TRUSS 1999; the result going back to Felgner's Doctoral Diss. 1972. Also the augmentation axiom is strictly stronger than the ordering axiom: see [167] MATHIAS 1974.

2.9.4 Augmentation and the set of initial intervals

Given a poset A, to every partially or totally ordered augmentation B of A there corresponds the ordering $\mathcal{J}(B)$ of the initial intervals of B. These initial intervals $(\text{mod}\, B)$ remain initial intervals $(\text{mod}\, A)$.

In other words **$\mathcal{J}(B)$ is a subset of $\mathcal{J}(A)$.**

More precisely to obtain the first set from the other: for each pair of elements x, y such that $x|y$ $(\text{mod}\, A)$ and $x < y$ $(\text{mod}\, B)$, suppress those sets which contain y yet not x.

Let A be a poset. If C is a totally ordered augmentation of A, then $\mathcal{J}(C)$ is a total ordering which is maximal among those totally ordered restrictions of $\mathcal{J}(A)$.

Conversely, **every maximal totally ordered restriction of $\mathcal{J}(A)$ is of the form $\mathcal{J}(C)$, where C is a totally ordered augmentation of A. Moreover this C is unique** ([18] BONNET, POUZET 1969).

• Starting with a totally ordered augmentation C of A, we already know that $\mathcal{J}(C)$ is closed under union and intersection, and has the separation property. Conversely, let $\mathbf{C}$ be a total ordering which is maximal among those totally ordered restrictions of $\mathcal{J}(A)$. By 2.5.2 this $\mathbf{C}$ is closed under union and intersection, and has the separation property. To obtain $\mathbf{C} = \mathcal{J}(C)$, define the totally ordered augmentation C of A by the condition that, given two elements x, y of E, we put $x \leq y$ $(\text{mod}\, C)$ iff every element of $\mathbf{C}$ which contains y as an element also contains

x. The antisymmetry of C follows from the fact that, for distinct x, y there exists an element of **C** which separates them. Moreover since **C** is a chain, C is also. The uniqueness of C follows from the fact that two distinct total orderings C and C' yield two distinct $\mathcal{J}(C)$ and $\mathcal{J}(C')$. •

2.9.5 Axiom of maximal chain on initial intervals

This axiom asserts that, **given a poset A, the partial ordering of inclusion on the initial intervals of A admits a maximal chain among its restrictions**. It is equivalent to the augmentation axiom (2.9.3).

Finally the augmentation axiom will allow us to define the conjunction of chains based on a same set, and the dimension of a poset: see 4.9 below.

2.10 Stratified poset, minimal stratified augmentation, partition in slices (Bonnet, Pouzet)

2.10.1 Stratified poset

A partial ordering or poset A is said to be **stratified** iff the union of identity and incomparability (mod A) forms a transitive relation, hence an equivalence relation. Consequently $x < y$ and $y|z$ imply that $x < z (\text{mod}\, A)$. Similarly with $>$ in the place of $<$. Several authors call it a **weak ordering** (or weak order); see [139] KONG, RIBENBOIM 1994.

The equivalence classes of incomparability-identity form a chain, by putting (Class of x) $<$ (Class of y) iff $x < y \,\text{mod}\, A$. This chain of the equivalence classes will be called the principal chain of the stratified partial ordering A. Every maximal chain in A is isomorphic with the principal chain.

2.10.2 Stratified augmentation, partition in slices

Given an arbitrary poset A, there exist stratified partially ordered augmentations of A. In particular there exist chains which are augmentations of A.

We obtain as follows a minimal stratified augmentation, in the sense that if $x < y$ (modulo the augmentation), then either $x < y$ (mod A) or there exist x', y' with $x < x'|y' \leq y$ or with $x \leq x'|y' < y$ (mod A).

Lemma for partition in slices.

Let A be a poset. There exists an equivalence relation R on $|A|$ for which the equivalence classes are free subsets (mod A), and a chain H on the set of equivalence classes, such that for any two elements x, y of $|A|$, we have: (Class of x) $\leq$ (Class of y) mod H iff there exist two elements x', y' which are equivalent (mod R) and satisfy $x \leq x'$ and $y' \leq y$ (mod A) (uses axiom of choice; [18] BONNET, POUZET 1969).

• First of all, denote by R any equivalence relation on the base $|A|$, and by H any chain on the set of equivalence classes of R, which satisfy the following condition: (Class of x) $\leq$ (Class of y) mod H) iff either x is equivalent to $y(\text{mod}\, R)$, or there exists an equivalence class U of R which is a free subset of $|A|$ and two elements x', y' of U with $x \leq x'$ and $y' \leq y(\text{mod}\, A)$.

Such couples (R, H) exist: it suffices to take R to be the trivial equivalence relation with one equivalence class $|A|$, and H the chain on the singleton of $|A|$.

A couple (R', H') is said to be finer than (R, H) iff every class $(\text{mod}\, R)$ which is free $(\text{mod}\, A)$ remains a class $(\text{mod}\, R')$, and moreover every class $(\text{mod}\, R)$ is a union of classes $(\text{mod}\, R')$ constituting an interval of H', and finally for any two of these intervals, the chain induced by H' is identical to the chain $(\text{mod}\, H)$ of the corresponding equivalence classes. The comparison "finer than" defines a partial ordering on the couples (R, H).

We shall prove that a couple (R, H) for which the classes are not all free $(\text{mod}\, A)$ admits a couple (R', H') which is strictly finer. For this, take a class U of R which is not free. Take a subset V of U which is maximal free $(\text{mod}\, A)$ (2.2.6), and partition U into three disjoint subsets: V, the set of elements having a greater element $(\text{mod}\, A)$ in V, and the set of elements having a smaller element $(\text{mod}\, A)$ in V. Define the equivalence classes $(\text{mod}\, R')$ as those $(\text{mod}\, R)$, excepting U which is partitioned into three equivalence classes $(\text{mod}\, R')$. The chain H' is the same as H, except for the substitution of U by the three classes with the obvious ordering. Using Hausdorff-Zorn axiom, consider a maximal couple obtained from a maximal chain of couples (R, H), totally ordered by the comparison "finer than" . For this maximal (R, H), the equivalence classes defined by R are all free $(\text{mod}\, A)$. •

The converse of the preceding lemma is false in the following sense. Given a stratified minimal reinforcement B of A, we may have either $x < y$ $(\text{mod}\, A)$, or the existence of x', y' with $x < x'|y' \leq y$ $(\text{mod}\, A)$ or $x \leq x'|y' < y$ $(\text{mod}\, A)$ and yet $x \not< y$ $(\text{mod}\, B)$.

2.11 Tree

2.11.1 Definition, examples, connection with graph theory

A poset A is called a **tree** iff, for each element x in the base, the set of all elements less than or equal to x is totally ordered modulo A.

For example, every chain or total ordering is a tree. The free poset reduced to the identity, is a tree.

Another example: beginning with a set a and a set A of subsets of a, where any two elements of A are either disjoint or one is included in the other, and no element of A is empty. Then reverse inclusion constitutes a tree based on A. Conversely, let A be a tree with base E. To each element $x \in E$, associate the set A_x of those elements $\geq x(\text{mod}\, A)$: the A_x ordered by reverse inclusion form a tree, isomorphic with A.

Consider the comparison graph of a tree A, i.e. the binary irreflexive and symmetric relation which takes the value $(+)$ for those couples (x, y) satisfying $x < y$ or $y < x$. Obviously this comparison graph has no finite cycle with $\geq$

3 elements. Conversely a finite connected graph without cycles is called a tree (in graph theory; see for instance [155] van LINT, WILSON 1992); such a finite "graph-tree" can be oriented to give a tree in our sense: start from an extremity a (a vertex a which is joined to only one other vertex), then take a as the minimum element and orient each edge by increasing distances to a.

2.11.2 Maximal chains and free subsets in a finite tree

If A is a finite tree, then every base of a maximal chain of A and every maximal free set $(\bmod A)$ have one and only one common element [145] (KUREPA 1952).

• Let F be the base of a maximal chain in A, and G a maximal free set $(\bmod A)$. Let u be the maximum of the chain A/F. Then either u is identical or incomparable $(\bmod A)$ to each element of G. In that case u belongs to G, since G is a maximal free set. Or there exists an element v of G, which is distinct from u and comparable with $u(\bmod A)$. Then $v < u$; for otherwise v could be added to F and so A/F would not be a maximal chain. Thus v is comparable with every element $< u$ and hence v belongs to F since A/F is maximal. Thus v is common to F and G. •

This result does not hold for an infinite tree.

• Take the chain of the integers and to each integer i, associate an element $i' > i$ but i' incomparable with integers $> i$. Finally the i' are set to be mutually incomparable. Then the set of integers defines a maximal chain, and the set of "primed" integers is a maximal free set; example due to KUREPA. •

The result no longer holds for an arbitrary finite partial ordering. Indeed take four elements and let $a < b, a < c, b' < c$ with $b|c, a|b', b|b'$. Then the chain (a, c) and the antichain (b, b') are maximal.

2.11.3 Chain associated with a tree

Let A be a tree and E its base. There exists a chain C based on E, such that for each element $x \in E$, the interval $\geq x$ $(\bmod A)$ becomes an interval $(\bmod C)$ with minimum x (uses ultrafilter axiom; ZF suffices if A is countable) .

• Suppose first that E is finite. Let $u_1, ..., u_h$ be the minimal elements $(\bmod A)$ and in an arbitrary manner order these elements, obtaining for example $u_1 < ... < u_h$. Then for each $i = 1, ..., h$ replace u_i by a sequence beginning with u_i and totally order in an arbitrary manner the immediate successors of $u_i(\bmod A)$. Continue in this fashion until all the elements of E are contained in this chain: thus we obtain the desired chain C.

Now consider the case where E is infinite. To each finite subset F of E, associate the non-empty set U_F of chains with base F which satisfy the proposition for the restricted finite tree A/F. Then the U_F satisfy the hypotheses of the coherence lemma 2.4.1 (equivalent to ultrafilter axiom). Then we obtain a chain

C with base E, all of whose finite restrictions satisfy the proposition. Hence A and C satisfy the proposition. •

Consequently, given a tree A, there exists a chain with the same base, and a set U of intervals (modulo the chain) which are mutually inclusive or disjoint, such that A is isomorphic with reverse inclusion on U.

Given a set E and a set U of subsets of E, which are mutually inclusive or disjoint, there exists a chain with base E, for which the elements of U are intervals (ultrafilter axiom; ZF suffices if E is countable).

• Add to U the singletons of elements of E, and apply the preceding proposition to the set thus obtained, which is ordered by reverse inclusion. Replace the chain by its restriction to the set of singletons; each singleton being identified with its unique element. •

2.11.4 Reduced tree or reduct

Let A be a tree. We introduce an equivalence relation called **reduction** (mod A). Two elements x, y of the base are said to be **equivalent** iff they are comparable and every element comparable to x is also comparable to y (mod A). Equivalently, every maximal chain (with respect to inclusion) restriction of A and which contains x, also contains y. The relation thus defined is reflexive, symmetric and transitive. An equivalence class is a totally ordered restriction of A.

We say that an equivalence class **precedes** another equivalence class, if the elements of the first precede (mod A) the elements of the second. The partial ordering on the equivalence classes thus defined, is a tree, and called the **reduced tree** or **reduct** of A.

Given a tree A, associate to each element x of the base the set A_x of maximal chains (with respect to inclusion) which contain x in their base. For x and y incomparable (mod A), the sets A_x and A_y are disjoint. For $x \leq y(\text{mod}\, A)$, we have $A_x \supseteq A_y$. Hence these sets, ordered by reverse inclusion, form a tree isomorphic with the reduct of A.

2.11.5 Initially maximal chain

Let A be a poset. We say that a chain (restriction of A) is **initially maximal** iff it is an initial interval of a maximal chain (with respect to inclusion). In other words, if x belongs to the base of an initially maximal chain U, then every element $< x(\text{mod}\, A)$ either belongs to the base U or is incomparable (mod A) to some element in this base.

The set of initially maximal chains, ordered by inclusion, forms a tree.

2.11.6 Non-amalgamation for trees

The amalgamation lemma 1.7.3, which holds for posets and leads to a poset, which also holds for chains and leads to a chain, no longer leads to a tree when starting with two trees.

• Take a tree on $\{a, b, c, d\}$ with a, b, c mutually incomparable , $d < a$ and $d < b$ and $d|c$; and another tree on $\{a, b, c, e\}$ with $e < b$ and $e < c$ and $e|a$. Then either $d < e < c$ or $e < d < a$: contradiction. •

2.12 Cofinality of a poset, cofinal height

2.12.1 Generalities

The **cofinality** of a poset A, denoted $\operatorname{Cof} A$ and introduced in 2.7.3, was studied above only in the case of a chain.

If A is non-totally ordered, then $\operatorname{Cof} A$ can be a singular aleph. Indeed construct A by taking, for each integer i, a chain of order type ω_i, and putting them together mutually incomparably: then $\operatorname{Cof} A = \omega_\omega$.

Given a poset A and a cofinal set $D(\operatorname{mod} A)$, there exists a cofinal subset U of D with $\operatorname{Card} U = \operatorname{Cof} A$. Consequently $\operatorname{Cof}(A/D) = \operatorname{Cof} A$ for each cofinal set D (uses axiom of choice; ZF suffices if A is countable or has well-orderable base).

• Take a set V cofinal $(\operatorname{mod} A)$ and with least cardinal, hence $\operatorname{Card} V = \operatorname{Cof} A$. Then replace each element $x \in V$ by an element in D which is greater than x. •

Let A be a poset and $u = \operatorname{Cof} A$. There exists a cofinal subset U of cardinal u, such that the restriction A/U is a well-founded poset with height $\operatorname{Ht}(A/U) \leq u$. A refinement of corollary 2.7.2, due to POUZET in 1979, published in ToR-86 p.50. Uses the axiom of choice; ZF suffices if A is countable or has well-orderable base.

• Let D be a cofinal subset with least cardinal $u = \operatorname{Cof} A$. Order D by its cardinality, which we assume to be an aleph, and call $a_i (i < u)$ this sequence (axiom of choice). Extract a sequence b_i by removing, for each i, the a_j such that $j > i$ and $a_j < a_i$ $(\operatorname{mod} A)$. Firstly the sequence of the b_j has length at most equal to u. Secondly, the inequality $i < i'$ implies $b_i <$ or $|b_{i'} (\operatorname{mod} A)$. Thirdly, the set U of values b is cofinal in A and A/U is a well-founded poset.

For each subscript $i < u$, we see by induction on i that the height of b_i $(\operatorname{mod} A/U)$ is $\leq i$. Indeed, firstly each b_i is $\geq$ or $|b_0$ $(\operatorname{mod} A)$. Secondly assume that there exists a least i with b_i of height $> i$ $(\operatorname{mod} A/U)$. Then there exists a b_k of height i, with $b_k < b_i$ $(\operatorname{mod} A)$. Hence $k < i$ by the previous alinea, which contradicts the minimality of i. Finally the height of A/U is $\leq u$. •

2.12.2 Cofinality of height ≤ cofinality

(1) **For every well-founded poset A, we have** $\operatorname{Cof} \operatorname{Ht} A \leq \operatorname{Cof} A$ (uses axiom of choice; ZF suffices if A is countable or with well-orderable base).

• Take a cofinal set D of least cardinality, so that $\operatorname{Card} D = \operatorname{Cof} A$. The heights $(\operatorname{mod} A)$ of the elements of D form a set H which is cofinal in the ordinal $\operatorname{Ht} A$. By the axiom of choice, H is subpotent with D, hence we have $\operatorname{Cof} \operatorname{Ht} A \leq \operatorname{Card} H \leq \operatorname{Card} D = \operatorname{Cof} A$. •

If A is a well-ordering, then $\operatorname{Ht} A$ is isomorphic with A, hence $\operatorname{Cof} \operatorname{Ht} A = \operatorname{Cof} A$.

On the other hand, take the well-founded poset A formed of a component ω, a component $\omega + 1$, the elements of one chain put incomparable with those of the other. Then $\mathrm{Cof}\, A = \omega, \mathrm{Ht}\, A = \omega + 1$ so that $\mathrm{Cof}\,\mathrm{Ht}\, A = 1$.

(2) **Let A be a well-founded poset. Then there exists a cofinal restriction C of A satisfying** $\mathrm{Ht}\, C \leq \mathrm{Card}\,\mathrm{Ht}\, A$ (POUZET in 1979, published in ToR-86 p.45).

• To each ordinal $i < \mathrm{Ht}\, A$ associate the class B_i of elements of height $i (\mathrm{mod}\, A)$. Order the set of the B_i by a well-ordering isomorphic with its cardinal, which is $\mathrm{Card}\,\mathrm{Ht}\, A$. Denote by B the well-founded stratified poset with base $|A|$, defined by putting $x < y (\mathrm{mod}\, B)$ iff "class of x" $<$ " class of y " according to the preceding well-ordering. Then $\mathrm{Ht}\, B = \mathrm{Card}\,\mathrm{Ht}\, A$.

By 2.7.2, there exists a cofinal restriction C of A such that any two elements of $|C|$ are never ordered in opposite senses by A and B. Hence if x, y belong to $|C|$ and $x < y$ $(\mathrm{mod}\, A)$, then $x < y$ $(\mathrm{mod}\, B)$, as incomparability $(\mathrm{mod}\, B)$ is impossible since x and y belong to two distinct classes B_i. Thus $B/|C|$ is an augmentation of $C = A/|C|$. By 2.9.2 proposition (3), we have $\mathrm{Ht}\, C \leq \mathrm{Ht}(B/|C|) \leq \mathrm{Ht}\, B = \mathrm{Card}\,\mathrm{Ht}\, A$. •

2.12.3 Cofinal height

Given a poset A, if there exists a well-founded cofinal restriction of A, then the least height of such restrictions is called the **cofinal height** of A, and denoted $\mathrm{Cofh}\, A$. With the axiom of choice, the cofinal height exists for every partial ordering A: see corollary in 2.7.2. Using only the axioms of ZF, the cofinal height exists at least for all well-founded posets.

(1) **For every poset A, we have** $\mathrm{Cofh}\, A \leq \mathrm{Cof}\, A$. Consequence of the preceding subsection; uses the axiom of choice; ZF suffices if A is countable or has well-orderable base. Strict inequality can happen: for example when A is reduced to the identity on its base E, then $\mathrm{Cofh}\, A = 1$ and $\mathrm{Cof}\, A = \mathrm{Card}\, E$.

(2) **If A is a total ordering, then** $\mathrm{Cofh}\, A = \mathrm{Cof}\, A$. Consequence of (1) and of the fact that the cardinal of a well-ordered restriction of A is at most equal to its height. Same conditions as in (1).

(3) **If A is a well-founded poset, then** $\mathrm{Cofh}\, A \leq \mathrm{Card}\,\mathrm{Ht}\, A$: another form of proposition (2) in the preceding subsection.

2.12.4 Two propositions on cofinal height

(1) **Given a poset A, the cofinal height is the least height of well-founded cofinal restrictions of A whose cardinal is equal to** $\mathrm{Cof}\, A$ (uses axiom of choice; ZF suffices if A is countable or has well-orderable base).

• Start with a cofinal subset D such that A/D is a well-founded poset of least height $\mathrm{Ht}(A/D) = \mathrm{Cofh}\, A$. Take a subset D' of D which is cofinal and of least cardinal $\mathrm{Cof}\, A$ (see 2.12.1, using the axiom of choice). Then $\mathrm{Ht}(A/D') \leq \mathrm{Ht}(A/D) = \mathrm{Cofh}\, A$ by 2.7.1 proposition (1). •

(2) **If B is a cofinal restriction of the poset A, then** $\operatorname{Cofh} B \leq \operatorname{Cofh} A$ (assuming that these cofinal heights exist). Indeed each cofinal restriction of B is a cofinal restriction of A.

The following example, due to [197] POUZET 1979, shows that **strict inequality is possible**, contrary to the situation for cofinality (see 2.12.1).

• Take as base the cartesian product $\omega \times \omega_1$. For each couple (i, j) with i integer, j countable ordinal, put $(i, j) \leq (i', j)$ for every integer $i' \geq i$. For each even integer i, put $(i, j) \leq (i, j')$ for all $j' \geq j$. For each odd integer i, we let the couples with first term i be mutually incomparable. Then we complete by transitivity. The poset A thus obtained has a cofinal subset of all couples with odd first term: thus we have $\operatorname{Cofh} A = \omega$. The restriction to couples with even first term has cofinal height ω_1. •

2.12.5 The cofinal height is a cardinal

For each poset A, the cofinal height of A is a cardinal (uses axiom of choice; ZF suffices if A is countable, or has well-orderable base, or is well-founded; POUZET in 1979, published in ToR-86 p.52).

• Take a well-founded cofinal restriction B of A with least height, so that $\operatorname{Ht} B = \operatorname{Cofh} A$: see corollary in 2.7.2. By the preceding subsection we have $\operatorname{Cofh} A \leq \operatorname{Cofh} B$. By 2.12.3 proposition (3), we have $\operatorname{Cofh} B \leq \operatorname{Card} \operatorname{Ht} B = \operatorname{Card} \operatorname{Cofh} A$. •

2.12.6 Comparison between cofinal height and cofinality of height

Let A be a well-founded poset; any of the possible comparisons "strictly less than", "strictly greater", "equal" can be obtained for the cardinals $\operatorname{Cofh} A$ and $\operatorname{Cof} \operatorname{Ht} A$.

(1) Start with the chain of integers. To each integer i associate an element i'. Put $i' > i$; put i' incomparable with integers $\geq i+1$; finally for all integers i and $j \neq i$, put i' incomparable with j'. Denote by A this partial ordering; then $\operatorname{Ht} A = \operatorname{Cof} \operatorname{Ht} A = \omega$ and $\operatorname{Cofh} A = 1$.

(1') In the above example 2.12.4 proposition (2), we have $\operatorname{Cofh} A = \omega$ and $\operatorname{Cof} \operatorname{Ht} A = \omega_1$.

(2) For each ordinal α, we have $\operatorname{Cofh} \alpha = \operatorname{Cof} \alpha = \operatorname{Cof} \operatorname{Ht} \alpha$.

(3) Construct A by starting with the ordinal ω_ω. For each integer i, add by a bifurcation another chain ω_i after the initial interval ω_i. Then we have $\operatorname{Ht} A = \omega_\omega$ so $\operatorname{Cof} \operatorname{Ht} A = \omega$ and $\operatorname{Cofh} A = \omega_\omega$.

We can also have $\operatorname{Cof} \operatorname{Ht} A < \operatorname{Cofh} A$ where $\operatorname{Cofh} A$ is a regular aleph. Consider the ordinal product $\omega_1.\omega$ and then the direct product $A = \omega_1 \times (\omega_1.\omega)$. Then $\operatorname{Ht} A = \omega_1.\omega$ so that $\operatorname{Cof} \operatorname{Ht} A = \omega$.

2.13 Net or directed poset, ideal

A poset is said to be a **net**, or **directed poset**, iff given any two elements in the base, there is a third element greater than both.

Examples: a chain (= total or linear ordering); a poset having a maximum.

A Sierpinski poset, in the sense of 2.2.7, **is a net**, provided that all real numbers are used in its construction (we use the continuum hypothesis). Indeed given two countable ordinals $u < v$ and their associated real numbers with the assumption $r(u) > r(v) (\mathrm{mod}\, \mathbf{R})$, there are ω_1 = continuum many countable ordinals $w > v$ with $r(w) > \mathrm{Max}(r(u), r(v)) (\mathrm{mod}\, \mathbf{R})$.

Given a poset A, an **ideal** in A is any initial interval which is directed.

For example, every initial interval of a chain is an ideal; or again, for an arbitrary poset A, the initial interval obtained by taking an element a and all elements $\leq a$ (mod A).

In the case of the partial ordering of inclusion among all subsets of a set E, an ideal is a set of subsets of E which is closed under taking a subset and taking the union of two subsets of E. Hence **an ideal on E is the complement of a filter on E**. Consequently **the traditional boolean prime ideal axiom is another designation of the ultrafilter axiom**.

Given a set U of ideals which are totally ordered by inclusion, or simply if U is a net with respect to inclusion, then the union of U is an ideal.

Maximal ideal axiom. Every ideal in a poset A is included in a maximal ideal of A (with respect to inclusion).

The maximal ideal axiom is equivalent to the axiom of choice.

• It follows from the maximal chain axiom, or HAUSDORFF-ZORN axiom: starting with an ideal U of A, take a maximal chain in the partial ordering of inclusion of ideals of A, and then take the union of the ideals in this chain, which is a maximal ideal.

Conversely, the maximal ideal axiom implies the well-ordering axiom: take a maximal ideal in the well-founded interval ordering: see 2.2.5. •

2.13.1 Characterization of an ideal

In a poset, an initial interval I is an ideal iff for any two initial intervals X, Y, the condition $X \cup Y = I$ implies that $X = I$ or $Y = I$.

• Suppose that I is an ideal which is the union of two initial intervals X, Y and that $X \neq I$ and $Y \neq I$. Thus there exists an element $x \in I - X$ and an element $y \in I - Y$. Consequently there exists an element z such that $z \in I$ and $z \geq x$ and $z \geq y$. Hence $z \notin X$ and $z \notin Y$: contradiction.

Conversely, suppose that I is an initial interval which is not an ideal. So there exist two elements u, v of I without any common upper bound in I. Define U to be the initial interval of those elements x for which there exists a common upper bound of x and u in I: hence $u \in U$ and $v \notin U$. Define V to be the initial interval of those y for which there exists a $z \geq y$ in I, such that there is no common upper bound of z and u in I: hence $u \notin V$ and $v \in V$. Finally we have $U \cup V = I$. •

2.13.2 Cofinal restriction of a net

(1) **For every net A, there exists a cofinal restriction of A which is a well-founded net** (uses axiom of choice; ZF suffices if A is countable).

• By 2.7.2 corollary (axiom of choice), there exists a cofinal F with A/F well-founded. For any two elements x, y of F, by hypothesis there exists an element $z \geq x$ and $\geq y (\bmod A)$. Since F is cofinal, there exists an element t in F with $t \geq z$ and so $t \geq x$ and $\geq y$: hence A/F is directed; in other words, a net. •

(2) **For every denumerable net A, there exists a totally ordered restriction of A which is cofinal and isomorphic either to ω or to 1.**

• Take an ω-sequence of the elements a_i(i integer) of the base. Then take $b_0 = a_0, b_1 =$ the element with least index which is greater than (mod A) both b_0 and a_1, then b_2 greater than b_1 and a_2, etc. •

On the other hand, the direct product $\omega \times \omega_1$, defined by using 4.8 below, has no totally ordered cofinal restriction. Indeed that would require a total ordering of order-type ω_1 formed of couples (i, j) with i, j increasing and i running through ω and j running through ω_1, which is impossible.

2.14 Computation of posets (Chaunier, Lygerōs)

Let n be the cardinal of a finite poset A; then we have $n(n-1)/2$ edges (or pairs). For each edge we can have either the incomparability $u|v$ between its two vertices (or elements) u, v or an arrow (or comparison) $u < v$ or $v < u \bmod A$.

We denote by $P_n(i)$ the number of isomorphism types of those posets of cardinal n which have exactly i arrows $(0 \leq i \leq n(n-1)/2)$. For $i = 0$ we have only the antichain, so that $P_n(0) = 1$ for every n. For $i = 1$ and $n \geq 2$ we have a unique poset (up to isomorphism) with exactly one arrow, so that $P_n(1) = 1$. For $i = n(n-1)/2$ we have only the chain, so that $P_n(n(n-1)/2) = 1$. For $i = (n(n-1)/2) - 1$, we have exactly two mutually incomparable elements; they can admit either 0, or 1, or 2, ... or $n-2$ strictly greater elements, so that P_n takes the value $n-1$. For $n = 3$ we have 5 isomorphism types with $P_3(0) = 1, P_3(1) = 1, P_3(2) = 2$ (we can have two "convergent" or two "divergent" arrows), $P_3(3) = 1$. For a given cardinal n, the best is to write the $P_n(i)$ as follows, by decreasing values of i.

The following table concerns cardinals 1 to 7 with a total number of posets equal resp. to 1, 2, 5, 16, 63, 318, 2045 (computed by WRIGHT in 1972).

1
1 1
1 2 1 1
1 3 3 4 3 1 1
1 4 6 9 12 10 10 6 3 1 1
1 5 10 17 28 35 44 46 43 36 25 16 7 3 1 1
1 6 15 29 54 83 123 168 204 239 249 243 209 167 113 74 38 18 7 3 1 1

Recent computations give 16 999 posets for the cardinal 8 (DAS 1977), 183 231 for 9 (MÖHRING 1984), 2 567 284 for 10 (CULBERSON, RAWLINS 1991), 46 749 427 for 11 (same authors), 1 104 891 746 for 12 (CHAUNIER, LYGERŌS 1991), 33 823 827 452 for 13 (same authors 1992).

In ([80] FRAÏSSÉ, LYGERŌS 1991) a conjecture of unimodality is proposed: each sequence $P_n(i)$ with n fixed and i variable would increase from 1 until a maximum and then decrease to 1.

This conjecture is positively checked for $n \leq 13$; see [31] CHAUNIER, LYGERŌS 1992. For $n = 13$ the maximum is $P_{13}(36) = 2\ 138\ 021\ 170$.

For $n = 14$, the value P_{14} is presently known for all integers (between 1 and $1/2(14.13) = 91$) except for 47. The biggest value is almost certainly $P_{14}(42) =$ 78 181 684 187, due to Paul ZIMMERMANN (april 2000). The previous conjecture is checked, unless an aberrant value would happen for 47.

2.15 Exercises

2.15.1 Poset and linear augmentation

Communicated by MILNER and POUZET.

1 - Given a denumerable poset A, if there exists a linear augmentation of A which is isomorphic with ω, then obviously (i) A is well founded, and (ii) for each element x of the base $|A|$ there exist only finitely many elements $\leq x \bmod A$.

Show that conversely (i) and (ii) imply the existence of such a linear augmentation of type ω.

Start with an arbitrary enumeration x_i(i integer) of all elements of the base $|A|$. Using 2.7.2, remove those xj such that there exists at least an x_i with $i < j$ and $x_j < x_i \bmod A$, and reindex as $y_0, y_1, y_2, .., y_n, ..$ the remaining x_i which form a cofinal w-sequence for A; we call them principal elements.

Note that $n < n'(n, n'$ integers) imply $y_n <$ or $|y_{n'}$; anticipating on chapter 4 below, the sequence of principal elements is bad for the converse of A.

More intuitively, let $y_0 = x_0$; say that an x_i is annexed to the principal element y_0 iff it is $\leq y_0 \bmod A$. Let y_1 be the first x_i which is not annexed to y_0; say that an x_i is annexed to the principal element y_1 iff it is not already annexed to y_0 and is $\leq y_1 \bmod A$. Let y_2 be the first x_i which is annexed neither to y_0 nor to y_1; say that an x_i is annexed to the principal element y_2 iff it is neither annexed to y_0 nor to y_1 and is $\leq y_2 \bmod A$; and so on. For each principal element y_n we denote by E_n the (finite) set of its annexed elements and by A_n the restriction A/E_n (whose maximum is y_n). Then we obtain a partially ordered augmentation of A by putting each element annexed to y_n before each element annexed to $y_{n'}$ for $n < n'$ (integers). Finally replace each A_n by a linear (= totally ordered) augmentation: we obtain an augmentation of A which is isomorphic with ω.

2 - More generally consider the case of a poset A whose cardinal is a regular aleph ω_α. If there exists a linear augmentation of A which is isomorphic with ω_α, then obviously (i) A is well-founded, and (ii) for each element x of the base $|A|$

the set of elements $\leq x(\mathrm{mod}\, A)$ has cardinality strictly less than ω_α. Show that conversely (i) and (ii) imply the existence of such a linear augmentation of A.

Generalize the preceding proof: start with an arbitrary ω_α-sequence $x_i(i < \omega_\alpha$ of all elements of the base $|A|$. Define as previously the principal and annexed elements; principal elements constitute an extracted sequence y_n with length wa because the regularity. Then for each $n < \omega_\alpha$ we define as precedently the set E_n of elements annexed to y_n (its cardinality is strictly less than $< \omega_\alpha$ and the restriction $A_n = A/E_n$ (whose maximum is y_n). Again we obtain a partial ordered augmentation of A by putting each element annexed to y_n before each element annexed to $y_{n'}(n < n' < \omega_\alpha)$. Finally replacing each A_n by a well-ordered linear extension, we obtain a linear extension of A which is isomorphic with ω_α.

3 - Consider now the case of a poset A whose cardinal is a singular aleph ω_α. If there exists a linear augmentation of A which is isomorphic with ω_α, then obviously the same conditions (i) and (ii) as in alinea 2 are satisfied.

In order to prove the converse proposition, the preceding proof in paragraphs 1 and 2 is not sufficient. For example take $\alpha = \omega$ so that $\omega_\alpha = \omega_\omega$ whose cofinality is ω. Then it could happen that the first ω-sequence of restrictions $A_0, A_1, A_2, ..$ lead to linear extensions of order types $\omega, \omega_1, \omega_2, ..,$ so that the complete sum would be strictly greater than ω_ω.

To avoid this inconvenient, we need the following lemma (communicated by POUZET). Given the well-ordered ω_α-sequence $x_i(i < \omega_\alpha$, regular or singular aleph), there exists a permutation f of ω_α which leads to an ω_α-sequence $y_j = x_{f(i)}$; moreover for each $j < \omega_\alpha$, the set of elements $< y_j \bmod A$ is at most equipotent with the index j.

Proof: define y_0 as the first x_i which is minimal $\bmod A$. Then define y_1 as the first x_i which is different from y_0 and has at most 1 element $< y_1 \bmod A$. Then define y_2 as the first element which is different from y_0 and y_1 and has at most 2 elements $< y_2 \bmod A$; and so on. To be sure that y_j exists for a given $j < \omega_\alpha$, remove from the base $|A|$ all the $y_{j'}$ $(j' < j)$ and consider any minimal element u in the obtained restriction: then the elements $< u(\mathrm{mod}\, A)$ form a set whose cardinality is at most $\mathrm{Card}\, j$.

In general f is a non-identical permutation of ω_α , so that we can have $i < i'$ with $f(i') < f(i)$. Moreover we can have $j < j'$ with $y_{j'} < y_j \bmod A$. Nevertheless, starting with the ω_α-sequence y_j instead of x_i, we define as previously the extracted sequence z_n of principal elements. To each z_n is associated the set E_n of its annexed elements and the restriction $A_n = A/E_n$. Finally the problem is solved in the singular case; indeed given $n < \omega_\alpha$, the union of all $E_{n'}(n' \leq n)$ is at most equipotent with the ordinal sum $1 + 2 + 3 + ... + j(n)$ where $j(n)$ is defined by $y_{j(n)} = z_n$: the preceding sum is equipotent with $j(n)$, hence is $< \omega_\alpha$.

2.15.2 The cardinal of the set of filters

1 - For each set X we denote by $\mathcal{F}(X)$ the set of finite subsets of X. By the axiom of choice, $\mathcal{F}(X)$ is equipotent with X for each infinite X. Let E be an infinite set with cardinal a; note that $\mathcal{F}(\mathcal{F}(E))$ is equipotent with E. In the following, we

obtain ultrafilters on $\mathcal{F}(\mathcal{F}(E))$.

Divide E into two complementary subsets A, B of the same cardinal a, and let f be a bijective mapping from A onto B. To each subset X of A associate $X^+ =$ the union of X and $(B - \overline{f}(X))$. Show that for two distinct subsets X, Y of A, neither of the two images X^+, Y^+ is included in the other.

2 - To the set E associate the set $E' = \mathcal{F}(E)$ of all finite subsets of E. To each subset X of A associate $X' = E' - \mathcal{F}(X^+)$. Note that the transformation from X into X' is injective, hence the cardinal of the set of X' is ${}^a2$. Moreover for X, Y distinct subsets of A, we have X', Y' neither included in the other.

3 - Let X be a subset of A and $Y_1, ..., Y_n (n$ integer) be a finite set of subsets Y of A, all distinct from X. Then the intersection of the images $Y' = E' - \mathcal{F}(Y^+)$ of the Y is not included in the image X' of X.

Indeed to each index $i = 1, ..., n$ associate an element $g(i) \in X^+ - (Y_i^+ \cap X^+)$. The finite set of the $g(i)$ is included in X^+ yet not included in any Y_i^+. Hence it does not belong to $X' = E' - \mathcal{F}(X^+)$ yet does belong to each $Y_i' = E' - \mathcal{F}(Y_i^+)$, hence belongs to the intersection of the Y'.

4 - Let $E" = \mathcal{F}(E') = \mathcal{F}(\mathcal{F}(E))$, so that $E"$ has cardinal a. Denote by $\mathbf{A}$ the set of preceding images X' of subsets X of A: hence $\mathbf{A}$ has cardinal ${}^a2$. Let $\mathbf{C}$ denote the set of all $\mathcal{F}(X')$. Hence $\mathbf{C}$ is the set of all $\mathcal{F}(E' - \mathcal{F}(X^+))$ where X is an arbitrary subset of A. Thus $\mathbf{C}$ is also of cardinal ${}^a2$. Moreover, each element of $\mathbf{C}$ is a subset of $E"$.

Let H, K be two finite non-empty disjoint subsets of $\mathbf{C}$; then the intersection $\cap K$ is not included in the union $\cup H$.

Indeed H is a set of $\mathcal{F}(X_i')(i = 1, ..., m)$ and K is a set of $\mathcal{F}(Y_j')(j = 1, ..., n)$ with m, n integers. Each X_i' and Y_j' is the image of a subset X_i or Y_j of A, all distinct. Fix an index $i \leq m$: by (3) above, the intersection of the $Y_j'(j = 1, ..., n)$ is not included in X_i'. Hence there exists an element $h(i)$ which belongs to this intersection and not to X_i'. The set of $h(i)(i = 1, ..., m)$ is a finite subset of each Y_j', hence an element of each $\mathcal{F}(Y_j')$, hence an element of the intersection $\cap K$. However this element is not included in any of the X_i', hence belongs to no $\mathcal{F}(X_i')$, and so is not an element of the union $\cup H$.

By the previous statement, each finite intersection of elements of $\mathbf{C}$ is non-empty. In particular, two complementary subsets of $E"$ cannot both belong to $\mathbf{C}$.

5 - Let U be a subset of $\mathbf{C}$. Associate to it $U^+ = U$ plus all complements $E" - S$, where S belongs to the difference set $\mathbf{C} - U$. For two distinct U, say U and V, there exists for example an element S of $\mathbf{C}$ which belongs to U, hence to U^+, and whose complement $E" - S$ belongs to V^+. It follows that U^+ and V^+ are distinct. For otherwise, two complementary elements would belong to U^+, these elements being obtained from two complementary elements of $\mathbf{C}$. Thus the set of all U^+, as the set of all U, has cardinality (2 to the power ${}^a2$).

Beginning with an arbitrary non-empty subset U of $\mathbf{C}$, consider an arbitrary positive finite number of elements of U^+, by distinguishing the elements $S_1, ..., S_m$ of U and the elements $E" - T_1, ..., E" - T_n$ where these T belong to the difference set $\mathbf{C} - U$.

By the previous (4), the intersection of the S is not included in the union of the T, hence the intersection of the S and the $E" - T$ is non-empty. Thus there exists an ultrafilter on $E"$ extending U^+. To two distinct U correspond two distinct ultrafilters, because of the existence of complementary elements (with respect to $E"$) belonging to the corresponding sets U^+. Thus the set of ultrafilters on $E"$, just as the set of subsets of **C**, has cardinality (2 to the power ${}^a2$).

6 - Modulo the generalized continuum hypothesis, we have a much simpler proof of the preceding, yielding, for a set E of cardinal a, that the set of ultrafilters on E has cardinality (2 to the power ${}^a2$).

To see this, consider all partitions of E into two disjoint subsets of cardinal a, and totally order this set of partitions by its cardinal $b = {}^a2$. Let u_0 be the first partition, whose two associated subsets shall be denoted $E_0(+)$ (associated to the 1-sequence $+$) and $E_0(-)$. For each of these subsets, $E_0(+)$ for example, take the first partition $u_1(+)$ such that the intersection of $E_0(+)$ with each of the subsets of E given by $u_1(+)$ has cardinal a. Denote by $E_1(++)$ and $E_1(+-)$ these two intersections. Do the same thing for the 2-sequences (-+) and (- -). In general, for each ordinal $i < b$, note that $\mathrm{Card}\, i \leq a$ by the generalized continuum hypothesis.

Fix i and fix an arbitrary i-sequence x with values (+) and (-). We have to define the partition $u_i(x)$. Assume that we have obtained, for each $j < i$, the sequence of the $E_j(x_{j+1})$ where x_{j+1} is the initial interval of x with length $j+1$. By 2.3.4, the set of the $E_j(x_{j+1})$ generates a filter which is not an ultrafilter. So there exists a partition $u_i(x)$ yielding for every intersection of finitely many of the $E_j(x_{j+1})$ a set of cardinal a.

Finally, for each b-sequence of values (+) and (-), we obtain a filter such that two distinct b-sequences give two distinct filters. Hence there are ${}^b2$ many such filters.

2.15.3 The possible cofinal restrictions of a denumerable poset

In the particular case of a denumerable directed poset, we know by 2.13.2 that there exists either a maximum, or a cofinal restriction which is isomorphic with ω.

In the general denumerable case, show that the possible cofinal restrictions are the following:

(1) an antichain with finite or denumerable cardinality;

(2) the union of components each of which is isomorphic with ω, with mutual incomparability for elements belonging to different components;

(3) the ω-tomic tree, i.e. the tree with denumerably many edges from each vertex;

(4) the union of two or three among the previous components, with mutual incomparability.

Chapter 3

Ramsey theorems, partition theorems, incidence matrix, combinatorial principles, profile

3.1 Ramsey's theorem, Ramsey number

3.1.1 Infinitary form

For historical information, see [213] RAMSEY 1930.

Partition the (unordered) pairs of integers into k classes (k finite) which we call **colors**. Then there exists an infinite set E of integers such that all pairs included in E have the same color. In the following, instead of integer (= element), pair, 3-element set, we often say **vertex, edge, triangle**.

The previous statement generalizes to the case of m-element sets, or sets with finite cardinal m which are assumed to be partitioned into k colors. There exists an infinite set E of integers such that all m-element subsets of E have the same color. Such a set E is called **monochromatic**.

Case of pairs. • Partition the non-zero integers x into k classes according to the color of the pair $\{0, x\}$: at least one of these classes is infinite. Let $u_0^0 = 0$ and let $u_1^0, u_2^0, \ldots$ be the elements of this class. Set $u_1^1 = u_1^0$ and partition the integers $x = u_i^0 (i \geq 2)$ into k classes according to the color of the pair $\{u_1^1, x\}$: at least one of these classes is infinite. Let $u_2^1, u_3^1, ..$ be the elements of this class. Set $u_2^2 = u_2^1$ and iterate.

At the end we obtain the infinite set with elements $v_0 = 0, v_1 = u_1^1, \ldots, v_i = u_i^i (i$ integer) satisfying the following condition. For every integer i the pairs $\{v_i, v_{i+1}\}, \{v_i, v_{i+2}\}, ..$ have the same color, which we say is associated with the integer i. Then at least one color is associated with infinitely many integers i: all pairs subsets of the set of v_i have the same color. •

Case of m-element sets ($m \geq 2$). • Assume that the statement is true for $m-1$. Put aside the integers 0, 1, 2, ... , $m-2$, and partition the integers $x \geq m-1$ into k classes, according to the color of the m-element set $\{0, 1, ..., m-2, x\}$: at least one of these classes is infinite. Let $u_0^0 = 0, u_1^1 = 1, ..., u_{m-2}^{m-2} = m-2$, and let u_{m-1}^{m-2}, u_m^{m-2}.. be the elements of this infinite class. Set $u_{m-1}^{m-1} = u_{m-1}^{m-2}$ and partition the integers $x = u_i^{m-2} (i \geq m)$ into a finite number of classes, by putting x and y into the same class iff for each $(m-1)$-element set I with last element u_{m-1}^{m-1}, the set I augmented with x and I augmented with y yield two m-element sets with the same color (which depends on I). As there are only finitely many I, one of these classes is infinite: let u_m^{m-1}, u_{m+1}^{m-1},.. be the elements of this class. Set $u_m^m = u_m^{m-1}$ and iterate this with the $(m-1)$-element sets having last element u_m^m.

At the end we obtain the infinite set with elements $v_0 = u_0^0 = 0, v_1 = u_1^1 = 1$ and in general $v_i = u_i^i$ for each integer i, satisfying the following condition. For each $(m-1)$-element set I formed of elements v and whose last element is v_i, the m-element sets obtained by adding a $v_j (j > i)$ to I all have the same color, which we call the color associated with the $(m-1)$-element set I. By the induction hypothesis, we apply the theorem for $m-1$: we obtain an infinite set of elements v all of whose $(m-1)$-element subsets have the same associated color. Hence an infinite set, all of whose m-element subsets have same color. •

3.1.2 Monotonic extracted sequence

Given a relation A, a sequence whose values are elements of the base shall be called a sequence in A. If A is a partial ordering, let u be an ω-sequence in A. Then there exists an ω-sequence extracted from u, which is either constant, or strictly increasing, or strictly decreasing, or consisting of elements which are pairwise incomparable (mod A).

• Partition the set of pairs $\{i, j\}$ of integers into four classes (put $i < j$ to fix things for discussion), by defining a first class by the equality $u_i = u_j$, a second by $u_i < u_j (\text{mod } A)$, a third by $u_i > u_j \text{ mod } A)$, and a fourth by $u_i | u_j (\text{mod } A)$; then apply RAMSEY's theorem. •

3.1.3 Finitary form of Ramsey's theorem; monochromatic set

Given three integers $m, k, p \geq m$, there exists an integer $p^+ \geq p$ such that, for every set of cardinal $\geq p^+$, whose m-element subsets are partitioned into k colors, there exists a p-element subset, all of whose m-element subsets have the same color. It is called a **monochromatic** p-element subset.

• Consider the case $m = 2$. Repeat the proof of 3.1.1, but instead of "one of the classes is infinite", say "one of the class is large", meaning that it contains at least $1/k$ of the original elements. It suffices to take $p^+ = (kp).k^{(kp-1)} = p.k^{kp}$ in order to obtain, after $kp-1$ operations, a sequence of length $\geq kp$ of elements v, analogous to those in 3.1.1. Thus we have a large class of v, of cardinality $\geq p$. •

3.1.4 Ramsey number

The least such p^+ in the preceding proposition is called a **Ramsey number**, denoted $(p)_k^m$ or indifferently $(p,p,...,p)^m$ where p is repeated k times (the number of colors). By the preceding proof, $(p)_k^2$ is at most equal to $p.k^{kp}$. Incidently this is a very bad majoration: $(3)_2^2 \leq 3.2^6$ far from the real value 6. Hereafter we give the smallest and best known values.

Case $m=1$. If each of the k classes had $p-1$ elements, the entire set would have $k.(p-1)$ elements. Hence it suffices to take $(p)_k^1 = k(p-1)+1$ to obtain at least one class with p elements. This argument is called the "pigeonhole principle": if $k(p-1)+1$ objects are partitioned into k pigeonholes, then at least one of the pigeonholes has p objects.

Case $k=1$: a single class, thus $(p)_1^m = p$.

Case $p=m$: a p-element set is necessarily monochromatic, thus $(m)_k^m = m$.

Calculation of $(3,3)^2 = (3)_2^2 = 6$.

• Consider the elements 1,2, ... ,6; partition the edges {1,2} to {1,6} into two colors. At least one contains three edges $\{1,a\},\{1,b\},\{1,c\}$. Either {a,b} or {b,c} or {c,a} has the same color, or these three edges have the opposite color: this shows that $(3)_2^2 \leq 6$. To see that $(3)_2^2 > 5$, take the usual pentagon with one color, and the starred pentagon with the opposite color. •

Calculation of $(3,3,3)^2 = (3)_3^2 = 17$ ([96] GLEASON, GREENWOOD 1955).

• Consider the elements 1,2, ... ,17 and partition the 16 edges (1,2) to (1,17) into three colors. At least one contains 6 edges, say $(1,a_1),...,(1,a_6)$. It remains to partition the edges $(a_i,a_j)(i,j =$ 1to 6) into two colors: hence we fall back to the case $(3)_2^2 = 6$; this shows that $(3)_2^2 \leq 17$.

The following counterexample shows that $(3)_2^2 > 16$. Consider the field composed of the integers 0 and 1 with 1+1 = 0 (the field of the integers modulo 2) and the ring of polynomials on this field with the identity $x^4 = x+1$. This ring is composed of 16 elements (0 or 1) + (0 or 1)$.x$ + (0 or 1)$.x^2$ + (0 or 1)$.x^3$. These elements are exactly $0,1,x,x^2,...,x^{14}$ (we have $x^{15} = 1$). Every non-zero element is a power $x^i(i=0,1,...,14)$ and has inverse x^{15-i}. Hence this ring is a field. Partition the pairs of polynomials into three colors, according to whether the difference of these two polynomials is a cube $x^{3u}(u=0,1,2,3,4)$ or is of the form x^{3u+1} or x^{3u+2}. It suffices to see that the sum of two non-zero cubes is not a cube. •

The only known, non trivial ternary number is $(4,4)^3 = (4)_2^3 = 13$: see [168] Mc KAY, RADZISZOWSKI 1991. The lower bound 13 goes back to [122] ISBELL 1969. The upper bound 15 goes back to [93] GIRAUD 1969, reduced to 13 after 22 years of stagnation.

3.1.5 Non-symmetric Ramsey numbers

Let E be a finite set; partition its m-element subsets into k colors $u_1, ..., u_k$. Given k integers $p_1, ..., p_k \geq m$, the least cardinal of E for which there exists either a p_1-element subset with color u_1, ... , or a p_k-element subset with color u_k, is said to be a **non-symmetric** Ramsey number and denoted by $(p_1, ..., p_k)^m$.

The value $(p_1, ..., p_k)^m$ does not depend of the ordering of arguments $p_1, ..., p_k$. Moreover, taking $p_1 = ... = p_k = p$, we obtain the **symmetric** Ramsey number $(p)_k^m = (p, ..., p)^m$.

Calculation of $(3,4)^2 = 9$.

• We show that this number is ≤ 9. Join up the integers 1 through 9 by edges. Either among the 8 edges (1,2) to (1,9) there exist 4 edges of color (+). This then yields either a 3-element set with color (+) or a 4-element set with color (-). Or there exist 6 edges with color (-), which then yields either a 3-element subset (-) or a 6-element subset (+). Or finally none of the preceding cases is realized for any of the points 1 through 9. Then from each point there emanate exactly 3 edges (+) and 5 edges (-). But this is impossible, since we would then have $3.(9/2) = 27/2$ edges (+).

We now show that $(3,4)^2 > 8$. Take the integers 0 to 7 , and give the color (+) to the edge (x, y) iff the absolute value of $y - x$ is 3, 4, or 5. •

Calculation of $(4,4)^2 = (4)_2^2 = 18$ (GLEASON, GREENWOOD 1955).

• Take the integers 1 to 18 . Among the edges emanating from 1 , there are at least 9 of the same color which we designate (+). They join 1 to the integers designated $a_1, ..., a_9$. By the preceding, in the set of $a_1, ..., a_9$ there exists either a 3-element set with color (+), or a 4-element set of the opposite color (-). Hence this ramsey number is at most 18.

The following example prove that the Ramsey number is not ≤ 17. Take the integers modulo 17, so 0 to 16. For any two distinct x, y in this set, we give to the edge (x, y) the color (+) iff x is congruent to y modulo a quadratic residue, so mod $\pm1, \pm2, \pm4$ or ±8; the color (-) in the opposite cases.

Suppose that there exist four integers a, b, c, d such that all their edges have the same color. We can replace these integers by $a - d, b - d, c - d$ and 0 , and thus can consider only the case $0, a, b, c$. We can require that the six integers $a, b, c, b - a, c - b, a - c$ be non-zero and either all residues or all non-residues. Multiplying by the inverse of a, we can reduce this to the case of $0, 1, b, c$ with the five integers $b, c, b-1, c-1, c-b$ which are all non-zero and all quadratic residues. Then the possibilities for b and c are reduced to -1, +2, -8 . For $b = -1$, we have $c \neq -1, c \neq 2$ since $c - b \neq 3$. Moreover $c \neq -8$ since $c - b \neq -7$. The same argument for $c = -1$: impossible. For $b = 2$ and $c = -8$ we obtain $b - c = -7$: impossible. •

Calculation of $(3,5)^2 = 14$.

• Take a set with 14 points and let a be in this set. Either from a there emanate at least 5 edges with color (+), which yields a 3-element monochromatic set with color (+), or a 5-element monochromatic set with color (-). Or from a there emanate at most 4 edges (+), hence at least 9 edges (-). Then since $(3,4)^2 = 9$, this yields either a (+)-monochromatic 3-element set, or a (-)-monochromatic 3-element set, or a (-)-monochromatic 5-element set. Thus our Ramsey number is at most equal to 14.

To see that it equals 14, take the 13 integers 0 through 12. Give the color (+) to the pair $\{x, y\}$ (where x and y are distinct elements among 0, 1, ... , 12) iff the absolute value of $y - x$ equals 2, 3, 10 or 11; color (-) in other cases. •

Other known values of binary Ramsey numbers:

$(3,6)^2 = 18$: see [131] KALBFLEISCH 1967; independently [133] KERY 1964.

$(3,7)^2 = 23$: see [100] GRAVER, YACKEL 1968.

$(3,8)^2 = 28$: see [212] RADZISZOWSKI 1994 (precedently 28 or 29 by [101] GRINSTEAD, ROBERTS 1982).

$(3,9)^2 = 36$: see [101] GRINSTEAD, ROBERTS 1982.

$(4,5)^2 = 25$; lower bound of 25 by KALBFLEISCH 1967; upper bound of 25 by [170] Mc KAY, RADZISZOWSKI 1995.

3.1.6 Inequalities

Below we list some inequalities for the smallest binary numbers whose exact value is not known.

$51 \leq (3,3,3,3)^2 = (3)_4^2 \leq 64$: lower bound of 51 by [33] CHUNG 1973; upper bound of 64 by [220] SANCHEZ-FLORES 1995; the precedent upper boud of 65 by [65] FOLKMAN 1974 is reduced to 64 after 21 years of stagnation.

$162 \leq (3,3,3,3,3)^2 = (3)_5^2 \leq 317$: for the upper bound, see 3.8.1;

lower bound 162 by EXOO 1994; precedent know value 159 by [89] FREDRICKSON 1979.

We easily get $(3)_k^2 \geq (3^k + 3)/2$: see 3.8.3 on Schur's numbers. Improved by $(3)_{k+1}^2 \geq 3.(3)_k^2 + (3)_{k-2}^2 - 3$. For example $(3)_4^2 \geq 3.(3)_3^2 = 3.17 = 51$ already mentioned.

$43 \leq (5,5)^2 = (5)_2^2 \leq 49$: lower bound 43 by [60] EXOO 1989; upper bound 49 by RADZISZOWSKI 1994.

$30 \leq (3,3,4)^2 \leq 31$; lower bound 30 by KALBFLEISCH 1967; upper bound 31 by PIWAKOWSKI, RADZISZOWSKI 1998.

$55 \leq (3,4,4)^2 \leq 79$: lower bound 55 by [141] KREHER, LI WEI, RADZISZOWSKI 1988; upper bound easily obtained via (3,3,4) $\leq$ 31 by the method of 3.8.1, which gives 2((3,3,4)-1) + ((4,4)-1) + 2 = 2.30 + 17 + 2 = 79.

$128 \leq (4,4,4)^2 = (4)_3^2 \leq 236$: lower bound 128 by [112] HILL, IRVING 1982; upper bound easily obtained via $(3,4,4) \leq 79$ by the method of 3.8.1, which gives 3((3,4,4)-1) + 2 = 3.78 + 2 = 236.

For further informations about Ramsey numbers, see [99] GRAHAM, ROTHSCHILD, SPENCER 1990.

3.1.7 Ramsey multiplicity function

Going back to $(3)_2^2 = 6$, the reader can easily see that, with 6 elements or vertices and 2 colors, one gets at least 2 monochromatic 3-element sets, also called monochromatic triangles: either two triangles of same color, or one of each color.

The **Ramsey multiplicity function** associates to each Ramsey number $(p)^m_k$ and each integer $n \geq m$ the smallest possible number of monochromatic p-element sets in a k-coloration of m-element sets inside a given n-element set. Denote this function by $[(p)^m_k](n)$. Obviously the value is zero from $n = m$ (except if $p = m$) as long as n is strictly less than the Ramsey number $(p)^m_k$. In particular, the function $[(3)^2_2](n)$ is defined for $n \geq 2$ and its first values are 0 for $n = 2, 3, 4, 5$ then jumping to value 2 for $n = 6$.

3.2 Lexicographically ordered set, Galvin's initial interval theorem, Nash-Williams' theorem

3.2.1 Lexicographically ordered set, lexicographic rank

Totally order the set of finite sets of integers lexicographically, by first difference: (set a) $\leq$ (set b) iff the least integer in a is strictly less than the least integer in b; or in the case of equality, compare the second least integer of a with the second least integer in b, etc. The empty set is defined to precede all other sets in this ordering. Finally if a is a proper initial interval of b, we put $a < b$.

A set $\mathcal{F}$ of finite sets of integers is said to be **lexicographically well-ordered** iff the lexicographic ordering of its elements is a well-ordering. The corresponding order type is called the **lexicographic rank** of $\mathcal{F}$; it is a countable ordinal. For example, the set of singletons of integers is lexicographically well-ordered with rank ω, the set of pairs with rank $\omega.2$. The set of all finite sets of integers is not lexicographically well-ordered. Indeed we obtain a lexicographically decreasing ω-sequence by taking the singleton $\{1\}$, then the pair $\{0,2\}$, then $\{0,1,3\}$, then $\{0,1,2,4\}$, etc.

Let $\mathcal{F}$ be a set of finite sets of integers, such that:

(1) **the elements of $\mathcal{F}$ are mutually incomparable under inclusion**;

(2) **every infinite set of integers includes a subset which belongs to $\mathcal{F}$**;

then $\mathcal{F}$ is lexicographically well-ordered (POUZET in 1980).

Note that (2) alone is not sufficient: take the set of all the finite sets of integers. Also (1) alone is not sufficient: take the above decreasing ω-sequence, replace each integer i by $2i$ then add the first odd integer which is greater than the maximum.

• Consider a non-empty subset $\mathcal{G}$ of $\mathcal{F}$, and show that there exists a minimum element in $\mathcal{G}$ for the lexicographic ordering. Let a_0 be the least integer such that there exists an element of $\mathcal{G}$ beginning with a_0. If the singleton $\{a_0\}$ belongs to $\mathcal{G}$, then it is the minimum of $\mathcal{G}$. Otherwise, take the elements of $\mathcal{G}$ which begin with a_0, and let a_1 be the least integer such that there exists an element of $\mathcal{G}$ beginning with a_0, a_1. If the pair a_0, a_1 belongs to $\mathcal{G}$, then it is the minimum of $\mathcal{G}$. Otherwise, if the procedure never terminates, then we obtain an infinite increasing sequence $a_0 < a_1 < ... < a_i < ..(i$ integer). By our hypothesis (2), there exists a finite set composed of certain of the a_i and belonging to $\mathcal{F}$: denote by a_h the last among these. There exists also a finite set beginning with $a_0, a_1, ..., a_h$ and belonging to $\mathcal{G}$, hence to $\mathcal{F}$. But this contradicts our hypothesis (1) of incomparability. •

3.2.2 Initial interval theorem

Let $\mathcal{F}$ be a set of finite sets of integers, such that every infinite set of integers includes as a subset at least one element of $\mathcal{F}$. Then there exists an infinite set E of integers such that every infinite subset of E has an element of $\mathcal{F}$ as an initial interval ([90] GALVIN 1968).

The following proof, using only the axioms of ZF, is due to POUZET in 1980, published in ToR-86, p.66-69.

Note that RAMSEY's theorem follows. Indeed, partition the pairs of integers into two colors (+) and (-). Then either there exists an infinite set of integers all of whose pairs belong to (+), or every infinite set includes an element of color (-). Then by the above statement, there exists an infinite set E such that every infinite subset of E begins with a pair belonging to (-): in other words every pair belongs to (-).

To simplify the proof, note that **it is always possible to assume that the elements of $\mathcal{F}$ are mutually incomparable with respect to inclusion**. Indeed, starting with an arbitrary $\mathcal{F}$, we obtain the subset $\mathcal{F}^*$ by taking those elements of $\mathcal{F}$ which are minimal with respect to inclusion. Every infinite set of integers includes at least one element of $\mathcal{F}^*$. The initial interval theorem, when restricted by the preceding condition, says that there exists an infinite set E, such that every infinite subset of E has an element of $\mathcal{F}^*$, hence of $\mathcal{F}$, as an initial interval.

By the previous subsection, we see that it suffices to prove the initial interval theorem for an arbitrary lexicographically well-ordered set (whose elements will no longer necessarily be incomparable under inclusion). Thus we are led to prove the following statement.

3.2.3 Case of a lexicographically well-ordered set

Let $\mathcal{F}$ be a lexicographically well-ordered set of finite subsets of integers ($\mathcal{F}$ could contain the empty set as an element); then:

(1) **either there exists an infinite set of integers which includes no element of $\mathcal{F}$**;

(2) **or there exists an infinite set of integers, each of whose infinite subset has an element of $\mathcal{F}$ as initial interval** (the empty set is considered as an initial interval of every set).

• We argue by induction on the lexicographic rank of $\mathcal{F}$. Suppose first that the rank is equal to 1, so that $\mathcal{F}$ is the singleton of a finite set F of integers. Then the infinite set of all integers not belonging to F satisfies our conclusion (1) if F is non-empty, our conclusion (2) if F is empty.

Let α be a countable ordinal. Suppose the statement is true for every set of lexicographic rank $< \alpha$. We shall prove it for every set $\mathcal{F}$ of rank α.

More strongly, in order to avoid use of the axiom of choice, or even a weakened form of choice, suppose that there exists a function h which, to each couple $(E, \mathcal{F})$ where E is an infinite set of integers, $\mathcal{F}$ a set of finite subsets of E of lexicographic ranks $< \alpha$, associates an infinite set $h(E, \mathcal{F})$ satisfying one of the conclusions (1) or (2). More precisely, either $h(E, \mathcal{F})$ includes as a subset no element of $\mathcal{F}$, or every infinite subset of $h(E, \mathcal{F})$ begins with an initial interval which belongs to $\mathcal{F}$. We will prove that there exists an analogous function for the couples whose second term $\mathcal{F}$ has lexicographic rank α. Thus h will be progressively extended to all countable lexicographic ranks, so to all couples $(E, \mathcal{F})$.

Start with an infinite set E of integers and a set $\mathcal{F}$ of finite subsets of E with lexicographic rank α. For each integer i of E, denote by $\mathcal{F}_i$ the subset of those elements of $\mathcal{F}$ which begin with i. Hence $\mathcal{F}$ is the union of the $\mathcal{F}_i$, and the lexicographic rank α of $\mathcal{F}$ is the sum along ω of the lexicographic ranks of the $\mathcal{F}_i$. If there exist infinitely many integers i of E with $\mathcal{F}_i$ empty, then the set of these i does not include as a subset any element of $\mathcal{F}$. This will be, by definition, $h(E, \mathcal{F})$ which then verifies our conclusion (1).

Consider the other case, and let $m(0)$ be the least integer of E after which $\mathcal{F}_i$ is never empty. Denote by M_0 the set of integers of E which are $\geq m(0)$ and by M_0^- the set M_0 with its minimum $m(0)$ removed. The $\mathcal{F}_i$ all have lexicographic ranks $< \alpha$, hence the function h is already defined for these. We put $\mathcal{G}_0 = \mathcal{F}_{m(0)}$ and then $\mathcal{G}_0^-$ is the set of the elements of $\mathcal{G}_0$, each of whose minimum $m(0)$ has been removed: the lexicographic rank of this $\mathcal{G}_0^-$ is still $< \alpha$.

Let $E_1 = h(M_0^-, \mathcal{G}_0^- \subseteq M_0^-$. Either there exist infinitely many integers i of E_1 for which E_1 includes no element of $\mathcal{F}_i$. Then the set of these i is by definition $h(E, \mathcal{F})$, which satisfies our conclusion (1) .

Or in the opposite case, let $m(1)$ be the least integer of E_1, from which point on every F_i restricted to its elements which are subsets of E_1 is never empty. Denote by M_1 the set of integers of E_1 which are $\geq m(1)$, and M_1^- the set M_1 with its minimum $m(1)$ removed. Let $\mathcal{G}_1 = \mathcal{F}_{m(1)}$ restricted to elements which are subsets of M_1. Then $\mathcal{G}_1^-$ is the set of the elements of G_1 each with its minimum removed: the lexicographic rank of $\mathcal{G}_1$ and of $\mathcal{G}_1^-$ is strictly less than α.

Let $E_2 = h(M_1^-, \mathcal{G}_1^- \subseteq M_1^-$, and iterate this procedure. Then either, at the end of a finite number of steps, we obtain an $E_r(r$ integer) with infinitely many integers i of E_r for which no element of $\mathcal{F}_i$ is a subset of E_r. Then by definition, we put $h(E, \mathcal{F})$ to be the set of these i, which satisfies (1). Or this described process continues indefinitely: we must consider two subcases.

First subcase. There exist infinitely many integers k for which the set $E_{k+1} = h(M_k^-, \mathcal{G}_k^-)$ contains as a subset no element of $\mathcal{G}_k^-$. Take the infinite corresponding set of minimums $m(k)$. Then $m(k') \in E_{k+1}$ included in M_k^- for all $k, k' > k$. So the set of minimums $m(k)$ contains as a subset no element of $\mathcal{G}_k$, hence no element of $\mathcal{F}$: we take it as our definition of $h(E, \mathcal{F})$, which satisfies (1).

In the second subcase, because of the definition of the function h, there exists an integer k_0 from which point on, every infinite subset of E_{k+1} begins by a possibly empty element of $\mathcal{G}_k^-$. A fortiori, if K denotes the set of minimums $m(k)$ for $k \geq k_0$, then every infinite subset of K begins by an element of a $\mathcal{G}_k$, hence by an element of $\mathcal{F}$. Thus the set K, which we take for $h(E, \mathcal{F})$, satisfies our conclusion (2). •

3.2.4 Separation theorem (Nash-Williams)

Consider two disjoint sets $\mathcal{F}, \mathcal{G}$ of finite sets of integers. Suppose that no element of $\mathcal{F}$ is an initial interval of an element of $\mathcal{G}$, and vice-versa. Then there exists an infinite set E of integers which contains as a subset no element of $\mathcal{F}$, or which contains as a subset no element of $\mathcal{G}$.

In [177] NASH-WILLIAMS 1965 only assumes that, for two distinct elements of $\mathcal{F} \cup \mathcal{G}$, one is never an initial interval of the other. The present stronger statement result from a remark by HODGES in 1984.

• Either there exists an infinite set of integers having no subset which is an element of $\mathcal{F}$. Or every infinite set of integers has a subset which is an element of $\mathcal{F}$. In the latter case, by GALVIN's theorem 3.2.2, there exists an infinite set E of integers, such that every infinite subset of E has an element of $\mathcal{F}$ as initial interval. Then E has no subset which is an element of $\mathcal{G}$. Indeed, if it contained as a subset an element G of $\mathcal{G}$, then take an infinite subset X of E with initial interval G. There exists an element F of $\mathcal{F}$ which is an initial interval of X: thus F is an initial interval of G, or G an initial interval of F: contradiction. •

Note that RAMSEY's theorem follows. Indeed, given two distinct pairs of integers, or in general, for p a fixed integer, given two distinct p-element sets of integers, one is never an initial interval of the other.

3.3 Uncountable case, partition theorems: Dushnik, Miller, Erdös, Rado

3.3.1 Sierpinski's counterexample

There exists a partition of the pairs of reals into two colors, such that every monochromatic set is countable (uses axiom of choice). See [227] SIERPINSKI 1933.

• Take a well-ordering A of the set of reals. Then to each pair of reals x, y with $x < y$ in the usual ordering, give the color (+) if $x < y \pmod A$ and the color (-) if $x > y \pmod A$. The proposition follows from the fact that every strictly increasing (or strictly decreasing) sequence of reals is countable. •

In other words, we have built a Sierpinski poset in the sense of 2.2.7.

We can summarize the situation by using the notation for Ramsey numbers with finite or infinite cardinal values. Then the usual RAMSEY theorem is written $(\omega)_k^m = \omega$ for all integers m, k.

The preceding proposition yields $(\omega_1)_2^2 > \omega_1$.

3.3.2 Partition lemma (Dushnik, Miller)

Let $A = \omega_\alpha$ be an infinite regular aleph. Partition the pairs of elements of A into two colors which we designate by (+) and (-). **Then either, for every subset B which is equipotent with A, there exists an element $a \in B$ and a set of elements $x \in B$ which is equipotent with A, where all the pairs $\{a, x\}$ have color** (-). **Or there exists a subset of A equipotent with A, all of whose pairs have color** (+). See [46] DUSHNIK, MILLER 1941.

• Assume there exists a subset B of A equipotent with A, which negates the first conclusion. Consider the elements of B ordered by the usual ordering on the ordinals. Let a_0 be the minimum of B; by hypothesis there exists an $a_1 > a_0$ in B, such that for every $x \geq a_1$ the pair $\{a_0, x\}$ has color (+). By induction, given $i < \omega_\alpha$, assume that for the $j < i$ we have a strictly increasing sequence of elements a_j of B, such that all pairs of the a_j have the color (+). For every $j < i$, there are less than wa many x in B such that the pair $\{a_j, x\}$ has color (-). Since ω_α is regular, the set of all such x for all $j < i$ has cardinality $< \omega_\alpha$. Thus there exists an a_i of B which is strictly above all the a_j, and such that for all $x \geq a_i$ and all $j < i$, the pair $\{a_j, x\}$ has the color (+). Finally we obtain an ω_α-sequence of elements, all of whose pairs have color (+). •

Note that the proposition is false for every singular aleph ω_α.

• Let $\gamma < \omega_\alpha$ be the cofinality of ω_α. For every $i > \gamma$ take a subset A_i of $A = \omega_\alpha$, such that the union of the A_i is A, but every A_i is strictly subpotent to A. For every pair contained in an A_i, give the color (-), and for pairs of elements belonging to distinct A_i's, the color (+). •

3.3.3 Partition theorem (Dushnik, Miller)

Let A be an arbitrary infinite set; partition the pairs of elements of A into two colors (+) and (-). **Then either there exists a denumerable subset of A which is** (-)**-monochromatic, or there exists a subset of A which is equipotent with A and** (+)**-monochromatic** (uses axiom of choice); [46] DUSHNIK, MILLER 1941.

Dushnik and Miller mention the influence of Erdős. A different proof of the theorem is given by [56] ERDÖS, RADO 1956. Using Ramsey numbers notation and replacing A by an aleph ω_α, we have $(\omega, \omega_\alpha)^2 = \omega_\alpha$.

• Replace A by an aleph which we designate by ω_α (axiom of choice), and assume first that this aleph is regular. By the preceding lemma, if our second conclusion is false, then there exists an $a_0 \in A$ for which the set A_0 of $x > a_0 (\mathrm{mod}\, A)$ such that $\{a_0, x\}$ has color (-) is equipotent with A. Take this a_0 minimum $(\mathrm{mod}\, A)$. Then replace A by A_0, thus yielding an element a_1 of A_0 satisfying the same condition and taken minimum. By iteration, we obtain an ω-sequence of elements a_i(i integer) , all of whose pairs of elements have color (-).

Assume now that $A = \omega_\alpha$ is singular. Then α is a limit ordinal (2.8.2, axiom of choice). Let γ be the cofinality of ω_α, so $\gamma < \omega_\alpha$ and $\gamma \leq \alpha$. Thus ω_α is the ordinal limit of the γ-sequence $\omega_{\alpha(i)}$ where $i < \gamma$ and $\alpha(i) < \alpha$. Moreover, we can choose the $\alpha(i)$ to be strictly increasing with i, and every $\alpha(i) > \gamma$. Finally every $\omega_{\alpha(i)}$ can be assumed to be regular, replacing if need be $\alpha(i)$ by its successor.

Suppose that the first conclusion fails: there is no denumerable subset of A all of whose pairs have color (-). Then by the preceding, there exists a subset B of A, equipotent with A, such that for every $x \in B$, there are strictly less than ω_α many $y \in B$ with $\{x, y\}$ having color (-).

For every subset X of B , denote by $M(X)$ the set of elements of $B - X$ which, together with at least one element of X, have color (-). Let U be any subset of B equipotent with A, and let i be an ordinal strictly less than γ. We shall show that there exists a subset W of U with cardinal $\omega_{\alpha(i)}$, satisfying the two following properties: every pair of elements of W has color (+); the set $M(W)$ has cardinal strictly less than ω_α.

Indeed, by our first paragraph and because for every $i < \gamma$ the cardinal $\omega_{\alpha(i)}$ is regular, there exists a subset V of U with cardinal $\omega_{\alpha(i)}$, all of whose pairs have color (+). For every $j < \gamma$, denote by V_j the set of elements x of V, such that there exist at most $\omega_{\alpha(j)}$ elements in $B - \{x\}$ which together with x have color (-). Then $V = \cup V_j (j < \gamma)$ since, by our third paragraph, no x together with ω_α many elements of B, has color (-), and ω_α is the limit of the $\omega_{\alpha(j)} (j < \gamma)$. Recall that the cardinal $\omega_{\alpha(i)}$ of V is regular and strictly greater than γ. It follows that there exists at least one ordinal $k < \gamma$ with V_k equipotent with V: put $W = V_k$. Then $\operatorname{Card} M(W) \leq \omega_{\alpha(i)} . \omega_{\alpha(k)} < \omega_\alpha$. Thus the two properties stated above for W are obtained.

It remains to construct a subset of B which is equipotent with B and thus with A, all of whose pairs have color (+). Let W_1 be a subset of B with cardinal $\omega_{\alpha(1)}$, all of whose pairs have color (+), with $\operatorname{Card} M(W_1) < \omega_\alpha$. Iterate by taking W_2 a subset of $B - (W_1 \cup M(W_1))$ with cardinal $\omega_{\alpha(2)}$, all of whose pairs have color (+), with $\operatorname{Card} M(W_2) < \omega_\alpha$.

Note that, in the union $W_1 \cup W_2$, all the pairs have color (+). Let $i < \gamma$ and assume that the W_j are defined for $j < i$. Then the union $\cup(W_j \cup M(W_j))$ (for all $j < i$) has cardinal $< \omega_\alpha$. For otherwise the i-sequence of cardinals $\operatorname{Max}(\omega_{\alpha(j)}, \operatorname{Card} M(W_j))$ would yield a sum $\geq \omega_\alpha$, with $\omega_{\alpha(j)} < \omega_\alpha$ and $\operatorname{Card} M(W_j) < \omega_\alpha$ for each $j < i$. Hence the cardinality of ω_α would be $\leq i < \gamma$, contradicting the fact that γ is the cofinality of ω_α. For U take the difference $B - \cup(W_j \cup M(W_j))$ (for all $j < i$): this has cardinality ω_α.

Hence there exists a subset W_i of this difference which satisfies the two properties: $\operatorname{Card} W_i = \omega_{\alpha(i)}$ with all pairs of W_i having color (+), and $\operatorname{Card} M(W_i) < \omega_\alpha$. It remains to note that the union of the $W_i (i < \gamma)$ has cardinal ω_α and that all its pairs have color (+). •

Example. For ω_1 we obtain either a denumerable (-)-monochromatic subset, or a (+)-monochromatic subset having cardinal ω_1. With the notation of ramsey numbers: $(\omega, \omega_1)^2 = \omega_1$.

3.3.4 Partition lemma (Erdös)

Let ω_α be an infinite aleph. Set $A = \omega_{\alpha+2}$ and assume the generalized continuum hypothesis in the form: (2 to the power ω_i) $= \omega_{i+1}$ for every $i \leq \alpha$. Partition the pairs of elements of A into the colors (+) and (-). **Then either there exists a (+)-monochromatic subset of A with cardinal $\omega_{\alpha+1}$, or there exists a (-)-monochromatic subset with cardinal $\omega_{\alpha+2}$.** See [52] ERDÖS 1942 or [56] ERDÖS, RADO 1956.

With Ramsey number notation: $(\omega_{\alpha+1}, \omega_{\alpha+2})^2 = \omega_{\alpha+2}$.

In particular, for every partition of the pairs of elements of a set of cardinal ω_2 into two colors, there is a monochromatic subset of cardinal ω_1 (uses axiom of choice plus the continuum hypothesis). With Ramsey number notation: $(\omega_1, \omega_2)^2 = \omega_2$; also $(\omega_1)_2^2 = \omega_2$.

Note that with only choice and without continuum hypothesis, there exists a model of ZF with $\omega_2 \leq$ continuum; in such a model SIERPINSKI's counterexample (3.3.1) shows that $(\omega_1)_2^2) > \omega_2$.

• We prove first that there exists a monochromatic subset of A with cardinal $\omega_{(\alpha+1)}$. Take $\alpha = 0$, so that $A = \omega_2$: we shall obtain a monochromatic set with cardinal ω_1. The proof will easily extend to the general case.

We say that a sequence of terms a_i in A(i ordinal) is **pre-monochromatic** iff for every index i, the color of the pairs $\{a_i, a_j\}$ remains the same for all $j > i$: we say that this color is **associated to the index** i. Construct as follows such a pre-monochromatic ω_1-sequence. We then immediately extract a monochromatic ω_1-sequence, by taking all those indices i with the same color, chosen to be cofinal in ω_1.

Let a_0 denote the minimum of A (in the usual well-ordering of $A = \omega_2$; thus $a_0 = 0$, the value is unimportant). Partition the elements $x \neq a_0$ of A into two classes: the class γ_1^+ of those x such that $\{a_0, x\}$ has color (+), and the class γ_1^- similarly defined with (-). Let a_1^+ be the minimum in the class γ_1^+, and then partition the elements $x \neq a_1^+$ of γ_1^+ into two subclasses: the class γ_2^{++} of those x for which $\{a_1^+, x\}$ has color (+), and the class γ_2^{+-} similarly defined with (-). Similarly, let a_1^- be the minimum of the class γ_1^-, and then partition the elements of γ_1^- distinct from this minimum, into two subclasses γ_2^{-+} and γ_2^{--}, defined as previously.

In general, let u be an ordinal strictly less than ω_1. If u has a predecessor $u-1$, assume that the classes γ_{u-1}^s are already defined, each characterized by a sequence s of + and -, with length $u-1$, hence defined on the indices strictly less than u.

Denote by a_{u-1}^s the minimum of γ_{u-1}^s, provided this class is non-empty. Partition the elements in this class which are distinct from the minimum (assuming of course that there are such), into two subclasses: the class γ_u^{s+}, characterized by the sequence s completed by the $(u-1)$-st term +, hence a sequence of length u of x for which $\{a_{u-1}^s, x\}$ has color (+). Analogous definition for the class γ_u^{s-}.

Suppose now that u is a limit ordinal. Given a sequence s of length u, hence of indices strictly less than u, consider for each $i < u$ the restricted sequence s/i

taking the same values as s but defined only for indices $< i$. Then we define the class γ_u^s as the intersection of the classes $\gamma_i^{s/i}$ for all ordinals $i < u$. Finally, whether u is a limit ordinal or not, we define the element a_u^s as the minimum of γ_u^s, provided that this class is non-empty.

Since the ordinals considered are at most countable, there are continuum many sequences s, hence ω_1 many, since we assume the continuum hypothesis. Hence there are ω_1 many classes γ and their minimums, for all indices u and all sequences s. Since A has cardinality w_2, there exist other elements besides the minimums.

Let r be one such. Beginning with a_0, pick either a_1^+ or a_1^-, depending on whether the pair $\{a_0, r\}$ has color (+) or (-). If we have chosen a_1^+, then choose either a_2^{++} or a_2^{+-} by the same consideration. Continue thusly: given an ordinal $u < \omega_1$, if we have already chosen the sequence s with length $u - 1$ and values (+) and (-), choose a_u^{s+} or a_u^{s-}, depending on whether the pair $\{a_{u-1}^s, r\}$ has the color (+) or (-). For every limit ordinal u, take the sequence s with length u which is the limit of the sequences already obtained, and take the corresponding a_u^s.

By the definition of the classes γ, whenever we reach a term a_u^s in the sequence defined by r, all following terms belong to the same class γ_{u+1}, i.e. that which contains r. Hence for every $v > u$ all pairs $\{a_u, a_v\}$ have the same color as $\{a_u, r\}$. Moreover, the class γ through which we pass is never empty, since r belongs to it. We thus obtain a pre-monochromatic ω_1-sequence from which we extract, as already explained, a monochromatic ω_1-sequence. •

• Let us now take up the proof of our stated lemma. Assume that the second conclusion does not hold, so that every (-)-monochromatic subset of A has cardinality at most ω_1.

Take a subset D_0 of A, which is maximal with respect to inclusion among the (-)-monochromatic subsets; hence D_0 has cardinality $\leq \omega_1$. For every element x of $A - D_0$, there exists at least one element y of D_0 such that the pair $\{x, y\}$ has color (+). Associate to each x of $A - D_0$ the minimum such y. Thus we partition the elements of $A - D_0$ into classes, each defined by an element a_0 of D_0. We denote by $\gamma(a_0)$ the class thus associated with a_0, with the color (+) for $\{a_0, x\}$ for all x in this class. If $\gamma(a_0)$ is non-empty, then take a subset $D_1(a_0)$ of this class which is maximal with respect to inclusion among (-)-monochromatic subsets; hence D_1 has cardinality $\leq \omega_1$.

As previously, partition the elements of $\gamma(a_0) - D_1(a_0)$ into classes which are defined by an element a_1 of $D_1(a_0)$. We denote by $\gamma(a_0, a_1)$ the class thus associated with the sequence (a_0, a_1), with the color (+) for $\{a_0, x\}$ and for $\{a_1, x\}$ for every x in this class.

The iteration can be continued in an obvious manner for all successor ordinals $u < \omega_1$. For a limit ordinal u and a sequence s with length u, having values $a_i (i < u)$, we define the class $\gamma(s)$ as the intersection of the $\gamma(s/i)$ for all $i < u$ (recall that s/i is the restriction of s to indices $< i$).

Of course, each such sequence $s = a_0, a_1, ..., a_i, ..$ must satisfy the following conditions. For every successor ordinal $i < u$, the term a_i belongs to $D_i(a_0, a_1, ..., a_{i-1})$, which is a maximal (-)-monochromatic subset of $\gamma_i(a_0, a_1, ..., a_{i-1})$. For every limit ordinal $i < u$, the term a_i belongs to $D_i(a_0, a_1, ..., a_j, ..)(j < i)$, which is a maximal

(-)-monochromatic subset of $\gamma_i(a_0, a_1, ..., a_j, ..)$, the latter being the intersection of the $\gamma_j(a_0, a_1, ..., a_{j-1})$($j$ successor ordinal $< i$). Note that the axiom of choice is used to define D: we do not have initially a well-ordering of the set of subsets of A.

For every ordinal $u < \omega_1$, there are at most ${}^{\omega}(\omega_1)$-many sequences $a_0, a_1, ..$ with length u, thus at most ω_1-many, since we assume the continuum hypothesis. Hence the set of all the a has cardinality ω_1 and likewise for the union of the D. Since A has cardinality ω_2, there exist elements which belong to no D.

Let r be one such. Pick a_0 such that r belongs to $\gamma(a_0)$. Then as r does not belong to D_0, pick a_1 such that r belongs to $\gamma(a_0, a_1)$; and in general, for every $u < \omega_1$ pick a_u such that r belongs to $\gamma(a_0, ..., a_u)$. The sequence of these a_u is (+)-monochromatic and has length ω_1, since the above considered classes γ all contain r as an element, and so are all non-empty. •

3.3.5 Partition theorem (Erdös, Rado)

(1) Let A be a regular limit aleph, B an aleph $< A$. Partition the pairs of elements of A into two colors (+) and (-). **Then either there exists a subset of A equipotent with A and (+)-monochromatic, or there exists a subset of A equipotent with B and (-)-monochromatic** (uses generalized continuum hypothesis)

(2) Let a be an infinite aleph, b the least aleph satisfying ${}^{b}a > a$. Let A be the aleph a_+, the successor of a. Partition the pairs of elements of A into two colors (+) and (-). **Then either there exists a subset of A equipotent with A and (+)-monochromatic, or there exists a subset of A equipotent with b and (-)-monochromatic**. (uses axiom of choice); [55] ERDÖS, RADO 1953.

• (1) Assume that every (+)-monochromatic subset of A has cardinality strictly less than A. We shall construct a subset of A equipotent with B and (-)-monochromatic. To every non-empty subset X of A, associate a subset $T(X)$ of X which is (+)-monochromatic and maximal with respect to inclusion, among (+)-monochromatic subsets of X (uses axiom of choice). Note that $T(X)$ is non-empty: at the worst $T(X)$ could be a singleton.

For every element x of A and every ordinal i, we define as follows the element $f_i(x) \in A$.

Fix x; take $f_0(x) = x$ if x belongs to $T(A)$. Otherwise set $f_0(x)$ to be the least element in $T(A)$ (with respect to the well-ordering of A), for which the pair $\{x, f_0(x)\}$ has color (-) (uses maximality of $T(A)$). Now let i be a non-zero ordinal, and assume that $f_j(x)$ is defined for each $j < i$. Either there already exists a $j < i$ for which $f_j(x) = x$ and in this case, set $f_i(x) = x$. Or all the $f_j(x)$ are distinct from x and distinct among themselves, and the pairs they form among themselves or with x all have the color (-).

Then let U be the set of the $y \in A$ for which $\{y, f_j(x)\}$ has color (-) for all $j < i$. In particular x belongs to U. If $x \in T(U)$, then set $f_i(x) = x$. Otherwise, if $x \in U - T(U)$, then by the maximality of $T(U)$ there exist elements $z \in T(U)$ with $\{x, z\}$ having color (-). Take $f_i(x)$ to be the least such z in the well-ordering of

A. By the preceding construction, for every $x \in A$, there exists an ordinal i_x such that, for all $i < i_x$, the $f_i(x)$ are distinct from x and distinct among themselves, and the pairs they form among themselves or with x have color (-); and $f_i(x) = x$ for $i \geq i_x$.

For every ordinal i, let M_i be the set of these $f_i(x)$, with i fixed and x running through A. We shall prove that M_i has cardinality strictly less than A for every $i < A$.

First of all, M_0 is included in $T(A)$ which is (+)-monochromatic, hence by hypothesis has cardinality strictly less than A. Consider an ordinal $C < A$ and assume that for every $i < C$ we have $\operatorname{Card} M_i < A$. Since A is regular by hypothesis, the maximum cardinal or the supremum cardinal of the $M_i (i < C)$ is strictly less than A. Otherwise A would be the union of ($< A$)-many sets, each strictly subpotent to A. Let D denote the maximum or supremum cardinal of the M_i. To every element $x \in A$, associate the sequence of the $f_i(x)(i < C)$. The number of distinct such sequences is less than or equal to ${}^{(\operatorname{Card} C)}D$.

Now since A is a limit aleph, by 2.8.5 proposition (2) (generalized continuum hypothesis), this cardinal is strictly less than A. Two identical sequences give the same set U, hence the same $T(U)$ of cardinality strictly less than A. For any x, the new element $f_C(x)$ belongs to $T(U)$. Hence as x varies, the possible sequences $f_i(x)(i > C)$ give ($> A$) many sets $T(U)$. Because of the regularity of A, the set M_C of all possible $f_C(x)$ has cardinality strictly less than A.

By hypothesis B is a cardinal strictly less than A. Thus there exists an r in A such that $i_r \geq B$. For otherwise M_B would be identical with A, contradicting the preceding. Hence the $f_i(r)$ are distinct for $i < B$, and their set is (-)-monochromatic and of cardinality B. •

• (2) For A of cardinality a_+, suppose that every (+)-monochromatic subset of A has cardinality $\leq a$. We shall construct a (-)-monochromatic subset of A with cardinality b.

For every x of A, construct as previously the $f_i(x)$ and the M_i for all ordinals i (axiom of choice). We shall prove that, for every i of cardinality $< b$, hence for every $i < b$, the set M_i has cardinality $\leq a$. M_0 is included in $T(A)$, which is (+)-monochromatic and hence has cardinality $\leq a$.

Take an ordinal $c < b$, hence $c < a$, and assume that $\operatorname{Card} M_i \leq a$ for every $i < c$. The maximum or the supremum of $\operatorname{Card} M_i (i < c)$ is $\leq a$. The sequences $f_i(x)(i < c)$ have cardinality at most ${}^{(\operatorname{Card} c)}a$ hence $\leq a$ since $(\operatorname{Card} c) < b$. As in the proof of (1) above, it follows that the cardinality of the set M_c of $f_c(x)$ for x running through A, is $\leq a \times a = a$.

Since $b \leq a$, the union of the $M_i(i < b)$ has cardinality $\leq b \times a = a$. Let r be an element of A not in this union. Then $r \neq f_i(x)$ for every x of A and every $i > b$. In particular $r \neq f_i(r)$ for every $i < b$: thus $i_r \geq b$. It follows that the set of the $f_i(r)(i < b)$ is (-)-monochromatic and has cardinality b. •

Note that, in the proof of (2), the cardinal $A = a_+$ is not a limit aleph. This condition, and also the use of generalized continuum hypothesis, is only necessary for (1) in order to apply 2.8.5 proposition (2).

The example of ω_1, given at the end of 3.3.1, can be expressed here by taking $a = \omega, b = \omega$ and $A = \omega_1$.

Recall that with generalized continuum hypothesis, the only regular limit alephs are ω and the inaccessible alephs (see 2.8.9). The statement (1) holds only for these.

On the other hand, with only the axiom of choice, the continuum can be regular or not, a limit or a successor aleph, with cofinality $\geq \omega_1$ (see 2.8.4).

3.4 Incidence matrix, linear independence lemma (Kantor), multicolor theorem (Pouzet)

3.4.1 Incidence matrix

Let p, q be two integers, E a finite set with cardinal h. Represent as "ordinate values" the set of p-element subsets of E, of cardinality $h!/p!(h-p)!$, and as "abscissa values" the set of $(p+q)$-element subsets, of cardinality $h!/(p+q)!(h-p-q)!$. To each couple (x, y) where x is a $(p+q)$-element set and y is a p-element set, attribute the value 1 iff $x \subseteq y$ and the value 0 otherwise. The rectangular table thus obtained shall be called the **incidence matrix** of E for p and q.

Note that if $h = \operatorname{Card} E \geq 2p + q$, then each row of the matrix, corresponding to a p-element set, is at least as long as each column, corresponding to a $(p+q)$-element set. Indeed we have $p!(h-p)! \geq (p+q)!(h-p-q)!$.

The reader is assumed to be familiar with the elementary theory of determinants and with the notion of linear dependence. If $\operatorname{Card} E < 2p + q$, then it is possible that a row of the incidence matrix depends linearly on one or several other rows. For example, for $p = q = 1$ and $\operatorname{Card} E = 2$, the matrix reduces to two rows and one column, with value 1. But for $\operatorname{Card} E \geq 2p + q$, we have the following result.

3.4.2 Linear independence lemma

If $\operatorname{Card} E \geq 2p+q$, the rows of the incidence matrix are linearly independent: no row is a linear combination of other rows. See [132] KANTOR 1972.

Equivalently, every non-zero determinant extracted from the matrix and depending on a finite number r of rows can be extended to a non-zero determinant based on the previous rows together with an arbitrary $(r+1)$-st row.

Consequently in the case that E is finite, there exists a non-zero determinant depending on all the rows. Hence there exists an injection which to each p-element set y associates a $(p+q)$-element set including y as a subset.

• To each permutation f of E associate the corresponding permutation $\overline{f}$ of p-element subsets of E. Hence $\overline{f}$ permutes the set of rows. There corresponds to f as well a permutation of the set of $(p+q)$-element subsets, hence of the set of columns, but it is unnecessary to consider this, since we are working with linear combinations of rows and reasoning by the coefficients attributed to each row in a given linear combination.

Assume that E has finite cardinal h; we argue ad absurdum. Assume that there exists a p-element subset, hence a row which is a linear combination of all the other rows, with positive, negative or zero rational coefficients, since these are quotients of determinants with values 0 and 1. Let us call b this p-element set and the corresponding row. Given an arbitrary permutation f of E which preserves the set b (but not necessarily each element of b), then $\overline{f}$ preserves the row b and permutes the set of the other rows. Two rows which are transformed one into the other represent two p-element sets y, y' such that $b \cap y$ and $b \cap y'$ are equipotent.

Transform the given linear combination by all possible $\overline{f}$, the number of such being $(h-p)!p!$, then take the combination which is the arithmetic average of the combinations thus transformed. By symmetry, all the rows which represent p-element sets disjoint from b will have the same coefficient. Similarly for all rows which represent p-element sets having a single element in common with b, and in general for all rows which represent p-element sets having equipotent intersections with b.

Consider a column a representing a $(p+q)$-element set disjoilnt from b: this a exists since $h \geq 2p+q$. In the column a, the p-element sets included in a are all disjoint from b, and so all have the same coefficient in our combination. Moreover, if we denote these p-element sets by y, these are the only ones yielding the value 1 in position (a, y) in the incidence matrix, while the matrix has the value 0 in position (a, b). It follows that their coefficient is zero, hence each row which represents a p-element set disjoint from b has coefficient zero. The problem is thus answered negatively for $p = 1$, since in this case the p-element sets distincts from b are disjoint with b, hence the above assumed linear combination does not exist.

Assume that $p \geq 2$, and consider a column a_1 representing a $(p+q)$-element set which intersects b in a unique element. then the rows y for which the matrix has value 1 in (a_1, y) are those which represent either a p-element set disjoint from b, hence with coefficient zero, or a p-element set intersecting b in a single point. By the preceding discussion, the latter have the same coefficient in the combination. Since the matrix has the value 0 in (a_1, b), this coefficient is zero. The problem is thus answered negatively for $p = 2$, since in this case the p-element sets distincts from b have at most one element in common with b. In the general case, by iterating the preceding argument, we prove that all the coefficients are zero, hence that the above assumed linear combination does not exist.

The result follows immediately in the case of E infinite.

Finally, for the conclusion concerning the extendibility of a non-zero determinant, assume on the contrary that there exists a non-zero determinant which is not extendible, and deduce that an arbitrary row of the matrix is a linear combination of rows of the submatrix which corresponds to this determinant. •

Degenerate case. In the "degenerate case" where $h = \text{Card}\, E < 2p+q$, the number of columns is strictly less than the number of rows. In this case the columns of the incidence matrix are linearly independent. In other words, there exists a non-zero determinant based on the columns.

• Interchange each p-element set y with the $(h-p)$-element set $E-y$, and each $(p+q)$-element set x with the $(h-p-q)$-element set $E-x$. Then the inclusion $y \subseteq x$ is equivalent to $E-x \subseteq E-y$. The role of p is played by $p' = h-p-q$; the role of $p+q$ is played by $p'+q' = h-p$, so that $q' = q$. We have $2p'+q' = 2h-2p-q > h$: hence we can apply the linear independence lemma with rows and columns interchanged. •

3.4.3 Multicolor theorem

Let E be a finite set, h its cardinal and p, q two integers. Partition the p-element subsets of E into a finite number k of classes which are called colors $u_0, u_1, \ldots, u_{k-1}$. For each $(p+q)$-element subset a of E, we call the **multicolor** of a the function which to each color $u_i (i < k)$ associates the number of p-element sets of color u_i which are included in a. When this number is non-zero, we say that the color u_i **figures** in the multicolor.

Theorem. If Card $E \geq 2p+q$, then the number of multicolors of $(p+q)$-element subsets of E is at least equal to the number of colors of p-element subsets. More precisely, there exists an injection which to each color u (to which at least one p-element set belongs) associates a multicolor in which u figures, and to which at least one $(p+q)$-element set belongs ([195] POUZET 1976).

• Assume first that E has finite cardinal $h \geq 2p+q$. Hence the number of $(p+q)$- element sets is at least equal to that of the p-element sets, and the rows of the incidence matrix are linearly independent. To each color there corresponds a finite set of rows of that color. Replace these by a unique row which is their sum, obtained by adding the values 0 or 1 in each column. Thus each new row represents a color u. Each column continues to represent a $(p+q)$ -element set, and indicates the number of p-element sets of color u which are included in this $(p+q)$-element set.

Note that, in the new matrix thus obtained, the rows are linearly independent. It suffices to see that, given a matrix with k independent rows ($k \geq 2$), the replacement of two rows b and b' by their sum yields a matrix with $k-1$ independent rows. Indeed, there exists a non-zero determinant based on the $k-2$ intact rows. So that the only other possibility would be that the row sum of b and b' is a linear combination of the $k-2$ intact rows. But then the row b, for example, would be a linear combination of the $k-2$ intact rows plus the row b', contradicting the hypothesis.

Thus, if k is now the number of colors, hence of rows, we have a non-zero determinant of order k. Take in this determinant a sequence of k couples (x, y) where x is a column and y a row, with non-zero value of the new matrix in each considered couple. We thus obtain the injective function in the theorem. This injection associates, to two distinct colors y, y' two $(p+q)$-element sets x, x' whose multicolors are distinct. Otherwise we would have two identical columns in the determinant. Thus this is an injection from the set of colors into the set of multicolors.

It remains to consider the case where E is denumerable. If we only have a finite number of colors, then we restrict E to a set of finite cardinality at least equal to $2p + q$ and including p-element subsets of each color. The rows, which represent the colors, are linearly independent, and remain so when one takes up the entire infinite set E. If there are infinitely many colors, then we still have linear independence. Then as mentioned for the linear independence lemma, every non-zero determinant is extendible to a non-zero determinant over one more row, hence one more color. The existence of the injective function in the theorem follows. •

3.5 Combinatorial lemmas, color and inclusion

3.5.1 First lemma

(1) Let E be a set, p, q two integers such that $p + q \leq \operatorname{Card} E$. Take a set of p-element subsets of E and call this the **color** A.

If every $(p + q)$-element subset includes the same number k of p-element subsets with color A, then every $(p + q + 1)$-element subset includes the same number $k(p + q + 1)/(q + 1)$ of p-element subsets with color A.

(2) Given two not necessarily disjoint sets of p-element subsets of E, call these the **colors** A and B.

If every $(p + q)$-element subset includes as many p-element subsets with color A as p-element subsets with color B, then the same is true for every $(p + q + 1)$-element subset of E.

• (1) Let F be a $(p + q + 1)$-element set. The cardinality of the set of $(p + q)$-element subsets of F is $p + q + 1$. Each includes k many p-element subsets with color A, which yields $k(p+q+1)$ couples, each formed with a p-element set having color A and with a $(p+q)$-element set which includes it. For every p-element subset with color A in F, there are $q + 1$ many $(p + q)$-element subsets which include it. This yields $k(p + q + 1)/(q + 1)$ as the number of p-element subsets included in F and having color A. •

• (2) Let F be a $(p + q + 1)$-element set. Every $(p + q)$-element subset of F includes as many p-element subsets with color A as with color B. Thus we have that in F, there are the same number of couples, each with first term a p-element set with color A and second term a $(p + q)$-element set including the first term, as of couples, each with first term a p-element set having color B and second term a $(p + q)$-element set including the first term. We obtain the number of p-element subsets with color A by dividing the preceding number by $q + 1$. Similarly for the color B. •

3.5.2 Second lemma

(1) Let E be a finite set, p, q two integers such that $p + q \leq \operatorname{Card} E$. Let A be a color of certain p-element subsets of E. Let $s \leq p$ and $s \leq (\operatorname{Card} E) - p - q$.

If every $(p+q)$-element subset includes the same number of p-element subsets with color A, then every s-element subset is included in the same number of p-element subsets with color A.

(2) Let A, B be two colors of p-element sets, not necessarily disjoint. **If every $(p+q)$-element set includes as many p-element subsets with color A as p-element subsets with color B, then every s-element subset is included in as many p-element sets with color A as p-element sets with color B** (communicated by POUZET in 1975).

• First we prove (2). For $s = 0$, this follows from the first lemma, statement (2) iterated from $p + q$ to $\operatorname{Card} E$. Assume that $s \geq 1$ and assume that the statement is true for $s - 1$ and $p + q$. In other words, for every E with finite cardinality $\geq s + p + q - 1$ and every $(s - 1)$-element subset of E. We shall prove this for s and $p + q$, hence for E with finite cardinality $\geq s + p + q$ and a s-element subset $H \subseteq E$.

Let u be an element of H. By the induction hypothesis, there exists a same number k of p-element subsets with color A as with color B, included in E and including $H - \{u\}$. Similarly, there exists a same number l of p-element subsets with color A as with color B, included in $E - \{u\}$ and including $H - \{u\}$ (the cardinality of these sets being respectively $\geq s + p + q - 1$ and equal to $s - 1$). By subtraction, there exist the same number $k - 1$ of p-element subsets with color A as with color B, included in E and including H. •

• Statement (1) follows from (2). Indeed, let H and H' be two s-element subsets of E. Take a permutation f of E which transforms H into H'. Take the p-element subsets with color A, and let A' be the color of their images via f. Then every $(p+q)$-element subset X includes as many p-element subsets with color A as with color A', as one can see by taking the $(p+q)$-element subset $\overline{f^{-1}}(X)$. By (2), the s-element subset H' is included in as many p-element subsets with color A as with color A'. But the latter are the images via f of p-element subsets with color A and including H. Thus H and H' are included in the same number of p-element subsets with color A. •

3.5.3 Third lemma

(1) Let E be a set, p an integer less than or equal to $\operatorname{Card} E$, and let A be a non-empty set, called color, of p-element subsets. **If there exists an integer q such that $2p + q \leq \operatorname{Card} E$, and if every $(p+q)$-element set includes the same number of p-element subsets with color A, then every p-element subset has color A.**

(2) Given E and p less than or equal to $\operatorname{Card} E$, let A, B be two sets, called colors, of p-element subsets. **If there exists an integer q such that $2p + q \leq \operatorname{Card} E$, and for which every $(p+q)$-element subset includes as many p-element subsets with color A as p-element subsets with color B, then the colors A and B are identical.**

• We can assume that E is finite, by replacing E if necessary by a finite subset of cardinality at least equal to $2p+q$. Now take the previous lemma with $s = p$. By statement (1), every p-element subset is included in the same number of p-element subsets with color A. In other words, every p-element subset has color A (since it is assumed that A is non-empty).

By statement (2), every p-element subset is included in as many p-element subsets with color A as p-element subsets with color B. In other words, the colors A and B are identical. •

If $\operatorname{Card} E < 2p + q$, then by taking $s \leq (\operatorname{Card} E) - p - q$, it is easy to give an example in which the color A does not extend to the entire set of p-element subsets. Take $E = \{a, b, c, d\}$ hence of cardinality 4, with $p = 2, q = 1$, and only the edges ab and cd with color A: then every 3-element subset contains such an edge. Note that $p + q \leq \operatorname{Card} E$ so that our second lemma works: each element belongs to an edge with color A.

Adding ac and bd with color B, every 3-element subset contains an edge with each color; yet $A \neq B$. In agreement with our second lemma, each element belongs to an edge of each color.

3.6 Profile increase theorem (Pouzet)

Let R be a relation with base E. To each non-negative integer p, associate the finite number $f(p)$ of isomorphism types of the p-element restrictions of R. The numerical function f thus defined is called the **profile** of R (notion due to [191] POUZET 1972). Note that $f(0) = 1$ and if $\operatorname{Card} E = h$ (finite), then $f(h) = 1$ and $f(p) = 0$ for all $p > h$.

Examples. If R is a chain, then the profile function has constant value equal to 1 (when $p \leq \operatorname{Card} E$).

If R is an infinite unary relation taking the value (+) for a finite number a of elements and (-) for all other elements, then the profile increases from $f(0) = 1$ to $f(a) = a + 1$, and then remains stationary.

3.6.1 Increasing profile

(1) **Theorem. Let p, q be two integers and R a relation with cardinality at least equal to $2p + q$. Then the number of isomorphism types of the restrictions of R to $(p+q)$ elements is at least as great as the number of isomorphism types of the restrictions of R to p elements.**

More precisely, there exists **an injective function which, to each isomorphism type u of a restriction of R to p elements, associates the type of a restriction to $p+q$ elements, which is an extension of u** ([195] POUZET 1976).

• This follows from the multicolor theorem 3.4.3, where the isomorphism types on p elements play the role of the colors of the p-element sets, and where two isomorphic $(p + q)$-element restrictions have the same multicolor. •

Consequently:

(2) If a relation has an infinite base, then the profile is increasing.

(3) If a relation has a finite base with even cardinality $2h$, then its profile is increasing for integers less than or equal to h. If a relation has odd cardinality $2h+1$, then its profile is increasing for integers less than or equal to $h+1$.

Note that proposition (1) is stronger than (2) and (3). For example, consider a relation on 7 elements. Not only does its profile increase for integers 0 to 4, but its value for 5 is greater than or equal to the value for 2 ; its value for 6 is greater than or equal to the value for 1 .

POUZET asked **if profile for finite relations satisfies unimodality** (increasing untill a maximum then decreasing). Negative answer by [236] STANTON 1990: take the equivalence relation of cardinal 24 with 4 equivalence classes: 2 with 8 elements and 2 with 4 elements. Then the profile is the following, starting from 1 restriction type of cardinal 0, then 1 type of cardinal 1, then 2 types of cardinal 2 (the two elements can belong either to a same equivalence class or to two different classes), and so until cardinal 24:

1 1 2 3 5 6 9 11 15 17 21 23 27 28 31 30 31 27 24 18 14 8 5 2 1

3.6.2 Isomorphic restrictions

Let p, q be two integers and R a relation with base E, where $\text{Card}\, E \geq 2p+q$, and let f be a permutation of E. **If for every $(p+q)$-element subset of E and its image under f, the restrictions of R to these two subsets are isomorphic, then for every p-element subset of E and its image under f, the restrictions of R are isomorphic.**

• Take an arbitrary restriction of R to p elements, and let U denote its isomorphism type. All p-element subsets of E with this isomorphism type shall have the color U . Now a p-element subset of E is said to have the color V iff its image under f has the color U.

By hypothesis, for each $(p+q)$-element subset a of E, the restrictions R/a and $R/\overline{f}(a)$ are isomorphic, hence a and its image $\overline{f}(a)$ have the same number of p-element subsets of color U. Thus a includes the same number of$p$-element subsets of color U as well as of color V. It follows that the colors U and V are identical, by 3.5.3 proposition (2). Hence f takes each p-element subset into another of the same color, i.e. with an isomorphic restriction of R. •

3.7 Ramsey sequence; another proof of Galvin's theorem (Lopez)

The following notion of **Ramsey sequence of conditions** is a form of the classical Ramsey set: see [54] ERDÖS, RADO 1952. The notion of Ramsey sequence and the connected proof of GALVIN's initial interval theorem are due to [161] LOPEZ 1983. As opposed with POUZET's proof in 3.2.2 and following, here we need neither lexicographic rank nor transfinite induction. As well as in 3.2.2, the axioms of ZF will be sufficient: see 3.7.4 below.

3.7.1 Ramsey sequence

Given two finite sets A, B of integers, put $A > B$ or $B > A$ iff every element of B is strictly greater than every element of A. We adopt the convention that the empty set is $<$ and $>$ any set; so that $<$ is irreflexive and transitive only for non-empty sets. Let H be a finite set, Z an infinite set of integers. A finite sequence of conditions $C_i(H, Z)(i = 1, ..., r)$ is said to be a **Ramsey sequence** iff we have the following:

$\forall_{H(fin)}\forall_{X(inf)}X > H \Rightarrow \exists_{Y(inf)}Y \subseteq X \wedge [(\forall_{Z(inf)}Z \subseteq Y \Rightarrow C_1(H, Z)) \vee ... \vee (\forall_{Z(inf)}Z \subseteq Y \Rightarrow C_r(H, Z))]$

(notations: fin = finite, inf = infinite set of integers; obvious logical symbols).

First example. Partition the pairs of integers into two colors (+) and (-). Take for $C(H, Z)$ the following statement: "for each integer $h \in H$, all pairs h, z where $z \in Z$, have same color (depending on h)" Then C alone constitutes a Ramsey sequence.

Second example. Take a condition C and define $\neg C$ as the negation of C. Then the sequence $(C, \neg C)$ is often Ramsey. It is the case, for instance, if $C(H, Z)$ means that the preceding pairs $\{h, z\}$ have color (+). In this case, the same infinite set Y as before satisfies the following condition. Either the color (+) is associated to each element of H (more precisely, any pair formed of an element of H and an element of Y has color (+)): then each infinite $Z \subseteq Y$ belongs to the set defined by C and H. Or there exists at least one element of H to which is associated the color (-) in the preceding sense; then each infinite $Z \subseteq Y$ belongs to the complement, defined by $\neg C$ and H.

Third example. Let H, F be finite sets of integers and $\mathcal{D}(H, F)$ be an arbitrary condition. Then the couple of the condition $C(H, Z) = \exists_{F(fin)}(F \subset Z \wedge \mathcal{D}(H, F))$ with its negation is Ramsey.

• Suppose the contrary. There exist a finite set H and an infinite set $X > H$ such that, for every infinite set $Y \subseteq X$, there exist two infinite subsets Z_1 and Z_2 with $\mathbf{C}(H, Z_1)$ and the negation $\neg\mathbf{C}(H, Z_2)$. Then each finite subset F of Z_2 satisfies $\neg\mathcal{D}(H, F)$. Now replace Y by Z_2: there exists an infinite subset Z_1' of Z_2 such that $C(H, Z_1')$. Thus there exists a finite subset F of $Z_1' \subseteq Z_2$ which satisfies $\mathcal{D}(H, F)$: contradiction. •

Remarks. In the case of two such opposite conditions, the above formula means that, given H, the set of all infinite $Z \subseteq Y$ satisfying $C(H, Z)$ is Ramsey in the sense of [54] ERDÖS, RADO 1952.

Among sets of infinite sets of integers, i.e. among sets of reals, it is known that the following are Ramsey: all open sets; Borel sets (see [91] GALVIN, PRIKRY 1973); analytic sets ([232] SILVER 1970). See also [47] ELLENTUCK 1974, who characterizes the "completely ramsey sets" by the Baire property.

3.7.2 Lemma on Ramsey sequences

Given a Ramsey sequence $C_1, ..., C_r$, we have the following statement (modulo the axiom of dependent choice):

$\exists_{A\ (inf)}\forall_{H\ (fin)}\ [\forall_{Z\ (inf)}(H \subset A\ \wedge\ Z \subseteq A \wedge\ Z > H) \Rightarrow C_1(H,Z))] \vee \ldots$
$\vee\ [\forall_{Z\ (inf)}(H \subset A\ \wedge\ Z \subseteq A\ \wedge\ Z > H) \Rightarrow C_r(H,Z))]$

• The proof generalizes the first part of RAMSEY's proof infinitary form, in obtaining elements v_i. Start from $u_0 = 0, H_0 = \{0\}$ and $X_0 =$ set of integers $\neq 0$. We get an infinite $Y \subseteq X_0$, called Y_0 and satisfying the above condition in brackets, where $H = H_0$ and $A = \{0\} \cup Y_0$. Then let u_1 be the first element of Y_0. Start from $H_1 = \{u_0, u_1\}$ and $X_1 = Y_0 - \{u_1\}$. We get an infinite $Y_1 \subseteq X_1$ which satisfies our above condition in brackets, where $H = H_0$ or H_1 and $A = \{u_0, u_1\} \cup Y_1$. Then start from $H'_1 = \{u_1\}$ and $X'_1 = Y_1$. We get an infinite $Y'_1 \subseteq Y_1$ which satisfies our condition, where $H = H_0$ or H_1 or H'_1 and $A = \{u_0, u_1\} \cup Y'_1$. Then let u_2 be the first element of $Y'_1.Y_2 \subseteq Y'_1$ which satisfies our condition, where $H = H_0$ or H_1 or H'_1 or H_2 and where $A = \{u_0, u_1, u_2\} \cup Y_2$.

Iterate, taking for $H'_2, H"_2, ..$ all the sets with last element u_2, and so getting $Y'_2, Y"_2, ..$ before defining u_3, H_3 and Y_3; and so on. Finally take for A the set of u_i(i integer). The axiom of dependent choice is used for choosing sets Y. •

3.7.3 A proof of Galvin's initial interval theorem

Let $\mathcal{F}$ be a set of finite sets of integers, assumed to be mutually incomparable under inclusion, and to satisfy GALVIN's hypothesis: **every infinite set of integers includes at least one element of $\mathcal{F}$ as a subset.**

Take $\mathcal{D}(H, F)$ to be the following condition: "the union $H \cup F$ admits an initial interval belonging to $\mathcal{F}$ "; more briefly "$H \cup F$ has $\mathcal{F}i.i.$". Then by 3.7.1 third example, the couple formed of the condition $\exists_{F(fin)}(F \subset Z \wedge H \cup F$ has $\mathcal{F}i.i.)$ with its negation, is Ramsey.

Consequently by the previous lemma, there exists an infinite set A such that, for every finite subset H of A:

either (1) $\forall_{Z(inf)}(Z \subseteq A \wedge Z > H) \Rightarrow \exists_{F(fin)}(F \subset Z \wedge H \cup F$ has $\mathcal{F}i.i.)$;

or (2) $\forall_{Z(inf)}(Z \subseteq A \wedge Z > H) \Rightarrow \forall_{F(fin)}(F \subset Z \Rightarrow H \cup F$ has no $\mathcal{F}i.i.)$.

Firstly we prove that, **assuming Galvin's hypothesis, there exists an infinite set A such the above (2) is false: so only (1) is true.**

• For H empty, the above conclusion (2) is false. Indeed (2) reduces to saying that, for every infinite set Z, there does not exist any finite subset of Z which belongs to $\mathcal{F}$.

Now let G be a finite set of integers. Assume that the above (2) is false for every subset H of G. Then it suffices to prove that there exists an integer $g > \mathrm{Max}\, G$ such that every $H \subseteq G \cup \{g\}$ falsifies (2). For this purpose, it suffices to prove that, for each subset H of G, there exist only finitely many integers $h > \mathrm{Max}\, G$ such that $\cup\{h\}$ satisfies (2): indeed it will suffice to choose g strictly greater than all such h.

Arguing ad absurdum, assume the existence of an infinite sequence $h_1, h_2, ..., h_i, ..$ with $H \cup \{h_i\}$ satisfying (2). Take Z to be the infinite set of these h_i: by hypothesis H falsifies (2), so H verifies (1). Thus there exists a finite subset F of Z such that $H \cup F$ has an initial interval which belongs to $\mathcal{F}$. Let h_p(p integer) be the first element of F; then $H \cup \{h_p\}$ falsifies (2): contradiction. •

Secondly **we obtain Galvin's conclusion**.

• Consider the infinite set A in the previous subsection, now assumed to satisfy only the above (1). Let B be an arbitrary infinite subset of A. Let K be a finite subset of B belonging to $\mathcal{F}$, and denote by H the initial interval of B which ends with $\operatorname{Max} K$. Then by (1) above, there exists a finite subset F of $B - H$ such that $H \cup F$ has an initial interval which belongs to $\mathcal{F}$. This initial interval cannot surpass $\operatorname{Max} K$, since elements of $\mathcal{F}$ are mutually incomparable under inclusion. Consequently our initial interval of $H \cup F$ reduces to an initial interval of H, thus of B. •

3.7.4 How avoiding the axiom of dependent choice

The axiom of dependent choice, used to prove the lemma 3.7.2, is avoidable in view of obtaining GALVIN's theorem.

• For the theorem we need only a particular case of the considered lemma, with only the condition $C(H, Z) = \exists_{F(fin}(F \subset Z \wedge H \cup F \textit{ has } \mathcal{F}i.i)$ and its negation $\neg C(H, Z) = \forall_{F(fin)}(F \subset Z \Rightarrow H \cup F \textit{ has no } \mathcal{F}i.i.)$. By 3.7.1 third example, the couple $(C, \neg C)$ is a Ramsey sequence; so that, given a finite set H and an infinite set $X > H$:

either (1) $\exists_{Y(inf)} Y \subseteq X \wedge \forall_{Z(inf)}[Z \subseteq Y \Rightarrow \exists_{F(fin)}(F \subset Z \wedge H \cup F \textit{ has } \mathcal{F}i.i.)]$

or (2) $\exists_{Y(inf)} Y \subseteq X \wedge \forall_{Z(inf)}[Z \subseteq Y \Rightarrow \forall_{F(fin)}(F \subset Z \Rightarrow H \cup F \textit{ has no } \mathcal{F}i.i.)]$.

Either (2) is false; in other words:

$\forall_{Y(inf)} Y \subseteq X \Rightarrow \exists_{Z(inf)}[Z \subseteq Y \wedge \exists_{F(fin)}(F \subset Z \wedge H \cup F \textit{ has } \mathcal{F}i.i.)]$

In such a case, take $Y = X$; then take any $Y \subseteq X$ but change the notations: writing Z instead of Y we get the following:

$\exists_{Y(inf)} Y = X \wedge \forall_{Z(inf)}[Z \subseteq Y \Rightarrow \exists_{T(inf)}[T \subseteq Z \wedge \exists_{F(fin)}(F \subset T \wedge H \cup F \textit{ has } \mathcal{F}i.i.)]]$

from which we immediately get:

$\exists_{Y(inf)} Y = X \wedge \forall_{Z(inf)}[Z \subseteq Y \Rightarrow \exists_{F(fin)}(F \subset Z \wedge H \cup F \textit{ has } \mathcal{F}i.i.)]$

So we obtain (1) strengthened by the unambiguous definition $Y = X$.

Or (2) is true, with (1) true or false, which is immaterial. In this case, take all the infinite sets Y which satisfy (2), and note that each infinite subset of a Y is still a Y satisfying (2). We proceed lexicographically: we take the least integer u_0 for which there exists a Y beginning with u_0; then the least $u_1 > u_0$ for which there exists a Y beginning with u_0, u_1; and so on. Finally we adopt the definition $Y = \{u_0, u_1, ..\}$.

Now we no more need 3.7.2 and we go directly to the proof of GALVIN's theorem as stated in the previous subsection. •

3.8 Exercises

3.8.1 A simple majoration for $(3)^2_k$

1 - Given two integers p, k, consider a p-element set whose pairs (or edges) are partitioned into k colors. This repartition is said to be a k-**good coloration** iff

no monochromatic triangle appear. Call $f(k)$ the maximum value p for which a k-good coloration exists: so that $(3)_k^2 = f(k) + 1$. From 3.1.4 we know that $f(1) = 2, f(2) = 5 = 2f(1) + 1, f(3) = 16 = 3f(2) + 1$.

Prove the inequality $f(k) \leq k.f(k-1) + 1$ (see [58] ERDÖS, RADO 1960).

• Let a be a vertex; given a color u, we have at most $f(k-1)$ vertices x such that the edge $\{a, x\}$ has color u. Indeed to avoid monochromatic triangles, every edge linking such two vertices x has color different from u. So that we dispose of $k-1$ colors; hence the upper bound $f(k-1)$. Now for all k colors, the whole number of vertices different from a is at most $k.f(k-1)$. •

2 - By iteration, get $f(k) \leq (k!)(1/0! + 1/1! + ... + 1/k!) < (k!).e$ (classical number e).

3 - Since $f(3) = 16$ we get for $f(4)$ the upper bound $4.16 + 1 = 65$. However [220] SANCHEZ-FLORES 1995 proved that $f(4) \leq 63$ (equivalently $(3)_4^2 \leq 64$). So that $f(5) \leq 5.f(4) + 1 \leq 316$; finally we find $(3,3,3,3,3)_5^2 \leq 317$.

3.8.2 Sperner's lemma

Let E be a set with finite cardinality $2h$ or $2h+1$. Then every set of subsets of E which are mutually incomparable with respect to inclusion has cardinality at most $(2h)!/(h!)2$ (even case) or $(2h+1)!/h!(h+1)!$ (odd case). In other words, the largest possible cardinality is obtained by taking the set of all h-element subsets ([235] SPERNER 1928).

Beginning with a set A of subsets of E, none of which is included in another, by replacing if necessary each set by its complement, we can always assume that the smallest cardinality of the elements of A is $p \leq h$. We shall prove that if $p < h$ (even case) or $p \leq h$ (odd case), then we can injectively substitute for every subset A of minimum cardinality p a superset B of A of cardinality $p+1$. Indeed, if this is possible, then the B will be distinct and of the same cardinal $p+1$, hence will be mutually incomparable with respect to inclusion. Moreover, no B can be included in an element of A of larger cardinality, since the p-element set A from which B was obtained would itself be included in that set.

Note that SPERNER's lemma follows from the linear independence lemma (see 3.4.2) in the case of E finite and $q = 1$. For further developments, see [205] POUZET, ROSENBERG 1982.

3.8.3 Schur numbers

1 - Given a partition of the strictly positive integers into a finite number of classes called columns, show that at least one of the columns contains three distinct integers a, b, c with $c = a + b$ ([224] SCHUR 1916).

Hint. To each column U associate the class of pairs of integers x, y such that the absolute value $|x - y|$ belongs to U; then apply RAMSEY's theorem.

2 - A set U of integers is said to be additively free iff the sum of any two integers belonging to U does not belong to U. Given an integer k, show that there exists an integer $k^+ > k$ such that, for each partition of the integers $1, 2, ..., k^+$

into k classes called columns, there is at least one non-additively free column. The smallest k^+ will be denoted by $s(k)$ and called the shur number relative to k. Show that $s(1) = 2, s(2) = 5, s(3) = 14$ (start with the column 5,6,7,8,9).

3 - In 1961, [5] BAUMERT, GOLOMB 1965 established that $s(4) = 45$. Here is the example he gave of a partition of the first 44 positive integers into four additively free columns:

1	2	4	9
3	7	6	10
5	8	13	11
15	18	20	12
17	21	22	14
19	24	23	16
26	27	25	29
28	33	30	31
40	37	32	34
42	38	39	35
44	43	41	36

Show that the Ramsey number $(3)^2_k \geq s(k)+1$: associate to each column U the set of pairs of integers from 1 to $s(k)$ for which the absolute value of the difference belongs to U ; thus $(3)^2_4 \geq 46$.

4 - Show that $s(k+1) \geq 3.s(k)-1$. Begin with the partition of the integers 1 to $p = s(k)-1$ into k columns. Add a $(k+1)^{st}$ column of the integers $p+1$ to $2p+1$. Then complete each column formed of integers u by the integers $2p+1+u$. Hence $s(5) \geq 134$ and $(3)^2_5 \geq 135$. [89] FREDRICKSON 1979 obtained $s(5) \geq 158$, thus $(3)^2_5 \geq 159$.

In general $s(k) \geq (3^k+1)/2$ and even $\geq (3^{k-4}.89/2) + 1/2$ for $k \geq 4$. This inequality is improved by [1] ABBOTT, HANSON 1972 who obtain, if one rectifies their numerical error: $s(k) \geq 89^{(k-7)/4}.1201 + 1$ for $k \geq 4$.

3.8.4 Polychromatic Ramsey numbers

1 - Let us start with a set E of 5 elements whose pairs are partitioned into three colors. From any element a in E there emanate at least two edges having a same color: consequently there exists at least one unicolor or bicolor triangle, that we call a $\leq$ **bicolor triangle**.

On another side starting with only 4 elements, which represent a quadrangle or a tetrahedron, we can put a same color on opposite edges, so that each triangular face wears three colors. In this sense 5 is the **polychromatic Ramsey number** from which, with 3 colors we are insured to get a $\leq$ bicolor triangle.

To our knowledge polychromatism appeared, for instance in a (unpublished) work of GALVIN, and in [41] DEVLIN 1979 (mimeog.), concerning Ramsey theory extended to embeddability: see below 5.12.2.

A more powerful lemma is the following (POUZET in 1999). **Consider a set E of cardinality 5 whose pairs are partitioned into three colors red, black, green. Then either there exists in E a quadrangle with possible**

colors black and green. Or there exists a triangle with possible colors red, black; or finally a triangle with possible colors red, green.

• Either all edges are black or green, and we are in the first case. Or there exists only one red edge: by deleting one of its vertices we find again a black and green quadrangle. Or finally there exist exactly two red, disjoint edges (in any other case we get a red and black triangle or a red and green triangle). Then to avoid our conclusion, we necessarily have a quadrangle a, b, c, d where ab, cd are red, bc, da are black and diagonals ac, bd are green. Now consider the fifth vertex u from which necessarily emanate ua, ub one of which is black and the other is green. Idem with uc, ud. Then we see that all possible colorations give at least a unicolor triangle, black or green: contradiction. •

2 - Let us start with a set E of 9 elements whose pairs are partitioned into three colors. Then let us group together two among these three colors: Since $(3,4)^2 = 9$, we are insured to get in E either a unicolor triangle or a $\leq$ bicolor quadrangle.

The polychromatic Ramsey number, starting with three colors, to get either a unicolor triangle or a $\leq$ bicolor quadrangle is either 8 or 9. It cannot be 7. Indeed starting with the set of integers 0 to 6, take color (1) for the edge (x, y) iff $|x - y| = 1$ or 6, take color (2) iff $|x - y| = 2$ or 5; finally take color (3) in the other case where the previous value is 3 or 4. Then we see that each triangle has at least two edges with different colors, and each quadrangle wears at least three colors.

We can prove that **the considered Ramsey number is** 8.

• Start with a set E of cardinality 8, whose pairs are partitioned into three colors red, black, green. Firstly suppose the existence of a vertex u from which emanate at least 4 red edges: then either we get a red triangle or a $\leq$ bicolor quadrangle with colors black and green.

Consequently we can suppose that from any vertex u there emanate at most three red edges. So that from u there also emanate at least two black edges for instance. Let us call a, b, c the extremities of the three red edges and d, e the extremities of two black edges emanating from u. We necessarily have on a, b, c a bicolor triangle black and green. On another side the edge (d, e) is either red or green. Let us use the lemma in alinea (1). Since we want to avoid any $\leq$ bicolor quadrangle in $\{a, b, c, d, e\}$, specially a black and green quadrangle, then we necessarily have either a red-black triangle (yet this is forbidden since by adding u we would get a $\leq$ bicolor red-black quadrangle); or necessarily we have a red-green triangle. Then we examine all possible cases and see that each case gives either a unicolor triangle or a $\leq$ bicolor quadrangle. •

3 - Consider **the polychromatic Ramsey number such that, partitioning all pairs into three colors, then there exists at least a $\leq$ bicolor quadrangle**. Firstly this number is at most 14 (POUZET, THOMASSÉ in 1999).

• Start from a set E of cardinality 14, whose pairs are partitioned into three colors red, black, green. From each vertex u there emanate at least 5 edges, say with color red, going from u to a, b, c, d, e. Then by the lemma in alinea (1), either we have a black and green quadrangle. Or a red and black triangle, which, by addition of u, gives a $\leq$ bicolor quadrangle. Or finally a red and green triangle which leads to the same conclusion. •

On another side, the following good coloration (obtained by the same authors) proves that the considered Ramsey number is > 9. Take 9 elements, each defined by its two coordinates (0,0), (0,1), (0,2), (1,0),..., (2,2). Let us give color (0) to the "small" triangle (0,0), (0,1), (0,2). Idem color (1) to the small triangle whose first coordinate is 1, and color (2) when the first coordinate is 2. Let us call "big" triangles with color (0) the three following ones. The big triangle (0,0), (1,0), (2,0) with 0 for all second coordinates. Idem the big triangle whose second coordinates are 1; then 2. Let us give color (1) to the big triangle formed of (0,0), (1,1), (2,2). Also the big triangle formed of (0,1), (1,2), (2,0): at each step the second coordinate increases by 1. Finally the big triangle with color (2) is defined by increasing by 2 (modulo 3) the second coordinate. We see that the total number of edges already considered is 4 times $3 = 12$ for each color: so that all the 36 edges have been colored.

We leave it to the reader to check that each quadrangle wears all the three colors (0), (1) and (2). To facilitate the task, note that the 126 quadrangles are easily classified into three classes. 18 "small" quadrangles are obtained from one of the 3 small triangles by adding any of the 6 vertices in the two other small triangles. On another side we have 27 trapezoidal quadrangles, each of which is obtained by joining a side of a small triangle to a side of another small triangle (9 times 6 possibilities, divided by 2). Finally we have 81 "big" quadrangles, each obtained from a side in a small triangle joined to vertices in the two other ones.

So we know that our polychromatic number is at most 14 and at least 10.

3.8.5 Approximation of a binary bicolor Ramsey number by a binomial coefficient

See for instance [11] BERGE 1970 p.418. Start from the obvious value of the binary Ramsey number $(2,a)^2 = a$ for $a \geq 2$.

Firstly show that, given two integers $a, b \geq 2$ then $(a,b)^2 \leq (a-1,b)^2 + (a,b-1)^2$.

• Put $r = (a-1,b)^2$ and $s = (a,b-1)^2$. Suppose we work on a set of cardinality $r+s$. Starting from a given vertex u, either from u there emanate at least r edges with color (+). Then by hypothesis either we get a (+)-monochromatic $(a-1)$-element set which gives with u a (+)-monochromatic a-element set. Or we get a (-)-monochromatic b-element set. Or from u there emanate at most $r-1$ edges with color (+), and consequently at least s edges with color (-). Then by hypothesis either we get a (-)-monochromatic $(b-1)$-element set which gives with u a (-)-monochromatic b-element set; or we get a (+)-monochromatic a-element set. In both cases our inequality $(a,b)^2 \leq r+s$ is proved. •

Finally $(a,b)^2 \leq \frac{(a+b-2)!}{(a-1)!(b-1)!}$. Indeed by induction: $(a,b)^2 \leq \frac{(a+b-3)!}{(a-2)!(b-1)!} + \frac{(a+b-3)!}{(a-1)!(b-2)!} = \frac{(a+b-3)!}{(a-2)!(b-2)!}(\frac{1}{b-1} + \frac{1}{a-1}) = \frac{(a+b-2)!}{(a-1)!(b-1)!}$.

Some numerical examples. $(3,3)^2 = 6$ equals $\frac{24}{2.2} = 6$. The number $(3,4)^2 = 9$ is near $\frac{120}{2.6} = 10$. The number $(4,5)^2 = 25$ is "near" $\frac{5.6.7}{6} = 35$.

Chapter 4

Good and bad sequence, well partial ordering, dimension, extraction property, directed well partial ordering

4.1 Less than relation, embedding between sequences

4.1.1 Generated initial interval

Given a poset A and a subset D of its base, the initial interval generated by $D \pmod A$ is the set of elements having an upper bound in D. An initial interval of A is said to be **finitely generated** iff it is generated by a finite subset of the base. We denote by $\mathcal{F}(A)$ the partial ordering (under inclusion) of finitely generated initial intervals of A.

If A is a well-founded poset, then so is $\mathcal{F}(A)$ (uses axiom of dependent choice; see [14] BIRKHOFF 1948).

• To each finitely generated initial interval X associate the set of its maximal elements $(\operatorname{mod} A)$, which is finite and non-empty (unless X is itself empty), and which is a free set, i.e. a set of mutually incomparable elements. Let us say that a set Y is **smaller than** Y' iff for each element $y \in Y$ there exists at least one upper bound of y in $Y'(\operatorname{mod} A)$. Then it suffices to prove that there does not exist any strictly decreasing ω-sequence of finite free sets, with respect to the preceding comparison (dependent choice). Given such a sequence Y_i(i integer), either each element of Y_0 belongs to finitely many sets Y_i: then remove one of these elements and keep only the Y_i to which it belongs. By iterating this procedure we eliminate all such elements from Y_0 but not all elements of Y_0. Finally we obtain a new, non-empty Y_0 which is disjoint from all other sets Y_i. Iterating this, we obtain a strictly decreasing ω-sequence of finite free, mutually disjoint sets.

Let Z_i be these disjoint sets. There exists at least one element z_0 in Z_0 with smaller elements in all other sets Z_i. Then at least one $z_1 < z_0$ in Z_1 with smaller elements in all sets $Z_i (i > 1)$; and so on. Thus, using dependent choice, we have a strictly decreasing ω-sequence of elements $(\bmod A)$: contradiction. •

4.1.2 Embedding between sequences

Given a sequence u in the poset A, the **initial interval generated by** u is that generated by the set of values of u: see previous subsection.

Let A be a poset and u, v be two sequences with values in A, both with the same length α. The sequence v is said to be **less than** $u(\bmod A)$ or **smaller than** u, or u is said to be **greater than** $v(\bmod A)$, iff $v_i \leq u_i(\bmod A)$ for each index $i < \alpha$. This comparison is reflexive, antisymmetric, transitive, thus defines a partial ordering on every set of sequences in A of a given length.

Given a poset A and two sequences u, v in A, we say that v is **embeddable in** u or that u admits an **embedding of** v $(\bmod A)$, iff there exists an extracted sequence of u (see 1.2.2) which is greater than v $(\bmod A)$. Consequently (Length of v) $\leq$ (Length of u).

Embeddability is reflexive, transitive but not antisymmetric, since the notion of extracted sequence is not itself antisymmetric: it defines a quasi-ordering on every set of sequences in A. Two sequences which are each embeddable in the other, are not necessarily extracted one from the other: in the total ordering of the positive integers, take the increasing sequences of even integers, respectively odd integers.

If a sequence v is embeddable in u, then the initial interval generated by v is included in the initial interval generated by u. Note also that, modulo the axiom of choice, starting with an initial interval A of a partial ordering, it suffices to well-order the base $|A|$ in order to obtain a sequence which generates A.

4.1.3 Partial ordering of words

Starting with a poset A, take the set of words, or finite sequences of elements of the base. Embeddability between words is antisymmetric and so defines a **partial ordering of words**.

If A is a well-founded poset, then the partial ordering of words in A is well-founded.

• Starting with a non-empty set of words in A, obtain a minimal word in this set. Firstly take the subset of words of minimum length n, and among such, those words beginning with a term u_0 of minimal value. Then among the latter, those words beginning with u_0, u_1 where u_1 has minimal value; and so on. •

4.1.4 Initial interval of a sequence

Given a sequence u with length α, an **initial interval** of u is any restriction of u to an ordinal domain smaller than or equal to α. The initial interval is called **strict** or **proper** iff its length is strictly less than α.

Let A be a poset, u, v be two ordinal-indexed sequences with values in A. If the length of u is a limit ordinal and if each proper initial interval of u is embeddable in $v(\operatorname{mod} A)$, then u is embeddable in v.

• Let γ be the length of u. Let $h(0)$ be the least ordinal such that $v_{h(0)} \geq u_0(\operatorname{mod} A)$. By induction, given an ordinal $i < \gamma$, assume that the strictly increasing sequence of the $h(j)(j < i)$ is defined, and let $h(i)$ be the least ordinal strictly greater than the $h(j)$ and such that $v_{h(i)} \geq u_i(\operatorname{mod} A)$. Clearly this $h(i)$ is the smallest possible, and by hypothesis for each $i < \gamma$, our $h(i)$ is strictly less than the length of v: our proposition follows. •

4.2 Good, bad, minimal bad sequence

4.2.1 Good sequence, bad sequence

Let A be a poset. A sequence u in $|A|$ with length (or domain) α is called **good** $(\operatorname{mod} A)$ iff there exist at least two indices i, j with $i < j < \alpha$ and $u_i \leq u_j(\operatorname{mod} A)$. Otherwise a sequence is called **bad**, i.e. iff every term is $<$ or $|(\operatorname{mod} A)$ with the terms of strictly smaller index.

Every sequence extracted from a bad sequence is bad.

Given a bad ω-sequence u, for every u^* obtained from u by a permutation of the set of indices, there exists a bad ω-sequence extracted from u^*. We can even require that the bad extracted sequence begin with, for instance, the first term u_0 of u.

4.2.2 Minimal bad sequence

Let A be a poset. A bad ω-sequence u in $|A|$ is said to be **minimal bad** iff, for any ω-sequence v extracted from u, every bad sequence less than $v(\operatorname{mod} A)$ is equal to v.

(1) **Every ω-sequence extracted from a minimal bad sequence is minimal bad.**

In a minimal bad sequence, all the terms are mutually incomparable.

• Let u be a minimal bad sequence in the poset A, and let $p < q$ be two integers with $u_p > u_q$ $(\operatorname{mod} A)$. Define the extracted sequence v by $v_0 = u_0, \ldots, v_{p-1} = u_{p-1}, v_p = u_q < u_p$, and $v_{p+i} = u_{q+i}$ for every $i \geq 1$. Then v is bad since it is extracted from u. Moreover v is a sequence less than $u_0, \ldots, u_{p-1}, u_p, u_{q+1}, u_{q+2}, \ldots$ which is extracted from u, and as v is distinct from this extracted sequence; this contradicts the minimality of u. •

(2) Let A be a poset, u a minimal bad sequence in A, and let p be an integer, and $x < u_p(\operatorname{mod} A)$. **Then $x \leq$ all terms of $u(\operatorname{mod} A)$, except possibly**

finitely many.

• Assume on the contrary that there exists a strictly increasing sequence of integers $p < f(1) < f(2) < ... < f(i) < ..$ with $u_{f(i)} <$ or $|x \pmod A$. Then the sequence $x, u_{f(1)}, ..., u_{f(i)}, ..$ is a bad sequence less than $u_p, u_{f(1)}, ..., u_{f(i)}, ..$ which contradicts the minimality of u. •

Note that a sequence with incomparable terms which satisfies the condition of the preceding proposition is not necessarily minimal.

• Take A to be a well-founded poset formed from denumerably many minimal elements which we denote by the integers 0, 1, 2, .. and with 0' > 0, 1' > 0 and > 1, and for every integer i an element $i' > 0, 1, ..., i$ and finally make the i' mutually incomparable. Then the sequence 0', 1', 2', .. has incomparable terms and satisfies the condition of the preceding proposition, but the sequence 0, 1, 2, .. is a bad sequence which is less than the considered sequence. •

4.2.3 Permutation of a minimal bad sequence

Let A be a poset. If u is a minimal bad ω-sequence $(\mathrm{mod}\, A)$, then the image of u by an arbitrary permutation of the set of indices is minimal bad.

• Denote by u^* the image of u under a permutation of the indices. By the preceding statement, the terms of u, hence of u^*, are mutually incomparable, so that u^* is bad. Suppose that u^* is not minimal bad. Then there exists an ω-sequence v extracted from u^* and a smaller ω-sequence w, which is bad and distinct from v. We can assume that the first terms of v and w are distinct, so $w_0 < v_0 \pmod A$.

Retransform u^*, v and w by the inverse of our initial permutation. Then we have again u and we obtain the images of v and w. Let w' denote the latter image. By 4.2.1, there exists a bad ω-sequence extracted from w', which begins by w_0 (more exactly by the term of w' whose value is w_0). This contradicts the minimality of u (proof of HODGES using only ZF). •

4.2.4 Initial interval generated by a minimal bad sequence

(1) **Given a poset A and a minimal bad sequence u in A, the initial interval of A generated by u is a well-founded poset** (uses dependent choice; ZF suffices if A is countable, or obviously if A is well-founded).

• Assume on the contrary that there exists a term u_p of u with a strictly decreasing ω-sequence $u_p > a_0 > a_1 > ... > a_i > ..(\mathrm{mod}\, A)$ (dependent choice). Then $a_0 \leq$ all except finitely many of the terms of u (see 4.2.2 proposition (2)). Hence there exists an integer q such that $a_0 \leq u_{q+i}$ for every integer i. Finally we have $a_i < u_{q+i}$ for every integer i, thus contradicting the minimality of the sequence u.•

(2) **Let A be a poset, u an ω-sequence in A with mutually incomparable terms. Then u is minimal bad iff every bad ω-sequence embeddable in u is extracted from u.**

• If u is minimal bad, our conclusion follows immediately from the definition.

Conversely, suppose that u has incomparable terms, so is bad, and yet not minimal bad. Then there exists a strictly increasing ω-sequence of integers $f(0) < f(1) < ... < f(i) < ..$ with a bad sequence $x = (x_0, x_1, ..., x_i, ..)$ which is a distinct sequence less than $u_{f(0)}, u_{f(1)}, ..., u_{f(i)}, ..$. Thus there exists an integer p such that $x_p < u_{f(p)} (\text{mod}\, A)$.

Consider the sequence of the $x_{p+j} (j = 0, 1, 2, ..)$. This sequence is embeddable in u and moreover is bad since x is bad. By hypothesis, this sequence is extracted from u, hence there exists a term $u_q = x_p < u_{f(p)} (\text{mod}\, A)$, which contradicts the hypothesis that the terms of u are mutually incomparable. •

4.2.5 Counterexamples

(1) In the preceding subsection, the hypothesis of incomparability of the terms of u is necessary.

• Take A to be a well-founded poset with $0 < 0'$ and denumerably many integers 1, 2, 3, .. which are set to be mutually incomparable, and also incomparable with 0 and 0'. Then the sequence 0', 0, 1, 2, 3, .. is bad yet not minimal bad (since $0' > 0$), and every bad ω-sequence embeddable in it is extracted from it. •

(2) Moreover, a bad non-minimal bad ω-sequence u can satisfy the condition that every bad sequence less than u is equal to u.

• Take an element a set to be minimal, and the even integers set to be mutually incomparable, with $0 > a$, $2 > a$, .. . Take the odd integers 1, 3, .. all to be minimal and incomparable with the preceding elements. Then the sequence of all integers 0, 1, 2, .. is not minimal since a is less than every even integer. Yet every distinct sequence less than it must take the value a in the position of an even integer, and hence is good. •

Problem. Let A be a poset, u an ω-sequence in A with incomparable terms. For u to be minimal bad, does it suffice that every ω-sequence with incomparable terms which is embeddable in u be extracted from u.

4.2.6 Existence of a minimal bad sequence

Theorem. Let A be a well-founded poset. For every bad ω-sequence u in A, there exists a minimal bad sequence which is embeddable in u (uses dependent choice; ZF suffices if A is denumerable).

• Replace the first term u_0 of u by $v_0 \leq u_0$ with v_0 minimal among those elements $x \leq u_0$ for which there exists a bad ω-sequence which begins by x and is embeddable in u. Denote by v_0, w_1 the first two terms in such a bad ω-sequence embeddable in u. Replace w_1 by $v_1 \leq w_1$ with v_1 minimal among those $x \leq w_1$ for which there exists a bad ω-sequence which begins by v_0, x and is embeddable in u. Denote by v_0, v_1, w_2 the first three terms in such a bad ω-sequence embeddable in u. By iteration, we obtain the sequence $v = v_0, v_1, v_2, ..$ which is bad and embeddable in u (dependent choice). Moreover, for every integer i and every $x < v_i (\text{mod}\, A)$, no

ω-sequence beginning by $v_0, v_1, ..., v_{i-1}, x$ is simultaneously bad and embeddable in u.

To see that v is minimal, suppose on the contrary that there exists a strictly increasing ω-sequence of integers $f(0) < f(1) < ... < f(i) < ..$ and a bad ω-sequence $x = (x_0, x_1, ..., x_i, ..)$ which is distinct and less than $v_{f(0)}, v_{f(1)}, ..., v_{f(i)}, ...$ Then there exists an integer p with $x_p < v_{f(p)} \bmod A)$. Consider the sequence $v_0, v_1, ..., v_{f(p)-1}, x_p, x_{p+1}, ...$ This sequence is embeddable in v and so embeddable in u. By the preceding inequality, this sequence is good.

As the sequences x and v are bad, there exists an integer $j < f(p)$ and an integer $k \geq p$, hence $j < f(p) \leq f(k)$ with $v_j \leq x_k \leq v_{f(k)} (\bmod A)$, so that v is good: contradiction. •

4.2.7 Strongly minimal bad sequence

Given a poset A, a bad ω-sequence u in A is said to be **strongly minimal bad** iff, for each integer i and each element $x < u_i (\bmod A)$, every ω-sequence in A beginning with $u_0, u_1, ..., u_{i-1}, x$ is good.

Every strongly minimal bad sequence is minimal bad.

• Suppose that u is strongly minimal bad but not minimal bad. There exists a strictly increasing ω-sequence of integers $f(0) < f(1) < ... < f(i) < ..$ and a distinct bad sequence x less than $u_{f(0)}, u_{f(1)}, ..., u_{f(i)}, ...$ Hence there exists an integer p with $x_p < u_{f(p)} (\bmod A)$. The ω-sequence $u_0, u_1, ..., u_{f(p)-1}, x_p, x_{p+1}, ..$ is good, since u is strongly minimal. As the sequences u and x are bad, there exists an integer $j < f(p)$ and also $k \geq p$ with $j < f(p) \leq f(k)$ and $u_j \leq x_k \leq u_{f(k)} (\bmod A)$, hence u is good: contradiction. •

There exist minimal, non-strongly minimal sequences.

More precisely, there exists an ω-sequence extracted from a strongly minimal bad sequence, which is not strongly minimal; however by 4.2.2 proposition (1) it is minimal bad. In addition, there exists a strongly minimal bad sequence and a sequence obtained from it by permutation of the set of indices, which is not strongly minimal bad; however by 4.2.3 it is minimal bad.

• Take two minimal elements a, b and put all even integers 0, 2, 4, .. to be mutually incomparable and incomparable with b, but all $> a$. Put all odd integers 1, 3, 5, .. to be mutually incomparable, also incomparable with the even integers and with a, yet all $> b$. Then the sequence $b, 0, 2, 4, ..$ is strongly minimal; but the sequence 0, 2, 4, .. is not; indeed for each integer i the sequence $0, 2, 4, ..., 2i, a, 1, 3, 5, ..$ is bad. •

• Take a minimum element a, then b, c mutually incomparable and $> a$. Put the even integers 0, 2, 4, .. to be mutually incomparable and incomparable with c, yet all $> b$. Put the odd integers to be mutually incomparable, also incomparable with the even integers and with b, yet all $> c$. Then the sequence $c, 0, 2, 4, ..$ is strongly minimal bad; but some of its permutations are not: indeed every strongly minimal bad sequence begins with b or c. •

There exists a minimal bad sequence which is not extracted from any strongly minimal bad sequence.

• Let A be the well-founded poset constructed in the following manner. Begin with the sequence of integers i in increasing order. To each i associate an element $i' > i$, the i' being mutually incomparable, each i' incomparable with all integers $> i$. Finally add infinitely many minimal elements $a_0, a_1, ..$ which are incomparable with the i and i'. Let u be the sequence of the i'. This sequence is minimal bad, since for each extracted ω-sequence, every distinct smaller sequence is good.

Suppose that u is extracted from a strongly minimal bad ω-sequence v. Then v is obtained from u by inserting elements a_i. Let p be an integer such that the p-th term of v is a term of u, in other words an i' (i integer $\leq p$). Hence v begins by $0', 1', 2', ..., i'$ between which can be inserted elements a_j. It suffices to replace i' by i, which is incomparable with $0', 1', ..., (i-1)'$ and which is $< i'$, and to add after i an ω-sequence of elements a_j which are not already inserted, in order to obtain a bad sequence. This contradicts the strong minimality of v.•

4.2.8 Existence of a strongly minimal bad sequence

Theorem 4.2.6 **as stated, does not extend to strongly minimal bad sequences.**

• Take the well-founded poset of the second counterexample 4.2.5, with the minimal element a followed by the even integers 0, 2, .. together with the odd integers taken as minimal and incomparable with the even integers. Then the sequence of even integers 0, 2, .. is bad and admits only itself and good sequences as embedded sequences. It is not strongly minimal bad, since for each integer i, the sequence $0, 2, 4, ..., 2i, a, 1, 3, ..$ is bad. •

However, **given a well-founded poset A, if there exists a bad ω-sequence in A then there exists as well a strongly minimal bad sequence** (uses dependent choice; ZF suffices for A countable).

• Take an element u_0 minimal among the first terms of bad ω-sequences, then an element minimal among those x for which there exists a bad ω-sequence beginning by u_0, x etc. •

Finally, by the method of the first paragraph of the proof in 4.2.6, we see that: given a well-founded partial ordering A and a bad ω-sequence u in A, there exists a bad ω-sequence v embeddable in u, such that for each integer p and each $x < v_p \pmod A$, every ω-sequence beginning by $v_0, v_1, ..., v_{p-1}, x$ and which is embeddable in u is good.

We can call this sequence v **strongly minimal with respect to embeddability in the sequence** u (as in 4.2.6, we use dependent choice; ZF suffices if A countable).

4.3 Finitely free poset; well partial ordering (w.p.o.) and well quasi-ordering (w.q.o.)

4.3.1 Finitely free poset

This is a poset such that any free subset of its base, or any antichain, is finite.

A poset can be well-founded without being finitely free: take an infinite base and the partial ordering of identity; then all elements are mutually incomparable.

A poset can be finitely free without being well-founded: take the natural ordering on the negative integers.

Every finite poset is finitely free.

Every restriction of a finitely free poset is finitely free.

(1) **Given a finitely free poset A with infinite base, there exists a totally ordered restriction of A which is equipotent with A** (uses axiom of choice; ZF suffices if A is countable or if Card A is a regular aleph).

• Partition the pairs of elements x, y of the base $|A|$ into two colors, according to whether x and y are comparable or incomparable (mod A). Assuming the axiom of choice, take an aleph equipotent with the base, and apply the DUSHNIK MILLER partition theorem 3.3.3. Then either there exists a denumerable subset with all its elements incomparable: contradiction. Or there exists a subset with same cardinality as the base, all of whose elements are comparable. •

(2) Let A be a well-founded poset and α its height. If for every ordinal $i < \alpha$, there are only finitely many elements of height i, then A is not necessarily finitely free.

• Take the chain of the integers, and for each integer i, add an i' set to be $> i$ and incomparable with integers $> i$, the i' mutually incomparable. •

4.3.2 Well partial ordering, well quasi-ordering

A poset which is well-founded and finitely free is called a **well partial ordering = w.p.o.**

For example, every finite poset is a well partial ordering.

Similarly for any well-ordering, or even a well-ordering where each element is replaced by a free finite set.

Every restriction of a well partial ordering is a well partial ordering.

A quasi-ordering (reflexive and transitive binary relation) is said to be a **well quasi-ordering = w.q.o.** iff the partial ordering of the equivalence classes is a well partial ordering (each class is formed by elements each of which is both greater and smaller than other elements).

(1) **A is a well partial ordering iff for any non-empty subset X of the base, the set of minimal elements (mod A/X) is finite and non-empty.**

• Let A be a well partial ordering and X a non-empty subset of the base. Since A is well-founded, there exist minimal elements (mod A/X): see 2.2.2. These elements are mutually incomparable (mod A), hence there are only finitely many such, since A is finitely free.

Conversely, if A is not a well partial ordering, then either there exists a subset X of the base without any minimal element, or there exists an infinite free subset X. In the latter case all the elements of X are minimal $(\text{mod}\, A/X)$. •

(2) **A necessary and sufficient condition for A to be a well partial ordering is that every ω-sequence in A be good** (the sufficiency uses dependent choice; ZF suffices if A is countable or has well-orderable base).

• Let A be a well partial ordering. If there exists a bad ω-sequence in A, then there is an extracted ω-sequence which is either strictly decreasing or with all its terms mutually incomparable: see RAMSEY's theorem; more precisely 3.1.2.

Conversely, if A is not a well partial ordering, then either A is not well-founded, so there exists a strictly decreasing ω-sequence (2.2.2, dependent choice). Or there exists an infinite free set, hence an ω-sequence of mutually incomparable elements: in both cases, a bad sequence. •

(3) **A necessary and sufficient condition for A to be a well partial ordering is that, for every ω-sequence in A with distinct values, there exists an extracted ω-sequence which is strictly increasing** $(\text{mod}\, A)$. Same conditions as for (2). The statement (3) results from (2) plus RAMSEY's theorem.

4.3.3 Characterization of a well partial ordering

A necessary and sufficient condition for A to be a well partial ordering is that, for each element u of the base $|A|$, the restriction of A to elements $<$ or $|u$ is a well partial ordering.

• If A is a well partiaml ordering, then every restriction of A is a well partial ordering; hence the necessity of our condition.

Conversely, assume that the condition holds. Let X be a non-empty free subset of the base. Take an element u of X. By hypothesis $X - \{u\}$ is finite, so X is finite: thus A is finitely free. Suppose now that X is an arbitrary non-empty subset of the base. Take an element u of X and let Y be the set of those elements not greater than or equal to $u(\text{mod}\, A)$. Then either Y is empty, so that u is minimal in X. Or Y is non-empty and so by hypothesis has a minimal element which is also minimal in X. Thus A is well-founded (proof of POIZAT using only ZF). •

4.4 Initial intervals of a well partial ordering; Higman: characterization of a w.p.o.; Rado's w.p.o.

4.4.1 Initial intervals of a well partial ordering

Theorem. For a partial ordering A to be a well partial ordering, it is necessary and sufficient that the partial ordering of inclusion of initial intervals of A be well-founded ([110] HIGMAN 1952; the necessity uses dependent choice; ZF suffices if A is countable).

• Let A be a well partial ordering, and suppose the statement of the proposition is false. By 2.2.2 (dependent choice), there exists an ω-sequence of initial intervals

A_i(i integer) which is strictly decreasing with respect to inclusion. For each i, choose an element u_i in $A_i - A_{i+1}$. Each u_i is $<$ or $|$(mod A) to the u with indices $< i$. Hence our ω-sequence is bad, contradicting 4.3.2 proposition (2).

Conversely, suppose that A is not a well partial ordering. Then either A is not well-founded or A is not finitely free. In the first case, there exists a subset D of the base $|A|$ such that A/D has no minimal element. To each element $x \in D$, associate the initial interval D_x of elements $\leq x(\text{mod}\, A)$. The set of the D_x has no element minimal with respect to inclusion; so the partial ordering of the initial intervals is not well-founded.

If A is not finitely free, then there exists an infinite set H of incomparable elements (mod A). By the definition of finiteness (1.1.1), there exists a set of subsets X of H, no one of which is minimal with respect to inclusion. Complete each X in the initial interval X^+ generated by X. Then $X^+ \cap H = X$ for every X, hence $X \neq Y$ implies $X^+ \neq Y^+$, and even $X \subset Y$ implies $X^+ \subset Y^+$ for every X, Y. The X^+ form a set of initial intervals of A, no one of which is minimal with respect to inclusion. Hence the partial ordering of the initial intervals is not well-founded (proof of sufficiency due to POUZET using only ZF). •

4.4.2 Rado's well partial ordering

If A is a well partial ordering, then the partial ordering of inclusion for the initial intervals of A is not necessarily a well partial ordering. In fact it is not necessarily finitely free.

The following example is called **Rado's well partial ordering** ([211] RADO 1954). It will be used in chapter 7 for the theory of better partial orderings. Consider the couples of natural numbers x, y and set $(x, y) \leq (x', y')$ iff $x = x'$ and $y \leq y'$ in the usual ordering, or $x' \geq x + y$ and y' arbitrary. The comparison relation thus defined is a well partial ordering. Yet there are infinitely many initial intervals which are mutually incomparable with respect to inclusion.

• The reader can verify reflexivity, antisymmetry and transitivity. Given a couple (x, y), there are only finitely many couples less than it, since either their first term is x and their second term is $\leq y$, or the sum of their terms is $\leq x$. Thus the partial ordering is well-founded.

Given (x, y), any couple which is incomparable with it, has first term strictly less than $x + y$. Hence there exist finitely many possible first terms. Moreover, two incomparable couples necessarily have distinct first terms. Hence the partial ordering is finitely free, and thus a well partial ordering.

Consider now, for each integer i, the initial interval A_i formed of the couples (i, y) with y an arbitrary integer, and the couples (x, y) with $x + y \leq i$: for an arbitrary integer $j \neq i$, the initial intervals A_i and A_j are incomparable with respect to inclusion. •

A second example of a well partial ordering with the same incomparable initial intervals: take again the couples of integers with either $x = x'$ and $y \leq y'$, or $x' > \text{Max}(x, y)$ and y' arbitrary.

A third example: take the couples of integers with either $x = x'$ and $y \leq y'$, or $x' > \mathrm{Max}(x, y)$ and $y \leq y'$.

Some of these posets will be reexamined and classified as α-better partial orderings: see 7.6.2 below.

4.4.3 Initial interval under a minimal bad ω-sequence

Let A be a poset and u a minimal bad ω-sequence in A. Then the initial interval of those elements x of $|A|$ such that there exists an integer i with $x < u_i(\mathrm{mod}\, A)$ is a well partial ordering (uses dependent choice; ZF suffices if A is countable). Compare with 4.2.4 proposition (1).

• We prove first of all that, given a term u_p(p integer) of the minimal bad sequence, the initial interval of those elements $< u_p \bmod A)$ is a well partial ordering. By 4.2.4 this interval is well-founded (dependent choice). Suppose that there exist infinitely many mutually incomparable elements, hence an ω-sequence of such elements $a_0, a_1, ..., a_j, ..$ all $< u_p$. Then each $a_j < u_p$, so each $a_j \leq$ all except a finite number of terms of u (see 4.2.2). Thus the ω-sequence of the a_j is embeddable in u, and hence by 4.2.4 proposition (2), it is extracted from u, so that there exist terms $u_q (q > p)$ which are $\leq u_p$. Yet all the terms of u are mutually incomparable by 4.2.2: contradiction.

Suppose now that the proposition is false. There exists a bad ω-sequence v such that for each integer i there exists a j with $v_i < u_j(\mathrm{mod}\, A)$: see 4.3.2 proposition (2) using dependent choice. At most a finite number of terms of v are $< u_0$, or $< u_1$, etc. Hence there exists an ω-sequence w extracted from v which is either strictly less than $u(\mathrm{mod}\, A)$ or strictly less than a sequence extracted from u. This contradicts the minimality of u.•

4.5 Well partial ordering: extraction theorem, words (Higman), augmentation

4.5.1 The extraction theorem for well partial orderings

Theorem. Let A be an infinite well partial ordering and ω_α a given aleph. Let u be a sequence in A with distinct values and length ω_α. Then there exists a sequence extracted from u which is strictly increasing (mod A) and has same length as u (uses axiom of choice if ω_α is singular; ZF suffices for ω_α regular).

• Partition the pairs of indices $i, j (i < j < \omega_\alpha)$ into two colors: the color (+) if $u_i < u_j(\mathrm{mod}\, A)$, and (-) if $u_i >$ or $| u_j(\mathrm{mod}\, A)$. Using the partition theorem 3.3.3, either there exists a set of indices equipotent with ω_α, all of whose pairs have color (+), and hence an ω_α-sequence extracted from u and strictly increasing. Or there exists a denumerable set of indices all of whose pairs have color (-), and hence a bad ω-sequence (mod A), thus contradicting 4.3.2 proposition (2). •

Conversely by 4.3.2 proposition (3), only well partial orderings satisfy our conclusion. However, the ordinal product $\omega_1.(\omega^-)$, and the partial ordering with denumerably many mutually comparable components, each isomorphic with ω_1, satisfy our conclusion, when restricted to sequences u of length $\omega_\alpha (\alpha \neq 0)$ (communicated by POUZET).

4.5.2 Well partial ordering of words

Theorem. If A is a well partial ordering, then the partial ordering of embeddability for words in A is a well partial ordering (uses dependent choice; ZF suffices if A is countable); [110] HIGMAN 1952.

• First of all, the partial ordering of words in A is well-founded by 4.1.3.

By 4.3.2 proposition (2) (dependent choice), it suffices to show that there does not exist any bad ω-sequence of words with respect to embeddability (defined in 4.1.2). If such a bad sequence exists, then there exists also a strongly minimal bad sequence U (see 4.2.8).

We shall show that this is impossible. Denote by $U_i (i$ integer) the ith term of U. By hypothesis, U_0 is such that no bad ω-sequence of words begins by a shorter word, i.e. a word obtained from U_0 by removing at least one element. No bad sequence beginning by U_0 continues with a word shorter than U_1, etc. Note that no U_i is empty, since the $U_j (j > i)$ are $<$ or $|U_i$ with respect to embeddability: **the first term** u_i of U_i **always exists**. The sequence of the u_i takes its values in the well partial ordering A and so (by 4.3.2 proposition (3)) has an extracted ω-sequence which is either constant or strictly increasing.

Denote by $h(i) (i$ integer) the indices corresponding to this constant or increasing sequence of first terms, and let $V_{h(i)}$ be the word $U_{h(i)}$ with its first term $u_{h(i)}$ removed.

Consider the sequence: $U_0, U_1, ..., U_{h(0)-1}, V_{h(0)}, V_{h(1)}, ...$ This sequence begins as U, but at the position $h(0)$, the word $U_{h(0)}$ is replaced by the shorter word $V_{h(0)}$. So by hypothesis, this latter sequence is good. Now, two words in position $< h(0)$ cannot admit an embedding of one in the other. Similarly for two words in position $\geq h(0)$, which are of the form $V_{h(i)}$ and $V_{h(j)}$ with $i < j$: given the fact that $u_{h(i)} \leq u_{h(j)} (\text{mod}\, A)$, such an embedding would yield $U_{h(i)} \leq U_{h(j)}$. Finally, there remains the possibility that a $U_i (i < h(0))$ is embeddable in a $V_{h(k)} (k$ integer) and hence embeddable in $U_{h(k)}$. However this contradicts the assumption that U is bad. •

The theorem is not extendible to ω-sequences. Indeed start from RADO's well partial ordering (see 4.4.2). Given two different integers p, q with for instance $q > p$, note that the two ω-sequences $(p, 0), (p, 1), ..., (p, i), ..(i$ integer), and $(q, 0), (q, 1), ..., (q, i), ..$ are incomparable with respect to embeddability.

4.5.3 Augmentation of a well partial ordering

(1) **Every partially ordered augmentation of a well partial ordering is a well partial ordering.**

• Let A be a well partial ordering, and B a partially ordered augmentation of A. If X is a free set $(\bmod B)$, i.e. a set whose elements are incomparable $(\bmod B)$, then X is free $(\bmod A)$ and hence finite; thus B is finitely free.

Now let X be a non-empty subset of the common base of A and B. Let Y be the set of those elements of X which are minimal $(\bmod A)$. Since these elements are incomparable $(\bmod A)$, there are only finitely many. Hence there exists in Y an element y which is minimal $(\bmod B)$. We shall show that y is minimal $(\bmod B)$ for the set X, which then implies that B is well-founded. If there exists an element x of X such that $x < y(\bmod B)$, then the non-empty set of elements of X which are $\leq x(\bmod A)$ has a minimal element z; so that $z \leq x < y(\bmod B)$ and z is an element of Y. But this contradicts the minimality of $y(\bmod B)$ in Y (proof communicated by POIZAT in 1976 and using only ZF). •

(2) **A necessary and sufficient condition for A to be a well partial ordering is that every totally ordered augmentation of A be a well-ordering** (sufficiency uses the reinforcement axiom plus dependent choice; ZF suffices if A is countable).

• If A is a well ordering, then by the previous proposition every totally ordered augmentation of A is a well-ordering.

Conversely, suppose that A is not a well partial ordering. Then either A is not well-founded, so that there exists a strictly decreasing ω-sequence $(\bmod A)$: see 2.2.2 proposition (1) (dependent choice). Or A has an infinite free subset, hence a denumerable free subset. In the latter case, take a total ordering ω^- (converse of ω) on this free subset. By the reinforcement axiom 2.9.3, there exists a chain which extends ω^- and is a reinforcement of A. •

(3) **If every totally ordered augmentation of A is a well-ordering, and if there exists such a total ordering, then A is a well partial ordering** (proof using ZF alone, POUZET 1979, published in ToR-86 p.105).

• Let C be a well-ordering augmentation of A. Then A is well-founded, for otherwise there would exist a subset D of the base with A/D having no minimal element, so C/D without any minimal element.

Suppose now that there exists an infinite free $(\bmod A)$ subset H. Let I denote the initial interval of those elements x for which there exists an element $y > x(\bmod A)$ with $y \in H$. Similarly denote by F the final interval of those x for which there exists an element $y < x$ with $y \in H$. Finally let L be the set of those elements which are incomparable $(\bmod A)$ with all elements of H. The four sets H, I, F, L are disjoint and form a partition of the base. Then the chain $C/I + C/L + (C/H)^- + C/F$ is an augmentation of A. However, as H is infinite, the well-ordering C/H is isomorphic with an ordinal $\geq \omega$, and hence its converse $(C/H)^-$ is not a well-ordering. •

(4) **Given a well partial ordering A, there exists a well-ordered augmentation of A**: consequence of (1) above plus the augmentation axiom 2.9.3 (uses ultrafilter axiom).

The result can be obtain using only the axiom of choice for finite sets. Indeed, to each ordinal u strictly less than $\operatorname{Ht} A$, associate the finite set F_u of those elements with height u in A; then the finite set of chains based on F_u. Then the axiom of choice for finite sets associates to each u one of these chains: it suffices to take the sum along the u's.

Note that with the considered axiom, a partial ordering A is a well partial ordering iff:

(i) every chain which is an augmentation of A is a well-ordering, and

(ii) there exists a totally ordered augmentation of A (use (3) above: ZF suffices).

More precisely, the preceding proposition is equivalent to the following weakening of the axiom of choice for finite sets: "for every well-orderable set of finite mutually disjoint sets, there exists a choice set" (POUZET in 1979).

4.6 Well-ordered restrictions

4.6.1 Well-ordered restriction meeting every height

Let A be a well-founded poset having at most finitely many elements of each height. Then there exists a well-ordered restriction C of A which is isomorphic with the height $\mathrm{Ht}\, A$.

More precisely, the base $|C|$ contains for each ordinal $i < \mathrm{Ht}\, A$ a unique element of height i (mod A). Uses ultrafilter axiom; ZF suffices if A is countable.

For $\mathrm{Ht}\, A = \omega$, this is just a form of KÖNIG's lemma.

Compare the present statement with 4.3.1 proposition (1).

• For every finite set F of ordinals $< \mathrm{Ht}\, A$, there exists at least one function h which associates to each i of F an element hi of height i (mod A). The hi ($i \in F$) are mutually comparable, so form a finite totally ordered restriction of A: it suffices to start with the largest i in F and to recall that for each $j < i$ there exists an element $hj < hi$ (mod A) ; see 2.7.1.

Consider the ordinal α and the base $|A|$ as disjoint sets. To each F associate $F^+ = F$ plus those elements in $|A|$ whose height belongs to F. The F^+ thus form a directed system with respect to ordering by inclusion.

Given an F, to each h with domain F associate the birelation with base F^+ formed by the unary relation taking the value (+) on F and (-) on $F^+ - F$, together with the binary relation taking the value (+) for couples (i, hi) where $i \in F$. Associate to each F the non-empty set U_F of these birelations for all h with domain F. The U_F satisfy the hypotheses in the coherence lemma 2.4.1 (equivalent to the ultrafilter axiom). Hence there exists a birelation on the union $|A| \cup \alpha$, which defines an isomorphism from α onto a totally ordered restriction of the poset A. •

4.6.2 Well-ordered restriction of maximum length

Let A be a well-founded poset. Then either there exist infinitely many elements, all of different heights (mod A) which are mutually incomparable, or there exists a well-ordered restriction of A with the same height (uses ultrafilter axiom; [199] POUZET 1979).

• Suppose the first conclusion does not hold, and argue by induction on the height of A. Assume first that $\mathrm{Ht}\,A$ is decomposable, hence of the form $\beta + \gamma$ with β non-zero and $\gamma < \beta + \gamma$. Let C denote the restriction of A to elements whose height $(\mathrm{mod}\,A)$ is greater than or equal to b. For each x in $|C|$ we have $\mathrm{Ht}\,x(\mathrm{mod}\,A) = \beta + \mathrm{Ht}\,x(\mathrm{mod}\,C)$ (proof by induction). Hence $\mathrm{Ht}\,C = \gamma$. By the induction hypothesis, there exists a well-ordered restriction C^* of C with height γ. Let u be the minimum of C^*. The restriction of A to elements $< u$ is a well-founded partial ordering B with height $\geq \beta$ (see 2.7.1). By the induction hypothesis, there exists a well-ordered restriction B^* of B with height β. Then $B^* + C^*$ is a restriction of A with height $\beta + \gamma$.

Assume now that $\mathrm{Ht}\,A = \alpha$, an indecomposable ordinal; and suppose that the first conclusion does not hold. By the previous subsection (ultrafilter axiom), it suffices to prove that there exists a restriction of A with the same height α, which is finitely free and hence a well partial ordering. Let k be the cofinality of α; decompose α into the sum $\Sigma\alpha_i(i < k)$ with the α_i increasing, all strictly less than α, and $\mathrm{Sup}\,\alpha_i = \alpha$ (see 2.7.4).

For each $i < k$, consider the restriction A_i of A to elements whose height is both $\geq \Sigma\alpha_j(j < i)$ and $< \Sigma\alpha_j(j \leq i)$. This A_i has height α_i. By the induction hypothesis, and since there do not exist infinitely many mutually incomparable $(\mathrm{mod}\,A_i)$ elements with different heights, there exists a well-ordered restriction C_i of A_i with the same height α_i.

Let D denote the restriction of A to the union of the bases $|C_i|$. Firstly $\mathrm{Ht}\,D$ is strictly greater than each α_i hence $\geq \alpha = \mathrm{Sup}\,\alpha_i$; so that $\mathrm{Ht}\,D = \alpha$. Secondly, there do not exist infinitely many mutually incomparable elements in D, since they would all have different heights $(\mathrm{mod}\,A)$, which contradicts our hypothesis. Hence D is finitely free, so that we can apply 4.6.1. •

As opposed to what happens in 4.6.1, **we cannot require that the well-ordering having the same height as A, have one and only one element of each height** $(\mathrm{mod}\,A)$. Consider the following counterexample due to POUZET.

• Take the points (x, y) of integer coordinates with $y \leq x$. Let $(x, y) < (x', y')$ iff either $x = x'$ and $y < y'$, or $x < x'$ and $y' \geq y + 2$. Then the height of each point (x, y) is y, hence $\mathrm{Ht}\,A = \omega$.

However, a totally ordered restriction of A isomorphic with ω can have at most a finite sequence of points with heights 0, 1, 2, .. . For example the points $(u, 0), (u, 1), .., (u, u) (u$ integer). After which one must pass to a point with ordinate $\geq u + 2$, hence with height $\geq u + 2$. •

4.6.3 Well-ordered restriction equipotent with the base

Let A be a poset whose base has cardinality an infinite regular aleph ω_α. Suppose that, for each element u of the base $|A|$, the set of elements $< u$ or $|u$ has cardinality $< \omega_\alpha$. Then there exists a totally ordered restriction of A which is isomorphic with the ordinal ω_α.

• Well-order the base $|A|$ in order-type ω_α and let C denote the chain thus obtained. Partition the pairs of elements x, y of the base into two colors. If

$x < y \pmod C$, we say that the pair is (+) iff $x < y \pmod A$ and (-) iff $x > y$ or $x|y \pmod A$. For a given element a in the base, there are $(< \omega_\alpha)$ many elements x for which the pair $\{a, x\}$ has color (-): so that either $x < a (\operatorname{mod} C)$ or $x < a$ or $|a (\operatorname{mod} A)$. Now apply 3.3.2 (partition lemma): there exists a subset equipotent to the base, in which all pairs have color (+). This yields a restriction isomorphic with the ordinal ω_α. •

The proposition is false for a singular aleph.

• For each integer i, take a set E_i with cardinality ω_i, the E_i being mutually disjoint. Define a partial ordering on the union, which has cardinality ω_ω, by taking each E_i as a free set, each element of E_i being less than each element of E_j for any two integers $j > i$. Then every totally ordered restriction has order-type at most ω. •

4.7 Ideal and finitely free poset; Bonnet's countable set of ideals

4.7.1 Well partial ordering and ideals

Every well partial ordering is a finite union of ideals (uses dependent choice; ZF suffices for a countable ordering).

• Let A be a well partial ordering; recall that the partial ordering of initial intervals of A is well-founded (HIGMAN's theorem in 4.4.1, dependent choice). Suppose that A is not a finite union of ideals. Among the initial intervals of A which are not finite unions of ideals, there exists a minimal such initial interval M, with respect to inclusion. As M is not an ideal, it is non-empty. By the preceding subsection, there exist two initial intervals U, V of M, which are distinct from M with $U \cup V = M$. By minimality of M, the intervals U, V are each a finite union of ideals; so M as well: contradiction. •

4.7.2 Characterization of a finitely free poset

(1) **For a poset to be finitely free, it is necessary and sufficient that every initial interval be a finite union of ideals** ([17] BONNET 1975; the necessity uses axiom of choice).

• Suppose that there exist infinitely many elements u which are mutually incomparable. Associate to each u the ideal of those elements $\leq u$: the union of these ideals is an initial interval which is not decomposable into a finite union of ideals.

Conversely, let A be a finitely free poset, and I an initial interval of A. By the corollary of 2.7.2 (axiom of choice), take a well-founded, hence a well partial ordering J which is a cofinal restriction of I. By the preceding proposition, J is a finite union of ideals. Complete each ideal X in J into an ideal of I by adding those elements in $|I|$ which are bounded above by an element of $|X|$. Thus I is a finite union of ideals. •

(2) **Every finitely free poset is a finite union of ideals**.

(3) **Every infinite finitely free poset has an infinite ideal**. Immediate consequences of (1).

The converse of (3) is false: take a non-finitely free poset which has a maximum. For further informations, see [171] MILNER 1981.

4.7.3 Set of mutually incomparable ideals

Theorem. Let A be a finitely free poset. Then every set of mutually incomparable ideals of A (with respect to inclusion) is countable ([17] BONNET 1975, uses axiom of choice).

• Suppose the existence of an uncountable set of mutually incomparable ideals of A, hence a set of cardinality ω_1 of such ideals H_i(i countable ordinal). To each H_i we associate a partial ordering C_i which is a well-founded cofinal directed restriction of H_i: see 2.13.2 (axiom of choice). Since A is finitely free, so is each C_i, and hence each C_i is a well partial ordering.

Given two distinct countable ordinals i, j, by using the incomparability with respect to inclusion of the bases of H_i and H_j, define an element $u_{i,j}$ belonging to $|C_i|$ but not to $|H_j|$. Thus $u_{i,j} >$ or $|(\bmod A)$ with every element of H_j.

For each countable ordinal i, let A_i be the set of the $u_{i,j}$ for all countable values j. More precisely, let A_i be the corresponding restriction of A: we have $\operatorname{Card} A_i \leq \omega_1$. In general, for each set U of countable ordinals and each countable ordinal i, let $A_i(U)$ be the set of the $u_{i,j}$ for all $j \neq i$ and $j \in U$. More precisely $A_i(U)$ will be the corresponding restriction of A. Each A_i or each $A_i(U)$ is a well partial ordering, as it is a restriction of the well partial ordering C_i.

(1) Suppose first the existence of a set U of countable ordinals which is cofinal in ω_1 and such that for each $i \in U$, the restriction $A_i(U)$ is countable; and obtain contradiction as follows.

Let U_0 designate this U and let $i(0)$ be its least element. Partition the elements $j \neq i(0)$ of U into classes, by putting j and j' in the same class iff $u_{i(0),j} = u_{i(0),j'}$. There are countably many such classes, since $A_{i(0)}(U)$ is countable. Since the cardinality of U is ω_1, at least one of these classes, say U_1, has cardinality ω_1. Let $k_0 = u_{i(0),j}$ for the j in U_1, and note that $k_0 >$ or $|(\bmod A)$ with all elements of every H_j for $j \in U_1$.

Iterate this by letting $i(1)$ be the least ordinal index belonging to U_1. Partition the elements j of $U_1 (j \neq i(1))$ into classes defined by the equality $u_{i(1),j} = u_{i(1),j'}$. There are countably many such classes, since $A_{i(1)}(U_1)$ is included in $A_{i(1)}(U)$ and the latter is countable. At least one of these classes, say U_2, has cardinality ω_1. Temporarily let $k_1 = u_{i(1),j}$ for the j in U_2. By the preceding, $k_0 >$ or $|k_1 \ (\bmod A)$.

We require that k_0 and k_1 be incomparable $(\bmod A)$. For this, note that $h = u_{i(1),i(0)}$ is an element of $A_{i(1)}$, and $h >$ or $|$ $(\bmod A)$ with all elements of $H_{i(0)}$, hence $h >$ or $|k_0 (\bmod A)$. Since $C_{i(1)}$ is directed, replace k_1 if necessary by a common upper bound of k_1 and h, belonging to $C_{i(1)}$. Note that the old value, hence also the new value of k_1 is $>$ or $|$ $(\bmod A)$ with all elements of every H_j for $j \in U_2$.

Iterating this yields an infinite sequence of elements $k_r (r$ integer) which are mutually incomparable $(\bmod A)$: this contradicts our hypothesis that A is finitely free.

(2) Now consider the case where, for each set U cofinal in ω_1, there exists an i in U with $A_i(U)$ of cardinality ω_1; and obtain a contradiction as follows. Starting with the set U_0 of all countable ordinals, take $i(0)$ such that $A_{i(0)}$ has cardinality ω_1. Since $A_{i(0)}$ is a well partial ordering, by 4.3.1 there exists a well-ordered restriction B_0 of $A_{i(0)}$ which is isomorphic with ω_1.

Let U_1 be the set of countable indices $j > i(0)$ such that $u_{i(0),j}$ belongs to the base $|B_0|$. This U_1 has cardinality ω_1, just as B_0. In the case now considered, we can iterate this as follows. Take an element $i(1)$ in U_1, such that $A_{i(1)}(U_1)$ has cardinality ω_1. Then take a well-ordered restriction B_1 of $A_{i(1)}(U_1)$, isomorphic with ω_1. Let U_2 be the set of countable indices $j > i(1)$, such that $u_{i(1),j}$ belongs to the base $|B_1|$, and so forth. We obtain, for each integer r, a set U_r, a countable ordinal $i(r)$ and a well-ordering B_r isomorphic with ω_1.

For each integer r, consider the set of those $u_{i(r),j}$ where j takes the values $i(r+1), i(r+2), ...$ This is a countable subset of the base $|B_r|$. Take the least upper bound v_r in the well-ordering B_r of this subset. Since $u_{i(r),j} >$ or $|(\bmod A)$ with each element of H_j, it follows that $v_r >$ or $|(\bmod A)$ with each element of $H_{i(r+1)}$ and of $H_{i(r+2)}$ and so forth. In particular, we have $v_r >$ or $|v_{r+1}, v_{r+2}, ..(\bmod A)$.

It remains to replace each v_r by an element w_r of $H_{i(r)}$ in such a manner that the w_r are mutually incomparable $(\bmod A)$. For this, let $w_0 = v_0$. Take an element x_0 in $H_{i(1)}$ which is $>$ or $|$ with all elements of $H_{i(0)}$, hence which is $>$ or $|w_0$. Replace v_1 by a common upper bound w_1 in $H_{i(1)}$ of v_1 and x_0. Then take in $H_{i(2)}$ an element, again called x_0, which is $>$ or $|$ with all elements of $H_{i(0)}$, and take an element x_1 in $H_{i(2)}$ which is $>$ or $|$ with all elements of $H_{i(1)}$, and replace v_2 by a common upper bound w_2 of v_2, x_0, x_1; and so forth. This contradicts our hypothesis that A is finitely free. •

A poset can be denumerable and finitely free, and can have continuum many mutually incomparable initial intervals.

• Take two chains, each isomorphic with the chain of rationals, every element of the first chain being incomparable with every element of the second. Then to each real number x, associate in the first chain the initial interval $I(x)$ which represents x as a cut. Similarly in the second chain, associate to x the initial interval which represents the opposite number $-x$ as a cut, say $J(-x)$. Finally in the given poset, the union $I(x) \cup J(-x)$ yields, when x runs through the reals, continuum many incomparable initial intervals (example due to [17] BONNET 1975). •

4.8 Direct product of posets

Given a set of posets $A_i (i \in I)$, the **direct product** of the A_i is defined to be the poset with base the cartesian product of the bases, hence with base the set of functions f with domain I taking a value $f(i) \in A_i$ for each $i \in I$. The comparison relation is defined by $f \leq g$ iff $f(i) \leq g(i)(\bmod A_i)$ for every $i \in I$.

This comparison relation is reflexive, transitive and antisymmetric.

In particular, for two posets A, B the direct product is denoted by $A \times B$. Up to isomorphism, the operation $\times$ is commutative and associative.

Another particular case: if all the posets A_i are identical to a fixed poset A, where i runs through an ordinal, then we find again the "less than" comparison relation between sequences in A with the same ordinal length: see 4.1.2.

4.8.1 Direct product of two well-founded posets

(1) **Let A, B be two well-founded posets; then the direct product $A \times B$ is well-founded.**

• Start with a non-empty subset D of the cartesian product of the two bases. Then take an element (x, y) of D, where $x \in |A|$ with minimal height (mod A), then $y \in |B|$ with minimal height (mod B): hence (x, y) is minimal in D. •

(2) **Let A, B be two well partial orderings; then the direct product $A \times B$ is a well partial ordering** (uses denumerable subset axiom: see 1.2.6; ZF suffices if A and B are countable).

• By the preceding (1), the direct product is well-founded. Suppose that there exist infinitely many incomparable elements, and hence an ω-sequence of incomparable elements. Then we have an extracted ω-sequence of elements (x, y) where the first terms x are increasing (mod A), and from this sequence, another extracted ω-sequence where the second terms y are increasing (mod B): against the incomparability. •

4.8.2 Natural sum and product for ordinals

From the unique decomposition of an ordinal into a sum of decreasing powers of ω, one defines the commutative operations of natural sum and natural product for ordinals: this goes back to [109] HESSENBERG 1906; see also [4] BACHMANN 1967 p. 107.

For the sum, we begin with $\alpha = \omega^{a(1)}.m_1 + ... + \omega^{a(h)}.m_h$, where the coefficients $m_1, ..., m_h$ are integers, and where $a(1) > ... > a(h)$ are ordinal exponents, and $\beta = \omega^{a(1)}.n_1 + ... + \omega^{a(h)}.n_h$. We can always assume that the ordinal exponents are the same for both decompositions, if necessary by inserting terms with coefficient zero. Then the **natural sum** $\alpha \oplus \beta$ is defined as:

$\omega^{a(1)}.(m_1 + n_1) + ... + \omega^{a(h)}.(m_h + n_h)$.

For the **natural product**, denoted $\otimes$, one first defines the product $\omega^\alpha \otimes \omega^\beta$ as being $\omega^{(\alpha \oplus \beta)}$. Then for two arbitrary ordinals, each written in the form of a sum of powers of ω, one multiplies them as with polynomials.

With respect to these notions, one defines the **positive and negative ordinals**, by taking for m positive or negative integers. Then one defines **rational ordinals** by taking the quotient field: see [231] SIKORSKI 1948, and again [4] BACHMAN 1967. More difficult and less "natural", one can define **transfinite reals**; see for instance [134] KLAUA 1959 ; also [135].

4.8.3 Height of a direct product of posets

(1) **Let A, B be two well-founded posets, x an element of the base $|A|$ and y an element of $|B|$. Then** $\text{Ht}(x, y)(\text{mod}\, A \times B) = \text{Ht}\, x(\text{mod}\, A) \oplus \text{Ht}\, y(\text{mod}\, B)$ **where $\oplus$ is the commutative, or natural product** defined in 4.8.2.

(2) **If A, B are non-empty well-founded posets, then**:
$\text{Sup}(\text{Ht}\, A, \text{Ht}\, B) \leq \text{Ht}(A \times B) < \text{Ht}\, A \oplus \text{Ht}\, B$.

• (1) If x is minimal $(\text{mod}\, A)$ and y minimal $(\text{mod}\, B)$, then the couple (x, y) is minimal $(\text{mod}\, A \times B)$: so the above equality (1) holds for $\text{Ht}\, x = \text{Ht}\, y = 0$.

Let γ be an ordinal $\neq 0$; assume that the equality holds for each x' and each y' with $\text{Ht}\, x' \oplus \text{Ht}\, y' < \gamma$. Assume moreover that the equality holds too if $\text{Ht}(x', y')(\text{mod}\, A \times B) < \gamma$. Let I be the initial interval of those (x', y') with height $< \gamma$, and J the complement of I. So (x, y) belongs to J iff $\text{Ht}\, x \oplus \text{Ht}\, y \geq \gamma$. By definition of height, a couple (x, y) is minimal in J iff $\text{Ht}(x, y)(\text{mod}\, A \times B) = \gamma$. Note that this implies $\text{Ht}(x)(\text{mod}\, A) \leq \gamma$ and $\text{Ht}(y)(\text{mod}\, B) \leq \gamma$. Finally it suffices to prove that (x, y) is minimal in J iff $\text{Ht}\, x \oplus \text{Ht}\, y = \gamma$.

For (x, y) minimal in J, either $\text{Ht}\, x \oplus \text{Ht}\, y > \gamma$. Then for instance $\text{Ht}(x) \leq \gamma$ and there exists an ordinal $j < \text{Ht}\, y$ with $(\text{Ht}\, x) \oplus j = \gamma$; and so an element $y' < y$ $(\text{mod}\, B)$ with $(\text{Ht}\, y') = j$ (see 2.7.1). Hence $(x, y') < (x, y)$ and (x, y') belongs to J, so that (x, y) is not minimal in J.

Or $\text{Ht}\, x \oplus \text{Ht}\, y = \gamma$. Then for any $(x', y') < (x, y)(\text{mod}\, A \times B)$ we have either $x' < x$ $(\text{mod}\, A)$ with $y' \leq y$ $(\text{mod}\, B)$, or conversely $x' \leq x$ with $y' < y$. By definition of commutative ordinal addition, this always gives $\text{Ht}\, x' \oplus Hty' < \gamma$, so that (x', y') belongs to I and (x, y) is minimal in J. •

• (2) The first inequality is obvious. For the second inequality, note that the height of a well-founded poset A equals $\text{Sup}(i + 1)$ where i denotes the height of any element in A. Similarly for B we obtain $\text{Sup}(j + 1)$; and finally for $A \times B$ we obtain $\text{Sup}(i \oplus j + 1)$, taking in account the preceding (1). Then the second inequality results from the following, which is a consequence of the definition of natural (commutative) sum, easily provable by the reader: $\text{Sup}(i \oplus j + 1) < \text{Sup}(i + 1) \oplus \text{Sup}(j + 1)$. •

4.9 Conjunction of posets, dimension: Dushnik, Miller, Hiraguchi

Given a poset A and a set of posets B_i which are all augmentations of A with common base $|A|$, we say that A is the **conjunction** of the B_i's iff, for any x, y in $|A|$, we have $x \leq y$ modulo A iff $x \leq y$ modulo each B_i.

In the interesting particular case where the B_i are chains, if $x < y \,\text{mod}\, A$ then $x < y(\text{mod}\, B_i)$ for each i and if $x | y(\text{mod}\, A)$ then there exists an i with $x < y(\text{mod}\, B_i)$ and a j with $x > y(\text{mod}\, B_j)$.

The **dimension** of a poset A is the least cardinal of a set of chains, each with base $|A|$ and whose conjunction is A. The notion of dimension goes back to [46] DUSHNIK, MILLER 1941. We shall denote $\text{Dim}\, A$ the dimension of a poset A.

Modulo the augmentation axiom (see 2.9.3), **every poset A is a conjunction of chains each of them augments A.** Indeed if $x|y(\mathrm{mod}\, A)$, with only axioms ZF there exist an augmentation of A with $x < y$ and another one with $x > y$. Since by SCOTT's definition 1.4.4, every set has a cardinal, the following definition of dimension depends only from augmentation axiom (in ZF at least every finite poset has a dimension).

Let A be a poset, B a restriction of A; if dimensions exist for A and B, then $\mathrm{Dim}\, B \leq \mathrm{Dim}\, A$.

Every chain has dimension 1.

Given a set E of cardinality ≥ 2, **the identity relation, or free poset, has dimension 2**: take a chain based on E and the converse chain (so we use the ordering axiom; ZF suffices if E is countable).

4.9.1 Two lemmas

(1) Let A be a poset with dimension h (positive integer). **Add a new element u which will be the minimum of a poset, extension of A to the new base $|A| \cup \{u\}$. Then we obtain a poset with the same dimension h.** Similarly if we add a maximum.

It suffices to start with the chains C_i whose conjunction is A; then add to each C_i the minimum element u.

(2) **Let h be an integer ≥ 2, and let A, B be two posets with disjoint bases, each one with dimension $\leq h$. Then the poset based on the union of the bases, common extension of A and B and in which every element of $|A|$ is incomparable with every element of $|B|$, has dimension $\leq h$.**

• Assume $h = 2$ (the proof immediately extends to any greater integer). Consider the two chains C, C' whose conjunction is A, and the two chains D, D' whose conjunction is B. Then, on the union of the two bases, take the chains $C + D$ and $D' + C'$. •

4.9.2 Every finite tree either is a chain or has dimension 2

• We construct the tree from its maximal elements, by a finite sequence of the two following operations.

(1) union of two trees with disjoint bases, each element of one base being incomparable with each element of the other;

(2) addition of a minimum;

finally we use the preceding subsection. •

4.9.3 Direct product of chains

Let a be a cardinal; every direct product of chains whose set has cardinality a, is a poset with dimension $\leq a$ (uses axiom of choice; augmentation axiom suffices if a is finite: see 2.9.3). This statement will be completed in 4.9.6 below.

• Denote the chains by $A_i (i \in I$ with cardinality $a)$, and let A be the direct product of the A_i. For each index i, consider the direct product D_i of the $A_j (j \neq i)$. By the augmentation axiom, there exists at least one totally ordered augmentation C_i of D_i. To each i associate a unique C_i (axiom of choice), then associate the chain B_i with base $|A|$, such that $f \leq g \pmod{B_i}$ iff $f(i) < g(i) \pmod{A_i}$ or $f(i) = g(i)$ with (Sequence of the $f(j)$ for $j \neq i$) $\leq$ (Sequence of the $g(j)$ for $j \neq i) \bmod C_i$. Then the direct product A is the conjunction of the chains B_i. •

4.9.4 Dimension and finite restrictions; every tree has dimension ≤ 2

Let A be a poset and p a positive integer. If every finite restriction of A has dimension $\leq p$, then A has dimension $\leq p$ (uses ultrafilter axiom; ZF suffices if A is countable).

• To each finite subset F of the base $|A|$, associate the set U_F of multirelations $(C_1, ..., C_p)$, a sequence of p chains with common base F, such that, for every $x, y \in F$, we have $x \leq y (\bmod A)$ iff $x \leq y (\bmod C_1)$ and ... and $x \leq y (\bmod C_p)$. By hypothesis U_F is non-empty for every F.

Moreover if $F' \subseteq F$ then every multirelation which is an element of U_F, when restricted to F', yields an element of $U_{F'}$. By the coherence lemma 2.4.1 (equivalent to the ultrafilter axiom), there exists a multirelation with base $|A|$ whose restriction to each F belongs to U_F. Hence this multirelation is a sequence of p chains, each of whose base is $|A|$, and whose conjunction is the poset A. •

In particular, it follows from 4.9.2 that **every tree, finite or infinite, either is a chain or has dimension** 2.

4.9.5 Dushnik-Miller dimension theorem

Let E be a set; to each element a of E associate the singleton $a' = \{a\}$ and the complement $a" = E - a'$. **Then for any given set of subsets of E which contains all the a' and $a"$ as elements, the partial ordering of inclusion has dimension equal to** $\operatorname{Card} E$ ([46] DUSHNIK, MILLER 1941; uses augmentation axiom).

• Let us denote by $<$ any chain which augments the partial ordering of inclusion. Then there exists at most one element a of E for which $a" < a'$. Indeed, if we have two distinct elements a, b with $a" < a'$ and $b" < b'$, hence we obtain $a" < a' \subseteq b" < b' \subseteq a"$ thus $a = b$.

For each element a of E, the two sets a' and $a"$ are incomparable with respect to inclusion. Hence among the chains whose conjunction is the partial ordering of inclusion, one at least satisfies the inequality $a" < a'$, with $x' < x"$ for all elements $x \neq a$. Associate this chain with a: then the set of those chains corresponding to all elements of E gives the desired ordering of inclusion. To see this, it remains to consider two subsets X, Y of E which are incomparable with respect to inclusion. So there exist an element x of E with $x \in X$ and $x \notin Y$, and an element y with

$y \in Y$ and $y \notin X$. Then the chain associated with x gives $Y \subseteq x" < x' \subseteq X$ so that $Y < X$. Similarly the chain associated with y gives $X < Y$. •

The preceding statement extends as follows.

Let A be a poset with base E. To each element a of E associate the initial interval a' of elements $x \leq a \pmod A$ and the initial interval $a"$ of elements $x <$ or $|a \pmod A$. **Then for any given set of subsets of E which contains all the a' and $a"$ as elements, the partial ordering of inclusion has dimension equal to the least cardinality of those sets of chains which are restrictions of A and whose bases cover the whole set E** ([188] POUZET 1969; uses augmentation axiom).

• For each a, b in E, if $a" < a'$ and $b" < b'$ (where $<$ denotes a chain which augments the partial ordering of inclusion), then the preceding argument proves that a and b are comparable: either $a \leq b$ or $b \leq a \pmod A$. Hence to each chain which augments the inclusion, there corresponds a totally ordered restriction I of A, such that the elements u of $|I|$ satisfy $u" < u'$, with however $x' < x"$ for those elements x in $E - |I|$.

Conversely, given an arbitrary totally ordered restriction I of A, the inequalities $u" < u'$ for each u in the base $|I|$ are mutually compatible, and are compatible with inclusion. We can always take each I to be maximal with respect to inclusion (ordering axiom: see 2.4.4), and then these conditions have to be completed by $x' < x"$ for every x in $E - |I|$. We end as in the previous proof. •

4.9.6 Dimension of a direct product of chains

Given a direct product of a-many chains (where a is a finite or an infinite cardinal), each chain being reduced to the elements 0 and 1, we obtain a partial ordering isomorphic with inclusion for subsets of a, by replacing each subset b of a by its characteristic function, taking the value 1 for each chain belonging to b and 0 for each chain belonging to $a - b$. Hence, by the preceding, the direct product just considered has dimension a.

Consequently **the direct product of a many chains, each having at least two elements, has dimension a.**

Indeed it has dimension $\geq a$ by the preceding argument and dimension $\leq a$ by 4.9.3.

Let us cite a result due to [113] HIRAGUCHI 1951, also [114] 1955: **for every poset with finite cardinality $p \geq 4$ the dimension is at most equal to $p/2$.**

4.10 Bound

4.10.1 Bounds of an initial interval

Let A be a poset and B an initial interval of A. An element u in the difference set $|A| - |B|$ is called a **bound** (more precisely a minimal strict upper bound) of B modulo A iff every element strictly less than $u \pmod A$ belongs to $|B|$.

The bounds of an initial interval are pairwise incomparable.

Consequently **if A is finitely free, then there are only finitely many bounds for each initial interval of A.**

However an initial interval is not in general defined by the set of its bounds. For instance take the chain **Q** of rationals: there are infinitely many distinct initial intervals of **Q** without any minimal upper bound.

Let B be an initial interval of a poset A. If $x \in |B|$, then $x <$ or $|u (\text{mod}\, A)$ for every bound u of B.

(1) **If A is well-founded, then the condition $x \in |B|$ is equivalent to the condition $x <$ or $|u$ for every bound u of B.**

• Assume that A is well-founded; if x does not belong to $|B|$, there exists at least one element $u \leq x$ which is minimal among those elements in the difference set $|A| - |B|$: this u is a bound. •

(1') If A is not well-founded, then the proposition is false.

• Take A to be the chain $\omega + \omega^-$ (where ω^- is the retro-ordinal converse of ω), and B to be the initial interval ω: then there does not exist any bound. •

(2) **Let A be a well-founded poset; then each initial interval of A is completely defined by its bounds.**

(3) **If A is a well partial ordering, then each initial interval of A is completely defined by finitely many bounds.**

Consequently if A is an infinite w.p.o., then the set of its initial intervals is equipotent with A (axiom of choice).

4.10.2 Bounds of an ideal

Let A be a poset and B an ideal in A.

For every finite sequence $u_i (i = 1, ..., h)$ of bounds of B, and for every sequence $v_i (i = 1, ..., h)$ of elements $v_i < u_i (\text{mod}\, A)$ for each i, there exists an element t of the base $|B|$ satisfying the condition $t \geq v_i$ and $t \not\geq u_i$ for each i.

Consequence of the definitions of ideal and bound, taking into account that $v_i < u_i$ implies $v_i \in |B|$ for each i.

The proposition is no longer true when B is an arbitrary initial interval of A.

• Take A with three elements a, b, c where $a < c$ and $b < c$ and $a|b (\text{mod}\, A)$. Let B be the restriction to $\{a, b\}$. Then c is a bound and it suffices to take $u_1 = u_2 = c$ with $v_1 = a$ and $v_2 = b$. •

Note also that the converse of the proposition is false.

• Take A with five elements a, b, c, d, e where $d < a < e$ and $d < b < e$ and $d < c < e$ and a, b, c mutually incomparable. Let B be the restriction to $\{a, b, d\}$, so that B is not an ideal. However the conclusion of the proposition holds. Indeed the only bound is c and the only element $< c$ is d: it suffices to take $t = a$ or $t = b$, or even $t = d$. •

4.11 Maximal augmented chain theorem for a well partial ordering: De Jongh, Parikh; extraction property

4.11.1 Heights in a well partial ordering

Let A be a well partial ordering, B an initial interval of A, thus a well partial ordering. Denote by $A-B$ the well partial ordering restriction of A to the difference set $|A| - |B|$. Recall that $\mathcal{J}(A)$ denotes the partial ordering of initial intervals of A, which is well-founded by 4.4.1 (dependent choice).

Let us denote by $\mathrm{Ht}^*(A)$ the height of the maximum A of $\mathcal{J}(A)$, i.e. the height of $\mathcal{J}(A)$ minus one; analogous notations with B and $A - B$. Then we have the following inequalities:

$\mathrm{Ht}^*(B) + \mathrm{Ht}^*(A - B) \leq \mathrm{Ht}^*(A) \leq \mathrm{Ht}^*(B) \oplus \mathrm{Ht}^*(A - B)$, where $+$ is the usual ordinal sum and $\oplus$ is the natural, commutative sum: see 4.8.2.

• Consider the isomorphism which, to each initial interval X of $A-B$, associates the initial interval $B \cup X$ of A. The partial ordering $\mathcal{J}(A)$ is an extension of $\mathcal{J}(B)$ followed by $\mathcal{J}(A - B)$: this yields the first inequality.

On another hand, to each initial interval X of A, associate the couple whose first term is the restriction $X/(|X| \cap |B|)$ and whose second term is the restriction $X/(|X| \cap |A-B|)$. This is an isomorphism which transforms $\mathcal{J}(A)$ into a restriction of the direct product $\mathcal{J}(B) \times \mathcal{J}(A-B)$: this yields the second inequality, by using 4.8.3. •

4.11.2 Maximal augmented chain theorem

Given a well partial ordering A and the associated well-founded partial ordering $\mathcal{J}(A)$, there exists a well-ordered restriction of $\mathcal{J}(A)$ which is isomorphic to the height of $\mathcal{J}(A)$, thus which has the maximum possible height.

Equivalently, there exists a well-ordering which is an augmentation of A, isomorphic to the height of $\mathcal{J}(A)$ minus one, thus isomorphic to the maximum possible ordinal ([39] DE JONGH, PARIKH 1977; uses axiom of choice).

• First of all, the equivalence of both preceding statements is an immediate consequence of 2.9.4 (see [18] BONNET, POUZET 1969): note that for any ordinal α, the well-ordering of initial intervals is isomorphic with $\alpha + 1$.

We shall prove the first statement, concerning the maximal chain (up to isomorphism) among restrictions of $\mathcal{J}(A)$. We procede by induction on the height $\mathrm{Ht}^*(A)$. Note that the statement is obvious for A empty, thus $\mathrm{Ht}^*(A) = 0$. It is obvious also if A is based on a singleton, thus $\mathrm{Ht}^*(A) = 1$. Assume first that $\mathrm{Ht}^*(A)$ is a decomposable ordinal $\alpha = \beta + \gamma$. Moreover we assume that γ is the last component of α in the Cantor normal form. Then the previous decomposition $\beta + \gamma$ holds as well with the natural, or commutative sum as with the usual ordinal sum.

Denote by B an initial interval of A with height β modulo $\mathcal{J}(A)$. Then the two inequalities in the previous subsection yield $\mathrm{Ht}^*(A - B) = \gamma$. By our induction hypothesis, there exists a chain of type β which is a restriction of $\mathcal{J}(B)$. Similarly with $\mathcal{J}(A - B)$ and γ. Hence there exists a chain of type γ whose elements are initial intervals of A including B. Finally the sum of these two chains has the isomorphism type α.

Secondly we assume that $\mathrm{Ht}^*(A)$ is an indecomposable ordinal of the particular form ω^{k+1}, where k is an ordinal. Take an initial interval B of A with height ω^k modulo $\mathcal{J}(A)$. Then by the previous subsection we have the following inequality: $\omega^{k+1} \leq \omega^k \oplus \mathrm{Ht}^*(A - B)$, thus $\mathrm{Ht}^*(A - B) = \omega^{k+1}$.

By iteration, we obtain an ω-sequence of initial intervals B_i(i integer) of A which is strictly increasing under inclusion, with $\mathrm{Ht}^*(B_{i+1} - B_i) = \omega^k$ for each i (obvious notation $B_{i+1} - B_i$ for the restriction of A to the difference set of the bases).

By our induction hypothesis, we associate with each i a chain of order type ω^k of initial intervals, each one situated between B_i and B_{i+1}. Hence the ω-sum of these chains yields a chain of initial intervals with type ω^{k+1}.

Thirdly we assume that $\mathrm{Ht}^*(A)$ is an arbitrary indecomposable ordinal α. Denote by $u = \mathrm{Cof}\,\alpha$ and take a u-sequence of increasing ordinals $\alpha_i (i < u)$ with $\mathrm{Sup}\,\alpha_i = \alpha$. For each $i < u$, take an initial interval A_i of A with height α_i modulo $\mathcal{J}(A)$ (axiom of choice). Hence $\mathrm{Ht}^*(A_i) = \alpha_i$ for each $i < u$. Let b_i be an element in the difference set $|A| - |A_i|$. Denote by b_i the height modulo $\mathcal{J}(A)$ of the initial interval B_i of elements $<$ or $|b_i$. Then $\alpha_i \leq \beta_i < \alpha$ thus $\mathrm{Sup}\,\beta_i = \alpha$. From the u-sequence of indices $i < u$, we extract another u-sequence for which all elements b_i and all ordinals β_i are distinct.

Since A is a well partial ordering, by 4.5.1 we may require that the u-sequence of the b_i be strictly increasing (mod A). Thus the sequence of corresponding intervals B_i is strictly increasing with respect to inclusion. By a new extraction, we may require that $\beta_{i+1} \geq \beta_i \oplus \beta_i$ (natural sum) for each index $i < u$. Then we have $\mathrm{Ht}^*(B_{i+1} - B_i) \geq \beta_i$ for each i, using the previous subsection. By the induction hypothesis, we can take in each difference $B_{i+1} - B_i$ a chain of initial intervals of A, situated between B_i and B_{i+1} and whose length is at least β_i. The sum of these chains yields a definitive chain whose length is $\Sigma\beta_i = \alpha$ by 2.7.4. •

4.11.3 The extraction property

Given an aleph ω_α, we shall say that a poset A has **the extraction property** for ω_α iff for every ω_α-sequence u of distinct elements in the base $|A|$, there exists a strictly increasing (mod A) sequence extracted from u and with the same length as u (ZAGUIA in 1983, published in ToR-86 p.124).

By 4.3.2 proposition (3), an infinite partial ordering A is a well partial ordering iff A has the extraction property for ω.

By 4.5.1, any well partial ordering has the extraction property for any infinite aleph.

Let A be a well partial ordering, $\mathcal{J}(A)$ the partial ordering of initial intervals of A (with respect to inclusion). Then $\mathcal{J}(A)$ has the extraction property for every regular aleph strictly greater than ω (see [207] POUZET, ZAGUIA 1985).

• For any $i < \omega_\alpha$ (regular aleph), let B_i be an initial interval of A. Denote by M_i the finite set of minimal elements (mod A) in the complement set $|A| - |B_i|$. For each integer p, denote by C_p the class of those M_i whose cardinal is p. Since ω_α is regular, then there exists an integer p such that C_p has cardinality ω_α. Reindexing the sequence of M_i, we consider it as an ω_α-sequence in the direct product $A \times .. \times A(p$ times) (choice for finite sets). This direct product being a well partial ordering by 4.8.1 proposition (2) (denumerable subset axiom); the proposition follows by 4.5.1. •

In the cited doct. dissert., it is proved that any distributive lattice which has the extraction property for every regular aleph strictly greater than ω, admits a chain isomorphic with its height: this is another proof for the theorem 4.11.2.

4.12 Cofinality of a finitely free poset

4.12.1 Increasing sequence lemma

Let A be a poset, and let $k = \operatorname{Cof} A$. Then a necessary and sufficient condition that A be generated by a strictly increasing sequence (i.e. that every element have an upper bound in the sequence) **is that every subset of the base with cardinal $< k$ have an upper bound in A.** Moreover k is necessarily regular (uses axiom of choice; POUZET in ToR-86 p.125).

• If A is generated by a strictly increasing ordinal-indexed sequence C, then $k = \operatorname{Cof} C$ thus k is regular. Moreover every subset of the base, with cardinal strictly less than k, has an upper bound in C.

Conversely suppose that every subset with cardinal $< k$ has an upper bound. Take a cofinal subset H of A with $\operatorname{Card} H = k$. Well-order H; consider k as an ordinal and denote by $a_i (i < k)$ the elements of H. Then define as follows the increasing k-sequence of elements b in H. Let $b_0 = a_0$. Let $1 \leq i < k$; by hypothesis there exists an upper bound of all a_j and all already defined $b_j (j < i)$. Then define b_i as being such an upper bound, belonging to H. •

4.12.2 Increasing sequence and regular aleph

Theorem. Let A be a partial ordering and $k = \operatorname{Cof} A$. If every proper initial interval of A has cofinality $< k$, then

(1) every subset of the base, with cardinality $< k$, has an upper bound;

(2) k is a regular aleph and A is generated by an increasing k-sequence (uses axiom of choice; communicated by POUZET).

• By the preceding lemma, (2) follows from (1). Moreover we can always reduce A to a cofinal restriction, still denoted by A, having cardinality k.

To each element x in the base, associate A_x = initial interval of elements $<$ or $|x(\mathrm{mod}\, A)$; then let $k_x = \mathrm{Cof}\, A_x < k$. For each cardinal $u < k$, denote by H_u the set of elements x such that $k_x \leq u$. First we see that, given $u < k$, every subset F of H_u with $\mathrm{Card}\, F < k$ has an upper bound. Indeed for all $x \in F$, the union of corresponding A_x has at most cofinality $u \times \mathrm{Card}\, F < k$. So that this union does not cover the entire base: hence the intersection of corresponding final intervals $\geq x$ is non-empty.

Now let us prove the regularity of k. For each $u < k$, since $\mathrm{Card}\, H_u \leq k$, consider H_u as the union of an arbitrary increasing $(\mathrm{Cof}\, k)$-sequence of subsets, each of which having cardinality $< k$: so that each has an upper bound. We obtain a set K_u of these upper bounds, with $\mathrm{Card}\, K_u \leq \mathrm{Cof}\, k$. Now when u varies, by taking $(\mathrm{Cof}\, k)$ many values, we obtain a cofinal set modulo A, having cardinality less than or equal to $(\mathrm{Cof}\, k)^2 = Cof k$: this proves that $k = \mathrm{Cof}\, k$.

We are finished provided we note that each set F with $\mathrm{Card}\, F < k$, is necessarily included in some $H_u (u < k)$. For otherwise F would be the union of $(\mathrm{Cof}\, k) = k$ many disjoint non-empty subsets: contradiction. •

4.12.3 Cofinality theorem for finitely free posets

Every finitely free poset, either has finite cofinality, or infinite regular cofinality ([197] POUZET 1979; uses axiom of choice).

• With previous notations, suppose that $k = \mathrm{Cof}\, A$ is an infinite singular aleph. We can always suppose that $\mathrm{Card}\, A = k$ and that A is well-founded (see 2.7.2, corollary). Then by the previous subsection, there exists a proper initial interval with cofinality k. Thus there exists a strictly decreasing ω-sequence of initial intervals (under inclusion), each of which having cofinality k. Take an element in each difference set: we obtain a bad ω-sequence $(\mathrm{mod}\, A)$. Since A is well-founded, we have infinitely many mutually incomparable elements. •

4.13 Cofinal restriction of a directed well partial ordering (Pouzet)

Given a poset A, an ordinal-indexed sequence γ in A without repetition is said to be **unbounded** $(\mathrm{mod}\, A)$ iff no element in $|A|$ is an upper bound $(\mathrm{mod}\, A)$ of the set of values of γ.

4.13.1 Some properties of a directed well partial ordering

Let A be a directed well partial ordering. Let $\gamma = \mathrm{Cof}\, A$ and let F be a cofinal subset of least cardinality, hence of cardinality γ. Then we have the following (modulo the axiom of choice).

(1) **There exists a γ-sequence $b_i (i < \gamma)$ without repetition, of elements forming a cofinal subset of F.** Moreover this sequence is bad with respect to the converse of A; i.e. if $i < j < \gamma$ then $b_j >$ or $|b_i (\mathrm{mod}\, A)$.

(2) **For each element $c \in F$ there exists an unbounded (mod A) strictly increasing γ-sequence with values in F, beginning with c.**

(3) **Given two unbounded strictly increasing γ-sequences in F, there exists a third strictly increasing γ-sequence in which the first two are embeddable** in the sense of 4.1.2.

(4) **The ideals generated by the unbounded strictly increasing γ-sequences in F constitute a directed partial ordering with respect to inclusion; moreover the union of these ideals includes F.**

• (1) Well-order F according to its cardinality. In the thus obtained γ-sequence of the $a_i (i < \gamma)$ for each index i, remove those a_j for which $j > i$ and $a_j < a_i (\text{mod } A)$. This extracted sequence yields a cofinal subset G of F. Thus G has cardinality γ and our extracted sequence is a γ-sequence satisfying (1).

(2) Let b_i be the preceding γ-sequence satisfying (1). By 4.5.1 (extraction theorem), there exists a strictly increasing (mod A) extracted γ-sequence. Suppose that there exists an upper bound u of the elements of this sequence. We can assume that u belongs to the cofinal set G of alinea (1); hence u is one of the b_i's : contradiction proving that our sequence is unbounded.

Now let c be an element of F and $u_i (i < \gamma)$ an unbounded strictly increasing γ-sequence in F. Since A is directed, to each i we associate an element v_i in the cofinal set F, which is a common upper bound of c and u_i. Moreover, we can require that the v_i be distinct. For this, let v_0 be a common upper bound of c and u_0. In general, given an index $i (1 \leq i < \gamma)$ and $v_j (j < i)$, note that the set of these v_j has cardinality strictly less than γ, and so is not cofinal in A. Thus there exists an x in F with $x >$ or $|$ with each v_j. Take v_i to be a common upper bound of c, u_i and x. This makes v_i distinct from the $v_j (j < i)$.

Finally, again using the extraction theorem, we extract a strictly increasing γ-sequence. Since the sequence of the u_i is embeddable in it, this strictly increasing γ-sequence is unbounded. Furthermore, it is formed of elements greater than c.

(3) Let u_i and v_i be two unbounded strictly increasing γ-sequences. Since A is directed, to each i associate a common upper bound w_i of u_i and v_i. By the procedure in the preceding (2), we can ensure that the w_i are distinct. Finally, extract a strictly increasing γ-sequence from the w_i.

(4) Follows immediately from (2) and (3). •

4.13.2 Embeddability between strictly increasing sequences

Let A be a directed poset and α a non-zero ordinal with ω_α regular aleph. Then the quasi-ordering of embeddability (see 4.1.2) **for the strictly increasing ω_α-sequences in A, is finitely free** (uses denumerable subset axiom).

• Suppose on the contrary that there exist denumerably many strictly increasing ω_α-sequences $u_1, ..., u_i, ..(i$ integer; this needs the denumerable subset axiom) which are mutually incomparable under embeddability. To each u_i and to each integer $j \neq i$, associate an element $a_{i,j}$ of u_i such that no term of u_j is greater than or equal to $a_{i,j} (\text{mod } A)$. Since ω_α is assumed to be regular and strictly greater than

ω, take an element b_i in u_i which is greater than all the $a_{i,j}$(i fixed, j variable). The b_i are mutually incomparable, hence A is not finitely free: contradiction. •

4.13.3 Increasing sequence of strictly increasing sequences

Let A be a poset and α a non-zero ordinal such that ω_α is a regular aleph. Then **for every increasing (with respect to embeddability) ω_α-sequence of strictly increasing ω_α-sequences in A, there exists a strictly increasing ω_α-sequence in A in which each is embeddable**.

• Let u_i be the given sequences and $a_{i,j}$ be the jth term in $u_i (i, j < \omega_\alpha)$. Let $b_0 = a_{0,0}$. Let b_1 be the first term in u_1 which is a common upper bound of b_0, $a_{0,1}$ and $a_{1,1}$. In general for each $i < \omega_\alpha$, let b_i be the first term in u_i which is a common upper bound of the $b_j (j < i)$ and the $a_{k,i} (k \leq i)$. Then the ω_α-sequence of the b_i satisfies our conclusion. •

4.13.4 Ideals generated by unbounded strictly increasing sequences

Let A be a directed well partial ordering whose cofinality is a regular aleph ω_α with $\alpha \neq 0$. Let F be a cofinal subset of least cardinality, hence of cardinality ω_α.

Consider the ideals which are generated, each of them by an unbounded strictly increasing ω_α-sequence with values in F (see 4.1.1). **Then there exists a set of at most ω_α many ideals which, under inclusion, constitute a directed well partial ordering of cofinality $< \omega_\alpha$. Moreover the union of the bases of these ideals includes F** (uses axiom of choice).

• By 4.13.2, the ideals considered in this statement form a finitely free partial ordering under inclusion, which furthermore is directed by 4.13.1 proposition (3). In this partial ordering, take a well-founded cofinal restriction, hence a directed well partial ordering (see 2.7.2, corollary). There are at most ω_α many ideals under consideration: see 4.10.1 proposition (3) (axiom of choice). Every element of F belongs to an unbounded strictly increasing ω_α-sequence: see 4.13.1 proposition (2). Thus the union of our ideals includes F and this holds as well when taking the set of ideals in a cofinal set.

The cofinality of the partial ordering under inclusion, for our ideals, is at most the cardinal of this set, hence at most ω_α. Yet this cofinality cannot be equal to ω_α, since by 4.13.1 proposition (2) there would then exist an unbounded strictly increasing ω_α-sequences of ideals, contradicting the previous subsection. Consequently the cofinality is strictly less than ω_α. •

4.13.5 Conclusion

Theorem. Let A be a directed well partial ordering; then there exists a cofinal restriction of A which is isomorphic with the direct product of finitely many distinct regular alephs, the largest of which is Cof A (uses axiom of choice). Due to POUZET in 1979, published in ToR-86 p.191; see also [173] MILNER, POUZET 1982).

• Firstly note that $\operatorname{Cof} A$ is a regular aleph by 4.12.3.

We already know that a directed well partial ordering of countable cofinality, either has a maximum element and so its cofinality is 1; or has a cofinal restriction isomorphic with ω: see 2.13.2 proposition (2).

Proceed by induction. Let ω_α(α non-zero) be an infinite regular aleph, and assume that, for each regular aleph $k < \omega_\alpha$ and each directed well partial ordering X with cofinality k there exists a cofinal restriction of X which is a direct product of finitely many distinct regular alephs, the largest of which is k.

Let B be the directed well partial ordering formed by the ideals in the previous subsection. By the induction hypothesis, the cofinality of B, which is strictly less than ω_α and regular, is either equal to 1 or to ω_β, with an ordinal $\beta < \alpha$. If $\operatorname{Cof} B = 1$, then there exists a strictly increasing ω_α-sequence which generates a cofinal subset of $|A|$, and we are finished.

Suppose now that $\operatorname{Cof} B = \omega_\beta$ with $\beta < \alpha$. Then by our induction hypothesis, ω_β is regular and there exists an integer n and a cofinal restriction C of B which is isomorphic with a direct product of n regular alephs, the largest of which is ω_β. Consequently, to each ideal in A which is an element of the base $|C|$, bijectively associate an n-tuple of coordinates $(t_1, ..., t_n)$, each of which runs through a distinct regular aleph, the first coordinate t_1 running through ω_β.

For each n-tuple, let $I(t_1, ..., t_n)$ designate an unbounded, strictly increasing ω_α-sequence in A which generates the ideal whose coordinates are $t_1, ..., t_n$ (obviously I is a choice function, several sequences yielding a same ideal).

Given two n-tuples, the inequality $(t_1, ..., t_n) \leq (t'_1, ..., t'_n)$ modulo the direct product, i.e. the set of inequalities $t_1 \leq t'_1, ..., t_n \leq t'_n$, is equivalent to the condition that the sequence $I(t_1, ..., t_n)$ is embeddable in $I(t'_1, ..., t'_n)(\operatorname{mod} A)$, in the sense of 4.1.2.

We are now going to replace each ω_α-sequence under consideration by an extracted sequence of the same length, with the aim in mind of bijectively associating to each $(n+1)$-tuple $(i, t_1, ..., t_n)$ where $i < \omega_\alpha$ and the t are as previously, an element of the base $|A|$ designated by $a(i, t_1, ..., t_n)$, which more precisely will be a term in the sequence $I(t_1, ..., t_n)$. We shall do this in a manner that the inequality $(i, t_1, ..., t_n) \leq (i', t'_1, ..., t'_n)$ modulo the new direct product, i.e. the set of inequalities $i \leq i', t_1 \leq t'_1, ..., t_n \leq t'_n$, is equivalent to the inequality $a(i, t_1, ..., t_n) \leq a(i', t'_1, ..., t'_n) \operatorname{mod} A$.

First of all, it is easy to insure that our elements a are distinct, since there are ω_β many n-tuples $(t_1, ..., t_n)$ and $\omega_\beta < \omega_\alpha$. We first take the $a(0, t_1, ..., t_n)$ to be all distinct. In general, given an ordinal $u < \omega_\alpha$ we ensure that the chosen values $a(u, t_1, ..., t_n)$ are all distinct, as well as being distinct from the already defined values $a(i, t_1, ..., t_n)$ for $i < u$. We do this while respecting the condition that $a(u, t_1, ..., t_n)$ belongs to the sequence $I(t_1, ..., t_n)$ and is strictly greater $(\operatorname{mod} A)$ than the already defined values $a(i, t_1, ..., t_n)$ for $i < u$. All that is easy because ω_α is regular, hence for each $u < \omega_\alpha$ no u-sequence extracted from the sequence I is cofinal.

Secondly, we shall ensure that, for two incomparable n-tuples $(t_1, ..., t_n)$ and $(u_1, ..., u_n)$, every value of a term in $I(t_1, ..., t_n)$ is incomparable $(\mathrm{mod}\, A)$ with every value of a term in $I(u_1, ..., u_n)$. To do this, for each n-tuple $(t_1, ..., t_n)$, designate by $(u_1, ..., u_n)$ all n-tuples which are incomparable with $(t_1, ..., t_n)$. Thus the sequence $I(t_1, ..., t_n)$ is incomparable, with respect to embeddability $(\mathrm{mod}\, A)$, with each sequence $I(u_1, ..., u_n)$. Take a term of the first, for which no term of the second is greater $(\mathrm{mod}\, A)$ than it, and designate it by $b(t_1, ..., t_n; u_1, ..., u_n)$. Then, fixing $(t_1, ..., t_n)$, let $(u_1, ..., u_n)$ vary, and replace the sequence $I(t_1, ..., t_n)$ by an extracted sequence of the same length ω_α, but beginning with an upper bound of all the preceding b. This is possible because there is at most ω_β many b.

Thirdly, we shall ensure that, for two distinct n-tuples satisfying $(t_1, ..., t_n) < (u_1, ..., u_n)$, there do not exist values $a \in I(t_1, ..., t_n)$ satisfying the opposite inequality $a > b \pmod{A}$: the only possibilities being incomparability $a|b$ or the inequality $a < b \pmod{A}$. To do this, for each n-tuple $(t_1, ..., t_n)$, let $(u_1, ..., u_n)$ designate any and all distinct lesser n-tuples: thus the sequence $I(t_1, ..., t_n)$ is not embeddable $(\mathrm{mod}\, A)$ in any $I(u_1, ..., u_n)$. Take a term of the first sequence, for which no term of the second is greater, and continue as in the preceding paragraph.

Fourthly and finally, we shall define by induction on i the elements $a(i, t_1, ..., t_n)$ as previously announced. To the $(n + 1)$-tuple (0,0,...,0) associate the beginning term of the sequence $I(0, ..., 0)$. Let $(t_1, ..., t_n)$ be an n-tuple and suppose that $a(0, u_1, ..., u_n)$ has already been defined for all n-tuples $(u_1, ..., u_n) < (t_1, ..., t_n)$. Then set $a(0, t_1, ..., t_n)$ to be the first term of $I(t_1, ..., t_n)$ which is strictly grater $(\mathrm{mod}\, A)$ than all $a(0, u_1, ..., u_n)$ for $(u_1, ..., u_n) < (t_1, ..., t_n)$. These values of a for first coordinate 0 shall be definitively kept.

To the $(n + 1)$-tuple (1,0,...,0) associate the first term of $I(0, ..., 0)$ which is strictly greater than $a(0, 0, ..., 0)$ and whichis not strictly less $(\mathrm{mod}\, A)$ than any value $a(0, u_1, ..., u_n)$ for any n-tuple $(u_1, ..., u_n)$. This is possible, since there is no element in the base $|A|$ which is greater $(\mathrm{mod}\, A)$ than every element of the sequence $I(0, ..., 0)$. Note that by the paragraph "thirdly", the value $a(1, 0, ..., 0)$ will be incomparable $(\mathrm{mod}\, A)$ with all those a with first coordinate 0 followed by a non-zero n-tuple $(u_1, ..., u_n)$. Indeed by construction $a(1, 0, ..., 0)$ is not strictly less $(\mathrm{mod}\, A)$ than these values; nor is it strictly greater, since otherwise the inequality of two terms would be in the opposite sense of the inequality of the n-tuples $(0, ..., 0) < (u_1, ..., u_n)$.

Now let $(t_1, ..., t_n)$ be an n-tuple, and suppose that $a(1, u_1, ..., u_n)$ has already been defined for all n-tuples $(u_1, ..., u_n) < (t_1, ..., t_n)$. Then for $a(1, t_1, ..., t_n)$ take the first term of $I(t_1, ..., t_n)$ which is strictly greater $(\mathrm{mod}\, A)$ than $a(0, t_1, ..., t_n)$ and which is strictly greater than $a(1, u_1, ..., u_n)$ for all n-tuples $(u_1, ..., u_n) < (t_1, ..., t_n)$; and finally which is not strictly lesser $(\mathrm{mod}\, A)$ than $a(0, v_1, ..., v_n)$ for any n-tuple $(v_1, ..., v_n)$. This last condition is possible by the fact that no element in $|A|$ is an upper bound of the sequence $I(t_1, ..., t_n)$.

Note that, by the paragraph "secondly", this value $a(1, t_1, ..., t_n)$ will be incomparable $(\mathrm{mod}\, A)$ with all the values a with first coordinate 0 or 1 followed by an n-tuple incomparable with $(t_1, ..., t_n)$. Our value $a(1, t_1, ..., t_n)$ will also be incomparable with the a having first coordinate 0 followed by an n-tuple

$(v_1, ..., v_n) > (t_1, ..., t_n)$. Indeed it is not strictly lesser $(\text{mod}\, A)$ by construction, nor is it strictly greater, since otherwise the inequality of the terms would be in the opposite sense of the inequality of the n-tuples.

In general, let k be an ordinal strictly less than ω_α, and suppose that all values of a with first coordinate $< k$ have been defined. To the $(n+1)$-tuple $(k, 0, ..., 0)$ associate the first term of $I(0, ..., 0)$ which is strictly greater than $a(i, 0, ..., 0)$ for all $i < k$, and which is not strictly less $(\text{mod}\, A)$ than any $a(i, u_1, ..., u_n)$. The value $a(k, 0, ..., 0)$ will be incomparable $(\text{mod}\, A)$ with all values a having first coordinate $i < k$ followed by a non-zero n-tuple.

Now let $(t_1, ..., t_n)$ be an n-tuple and suppose that the values $a(k, u_1, ..., u_n)$ have been defined for all n-tuples $(u_1, ..., u_n) < (t_1, ..., t_n)$. Then take $a(k, t_1, ..., t_n)$ to be the first term of $I(t_1, ..., t_n)$ which is strictly greater than $a(i, t_1, ..., t_n)$ for all $i < k$ and strictly greater than $a(k, u_1, ..., u_n)$ for all n-tuples $(u_1, ..., u_n) < (t_1, ..., t_n)$, and finally which is not strictly less $(\text{mod}\, A)$ than $a(i, v_1, ..., v_n)$ for any $i < k$ and any n-tuple $(v_1, ..., v_n)$. The reader can verify the desired incomparabilities. •

4.14 Exercise

4.14.1 Boundedly finitely free poset

Let A be a poset with base E, whose free subsets are assumed to be finite and all of cardinalities less than a given positive integer. Let D be one of the free subsets with maximum cardinal. Then to each element i of D there corresponds a totally ordered restriction A_i of A whose base intersects D with the singleton of i, such that the union of the bases $|A_i|$ is E (see [43] DILWORTH 1950; the proof given here is due to [184] PERLES 1963); see also [185].

We assume that the bases of these chains, or totally ordered restrictions, are disjoint, which is obviously possible.

1 - First assume that E is finite, and argue by induction on $\text{Card}\, E$. Fix A and D, and examine the first case where D is different from the set of minimal elements $(\text{mod}\, A)$ and from the set of maximal elements. Then consider the restriction A^+ of A to those x for which there exists in D an element $i \le x$; analogous restriction A^- with $i \ge x$. By the induction hypothesis, the proposition is true for A^+ and A^-, which are proper restrictions of A. Then to each i in D, there is associated a chain A_i^+ and a chain A_i^-: define A_i as their common extension. In this first case, note that the set of minimal elements modulo A is always a subset of the set of minimums for all chains A_i.

Examine the second case where D is the set of minimal elements $(\text{mod}\, A)$, and moreover there exists another free set D' with same cardinality, which is neither the set of minimal, nor the set of maximal elements. Then by our last remark in the first case, we can associate with D the same set of chains A_i as with D'. Analogous argument with maximal elements.

Now examine the third case where D is, for instance, the set of minimal elements, but the only other possible free set with maximum cardinality is the set of maximal elements. Take i in D, then a maximal chain (under inclusion) whose minimum is i. Then remove from E the elements of this maximal chain. Then use the induction hypothesis on this proper restriction of A and the subset $D - \{i\}$.

2 - Suppose that E is infinite and assume the ultrafilter axiom. Let D be a free set with maximum (finite) cardinality. To each finite subset F of E including D, associate as previously all possible partitions of F yielding for each i in D a totally ordered restriction of A/F containing i as an element. Consider each partition as an equivalence relation, and call U_F the set of these equivalence relations verifying our proposition: it suffices to apply the coherence lemma 2.4.1.

Note that the proposition does not extend to the case of an arbitrary finitely free partial ordering. For example, take A to be the direct product of ω_1 with itself, which is a well partial ordering by 4.8.1 proposition (2) (uses denumerable subset axiom). Yet every decomposition into a sum of chains consists of ω_1 many chains (PERLES 1963).

Chapter 5

Embeddability between relations, posets, chains; Suslin chain and tree, universal class

5.1 Embeddability, immediate extension (Hagendorf)

5.1.1 Embeddability, equimorphism

Let R, S be two relations of the same arity. We say that R is **embeddable** in S or is **smaller than** S **under embeddability**, or that S **admits an embedding** of R or is **greater** than R, iff there exists a restriction of S isomorphic with R; we write $R \leq S$ or $S \geq R$.

We say that R is **strictly embeddable** in S or **strictly smaller** than S, or that S **admits a strict embedding** of R or is **strictly greater** than R, denoted by $R < S$ or $S > R$, iff $R \leq S$ but $S \nleq R$.

We say that R is **equimorphic** with S, denoted $R \simeq S$, iff $R \leq S$ and $S \leq R$. The comparison relation $\leq$ is reflexive and transitive, hence defines a quasi-ordering on each set of relations. Equimorphism is symmetric and hence defines an equivalence relation. Embeddability is not antisymmetric, even up to isomorphism: see the following examples.

Let $\mathbf{Q}$ be the chain of the rationals, and $\mathbf{Q}+1$ the extension obtained by adding a last element: then $\mathbf{Q} \simeq \mathbf{Q}+1$.

Let ω be the chain of the natural numbers, and ω^- the converse chain (for instance the chain of the negative integers). Then $\omega^-.\omega \simeq 1 + (\omega^-.\omega)$.

In the chain of natural numbers, replace each even number by $\mathbf{Z}$ (the chain of positive and negative integers) and each odd number by a finite chain. We obtain continuum many mutually non-isomorphic chains, all of which are equimorphic.

5.1.2 Bernstein-Schröder statement for equimorphism

Let R, S be two equimorphic relations. Then there exists a partition of the base $|R|$ into two disjoint subsets D, D' and a partition of $|S|$ into two disjoint subsets E, E' with R/D isomorphic with S/E and R/D' isomorphic with S/E'. Repeat the proof of BERNSTEIN-SCHRÖDER's theorem where f and g become isomorphisms from one relation onto a restriction of the other.

The converse is false, even for chains. Indeed the chain ω of the natural numbers and the chain $\omega + 1 > \omega$ (with respect to embeddability) give rise to partitions satisfying the above conditions. Similarly for the incomparable chains $\omega + \omega^-$ and $\mathbf{Z} = \omega^- + \omega$.

Problem proposed by THOMASSÉ in 1995. Given two equimorphic denumerable relations R, S does it always exist a partition of the base $|R|$ into two disjoint subsets D, D' and a similar partition of $|S|$ into E, E' with the preceding isomorphisms, and moreover $R/D \simeq R$ or $R/D' \simeq R$ (and consequently a similar equimorphism for S).

Note that such a partition may coexist with another partition with strict embeddings. Take $R = \mathbf{Z}.\omega$ and $S = 1 + \mathbf{Z}.\omega$: we have an obvious partition into $\omega.2$ and $\omega.(\omega^-)$ which are strictly embeddable in R and S; nevertheless we have also a partition into 1 and $\mathbf{Z}.\omega$.

5.1.3 Immediate extension

Given a relation R, we say that S is an **immediate extension** of R iff S is an extension, and furthermore there does not exist any strictly intermediate relation T such that $R < T < S$ with respect to embeddability. We say also that S is an immediate successor of R with respect to embeddability.

For each relation R, there exists an immediate extension of R. Moreover if R has arity ≥ 1, then there exist at least two immediate extensions ([104] HAGENDORF 1977, prop. VI.5.6).

• Suppose first that R is a 0-ary relation, say $R = (E, +)$: then it suffices to replace the base E by a set with immediately greater cardinality in the weak sense (inexistence of a strictly intermediate cardinality): see 1.6.5.

Suppose that R is a unary relation. Let a be the cardinality of the set of elements giving the value (+) to R, and b the analogous cardinality for (-). Then it suffices to replace either a or b by an immediately greater cardinal.

Suppose now that R has arity $n \geq 2$. Add to the base E of R a set D disjoint with E, and define R^+ to have base $E \cup D$ with $R^+/E = R$ and R^+/D always (+) , and finally with R^+ taking the value (+) for those n-tuples containing at least one term in D. Finally choose for $d = \operatorname{Card} D$ the least aleph for which R^+ is not embeddable in R, hence $R^+ > R$. Let us prove that R^+ is an immediate extension of R; the relation R^- similarly defined by exchanging (+) and (-), being another immediate extension, obviously incomparable with R^+ with respect to embeddability.

Suppose first that d is an infinite aleph, and that there exists a strictly intermediate relation T with $R < T < R^+$. Consider T as a restriction of R^+. Then the intersection $D \cap |T|$ has cardinality d. Indeed if it had cardinality $< d$, then T would be embeddable in R: more precisely, in a restriction of R^+ to E increased with ($< d$)-many elements of D. Now partition $D \cap |T|$ into two disjoint subsets, each with cardinality d , say D' and D". Then T is isomorphic with its restriction to $|T| - D$" so that R is embeddable in this restriction, and finally R^+ is embeddable in T: contradiction.

Now it remains to consider the case where $d = \operatorname{Card} D$ is finite. We first see that $d = 1$. Indeed assume that d finite and ≥ 2. Then by hypothesis, the extension of R obtained by adding only one element u to the base $|R|$, with value (+) for all n-tuples containing u, is embeddable in R, thus equimorphic with R. Iterating this, the similar extension obtained by adding 2 elements is still equimorphic with R, and so until we add d elements: contradiction.

Now examine the case where $d = 1$. Call u the supplementary element, whose singleton constitutes D. Consider again the intermediate relation T as a restriction of R^+; obviously u belongs to the base $|T|$. We say that an element x in the base is a (+)-element (mod R) iff every n-tuple which contains x gives value (+) to R. Analogous definition for a (+)-element (mod T); in particular, u is a (+)-element (mod T). Every (+)-element (mod R) belongs to the base $|T|$ and is a (+)-element (mod T). Indeed otherwise, if x is a (+)-element (mod R) and does not belong to $|T|$, then by replacing u by x we could embed T in R: contradiction.

Either there exist Dedekind-infinitely many (+)-elements (mod R). Then R^+ is isomorphic with R: contradiction. Or the set of (+)-elements (mod R) has Dedekind-finite cardinality, say h. Then there exist at least ($h+1$)-many (+)-elements (mod T). Consider a restriction R' of T which is isomorphic with R. Since we have exactly h-many (+)-elements (mod R') and at least (h+1) - many (+)-elements (mod T), there exists at least one (+)-element (mod T), say v, which does not belong to the base $|R'|$. Then the restriction of T to the base $|R'|$ plus the element v is isomorphic with R^+: contradiction. •

5.2 Embeddability between posets; Cantorian theorem (Dilworth, Gleason)

(1) **There exist infinitely many finite posets which are mutually incomparable with respect to embeddability.**

• Let A_1 be the poset on five elements a, b, a', b', v with $a < b < v, a' < b' < v$ and incomparability elsewhere. Now more generally for each integer i, let A_i be the poset based on $2i+3$ elements $a, b, a', b', u_1, ..., u_{i-1}, v_{0,1}, v_{1,2}, ..., v_{i-1,i}$ with $a < b < v_{0,1}$ and $a' < b' < v_{i-1,i}$ and comparabilities $u_1 < v_{0,1}$ and $u_1 < v_{1,2}$ and then $u_2 < v_{1,2}$ and $u_2 < v_{2,3}$ and so until $u_{i-1} < v_{i-2,i-1}$ and $u_{i-1} < v_{i-1,i}$; and incomparability elsewhere. The posets thus defined are mutually incomparable with respect to embeddability (example communicated in 1969 by JULLIEN). •

(2) **There exists a strictly decreasing (under embeddability) ω-sequence of denumerable posets.**

• To each set I of integers, associate the poset A_I obtained as an extension of the finite posets $A_i (i \in I)$, taken to be mutually incomparable. Then take an infinite strictly decreasing sequence of sets I. •

(3) **There exist continuum many denumerable posets which are mutually incomparable under embeddability.**

• Take the preceding A_I and note that, for two sets I, J of integers neither of which includes the other, then A_I and A_J are incomparable. •

Problem posed by HAGENDORF. Existence of a strictly decreasing ω_1-sequence of denumerable posets; an affirmative answer is given by Kenneth KUNEN and Arnold MILLER, unpublished; see 5.11.2.

5.2.1 Characterization of posets, chains, trees

A binary relation A is a poset iff A does not admit an embedding of the following finite relations: The binary relation with cardinality 1 and value (-) (this ensures reflexivity); the relation always (+) with cardinality 2 (antisymmetry); the reflexive binary cycle with cardinality 3, and finally the consecutivity relation on 3 elements associated with the chain of cardinality 3: these two non-embeddabilities ensure transitivity.

A binary relation A is a chain iff A does not admit an embedding of the preceding finite relations, plus the identity relation of cardinality 2. Indeed this last condition implies that, given two elements x, y in the base $|A|$, either $x < y$ or $x > y \bmod A$.

A binary relation A is a tree iff A does not admit an embedding of the four first finite relations, plus the ordering on 3 elements a, b, c with $b < a, c < a$ and $a|b$.

5.2.2 Either ω or its converse or an identity relation

Every denumerable poset admits an embedding of the ordinal ω or its converse ω^- or the denumerable free poset (= identity relation).

In particular, every denumerable finitely free poset, hence every denumerable well partial ordering (see 4.3.2), hence every denumerable chain, admits an embedding of ω or ω^-. This is KÖNIG's lemma.

• Enumerate the elements $a_i (i$ integer) of the base, and partition the pairs of integers i, j where we assume $i < j$, into 3 colors, according to whether $a_i < a_j$ or $> a_j$ or $|a_j$ (modulo the given poset). By RAMSEY's theorem, there exists a denumerable set of integer indices in which all pairs have the same color. According to whether it is the first, second or third color, the given poset admits an embedding of ω or ω^- or a denumerable relation of identity. •

Modulo the denumerable subset axiom (1.2.6), every infinite poset admits an embedding of ω or ω^- or the denumerable identity relation.

A well-founded poset does not admit an embedding of ω^-. Conversely, a denumerable poset (or more generally a poset with well-orderable base) which does not admit an embedding of ω^-, is well-founded.

With the axiom of dependent choice, every poset which does not admit an embedding of ω^- is well-founded.

5.2.3 Cantorian theorem for posets

Theorem. Given a poset A, the partial ordering of inclusion among initial intervals of A admits a strict embedding of A ([44] DILWORTH, GLEASON 1962).

In particular, if A is a chain, then **the chain of cuts of A admits a strict embedding of A.**

• Let B denote the partial ordering of initial intervals of A (under inclusion). To see that $A \leq B$, it suffices to associate to each element x of the base $|A|$ the initial interval of elements $\leq x (\operatorname{mod} A)$.

Now suppose that $B \leq A$ and let f be an isomorphism of B into A. Some initial intervals X of A satisfy the relation $f(X) \in X$: for example the entire interval $X = |A|$. Let U be the intersection of all these intervals. We shall prove that $f(U) \in U$. Indeed if $f(U) \in U$, then the interval V of those elements $x < f(U)(\operatorname{mod} A)$ satisfies $V \subset U$ (strict inclusion). Hence by the isomorphism $f(V) < f(U)(\operatorname{mod} A)$, hence $f(V) \in V$ and so $V \supseteq U$: contradiction.

Thus $f(U) \in U$ and since U is the intersection of those X such that $f(X) \in X$, there exists one of these X, say X_0, which does not contain the element $f(U)$. Thus we have $X_0 \supseteq U$ and $\neq U$, hence by isomorphism $f(U) < f(X_0)(\operatorname{mod} A)$. Moreover $f(U) \in X_0$ and so $f(X_0) \notin X_0$: contradiction. •

5.3 Dense chain; embeddability conditions

5.3.1 Dense chain

A chain is said to be **dense** iff its base is infinite, and between any two distinct elements $x < y$ there exists an element $z : x < z < y$.

Hence between any two elements x and $y > x$ there exist infinitely many elements (in the sense of TARSKI: see 1.1.1).

• To each $z (x < z < y)$ associate the open interval $)x, z($ which by hypothesis is not empty. So that for x, y fixed and z variable there is no minimal interval $)x, z($ with respect to inclusion. •

Every denumerable dense chain without any minimum or maximum, is isomorphic with the chain Q of the rationals.

Every countable chain is embeddable in Q.

Every dense chain admits an embedding of Q (uses dependent choice; ZF suffices if the base is denumerable, or only well-orderable).

• Consider the set of finite strictly increasing sequences of elements of the chain, say $u_0 < u_1 < ... < u_h$(h integer), and the relation R which to each of these sequences with arbitrary length h associates any sequence with length $2h+1$ of the form $v_0 < u_0 < v_1 < u_1 < ... < v_h < u_h < v_{h+1}$. Then apply dependent choice to the relation R. The obtained ω-sequence yields a restriction isomorphic with $\mathbf{Q}$. •

5.3.2 Denumerable chain embedding every countable ordinal

Let A be a denumerable chain. If A admits an embedding of every countable ordinal, then A admits an embedding of the chain $\mathbf{Q}$ (uses countable axiom of choice).

• To each countable ordinal α, associate $\alpha + 1 + \alpha$ which by hypothesis can be embedded in A. Hence associate an element $a(\alpha)$ such that α is embeddable in the lower interval and in the interval above $a(\alpha)$. The chain A is supposed to be denumerable, yet there are ω_1-many countable ordinals α. So there exists at least one element a such that every countable ordinal is embeddable both below and above a (countable axiom of choice: we use the fact that every denumerable union of denumerable ordinals is a denumerable ordinal: see 1.2.5. By iteration, we obtain a restriction of A which is isomorphic with $\mathbf{Q}$. •

5.3.3 Auto-embeddability in two disjoint intervals

Let A be a non-empty chain with an initial interval and a final interval, both disjoint and in each of which A is embeddable. Then $\mathbf{Q}$ is embeddable in A (uses dependent choice; ZF suffices if A is denumerable or with well-orderable base).

• By iteration, the hypothesis allows to divide A into three disjoint intervals, in each of which A is embeddable. Hence there exists an element a such that A is embeddable both before and after a. By iteration using dependent choice, we obtain a restriction of A which is isomorphic with $\mathbf{Q}$. •

5.3.4 Countable dense set

A chain A is embeddable in the chain of reals iff there exists a countable subset of $|A|$ which is dense in A (uses countable axiom of choice).

• If A is a restriction of the chain of reals, then for each integer n take a partition of the reals into intervals of length $1/n$, and choose an element of $|A|$, if there are some, in each interval (countable axiom of choice).

Conversely, denote by D a restriction of A which is denumerable and dense in A. Using 2.6.5, there exist countably many trivial D-cuts, each having either one right bound or one left bound (or both) belonging to the base $|D|$. Moreover because the density (see 2.6.6), there exists at most one element of $|A|$ inside each D-cut. We can first embed in the chain $\mathbf{Q}$ of rationals all elements of $|D|$ plus those elements of $|A|$ which are inside trivial cuts. Finally each element which is inside a non-trivial D-cut is mapped onto an irrational number. •

5.4 Well partial ordering of finite trees (Kruskal)

Theorem. Embeddability between finite trees is a well partial ordering ([142] KRUSKAL 1960; uses dependent choice).

• We can always assume that each of the considered finite trees has a minimum element: it suffices to add a minimum, and even a new minimum if there already exists a minimum, to each finite tree; then embeddability or non-embeddability is preserved.

Suppose that embeddability between finite trees with a minimum is not a well partial ordering. Then there exists an ω-sequence of such trees, which is bad with respect to embeddability: see 4.3.2 proposition (2), countable case. Hence there exists a strongly minimal bad ω-sequence: see 4.2.8. In this sequence, the terms are mutually incomparable with respect to embeddability. Let U denote this sequence, and U_i($i$integer) each term. Finite trees with a minimum which are strictly embeddable in a U_i form a well partial ordering: see 4.4.3 (dependent choice).

Let H denote this countable set (up to isomorphism) of finite trees with a minimum. Embeddability between words, or finite sequences of elements of H, is a well partial ordering: see 4.5.2 (HIGMAN). To each tree U_i associate one of the finite sequences obtained by totally ordering in an arbitrary manner the immediate successors of the minimum, then replacing each of them by the sub-tree of those greater or equal elements. Then the preceding sequence U becomes a sequence of words in H, hence is good. Moreover, if the word thus substituted for the tree U_i is embeddable in the word substituted for $U_j (i, j$ integers), then the tree U_i itself is embeddable in U_j. Hence the sequence U is good: contradiction. •

5.5 Decreasing sequences and sets of incomparable chains of reals: Dushnik, Miller, Sierpinski

5.5.1 Isomorphic restrictions of R

Let A, B be two chains, each of which is embeddable in the reals. Then there are at most continuum many restrictions of B isomorphic with A.

• Consider the base $|B|$ as a subset of the set E of reals, and let F be a subset of $|B|$ such that B/F is isomorphic with A (if there exists such). For every subset X of $|B|$ such that B/X is isomorphic with A, there exists a strictly increasing map f_X from F onto X, hence from F into E. Now there are continuum many strictly increasing maps from F into E: see 2.1.4; and finally $X = \overline{f_X}(F)$ is determined by f_X (notation $\overline{f}$ in 1.1.2). •

5.5.2 Strictly decreasing sequence from R

Let A be a chain of continuum cardinality, which is embeddable in the chain of the reals.

(1) **There exists a strictly smaller (with respect to embeddability) restriction of A which has continuum cardinality** ([45] DUSHNIK, MILLER 1940; uses the axiom of choice).

(2) Even stronger, **there exists a restriction B of A with continuum cardinality, such that no interval of A, with continuum cardinality, is embeddable in B** (HAGENDORF 1977, published in ToR-86 p. 147).

(3) **For every denumerable chain U we have $A >$ or $|B.U$**, where B satisfies the preceding (2).

• (2) Let X be any subset of the base $|A|$ such that A/X is isomorphic with an interval of A of continuum cardinality. For each such interval, by the preceding proposition, there are at most continuum many corresponding sets X. Moreover, if we consider A as a restriction of the chain $\mathbf{R}$ of reals, each interval of A is the restriction to $|A|$ of an interval of $\mathbf{R}$, which is itself defined by its endpoints. Consequently there are continuum many such intervals. Finally, the set of all the X has at most continuum cardinality.

Apply 2.3.3 (axiom of choice). There exists a set C included in $|A|$, where C and $D = |A| - C$ both are equipotent with the continuum, as well as all the intersections $C \cap X$ and $D \cap X$. Consequently no X is included in C; moreover the restriction $B = A/C$ admits no embedding of any interval of A which has continuum cardinality. •

• (3) Suppose $A \leq B.U$. Then the base $|A|$ is partitioned into countably many intervals, each corresponding with an element of U. At least one of these intervals is equipotent with the continuum: see 1.5.3 (axiom of choice). Moreover it is embeddable in B, contradicting the preceding (2). •

Note that (1) implies the existence, starting from the chain of the reals, of a strictly decreasing ω-sequence of chains. We shall see that such a sequence does not exist for scattered chains, i.e. those in which the chain $\mathbf{Q}$ of the rationals is not embeddable: see ch.7.

5.5.3 Separation by injective functions

(1) **Let E be an infinite set, I a set of injective functions $f_i (i \in I)$ from E into E where I is subpotent to E, and each f_i satisfies $f_i(x) \neq x$ on a subset of E equipotent with E. Then there exists a subset C of E, equipotent with E, with the non-inclusion $\overline{f_i}(C) \not\subseteq C$ for each i** (notation $\overline{f}$ from 1.1.2).

(2) **Under the same hypotheses, there exist two subsets C, D of E which are disjoint and equipotent with E, with the non-inclusion $\overline{f_i}(C) \not\subseteq C \cup D$** and similarly by interchanging C and D (uses axiom of choice; [229] SIERPINSKI 1950; see also [217] ROSENSTEIN 1982).

• (1) Well-order E by its cardinality α; well-order I first by its cardinality, then with repetitions according to the ordinal α, in the case where $\mathrm{Card}\, I < \mathrm{Card}\, E$. Put the first x_0 of E satisfying $f_0(x_0) \neq x_0$ into C, and put $x'_0 = f_0(x_0)$ into $E - C$. In general, for each $i < \alpha$, let x_i be the first element of E such that $f_i(x_i) \neq x_i$ and where x_i and $x'_i = f_i(x_i)$ are distinct from all x_j and $x'_j (j < i)$. Such an x_i exists, because the number of $j < i$ is strictly less than the cardinal α, and there are $(\mathrm{Card}\, \alpha)$-many x which are different from $f_i(x)$, as f_i is injective. Finally put x_i into C and x'_i into $E - C$. •

• (2) The preceding proof modified as follows. Firstly, put the first $x_0 \neq f_0(x_0)$ into C and $x'_0 = f_0(x_0)$ into $E-(C \cup D)$. Secondly, put the first element $y_0 \neq f_0(y_0)$ where y_0 and $f_0(y_0)$ are both distinct from x_0 and x'_0, into D, and $y'_0 = f_0(y_0)$ into $E - (C \cup D)$. In general, for each $i < \alpha$, firstly find the first element $x_i \neq f_i(x_i)$ where both x_i and $x'_i = f_i(x_i)$ are distinct from all $x_j, x'_j, y_j, y'_j (j < i)$. Then put x_i into C and x'_i into $E - (C \cup D)$. Secondly, consider the first $y_i \neq f_i(y_i)$ with y_i and $f_i(y_i)$ distinct from all $x_j, x'_j, y_j, y'_j (j < i)$ as well as from x_i, x'_i. Then put y_i into D, and put $y'_i = f_i(y_i)$ into $E - (C \cup D)$. •

5.5.4 Strictly decreasing sequence indexed by the continuum; mutually incomparable chains

Let E be the set of reals and $\mathbf{R}$ their usual chain. There exist two subsets C, D of E which are disjoint, equipotent with the continuum, dense in $\mathbf{R}$ and such that, for each subset X of D, every restriction of $\mathbf{R}$ isomorphic with $\mathbf{R}/(C \cup X)$, either has base $C \cup X$, or its base contains at least one element of $E - (C \cup D)$.

Consequently **for $Y \subset X \subseteq D$, we have the strict embeddability $\mathbf{R}/(C \cup Y) < \mathbf{R}/(C \cup X)$.**

For X and Y subsets of D which are incomparable with respect to inclusion, the preceding two restrictions are incomparable with respect to embeddability ([229] SIERPINSKI 1950; see also [217] ROSENSTEIN 1982; uses axiom of choice).

• Take the sets C, D in 5.5.3, proposition (2) (axiom of choice), where the f_i designate all isomorphisms from $\mathbf{R}$ into itself, distinct from the identity. For such an isomorphism f, if a real x is mapped to $fx \neq x$, for example if $fx > x (\mathrm{mod}\, \mathbf{R})$, then every real in the interval (x, fx) is mapped to a strictly greater real, hence $fy \neq y$ for continuum many reals y. Note that C and D are disjoint and each equipotent with the continuum. Moreover by the same proposition we have $\overline{f}(C) \neq C$ for each considered isomorphism f; similarly with D. Thus by 2.1.4 proposition (2), the sets C and D are both dense in $\mathbf{R}$.

Take an arbitrary subset X of D and an isomorphism g from $\mathbf{R}/(C \cup X)$ into $\mathbf{R}$, which is distinct from the identity. Since C, hence $C \cup X$, is dense, there exists an isomorphism g^+ from $\mathbf{R}$ into itself, which extends g to the domain E of all reals: see 2.1.3. Hence g^+ is one of the previously considered isomorphisms f. By 5.5.3, the set of images $\overline{g}(C) = \overline{g^+}(C)$ is not included in $C \cup D$. •

This immediately implies **the existence of a strictly decreasing sequence (w. r. to embeddability) indexed by the continuum, of chains of reals.**

Also the existence of a set, equipotent with the continuum, of mutually incomparable chains of reals ([229] SIERPINSKI 1950).

5.6 Immediate extension between chains (Hagendorf)

An ordinal U is indecomposable iff every ordinal $V < U$ satisfies $V.2 < U$ (indecomposable ordinal is defined in 1.3.6).

• Let U be an indecomposable ordinal and $V < U$. Then $V + U = U$ hence $V.2 < U$. Conversely, suppose U decomposable: there exist $V < U$ and $W < U$ with $V + W = U$. Let $T = \text{Max}(V, W)$; then $T < U$ and $T.2 \geq V + W = U$. •

For an analogous statement concerning chains, see below 6.10.2.

5.6.1 Condition of indecomposability

Let A, A' be two chains and U the least ordinal such that $A + U + A' \not\leq A + A'$; then U is indecomposable ([102] HAGENDORF 1971).

• Let $V < U$, so that $A + V + A' \simeq A + A'$ (equimorphism defined in 5.1.1). Take an isomorphism from $A + V + A'$ into $A + A'$. Let $V = W + W'$ with W embeddable in A and W' embeddable in A'. In order to fix our ideas, suppose that $W \leq W'$. Then $W' + A' \leq A'$ so $(W'.2) + A' \leq A'$ and so $(W'.4) + A' \leq A'$, hence $(V.2) + A' \leq A'$ and so $A + (V.2) + A' \leq A + A'$ and finally $V.2 < U$. Argue analogously to obtain the same conclusion if $W \geq W'$. Thus U is indecomposable, by the preceding proposition. •

5.6.2 Immediate extension of a chain

We say that a chain B is an **immediate extension** of A (up to isomorphism) or is **immediately greater** than A (with respect to embeddability) iff $B > A$ and there does not exist any chain X satisfying $A < X < B$.

Let A be a chain and U the least ordinal such that $A + U \not\leq A$. Then $A + U$ is an immediate extension of A ([103] HAGENDORF 1972).

Note that the proposition no linger holds if we replace U by a retro-ordinal. For example, let $A = \omega^-$, the converse of ω. Then we find $U = \omega^-$; yet between A and $A + U = \omega^-.2$ we have $1 + \omega^-, 2 + \omega^-$ etc.

• The ordinal U of the proposition is non-zero and by the preceding subsection it is indecomposable. In order to obtain a contradiction, let us suppose that there exists a chain B satisfying $A < B < A + U$. Consider an isomorphism of B onto a restriction of $A + U$, which thus decomposes B into an initial interval B' and a final interval $B"$ such that $B' \leq A$ and $B" \leq U$. The chain $B"$ is an ordinal. If $B" < U$ it follows that $B = B' + B" \leq A + B" \leq A$ thus contradicting the definition of B. Hence $B" = U$ and $B = B' + U$ (by $=$ we mean "isomorphic to").

Consider an isomorphism from A onto a restriction of $B = B' + U$, which decomposes A into an initial interval A' and a complementary final interval $A"$ such that $A' \leq B'$ and $A" \leq U$. The chain $A"$ is an ordinal. If $A" < U$, since U is indecomposable we have $A" + U = U$ so $A + U = A' + A" + U = A' + U$ hence $A + U \leq B' + U = B$, thus contradicting the definition of B. Hence $A" = U$ and $A = A' + U$.

Then either U is an infinite indecomposable ordinal, in which case $1 < U$ so $A+1 \simeq A$ (equimorphy). The preceding argument, with A replaced by $A+1$, proves the existence of a final interval of $A+1$ which is isomorphic with U: contradiction.

Or $U = 1$ and so $A = A'+1$ and $B = B'+1$, hence $A'+1 < B'+1 < A'+2$, so that $A' < B' < A' + 1$. Iterate the preceding argument: we see that A is the sum of a certain initial interval and of the final interval ω^-. Hence $A + 1$ is isomorphic with A, which contradicts the hypothesis that $U = 1$. •

5.6.3 Existence of two immediately greater chains

Every infinite chain admits at least two immediately greater chains (with respect to embeddability) which are mutually incomparable ([103] HAGENDORF 1972).

• Let A be an infinite chain and U the least ordinal such that $A + U$ is not embeddable in A. Similarly let U' be the least retro-ordinal such that $U' + A$ is not embeddable in A. By the preceding statement, these sums are immediately greater than A. It remains to prove that they are mutually incomparable with respect to embeddability. We shall argue ad absurdum: to fix the ideas, suppose $U' + A \leq A + U$. Take an isomorphism from $U' + A$ onto a restriction of $A + U$. This isomorphism cannot embed $U' + A$ in A, nor in $A + V$ for any $V < U$. Thus there exists a final interval of A which is cofinally embeddable in U. This final interval is isomorphic with U: otherwise there would exist $V < U$ such that $U' + A \leq A + V \leq A$, which contradicts the definition of U'. Finally by 5.6.1, the ordinal U is indecomposable.

If U is an infinite indecomposable ordinal, then $1 < U$ so $A + 1 \simeq A$. The preceding argument, where A is replaced by $A + 1$, proves the existence of a final interval of $A + 1$ which is isomorphic with U: contradiction. Hence $U = 1$ and $U' + A \leq A + 1$. Thus every isomorphism of $U' + A$ into $A + 1$ has in its range the maximum element of $A + 1$; for otherwise $U' + A$ would be embeddable in A. Thus A has a maximum. By iteration, we see that A has a final interval which is isomorphic with the retro-ordinal ω^-. Thus $A + 1$ is isomorphic with A, contradicting the inequality $A + 1 = A + U > A$. •

5.7 Dense chain for an infinite cardinal

Given an infinite cardinal a, we say that a chain is a-**dense** iff its base is infinite and between any two distinct elements, the interval has cardinality equal to a. Analogous definition for a chain ($\geq a$)-**dense**.

A dense chain, in the sense of 5.3.1, is $(\geq \omega)$-dense (uses denumerable subset axiom).

5.7.1 Existence

Given an infinite cardinal a, there exists a chain with cardinal a which is a-dense (uses axiom of choice).

• Start with the ordinal product $\mathbf{Q}.a$ where $\mathbf{Q}$ is the chain of the rationals, and a is an ordinal, more precisely an aleph (recall that, with choice axiom, every cardinal is an aleph). The chain $\mathbf{Q}.a$ has cardinality a and is obviously dense without any minimum or maximum. Also the set of finite sequences of elements in $\mathbf{Q}.a$ has cardinality a. We order this set lexicographically.

Now it suffices to see that it is an a-dense ordering. Indeed, consider two finite sequences u and v strictly after u with respect to lexicographical comparison. Denote by u_i and v_i(i integer) the i-th terms. Either there exists a least index i with $u_i < v_i (\bmod \mathbf{Q}.a)$; then we take w_i which is strictly between u_i and v_i, and we have a-many finite sequences beginning with $u_0, ..., u_{i-1}, w_i$: they all are situated between u and v. Or u is an initial interval of v, and we have the same conclusion by taking sequences beginning with u followed by an element $w_h < v_h (h =$ length of u). •

5.7.2 Either ω_α or its converse or an ω_α-dense chain

Let ω_α be an infinite regular aleph. Every chain with cardinality ω_α admits an embedding either of the ordinal ω_α, or its converse $(\omega_\alpha)^-$, or an embedding of an ω_α-dense chain.

• Let A be a chain with cardinality ω_α, in which neither ω_α nor its converse is embeddable. Let x, y be elements of the base $|A|$; we put x equivalent with y iff the interval between them has cardinality $< \omega_\alpha$. The equivalence classes are intervals of A.

Every equivalence class has cardinality $< \omega_\alpha$. Indeed let T be an equivalence class. Take an ordinal indexed sequence of elements of T which is strictly increasing (mod A) and cofinal in T. This sequence has length strictly less than ω_α, since ω_α is not embeddable in A. The subset of T above the first term of the sequence, is a union of $(< \omega_\alpha)$- many intervals, each with cardinality $< \omega_\alpha$. By regularity of ω_α this union has cardinality strictly less than ω_α. Same argument for the subset of T before the first term of the sequence, by constructing a decreasing ordinal-indexed sequence.

Since the equivalence classes all have cardinality strictly less than ω_α and ω_α is regular, there exist ω_α-many equivalence classes. Take a representative element from each class. Again by regularity, between two representatives there necessarily exist ω_α-many equivalence classes, hence as many representatives. Finally, the restriction of A to the set of representatives is an ω_α-dense chain.•

The proposition is no longer true for a singular aleph.

• Take the sum $\omega^- + \omega_1^- + ... + \omega_i^- + ..$ where i is an arbitrary integer and ω_i^- designates the retro-ordinal, converse of ω_i. Our sum is a chain with cardinality ω_ω. Firstly, the only well-ordered restrictions of our sum are the finite chains and ω. Secondly, no dense chain is embeddable in the sum. Thirdly, a retro-ordinal is embeddable in the sum only if it is embeddable in one of the ω_i^-. •

5.7.3 Embedding and regular alephs

Theorem (1) **Let ω_α be a regular limit aleph. Then each ω_α-dense chain admits either an embedding of ω_α or its converse, or an embedding of all ordinals strictly less than ω_α and their converses.**

(2) **Let ω_α be a regular aleph. Then each $\omega_{\alpha+1}$-dense chain admits either an embedding of $\omega_{\alpha+1}$ or its converse, or an embedding of all ordinals equipotent with ω_α and their converses** (ERDÖS, RADO 1953; (1) and (2) use the generalized continuum hypothesis; for $\alpha = 0$, ZF plus choice suffices; for $\alpha = 1$, ZF plus choice plus continuum hypothesis suffices).

• (1) Let A be an ω_α-dense chain. By restricting to one of its intervals, we can assume A to have cardinality ω_α. Let C denote a well-ordering with base $|A|$ and order-type ω_α. Partition the pairs $\{x, y\}$ in the base into two colors: assuming that $x < y (\text{mod}\, C)$, the pair will have color (+) if $x < y (\text{mod}\, A)$ and (-) if $x > y (\text{mod}\, A)$. By ERDÖS-RADO partition theorem 3.3.5 proposition (1) (generalized continuum hypothesis), either there exists a (-)-monochromatic subset equipotent with the base: thus A admits an embedding of the retro-ordinal ω_α^-

Or for each cardinal $b < \omega_\alpha$, there exists a (+)-monochromatic subset with cardinality b thus A admits an embedding of every ordinal strictly less than ω_α . The proof ends by interchanging the colors. •

• (2) Let A be an $\omega_{\alpha+1}$-dense chain. Since ω_α is regular by hypothesis, the least cardinal b satisfying ${}^b(\omega_\alpha) > \omega_\alpha$ is $b = \omega_\alpha$: see 2.8.5 proposition (1) using generalized continuum hypothesis. In order to prove (2), we now denote by A those $\omega_{\alpha+1}$-dense chains which embed neither the ordinal $\omega_{\alpha+1}$ nor its converse.

Partition the pairs of elements of $|A|$ into two colors, exactly as in (1) above. By ERDÖS-RADO partition theorem proposition (2), either there exists a (-)-monochromatic subset equipotent with the base: thus A admits an embedding of $\omega_{\alpha+1}$. Or a (+)-monochromatic subset with cardinality ω_α: thus A admits an embedding of ω_α. Hence in every case A admits an embedding of the ordinal ω_α and its converse.

Let $a_0, a_1, ..., a_i, ..(i < \omega_\alpha$) be a strictly increasing ω_α-sequence (mod A). Let c be the least ordinal for which there exists an $\omega_{\alpha+1}$-dense chain which admits no embedding of c. By the preceding we have $c \geq \omega_\alpha$.

Assume, in order to obtain a contradiction, that c has cardinality ω_α. Then c is the limit of a strictly increasing sequence of ordinals $c_i < c$, indexed by i running through at most ω_α. Each c_i is embeddable in every $\omega_{\alpha+1}$-dense chain. For each i, the ordinal, or the aleph equal to c_i, is embeddable in the interval (a_i, a_{i+1}). Hence the sum of the c_i is embeddable in A and so c is embeddable in A. This contradiction shows that A admits an embedding of every ordinal equipotent with ω_α. The same argument proves that A admits as well an embedding of the corresponding retro-ordinals. •

Corollary. Any chain of cardinal ω_2 admits an embedding either of every ordinal, or every retro-ordinal equipotent with ω_1 (axiom of choice plus continuum hypothesis). Consequence of 5.7.2 (with $\alpha = 2$) and the previous statement alinea (2) (with $\alpha = 1$).

Note that we find again the Ramsey value $(\omega_1)^2_2 = \omega_2$ (see 3.3.4).

5.8 Suslin chain and Suslin tree

Given a chain A, we already defined a set dense in A: see 2.6.6.

The chain of reals, and more generally any chain A which is embeddable in the chain of reals, satisfies the two following conditions.

(1) **There exists a countable set which is dense in A;**

(2) **Every set of mutually disjoint intervals of A, none of which are singletons, is countable.**

The condition (2) follows from (1).

• If D is countable and dense in A, then every non-singleton interval contains at least one element of D Two disjoint intervals cannot contain a same element of D, so there are countably many intervals. •

Suslin's hypothesis ([239] SUSLIN 1920). The axiom called "Suslin's hypothesis" asserts that the preceding condition (2) implies (1), hence that (1) and (2) are equivalent.

This axiom is neither provable nor refutable in ZF, even with the axiom of choice and even with the generalized continuum hypothesis. More precisely JECH and TENNENBAUM have proved the consistency of the existence of a Suslin tree (i.e. the negation of the axiom) with ZF (modulo the consistency of ZF). Whereas SOLOVAY and TENNENBAUM have proved the relative consistency of the axiom: see [126] JECH 1978. For a detailed discussion of Suslin chains and Suslin trees, as well as for the advanced results of JENSEN, see for example [42] DEVLIN, JOHNSBRĂTEN 1974.

5.8.1 Suslin chain

It is more convenient to work with the negation of SUSLIN's hypothesis, rather than the hypothesis itself. We say that **a chain is Suslin** iff it satisfies (2) and not (1), i.e. **iff every set of non-singleton mutually disjoint intervals is countable, yet there exists no countable set which is dense in the chain.**

A Suslin chain is uncountable; moreover it admits an embedding of the chain of rationals (uses axiom of choice).

• The inexistence of any countable dense set implies that the chain itself be uncountable: its cardinality is at least ω_1 (axiom of choice). If the chain of rationals is not embeddable in it, then either the ordinal ω_1 or its converse is embeddable in it: see 5.7.2 with $\alpha = 1$. Hence there exist uncountably many non-singleton mutually disjoint intervals. •

Every Suslin chain has cardinality exactly ω_1 (uses axiom of choice and continuum hypothesis).

• Let A be a Suslin chain; we already know that A is uncountably infinite, so has cardinality $\geq \omega_1$ (axiom of choice). Suppose that A has cardinality at least ω_2. Then by 5.7.3, corollary, A admits an embedding either of ω_1 or its converse (choice plus continuum hypothesis). Hence there exist ω_1 many non singleton mutually disjoint intervals, which contradicts the definition of a Suslin chain. •

5.8.2 Suslin tree

We say that **a tree is Suslin** iff it has cardinality ω_1 and every chain (or totally ordered restriction) and every antichain is countable.

The existence of a Suslin chain of cardinality ω_1 **implies the existence of a Suslin tree** (the additional assumption of cardinality ω_1 allows us to avoid using the axiom of choice: ZF suffices).

• Let A be a Suslin chain and remember that it has cardinality ω_1. To each countable ordinal i, associate an interval A_i defined by its two endpoints $u_i < v_i (\mathrm{mod}\, A)$, where all u_i and v_i are distinct. To do this, begin with $A_0 = (u_0, v_0)$ an arbitrary interval. Let i be a non-zero countable ordinal, and suppose that the A_j for $j < i$ have already been defined so that they are mutually either disjoint or one included in the other. The set of endpoints $u_j, v_j (j < i)$ is countable: by hypothesis it is not dense in A, hence there exist two elements u, v in the base $|A|$ between which there is neither an initial endpoint u_j nor a final endpoint $v_j (j < i)$. Set $A_i = (u, v)$ so that $u_i = u$ and $v_i = v$: this interval is either disjoint or included in each $A_j (j < i)$.

The set of intervals A_i thus obtained has cardinality ω_1. Reverse inclusion defines a tree on the set of these A_i. Every antichains, i.e. every set of intervals A_i which are mutually disjoint, is countable. Finally a chain, or set of intervals A_i which are mutually comparable with respect to inclusion, is well-ordered by the ordinal indices, with $A_j \subset A_i$ for every pair of countable ordinals $i, j (i < j)$. Such a chain is countable; for if it had cardinality ω_1 , then using the endpoints of preceding intervals, we could obtain ω_1 many mutually disjoint intervals. •

5.8.3 Suslin tree and Suslin chain: the equivalence

Theorem. The existence of a Suslin tree implies, and hence is equivalent with the existence of a Suslin chain (we use the regularity of ω_1, thus for example the countable axiom of choice).

• Let A be a Suslin tree; we can assume that A is a well-founded poset, if necessary by replacing A by a cofinal well-founded restriction. To see this, apply 2.7.2, corollary. Note that, by hypothesis, the base $|A|$ is well-orderable with cardinality ω_1. More precisely let $u_i (i < \omega_1)$ be an indexation of the base; then remove each u_j for which there exists an $i < j$ with $u_i > u_j (\mathrm{mod}\, A)$.

To see that what remains is still a Suslin tree, note that for each index i, the removed elements u_j are all $< u_i (\mathrm{mod}\, A)$. Since A is a tree, then they are totally ordered; since A is Suslin, then there are countably many such; thus there remain ω_1 many elements.

We can further assume that every non-empty final interval of A has cardinality ω_1. To see this, let x be an element of the base which has only countably many successors (mod A). Those x of minimal height are mutually incomparable, and so there are countably many such: it suffices to remove these x and their successors.

Neither A nor any final interval of A is finitely free. For otherwise, by 4.3.1 proposition (1) (using the regularity of ω_1), there would exist a totally ordered restriction of A with cardinality ω_1, contradicting our hypotheses.

We can assume that A has a minimum, whose singleton will be denoted by E_0. For each countable non-zero ordinal i, let E_i be the denumerable set of elements of height i: we require that this set be infinite. For each element x of E_i we require that there exist denumerably many elements which are immediate successors of $x (\mathrm{mod}\, A)$. This set is denoted $E_{i+1,x}$ and the union of these sets must be E_{i+1}. For each countable limit ordinal i and each x of height $< i$, we require that there exist infinitely many elements in E_i which are $> x$ (mod A). Finally, for each countable limit ordinal i and each totally ordered restriction X of A containing elements of all heights $< i$, we require that there exist either a unique element of E_i which is a successor of all elements of X, or none such.

These requirements are easy to satisfy. For example for the minimum, take an arbitrary element having ω_1-many successors. Then having obtained E_i, the set of elements x with height i, note that for each $x \in E_i$, the final interval $> x$ has cardinality ω_1 and is not finitely free. Hence take a denumerable free subset as $E_{i+1,x}$ and among the successors of x, retain only those which are identical to or successors of an element of $E_{i+1,x}$. Finally, for each countable limit ordinal i and each chain X containing elements of all heights $< i$, if there exist elements above X, then decide to retain one such plus the ω_1 many successors of this element.

For each element x of the base $|A|$ with height i, totally order the denumerable set $E_{i+1,x}$ with the order type of a dense chain $C_{i+1,x}$. Now consider the set of all maximal totally ordered restrictions, or maximal chains of A. This set is totally ordered by the preceding dense chains. Indeed, given two distinct maximal chains U and V: none of the two bases is included in the other. Moreover there exists a least element u among those elements of $|U|$ which do not belong to $|V|$ and a least element v among those elements of $|V|$ which do not belong to $|U|$. By the preceding, there exists a last element x whose height will be denoted i, common to both bases $|U|$ and $|V|$ and having u and v as immediate successors. Let $U < V$ iff $u < v (\mathrm{mod}\, C_{i+1,x})$: this totally orders the set of maximal chains.

Let H be the chain thus obtained. We shall prove that H is Suslin. First of all a set D which is dense in H cannot be countable. For if it were, then take an ordinal k greater than the heights of all elements of those maximal chains constituting D. Then for an arbitrary element z with height k, the interval of maximal chains passing through z does not contain any element of D.

Now suppose that there exist w_1 many non-singleton mutually disjoint intervals of H. In each interval take two elements, or maximal chains U and V. As before, take an element x whose height is denoted i, and the elements u, v immediate successors of $x(\bmod A)$; and take w between u and v modulo the chain $C_{i+1,x}$. Then these w thus associated with our disjoint intervals of H, are mutually incomparable $(\bmod A)$; they must be countably many: contradiction. •

5.9 Aronszajn tree, Specker chain

5.9.1 Construction of an Aronszajn tree

An **Aronszajn tree** is a well-founded tree of cardinality ω_1 whose chains and height levels are countable. It is not required that every antichain be countable. Hence every well-founded Suslin tree is an Aronszajn tree; but the converse possibly depends on set-theoretic axioms.

The following construction of an Aronszajn tree, using ZF plus choice, goes back to [144] KUREPA 1935 p. 96, citing a letter from ARONSZAJN in 1934.

The elements of the tree will be ordinal sequences of integers $a_i (i < \alpha)$, without repetition, where α varies over all countable ordinals. We say that such a sequence u **precedes** v or that v **follows** u iff u is an initial interval of v. Moreover we require the following conditions of convergence and denumerability.

Convergence. For each sequence $a_i (i < \alpha)$, the sum of the inverses $1/a_i$ is finite. Furthermore for each sequence u with length α, each countable ordinal β and each positive real number r, there must exist, in our set, a sequence with length $\alpha + \beta$, following u, and such that the sum of the inverses $1/a_i$ for $a \leq i < \beta$ be less than r.

Denumerability. For each non-zero countable ordinal α, we take in our set denumerably many α-sequences (uses axiom of choice). Using the preceding convergence condition, we see that, given a limit countable ordinal α, we can retain only α-sequences u such that, for each $i < \alpha$, the initial interval of u with length i already belongs to our set; more exactly, we retain denumerably many such α-sequences.

Finally for each sequence u in our set, every initial interval of u belong to our set.

These conditions allow us, for each α, to construct a "definitive" denumerable set of α-sequences which will not be unduly increased by the ulterior construction of longer sequences.

Because the non-repetition of values a_i, every totally ordered restriction, or chain, of the preceding Aronszajn tree, is countable.

In the preceding tree, for each countable ordinal α and each α-sequence u, there exist ω_1-many sequences which are successors of u. However this is not true for all Aronszajn trees; for instance we can add to the preceding tree new ordinal sequences without successors, or with countably many successors.

5.9.2 Specker chain

This is an uncountable chain A such that neither ω_1 nor its converse are embeddable in A; moreover any chain which is embeddable both in A and in the chain of reals, is countable.

A Specker chain has no countable dense subset. For if so, by 5.3.4 above, it would be embeddable in the chain of reals as well as in itself, and consequently it would be countable.

Let A be an uncountable chain such that neither ω_1 nor its converse are embedded in A; then A has cardinality ω_1 (uses choice plus continuum hypothesis).

In particular **every Specker chain has cardinality ω_1** (same conditions).

• Suppose our conclusion is false: then A has cardinality $\geq \omega_2$ (axiom of choice). By the corollary in 5.7.3 (choice plus continuum hypothesis), A admits an embedding either of ω_1 or its converse: contradiction. •

Say that a chain A is **weakly Specker** iff

(i) neither ω_1 nor its converse are embedded in A; and

(ii) A has no countable dense subset.

Every Specker chain is weakly Specker. However take a specker chain A and consider the ordinal sum $A + \mathbf{R}$ where $\mathbf{R}$ is the chain of reals. Then this sum is weakly specker yet not specker, since $\mathbf{R}$ is embeddable in $A + \mathbf{R}$ and in itself without being countable.

Note that **any Suslin chain is weakly Specker**.

5.9.3 Construction of a Specker chain

Start with the Aronszajn tree of 5.9.1, defined on a set of ordinal sequences of integers, without repetition. Apply 2.11.3 (using ultrafilter axiom) which associates to the given tree A, a chain C with the same base. For each element u (ordinal sequence of integers without repetition), the final interval $\geq u (\operatorname{mod} A)$ becomes an interval $(\operatorname{mod} C)$ with minimum u.

We show first of all that **the chain C thus defined does not admit an embedding of ω_1 nor of its converse**.

• By 2.11.3, C can be decomposed in a denumerable sum of disjoint intervals, each corresponding to an integer a, and formed from some sequences which begin with a. Consequently, given a strictly increasing $(\operatorname{mod} C)$ ω_1-sequence whose terms are denoted u_i(i countable ordinal), there exists an integer a_1 such that, from a certain ordinal on, the sequence u_i begins with a_1. By iteration, for every countable ordinal k, there exists a k-sequence of integers $a_1, a_2, ..., a_j, ..(j < k)$ such that, from a certain ordinal index on, the sequence u_i begins with the above k-sequence. Finally these $a_1, a_2, ..., a_j, ..$ form a strictly increasing ω_1-sequence of integers without repetition: contradiction. •

Let us now show that **if H is an uncountable chain which has a denumerable restriction D dense in H, then H is not embeddable in C.**

• Since D is denumerable, it is formed of ordinal sequences of integers without repetition, all of length less than a certain countable ordinal. Let a be a countable ordinal strictly greater than all these lengths. Since H is uncountable, there exists at least one sequence u with length a having ω_1-many sequences in $|H|$ which extend u. By 2.11.3, there exists an interval $(\bmod C)$ formed of all those elements, or sequences extending u. Thus any two of them are not separated $(\bmod C)$ by any sequence belonging to $|D|$: this contradicts the density of D in H. •

5.10 Universal class (Tarski, Vaught)

5.10.1 Definition; bound; examples

Given an arbitrary finite set U of finite relations of given arity, the **universal class** defined by U is formed of those relations (up to isomorphism) which do not admit an embedding of any element of U.

A universal class is closed under restriction and isomorphism, hence under embeddability. We can always suppose that all elements of U are mutually incomparable under embeddability: it suffices to remove each relation in U which has a proper restriction belonging to U.

A finite relation A with a non-empty base is said to be a **bound** of the universal class U iff A does not belong to U yet every proper restriction belongs to U. By the previous removing, we reduce U to the bounds and we see that **a universal class is completely defined by the set of its bounds**.

If we consider the partial ordering of embeddability between finite relations (considered up to isomorphism), then a universal class, reduced to its finite relations, constitutes an initial interval; then we find the bounds of this initial interval, in the sense of 4.10.

Note that a universal class is completely defined by the set of finite relations (up to isomorphism) which belong to it.

Let us mention the original, logical definition due to TARSKI: there exists a prenex formula (in the first order predicate calculus) having only universal quantifiers, and taking the value (+) only for the relations belonging to the class. Note that the minimal possible number of quantifiers is equal to the maximum cardinality of bounds.

The reader will see the equivalence between our definition and the following **characterization of universal classes by** [248] VAUGHT 1953:

U is universal iff there exists an integer p (in fact p is the maximum cardinality of bounds) such that:

if a relation R belongs to U and S is a relation all of whose restrictions to at most p elements are embeddable in R, then S itself belongs to U.

Examples of universal classes. The reflexive binary relations constitute a universal class defined with one universal quantifier by the formula $\forall_x \rho xx$ where ρ is a binary predicate, substituable by any binary relation R which gives to the formula either the value (+) or (-), according to whether R is reflexive or not.

Equivalently the universal class of reflexive binary relations is defined by a unique bound, which is the binary relation based on one element and taking value (-).

The symmetric binary relations constitute a universal class defined with 2 universal quantifiers by the formula $\forall_{x,y} \rho xy \iff \rho yx$. Equivalently this class is defined by the following bounds based on 2 elements: the chain of cardinality 2, arbitrarily modified on its diagonal where it takes indifferently value (+) or (-).

The antisymmetric binary relations constitute a universal class defined by $\forall_{x,y}(\rho xy \wedge \rho yx) \Rightarrow (x = y)$. Equivalently this class is defined by the following bounds based on 2 elements: the identity relation (=) and the difference relation ($\neq$) arbitrarily modified on the diagonal.

The transitive binary relations constitute a universal class defined with 3 universal quantifiers by the formula $\forall_{x,y,z}(\rho xy \wedge \rho yz) \Rightarrow \rho xz$ where ρ is again a binary predicate.

Equivalently this class is defined by the following bounds based on at most three elements. No bound on one element: indeed the binary relations on one element with either value (+) or (-) are both transitive. The bounds on 2 elements are again the identity relation and the difference relation, arbitrarily modified on the diagonal. Finally the following bounds on 3 elements: the binary 3-cycle and the consecutivity relation associated with the chain of cardinality 3.

For the meticulous reader, note that, when extending the usual truth-value to the case where the base of relations can be empty, then the logical definition for universal classes fails. The exceptional case concerns this universal class which reduces to the singleton of the 0-ary relation (0,+) with empty base (and similarly with the value (-)). This is a case where the representing formula cannot be written under prenex form: see 10.10.4.

5.10.2 Finite intersection or union of universal classes

Every finite intersection of two universal classes (with same arity) is universal. It suffices to take the union of the sets of bounds, and then to remove those relations which have a proper restriction in this union.

Every finite union of universal classes (of a given arity) is universal.

• Let V and W be two universal classes. A finite relation X belongs to the union iff, for each bound A of V and each bound B of W, either X does not admit any embedding of A, or of B. Equivalently, for each minimal common extension C of A and B, the relation X does not admit any embedding of C. Finally note that $\mathrm{Card}\, C$ is at most equal to $\mathrm{Card}\, A + \mathrm{Card}\, B$, so that there exist only finitely many such C. •

Now we have other examples of universal classes, by taking finite union and finite intersections of preceding universal classes.

5.10.3 Universal class and finite relations

Given a finite relation R and a universal class U, **if there exist in U infinitely many extensions of R with arbitrary large finite cardinalities, then there**

exists in U a denumerable extension of R.

• We can assume that R is defined on the integers $1, ..., p$ and that, for each integer i, there exists in U an extension R_i of R based on the integers 1 to $p+i$. For infinitely many integers i, the R_i have a same restriction S_1 to $\{1, ..., p+1\}$. Among these, there are infinitely many integers i for which the R_i have a same restriction S_2 to $\{1, ..., p+2\}$. Iterating this, we thus define S_j for each integer j. It now suffices to take the common extension of the S_j, based on the set of all integers: by the preceding VAUGHT's characterization, R belongs to U. •

For further informations about universal classes, see [140] KRAUSS 1971

5.10.4 Maximal element in a universal class; Malitz' universal class

Start with a finite relation A; then the set of all restrictions of A is a universal class: denote by p the cardinal of $|A|$ and take for the bounds: each extension of A having cardinality $p+1$, plus each relation B having cardinal 1 to p, such that $B \not\leq A$ yet every proper restriction of B is embeddable in A.

In such a universal class, A is the maximum with respect to embeddability.

In the general case, **we can have infinitely many maximal elements** (which are finite relations) in our universal class. The following example is due to [166] MALITZ 1967.

Consequently, **in general a universal class is not directed with respect to embedding**; this is an important difference with ages (see 10.2.5).

• Take the base of integers from 0 to $n-1$. Let I_n be the usual chain of these integers; let C_n be the consecutivity relation ($y = x+1$); let 0_n be the unary relation called the' **singleton of zero**, i.e. the relation taking (+) for 0 and (-) elsewhere; and let U_n be the relation singleton of $n-1$. Finally let R_n be the quadrirelation $(I_n, C_n, 0_n, U_n)$. From $n = 7$ on, all the R_n have the same restrictions of cardinalities 1, 2 and 3, up to isomorphism. Let $A_1, ..., A_h$ be those quadrirelations of the same arity and cardinalities 1, 2, 3 which are not embeddable in R_7, and hence in $R_n (n \geq 7)$. We see that every extension of an R_n to a new element added to its base admits an embedding of one of the $A_1, ..., A_h$.

Using VAUGHT's characterization (5.10.1), take the universal class formed of those quadrirelations R, all of whose restrictions to at most 3 elements are embeddable in R_7, and consequently embeddable in each $R_n (n \geq 7)$. Then each R_n for $n \geq 7$ is a maximal element of our universal class. •

For further informations, see [74] FRAÏSSÉ 1971 vol.1 p.102-106 (english translation 1973 p.98-103).

5.11 Decreasing ω_1-sequence of denumerable posets: K. Kunen and A. Miller

5.11.1 Existence of a strictly decreasing ω_1-sequence of sets of integers, under inclusion up to finite differences

Let A, B two denumerable sets of non-negative integers. Set $A \subseteq B$ (inclusion up to finite difference) iff there exist only finitely many elements which belong to A yet not to B. This inclusion relation is reflexive and transitive, thus it defines a quasi-ordering on denumerable sets of integers. The corresponding equivalence relation means that the symmetric difference between A and B is finite. We denote $A \subset B$ the corresponding strict inclusion, which is irreflexive and transitive.

The existence of a strictly decreasing ω_1-sequence of sets of integers goes back to Du BOIS-REYMOND (end of 19-th century). To see this, let α be a non-zero countable ordinal, and let $A_i (i < \alpha)$ be a strictly decreasing α-sequence of sets of integers. If α is a successor ordinal, then $A_{\alpha-1}$ is the last term of our decreasing sequence, and by hypothesis it is a denumerable set of integers. Then we complete our sequence with a last term A_α by removing denumerably many elements (yet keeping denumerably many) in $A_{\alpha-1}$.

Suppose now that α is a limit ordinal; then take a cofinal ω-sequence B_j extracted from the A_i's. Let $c_0 \in B_0$, then $c_1 \in B_0 \cap B_1$ with $c_1 \neq c_0$, ... , then $c_j \in B_0 \cap B_1 \cap ... \cap B_j$ with $c_j \neq c_0, ..., c_{j-1}$. Then the set $C = \{c_0, c_1, c_2, ..\}$ is included in each B_j up to the eventual addition of finitely many elements $c_0, ..., c_{j-1}$.

Using transitivity we see that the set C satisfies $C \subset B_j$ for each integer j, hence also $C \subset A_i$ for each $i < \alpha$.

5.11.2 Existence of a strictly decreasing ω_1-sequence of denumerable posets

The following construction is due to Kenneth KUNEN and Arnold MILLER (1986). Start from the existence of a strictly decreasing ω_1-sequence of sets of integers with respect to inclusion, up to finite differences: see 5.11.1. To each integer n associate the finite poset classically called the "crown" on $2n$ elements. To each infinite set A of integers associate the poset P_A with a unique first element immediately followed by all those crowns on $2n$ elements, where $n \in A$. Finally to the (denumerable) equivalence class $\mathcal{A}$ formed of an infinite set A of integers and all sets having only a finite difference with A, associate the denumerable poset $P_\mathcal{A}$ formed of the preceding P_A's which are taken mutually incomparable. Then the strict embedding $P_\mathcal{A} < P_\mathcal{B}$ is equivalent with the following condition: for every $A \in \mathcal{A}$ and $B \in \mathcal{B}$ we have the strict inclusion $A \subset B$ up to finite differences. Consequently the existence of astrictly decreasing ω_1-sequence of denumerable posets follows from the Du BOIS-REYMOND theorem.

5.12 Exercises

5.12.1 Chain with several immediately greater chains

1 - An infinite chain can have 3, 4,.. immediately greater chains. For example the product $A = \omega^-.\omega$ has the 3 immediately greater chains $B = A+1, C = \omega + A, D = (\omega^2)^- + A$, where $(w^2)^-$ is the converse of ω^2. These are the only chains immediate extensions of A.

• The reader easily sees that there does not exist any chain either strictly between A and B, or between A and C, or between A and D.

Let us see that any chain $> A$ admits an embedding either of B or C or D. Indeed every chain which is an extension of A can be obtained as follows. Either by adding news elements at the end, so embeddinging B. Or by adding a chain at the beginning, in which either ω is embeddable (embedding of C), or in which the converse of an ordinal is embeddable (from ω^2 we get an embedding of D). Or by partitionning the new elements into a finite number of components ω^- of A, which is the same as adding them at the beginning. Or by modifying infinitely many components, so adding an ordinal ω (embedding of C), or finally by replacing these components by a retro-ordinal (so getting an embedding of D from the $(\omega^2)^-$). •

2 - Other examples. The chain $\omega_1 + \mathbf{Q}$, where $\mathbf{Q}$ is the chain of rationals, has the 5 following immediate extensions. The addition of ω_1 or its converse at the end or between ω_1 and $\mathbf{Q}$; the addition of ω^- at the beginning.

For each integer $p \geq 2$, the chain $\omega.(p-1)$ has exactly p immediate extensions (among chains).

There exist chains having infinitely many immediate greater chains. For example every infinite rigid chain A (i.e. a chain A without any strict restriction isomorphic with A: see below 6.9). Use the method of 5.5.4.

3 - Examples due to HAGENDORF in 1983: see ToR-86 p.158.

Prove that $A = \omega^- + \omega_1$ has 3 immediate extensions among chains: $1 + A$ and $A + 1$ and $\omega.\omega^- + \omega_1$ which is not obtained by simply adding an interval.

Prove that $B = \mathbf{Q} + \omega_1$ has 5 immediate extensions: $B + 1, \omega_1 + B, (\omega_1)^- + B, \mathbf{Q} + (\omega_1)^- + \omega_1, \mathbf{Q} + \omega^-.\omega_1$. The last one is not obtained by adding an interval.

5.12.2 Ramsey theory extended to embeddability

Given two order types α, β and two integers m, k, let us write $\alpha \rightarrow (\beta)^m_k$ iff, partitioning all m-element sets in the base $|\beta|$ into k colors, then α is embeddable in at least one unicolor restriction of β. In the contrary case we write $\alpha \not\rightarrow (\beta)^m_k$.

The usual RAMSEY's theorem takes the form $\omega \rightarrow (\omega)^m_k$ for any two positive integers m, k. Yet already $\omega + 1 \not\rightarrow (\omega)^2_2$.

Let $\mathbf{Q}$ be the chain of rationals. Let us take an ω-enumeration of $\mathbf{Q}$. Then in SIERPINSKI's manner (see 3.3.1), let us partition all pairs of rationals $\{a, b\}$ into two colors, by assuming that $a < b \pmod{\mathbf{Q}}$, according to whether $a < b$ or $b < a$ modulo ω.

Then we see that an infinite unicolor subset has either the order-type ω or the opposite order-type ω^-: so that $\mathbf{Q}$ itself cannot be embedded in a unicolor subset. We express this non-embedding by the Ramsey notation $\mathbf{Q} \not\to (\mathbf{Q})^2_2$ with the superscript 2 denoting pairs, the subscript 2 denoting the number of colors. Nevertheless it remains that $\omega \to (\mathbf{Q})^2_2$.

Let us start with the chain $\mathbf{Q}$ of rationals, whose pairs are partitioned into two colors (+) and (-). Then **either we have a (+)-monochromatic restriction isomorphic with $\mathbf{Q}$, or a denumerable (-)-monochromatic restriction (thus isomorphic either with ω or with its converse)**; due to [57] ERDÖS, RADO 1959.

• Either for any set X of rationals giving a restriction $\mathbf{Q}/X$ equimorphic with $\mathbf{Q}$, there exists at least one element $x \in X$ such that the set of those rationals in X which are (-)-connected with x is itself equimorphic with $\mathbf{Q}$. Then by iteration we construct a denumerable (-)-monochromatic set.

Or for any rational x the set of rationals (-)-connected with x is scattered. Assume that for a given integer n we have a n-sequence of rationals $x_1, ..., x_n$ (whose order is not important) which are mutually (+)-connected. Then in any one of the $n+1$ intervals delimited by these n rationals plus ∞ and $-\infty$, there exists at least one rational which is (+)-connected with $x_0, ..., x_n$ (use the fact that any finite union of scattered sets is scattered).

Consequently for any enumeration of all rationals we can construct an isomorphic ω-sequence of rationals mutually (+)-connected: this gives a (+)-monochromatic, denumerable restriction of $\mathbf{Q}$ isomorphic with $\mathbf{Q}$. •

A deeper result is due independently to GALVIN (not published) and to [41] DEVLIN 1979 (see also A.5.5 below):

Let us start with the chain $\mathbf{Q}$ of rationals and partition pairs into three colors. Then there exist two colors and a subset F of the base such that any pair in F has one of the two chosen colors, and the restriction $\mathbf{Q}/F$ is equimorphic with $\mathbf{Q}$.

To our knowledge, this is the first apparition of polychromatism in Ramsey theory; see 3.8.4.

Chapter 6

Scattered chain, right or left indecomposable chain, scattered poset, topology on initial intervals

6.1 Scattered chain

A chain is said to be **scattered** iff the chain **Q** of rationals is not embeddable in it. For example, every well-ordering or converse well-ordering, the chain Z of positive and negative integers, the product $\omega^{-}.\omega$ or its converse.

If a chain A is scattered, then every restriction of A is scattered.

If a chain has no dense restriction, then it is scattered. Conversely by 5.3.1 using dependent choice, a scattered chain has no dense restriction.

Given a scattered chain A with at least two elements in its base, there exist at least two consecutive elements (mod A). Indeed in the contrary case, A would be dense, hence would embed **Q** (uses dependent choice).

Finally **for A (with at least two elements) being scattered, it is necessary and sufficient that each restriction of A has at least two consecutive elements**.

6.1.1 Sum of scattered chains, initial intervals

(1) **Every sum of scattered chains, ordered along a scattered chain, is scattered**.

In particular, **the ordinal sum and the ordinal product of two scattered chains is scattered**.

• Let A_i be the chains in consideration, and A their sum along the chain I. If **Q** is embeddable in A, then either there exist two elements of $\mathbf{Q}$ in the same interval A_i, in which case $\mathbf{Q}$ is embeddable in it, or there exists at most one element of $\mathbf{Q}$ in each A_i, in which case $\mathbf{Q}$ is embeddable in I. •

(2) Let A be a chain; **if every strict initial interval of A is scattered, then A is scattered**. Same statement with "final interval".

6.1.2 Construction of strictly greater scattered chains

(1) **Given a scattered chain A, the chain $A.2 = A + A$ is scattered and strictly greater than A with respect to embeddability** (uses dependent choice; ZF suffices if A is denumerable or with well-orderable base). This follows from the preceding subsection and 5.3.3.

(2) **For every countable set of denumerable scattered chains, there exists a denumerable scattered chain in which all are embeddable**. Consequence of the previous subsection, proposition (1).

6.1.3 Initial intervals of a denumerable scattered chain

Let A be a denumerable scattered chain; then there are only countably many initial intervals of A (uses dependent choice).

If A is denumerable and not scattered, then there are exactly continuum many initial intervals.

• For the second statement, it suffices to note that $\mathbf{Q}$ is embeddable in A and there are exactly continuum many initial intervals of $\mathbf{Q}$.

For the first statement, suppose that A has uncountably many initial intervals: we shall prove that $\mathbf{Q}$ is embeddable in A. For this, we shall prove that there exists an element u of the base $|A|$, such that uncountably many initial intervals do not contain u, and uncountably many initial intervals do contain u.

Suppose the contrary. We say that an element of the base is "on the right" if there are uncountably many initial intervals which do not contain it. Analogously define an element to be "on the left" (uncountably many intervals contain it). The elements on the left form an initial interval B, and the elements on the right form a final interval C. Every element is either on the left or right. Moreover by our assumption of the non-existence of u, no element is both on the left and right. Thus B and C are complements. At least one of the two, say B, has uncountably many initial intervals. Yet B is countable: enumerate the elements of its base. Now B has no maximum element, for otherwise it would be preceded by uncountably many initial intervals. Hence there exists a strictly increasing ω-sequence cofinal in B. Partition B according to this sequence. We then have denumerably many intervals, each of which has countably many initial intervals, and yet their union has uncountably many initial intervals. This contradicts the countable axiom of choice: see 1.2.5.

The existence of u is then proved: call it u_0 . Iterating this, we obtain an analogous element $u_1 < u_0 (\text{mod}\, A)$ with uncountably many initial intervals which do not contain u_1 and uncountably many ones which contain u_1 but not u_0. And an analogous element $u_2 > u_0$, and so forth. Using dependent choice, this yields a restriction of A which is isomorphic with **Q**. •

6.1.4 Chain of initial intervals

(1) **Given an infinite scattered chain A, the chain of inclusion of the initial intervals of A is scattered** (uses dependent choice; ZF suffices if the base is well-orderable).

• Let B be the chain of the initial intervals of A, and suppose that the chain **Q** of rationals is embeddable in B. We shall prove first that there exists an element u of the base $|A|$, for which the chain of those initial intervals containing u, and the chain of those initial intervals not containing u, both admit an embedding of **Q**. It suffices to start with a restriction of B which is isomorphic with **Q**, and to partition it into a sum of three terms, each isomorphic with **Q**, then to take two elements, i.e. two initial intervals of A, in the middle term; and finally to take an element u in the difference of these two initial intervals of A.

Having proved the existence of u, call it u_0. Iterating this, we obtain an analogous element $u_1 < u_0 (\mathrm{mod}\, A)$ and $u_2 > u_0 (\mathrm{mod}\, A)$, and so forth (here we use dependent choice; ZF sufficient if A is countable or has well-orderable base). This yields a restriction of A isomorphic with **Q**. •

(2) **Given an infinite scattered chain A, the set of initial intervals of A is equipotent with A** (uses axiom of choice).

• Let the aleph ω_α be the cardinal of the base $|A|$, and suppose that the set of initial intervals of A has cardinality $\geq \omega_{\alpha+1}$. By the preceding proposition, the chain **Q** of rationals is not embeddable in the chain B of these initial intervals. By 5.7.2, and since $\omega_{\alpha+1}$, like every successor aleph, is regular (2.8.2 using axiom of choice), the chain B admits an embedding of either the ordinal $\omega_{\alpha+1}$ or its converse. To fix ideas, suppose that $\omega_{\alpha+1}$ is embeddable in B, hence that there exists a strictly increasing $\omega_{\alpha+1}$-sequence of initial intervals $H_i (i < \omega_{\alpha+1})$ of A. Take an element in each difference of successive H_i. We then obtain a strictly increasing $\omega_{\alpha+1}$-sequence of elements in the base of A, contradicting the hypothesis that $\mathrm{Card}\, A = \omega_\alpha$. •

6.2 Hausdorff decomposition of a chain; neighborhood, decomposition into lesser chains

6.2.1 Hausdorff decomposition

Every non-scattered chain can be uniquely decomposed into a dense sum of scattered chains ([107] HAUSDORFF 1914; uniqueness uses dependent choice)

• Let two elements x, y of the base be equivalent iff the interval (x, y) is scattered. This condition yields equivalence classes which are themselves scattered intervals. No two of these intervals can ever be consecutive; thus they constitute a dense chain. If there were two distinct decompositions of a chain into a dense sum of scattered chains, then one of these scattered chains would admit an embedding of a dense sum, hence an embedding of $\mathbf{Q}$: see 5.3.1 using dependent choice. •

6.2.2 α-neighborhood

Two elements x, y of the base of a chain A are said to be 0-**neighbors** (mod A) iff the interval with endpoints x and y is finite. The equivalence classes of this relations are called 0-**neighborhoods**. These are intervals which are either finite, or isomorphic with ω or its converse, or isomorphic with $\mathbf{Z}$.

If A is scattered, at least one 0-neighbor has at least two elements (modulo dependent choice). Indeed we have at least two consecutive elements: see 6.1.

Let α be an ordinal. If α is a successor, then x and y are said to be α-**neighbors** iff there exist only finitely many $(\alpha - 1)$-neighborhoods between x and y. If α is a limit ordinal, then x and y are said to be α-neighbors if there exists an ordinal $\beta < \alpha$ for which they are β-neighbors.

The equivalence classes, for this equivalence relation, are intervals and are called α-**neighborhoods**.

6.2.3 Strongly scattered chain

We say that a chain A is **strongly scattered** iff there exists no dense set of non-empty mutually disjoint intervals of A. Here, if I and J are two disjoint intervals, then we write $I < J$ iff every element of I is less than (mod A) every element of J; this yields the notion of a **dense chain of intervals**.

If A is strongly scattered, then no dense chain is embeddable in A. Hence every strongly scattered chain is scattered.

Every scattered chain is strongly scattered, then both notions are equivalent, modulo the axiom of dependent choice (ZF suffices for a denumerable chain).

• Suppose that A is not strongly scattered, and take a dense set of mutually disjoint intervals. By 5.3.1 (using dependent choice), there exists a subset formed of intervals which constitutes a chain isomorphic with $\mathbf{Q}$. Then take an element in each of these intervals (countable axiom of choice): this yields a restriction of A isomorphic with $\mathbf{Q}$. •

6.2.4 Neighborhood rank

Given a strongly scattered chain, there exists a minimum ordinal α such that from that ordinal on, all elements of the base are α-neighbors. Moreover for all $i, j (i < j < \alpha)$, there exist j-neighbors which are not i-neighbors. Indeed, the i-neighborhoods form a non-dense chain: hence there exist two consecutive i-neighborhoods.

The ordinal α thus defined is called the **neighborhood rank** of the chain. In particular this rank is 0 iff the chain is either finite or isomorphic with ω or its converse or $\mathbf{Z}$.

If A is a strongly scattered chain and if B is embeddable in A, then B is strongly scattered and the neighborhood rank of B is at most equal to the neighborhood rank of A.

If B is equimorphic with A, then the neighborhood ranks are equal.

The converse of A has the same neighborhood rank as A.

6.2.5 Decomposition of a strongly scattered chain having a minimum

Let A be a strongly scattered chain having a minimum element, and let α be the neighborhood rank of A.

(1) **If $\alpha = 0$, then A is finite or isomorphic with ω.**

If $\alpha \geq 1$, then either α is a successor ordinal and A is a finite sum or the sum of an ω-sequence of chains with neighborhood ranks $< \alpha$. Or α is a limit ordinal; letting $\gamma = \mathrm{Cof}\,\alpha$, then A is the sum of a γ-sequence of chains with neighborhood ranks $< \alpha$.

(2) If $\alpha = 0$, then every proper initial interval of A is finite; otherwise, every proper initial interval is a finite sum of chains with neighborhood ranks $< \alpha$.

(3) **If $\mathrm{Cof}\,\alpha$ is infinite, then** $\mathrm{Cof}\,A = \mathrm{Cof}\,\alpha$ (this conclusion (3) uses axiom of choice).

We have the analogous statement if A has a maximum element: replace initial interval by "final interval of A " and cofinality by "co-initiality of A ".

• (1) Let a be the minimum element of A. If α is a successor ordinal, then A is a finite sum or the sum of an ω-sequence of $(\alpha - 1)$-neighborhoods. If α is a limit ordinal, let $b_i (i < \gamma = \mathrm{Cof}\,\alpha)$ be a γ-sequence of strictly increasing ordinals whose supremum is α. The base $|A|$ is then the union of the b_i-neighborhoods of a. Call A_i these neighborhoods: the chain A is the sum of the difference intervals $A_{i+1} - A_i$.

(2) Immediate consequence of (1).

(3) Using the axiom of choice, take an element in each difference $A_{i+1} - A_i$. •

6.2.6 Decomposition into a sum of strictly lesser chains

Every strongly scattered chain is either the sum of two strictly lesser chains (with respect to embeddability); or the sum of strictly lesser chains along an ordinal which is a regular aleph, or the sum of strictly lesser chains along the converse of a regular aleph.

• Either A is the sum of two strictly lesser chains. Or, taking an element a in the base, A is equimorphic with the final interval with minimum a, or with the initial interval with maximum a. Now apply the preceding proposition, and note that, in the case of a finite sum of intervals with neighborhood ranks strictly less than that of A, we can still decompose A into two intervals, neither of which admits an embedding of A. •

6.2.7 Induction scheme for scattered chains

For historical information, see [107] HAUSDORFF 1914.

Suppose that a condition C holds for the empty chain and for every singleton chain. If, assuming that C holds for each chain $A_i (i < \gamma =$ an integer or an infinite regular aleph), then C holds for the sum of the A_i along γ and along its converse γ^-. Then C holds for every strongly scattered chain. Consequently C holds for every scattered chain (modulo the axiom of dependent choice).

Consequence of 6.2.5, by induction on the neighborhood rank.

Compare with the "definition scheme" of the hereditarily indecomposable chain (7.4 below).

6.3 Indecomposable chain, right or left indecomposable chain

We say that a chain A is **indecomposable** iff for every decomposition $A = A' + A"$ where A' is an initial interval and $A"$ a final interval, A is embeddable either in A' or $A"$; so that A is equimorphic with either A' or $A"$. A chain is said to be **decomposable** otherwise. Some authors say "additively indecomposable" and "additively decomposable".

6.3.1 Right or left indecomposable chain, strictly or not

We say that A is **right indecomposable** iff A is embeddable in every non-empty final interval. We say that A is **strictly right indecomposable** iff, for every decomposition of A into an initial and a non-empty final interval, A is embeddable in the final interval but not in the initial interval. Analogously, we define **left indecomposable** and **strictly left indecomposable** chain.

Every right indecomposable or left indecomposable chain is indecomposable.

However, starting with the chain $\mathbf{Q}$ of rationals, the sum $1 + \mathbf{Q} + 1$ is indecomposable yet neither right nor left indecomposable. There exists, though, an interval of this sum, namely $\mathbf{Q}$, which is equimorphic with the entire chain and which is both right and left indecomposable: see 6.3.4 below.

(1) **If A is an infinite chain, both right and left indecomposable, then A is equimorphic with $A.2 = A + A$.**

(2) **If A is infinite and equimorphic with $A.2$, then the chain $\mathbf{Q}$ of the rationals is embeddable in A** (uses dependent choice, ZF suffices if A is denumerable). This is another form of 5.3.3.

(3) **Every non-empty, scattered, right indecomposable chain is strictly right indecomposable**; similarly with "left" (uses dependent choice; consequence of (1) and (2)).

However $\mathbf{Q}.\omega_1$ is strictly right indecomposable yet not scattered.

6.3.2 Two lemmas on right (or left) indecomposable chains

(1) **If A is a strictly right indecomposable chain, then every chain which is equimorphic with A is strictly right indecomposable**; similarly with "left".

• Let B be equimorphic with A. Decompose $B = B' + B"$ with $B"$ non-empty. Then $A = A' + A"$ with $A' \leq B'$ and $A" \leq B"$ and $A"$ non-empty, for otherwise A would be embeddable in a proper initial interval. Then $B \leq A \leq A" \leq B"$ and $B \not\leq B'$. •

(2) **Given a chain A which is not right indecomposable, there exists a decomposition $A = A' + A"$ with $A"$ non-empty and $A' \not\leq A"$** (axiom of choice).

• Set $\alpha = \text{Cof}\, A$; start with a decomposition $A = \Sigma A_i (i < \alpha)$ and note that, if our conclusion is false, then each A_i is non-cofinally embeddable in any final interval of A. •

6.3.3 The maximum right indecomposable initial interval

Theorem. Every chain A admits a maximum (with respect to inclusion) right indecomposable initial interval I possibly empty. Moreover if $I \neq 0$ and $A = I + B$ then B is the maximum final interval strictly less than A.

• By DEDEKIND's generalized statement 2.6.4, there exists a maximum cut (I, B) of A such that A is embeddable in each final interval which is strictly larger than B. If $I \neq 0$, then necessarily $B < A$. Let $X < I$: then $X + B < A$, since an embedding from $I + B$ into $X + B$ would map I into X. Finally let J be a final interval of I: if $J < I$ and J non-empty, then $J + B \simeq A$ and $J + B < A$: contradiction proving that I is right indecomposable. Moreover I is maximum since $B < A$. •

6.3.4 Unique sum-decomposition of an indecomposable chain

Theorem. Let A be an indecomposable chain. Then either A is strictly right indecomposable, or strictly left indecomposable, or A is uniquely decomposable into a sum $A = B + H + C$ where $B < A$ and $C < A$ and H is equimorphic with A and both right and left indecomposable (see [104] HAGENDORF 1977).

These three cases are mutually exclusive; in the third case only, we have A equimorphic with $A.2$.

• We say that a cut in A is a **right cut** if A is embeddable before it but not after; **left** if A is embeddable after but not before; **bilateral** if A is embeddable both before and after. Suppose first that there only exist right and left cuts. Again by DEDEKIND's generalized statement, there exists either a maximum left cut or a minimum right cut. In the first case A is embeddable in a final interval which has only right cuts, so A is strictly left indecomposable. Analogous conclusion in the second case.

Now consider the case where there exist bilateral cuts. Then each bilateral cut has other bilateral cuts both before and after it. Hence there can be neither a minimum nor a maximum bilateral cut. Thus there exists a maximum left cut (or no left cut) and a minimum right cut (or none of them). So that A is decomposable into a sum $A = B + H + C$, with $B < A$ and $C < A$ and H equimorphic with A; the intervals B and C may be empty. Every cut in H is bilateral $(\bmod A)$, so is again bilateral $(\bmod H)$. Hence H is both right and left indecomposable. Finally the existence of bilateral cuts gives A equimorphic with $A.2$.

The uniqueness of the decomposition follows from the fact that a lengthening of B to the right would give $B \geq A$; similarly for a lengthening of C to the left. Finally a lengthening of H either to the left or to the right would destroy its bilateral indecomposability. •

Corollaries (1) **For every indecomposable chain A, there exists an interval which is left or right indecomposable (or both) and equimorphic with A.**

(2) **If A is infinite and equimorphic with $A.2$, then A is equimorphic with a chain both right and left indecomposable.**

(3) **Every non-empty, scattered and indecomposable chain is strictly right or strictly left indecomposable**: reinforcement of 6.3.1 proposition (3); uses dependent choice ([130] JULLIEN 1969).

6.3.5 Strongly scattered right indecomposable chain

Let A be a strongly scattered right indecomposable chain, and α its neighborhood rank.

If $\alpha = 0$, then every proper initial interval of A is finite.

If $\alpha \geq 1$, then every proper initial interval is a finite sum of chains with neighborhood ranks strictly less than α.

Consequently we find again 6.3.1 proposition (3): **A is strictly right indecomposable. Moreover in the case where $\alpha \geq 1$: if a proper initial interval of A is indecomposable, then it has rank strictly less than α.**

• If A has a minimum, then we are in the case of 6.2.5.

Otherwise, take an element u of the base. Let B be the initial interval of elements strictly less than $u(\bmod A)$ and C the final interval beginning with u. By hypothesis A is equimorphic with C, so the neighborhood rank of C is α. The initial interval B is embeddable in a proper initial interval of C, hence by 6.2.5 if $\alpha = 0$ then B is finite, and if $\alpha \geq 1$ then B is embeddable in a finite sum of chains with neighborhood ranks strictly less than α. Thus B itself is such a finite sum. •

6.3.6 Lemma on indecomposable chains

Lemma. Let A be an infinite right indecomposable chain, B an infinite left indecomposable chain, and C an arbitrary chain. If $C \geq A$ and $C \geq B$, then either $C \geq A + B$ or $C \geq B + A$ (see [130] JULLIEN 1969; also the corollary).

• Let f be an isomorphism of A onto a restriction of C, and let C_A be the initial interval of C formed of those elements less than (mod C) or equal to the images fx as x runs through $|A|$.

Then A is embeddable in every non-empty final interval of C_A. Similarly let C_B be the final interval of C such that B is embeddable in every non-empty initial interval of C_B. Then either C_A and C_B are disjoint, and hence $C \geq A + B$. Or there exists an element u common to the bases $|C_A|$ and $C_B|$. Then B is embeddable before u in the intersection $|C_A| \cap |C_B|$, and A is embeddable after u: hence $C \geq B + A$. •

Corollary. The chain $\omega + \omega^-$, which is embeddable neither in $A = \omega^-.\omega$ nor in its converse $A^- = \omega.\omega^-$, is however embeddable in every chain which is a common extension of A and A^-.

• The chain A is right indecomposable and A^- left indecomposable. Thus if a chain $X \geq A$ and $\geq A^-$, then either $X \geq A + A^- \geq \omega + \omega^-$ or $X \geq A^- + A \geq \omega + \omega^-$. •

Problem posed by [104] HAGENDORF 1977. Let A be a strictly right indecomposable chain. If for every chain $X < A$ we have $X + 1 \leq A$, then is A a well-ordering. Affirmative answer for scattered chains: see [149] LARSON 1978.

6.4 Covering by indecomposable chains or by doublets

6.4.1 Union of two right indecomposable chains

Let A be a chain which is the union of an initial interval and a final interval, both having at least one common element and both of which are right indecomposable; then A is right indecomposable. Same statement for "left".

• Let B be the initial interval, C the final interval and D their intersection. Then B is embeddable in D. Either $D = C$ so that $A = B$ is right indecomposable. Or C has the form $D + E$ and so $A = B + E$ is embeddable in $C = D + E$, hence A is again right indecomposable. •

6.4.2 Intersection of two indecomposable chains

Consider a chain which is the union of a right indecomposable initial interval B and a left indecomposable final interval C, both infinite and having at least one common element; then the intersection $B \cap C$ is both left and right indecomposable and admits an embedding of the chain Q

of rationals (uses dependent choice). This is a consequence of 6.3.1 propositions (1) and (2).

6.4.3 Covering by right or left indecomposable chains

Let A be a chain. We say that two elements u, v of the base are **equivalent with respect to right indecomposable chains** iff there exists an interval of A which is right indecomposable and contains both elements u and v.

The condition thus defined is reflexive and symmetric. Moreover by 6.4.1 it is transitive.

Analogously we define **the equivalence relation with respect to left indecomposable chains**; we call these **covering by right (resp. left) indecomposable chains**.

There can be infinitely many equivalence classes of this covering relation. For instance, take the converse ω^- of ω: the equivalence classes for covering by right indecomposable chains are singletons.

An equivalence class for covering by right indecomposable chains is not necessarily a right indecomposable chain: take the chain $\mathbf{Z}$ of the positive and negative integers.

6.4.4 Doublet of indecomposable chains

A **doublet** is a chain which is the union of a left indecomposable interval and a right indecomposable interval, both having at least one common element. For example the chain $\mathbf{Z}$ of positive and negative integers is a doublet, being the union of the final interval ω and the initial interval ω^-, which can be chosen to have one or several common elements.

On the other hand, although the product $\omega^-.\omega$ is right indecomposable and its converse is left indecomposable, the sum $\omega.\omega^- + \omega^-.\omega$ is not a doublet. Nor even is the sum $\omega.\omega^- + \mathbf{Z} + \omega^-.\omega$. The latter example is as well the union of $\omega^-.\omega$ and its converse, with some possible common elements: if we decompose $\mathbf{Z}$ into ω^- and ω and then attach ω^- to $\omega^-.\omega$, we have a sum isomorphic with $\omega^-.\omega$, and similarly for the converse chains; yet the considered indecomposable chains are no longer intervals of the final constructed sum.

Every right or left indecomposable chain is a particular kind of doublet, in which one of the intervals reduces to a singleton. Note that it is not required that our indecomposable chains be initial or final intervals: one of them may be a middle interval.

In the case of a scattered chain which is a doublet, either one of the indecomposable chains is a restriction of the other, or the left indecomposable chain is an initial interval and the right indecomposable chain is a final interval of the doublet.

On the other hand $1+\mathbf{Q}+1$, where $\mathbf{Q}$ is the chain of rationals, is a non-scattered doublet having a decomposition into $1+\mathbf{Q}$, a right indecomposable initial interval, and $\mathbf{Q}+1$, a left indecomposable final interval.

6.4.5 Covering by doublets

Consider a scattered chain and let u, v be two elements in the base. We say that u and v are **equivalent with respect to doublets** iff there exists a doublet of indecomposable chains covering u and v.

This condition is reflexive and symmetric; let us show that it is transitive (uses dependent choice); the considered equivalence relation will be called **covering by doublets**.

• Take three elements $u < v < w$ (modulo the chain). If u and v, on the one hand, v and w on the other hand, are covered by two right or two left indecomposable chains, then apply 6.4.1. If u, v are covered by a right indecomposable chain, necessarily infinite and having no maximum, and if v, w are covered by a left indecomposable chain, infinite and having no minimum, then either one of these indecomposable chains covers u and w, or their intersection is infinite and both right and left indecomposable, hence admits an embedding of $\mathbf{Q}$ (see 6.4.2, dependent choice): contradiction.

If u, v are covered by a doublet and v, w by a right indecomposable chain, then by 6.4.1 we are in the case of a doublet covering u and w. If u, v are covered by a doublet and v, w by a left indecomposable chain, necessarily infinite, and if the second term of the doublet is infinite, then we obtain a contradiction again using 6.4.2. If the second term reduces to a singleton, then by 6.4.1 we have a left indecomposable chain covering u and w .

Now assume that u, v are covered by a doublet formed of A, a left indecomposable, and B, a right indecomposable chain; and similarly v, w are covered by a doublet formed of C , a left indecomposable, and D, a right indecomposable chain. Then either the cut situated to the right of B falls in or after D, and so by 6.4.1 we have a doublet covering u and w. Or the cut situated to the left of C falls in or before A, and we have the same conclusion. Or finally $\mathbf{Q}$ is embeddable in the intersection of B and C: contradiction. •

Transitivity no longer holds for a chain in which Q is embeddable.

• Take $A = 1+\mathbf{Q}+\omega_1.\omega$ and let u be the minimum of A. Let v be the minimum of the final interval $\omega_1.\omega$; let w be separated from v by the first component ω_1. Then u and v are covered by the doublet $1+\mathbf{Q}+1$ formed of a right indecomposable $1 + \mathbf{Q}$ and a left indecomposable chain $\mathbf{Q} + 1$, and v and w are covered by the indecomposable chain $\omega_1.\omega$. However no common doublet covers u and w, since 1+Q+1 is neither left nor right indecomposable. •

Even if we would reinforce the definition of the doublet, by requiring that the left indecomposable chain always be an initial interval, and the right indecomposable chain be a final interval of the doublet, both with a common element, then consider $A = 1 + \mathbf{Q} + 1$ and let u be the minimum, w the maximum of A, and v an element of the middle interval $\mathbf{Q}$. Then u is equivalent to v by a right indecomposable chain, v equivalent to w by a left indecomposable chain, yet u and w are no longer equivalent elements, since $1 + \mathbf{Q} + 1$ is no longer a doublet.

For a scattered chain, the equivalence relation of covering by doublets is the union of the equivalence relation for right indecomposable

chains and the equivalence relation for left indecomposable chains. In other words, x and y are equivalent with respect to doublets iff there exists a finite sequence of elements, from x until y, where two consecutive terms are equivalent with respect to right or left indecomposable chains.

In chapter 7 we shall see that, for a scattered chain, there are only finitely many equivalence classes with respect to covering by doublets. Moreover each class is a doublet (possibly reduced to a right or left indecomposable chain).

6.5 Scattered poset; Bonnet-Pouzet's conditions

We say that a poset A is **scattered** (with respect to embedding) iff A does not embed the chain $\mathbf{Q}$ of rationals. In the particular case of a chain, we find again the classical scattered chain: see 6.1.

Another example. Every well-founded poset is scattered; in particular every well partial ordering is scattered.

6.5.1 Four equivalent conditions on posets

Let A be a poset; then the following four conditions are equivalent ([18] BONNET, POUZET 1969; uses axiom of choice).

(1) **A is finitely free and scattered.**

(2) **Every totally ordered augmentation of A is a scattered chain.**

(3) **The partial ordering of inclusion among initial intervals of A is scattered.**

(4) **For every function f mapping the rationals into the base $|A|$, there exist two rationals x, y with $x < y \pmod{\mathbf{Q}}$ and $fx \geq fy \pmod{A}$.**

• Condition (2) implies (1). Indeed if $\mathbf{Q}$ is embeddable in A, then $\mathbf{Q}$ is embeddable in every augmentation of A. On the other hand, if A has an infinite, hence a denumerable free subset, then take a chain B isomorphic with $\mathbf{Q}$ on this subset, and then by 2.9.1 take a partially ordered augmentation of A which extends B. By the augmentation axiom in 2.9.3, there exists a totally ordered augmentation of A in which $\mathbf{Q}$ is embeddable. •

• Conditions (2) and (3) are equivalent. First suppose that there exists a totally ordered augmentation C of A in which $\mathbf{Q}$ is embeddable. Then the chain of inclusion for initial intervals of C is a restriction of the partial ordering of inclusion for initial intervals of A, and admits an embedding of $\mathbf{Q}$.

Conversely, suppose that $\mathbf{Q}$ is embeddable in the partial ordering of inclusion for initial intervals of A. Take a chain U, in which $\mathbf{Q}$ is embeddable, and which is maximal with respect to inclusion, and formed of initial intervals of A (see 2.2.4, axiom of choice). Then U is the chain of initial intervals of a certain totally ordered augmentation V of A: see 2.9.4. Moreover by 6.1.4 proposition (1), the chain $\mathbf{Q}$ is embeddable in V (dependent choice). •

• Conditions (2) and (4) are equivalent. Suppose that there exists a totally ordered augmentation of A in which $\mathbf{Q}$ is embeddable. Then there exists a bijection

f of $\mathbf{Q}$ onto a restriction of A, which maps any two rationals $x < y \pmod{\mathbf{Q}}$ into $fx <$ or $|fy \pmod{A}$.

Conversely, if there exists such a (necessarily injective) function, then the image $\overline{f}(Q)$ is an augmentation of the restriction $A/\overline{f}(|\mathbf{Q}|)$ (restriction of A to the set of images of rationals). By 2.9.1 and the augmentation axiom 2.9.3, there exists a totally ordered augmentation of A in which $\mathbf{Q}$ is embeddable. •

• Condition (1) implies (4). Suppose that A is finitely free and that $\mathbf{Q}$ is not embeddable in A, yet there exists a function f from $\mathbf{Q}$ into A, such that for $x < y (\mathrm{mod}\, \mathbf{Q})$ we have $fx <$ or $|fy \pmod{A}$ hence f is injective. Then the restriction $A/\overline{f}(|\mathbf{Q}|)$ is a denumerable finitely free poset in which $\mathbf{Q}$ is not embeddable. By applying RAMSEY, more exactly 3.1.2, there exists a denumerable, maximal, totally ordered and scattered restriction of $A/\overline{f}(|\mathbf{Q}|)$. Thus there exist two rationals x_0 and $y_0 > x_0 \pmod{\mathbf{Q}}$ with $f(y_0)$ an immediate successor of $f(x_0)(\mathrm{mod}\, A/\overline{f}(|\mathbf{Q}|)$.

Iterate this, by replacing $\mathbf{Q}$ by its restriction to the open interval (x_0, y_0). We obtain, for each integer i, an interval (x_i, y_i) with $y_i > x_i (\mathrm{mod}\, \mathbf{Q})$ and each interval (x_{i+1}, y_{i+1}) properly included in $(x_i, y_i)(\mathrm{mod}\, \mathbf{Q})$. For each i we have $f(x_i) < f(y_i)(\mathrm{mod}\, A)$, without any element of $\overline{f}(|\mathbf{Q}|)$ between $f(x_i)$ and $f(y_i)(\mathrm{mod}\, A)$: it suffices to consider a rational x and all possible positions of fx, either for x between x_{i-1} and $y_{i-1}(\mathrm{mod}\, \mathbf{Q})$, or for x elsewhere in the chain $\mathbf{Q}$.

Again using RAMSEY, we see that there exist denumerably many integers i with all values $f(x_i)$ and $f(y_i)$ mutually comparable $(\mathrm{mod}\, A)$. Hence there exist i and $j > i$ with, for instance, $f(x_j)$ situated between $f(x_i)$ and $f(y_i)(\mathrm{mod}\, A)$: contradiction. •

6.5.2 A maximal initial interval

Recall that a poset A is said to be **stratified** (see 2.10.1) iff incomparability $(\mathrm{mod}\, A)$ together with identity, forms an equivalence relation. The equivalence classes form a chain, called the **principal chain**.

Using the previous definition of a scattered poset, **a stratified poset is scattered iff its principal chain is itself scattered**.

Given a poset A, there exists an initial interval of A which is maximal (with respect to inclusion) among those initial intervals having a scattered stratified augmentation. Analogous statement for final intervals. Uses axiom of choice.

• Consider those initial intervals of A having a scattered stratified augmentation. There exist such, for example the empty interval. The set of these intervals is ordered by inclusion: let C be a maximal chain of inclusion (HAUSDORFF-ZORN axiom equivalent to the axiom of choice). Let X be the initial intervals of A which belong to C, and let I be the initial interval of A which is the union of the X. No initial interval of A which is a proper extension of I has a scattered stratified augmentation.

Given two initial intervals X, Y, a scattered stratified augmentation Y^+ of Y is said to be **finer** than a scattered stratified augmentation X^+ of X iff the base

$|X|$ is included in $|Y|$, and each incomparability class of X^+ is an incomparability classs of Y^+, and finally the principal chain of X^+ is an initial interval of the principal chain of Y^+.

If X, Y belong to C and if the base $|X|$ is included in $|Y|$, then for each scattered stratified reinforcement X^+ of X, there exists a scattered stratified augmentation of Y which is finer than X^+. Indeed, take an arbitrary scattered stratified augmentation Z of Y. Keep the incomparability classes of X^+ and replace each class of Z by its intersection with $(|Y| - |X|)$ (provided this intersection is non-empty). Then while preserving their order, place these classes coming from Z after all the classes of X^+. This is possible, since every element of $|Y| - |X|$ is greater than or incomparable $(\mathrm{mod}\, A)$ with every element of the base $|X|$.

The set of scattered stratified augmentations of the elements of C is partially ordered by the comparison "finer than". Take a maximal chain, again using the maximal chain axiom. It suffices to prove that this maximal chain terminates with a scattered stratified augmentation of I. For this, note that this maximal chain cannot terminate with an augmentation X^+ of an X in C, distinct from I. Indeed, take a Y in C, a proper extension of X. The preceding alinea shows the existence of a scattered stratified augmentation of Y which is finer than X^+: contradiction.

Furthermore, this chain necessarily has a maximum element, which is the augmentation of an element of C. Indeed, given a set composed of augmentations X^+ of elements X of C, without any X maximum, and totally ordered by the comparison "finer than", take the initial interval J formed by the union of the X being considered, and then the augmentation of J which is the common extension of the X^+. This augmentation is finer than each X^+, and is scattered, since every chain which is the union of scattered proper initial intervals is scattered: see 6.1.1 proposition (2). •

6.5.3 Three equivalent conditions on posets

The following three conditions are equivalent ([18] BONNET, POUZET 1969; uses axiom of choice).

(1) A **is scattered.**

(2) **There exists a scattered totally ordered augmentation of** A.

(3) **There exists a scattered stratified augmentation of** A.

• Condition (2) obviously implies (1) and (3).

Condition (3) implies (2). Indeed, starting with a stratified augmentation of A with a scattered principal chain, it suffices to choose a well-ordering (hence a scattered chain) based on each incomparability class of the stratified partial ordering (axiom of choice); and then to take the ordinal sum of these well-orderings along the principal chain.

We now show that (1) implies (3). For this, suppose that every stratified augmentation of A admits an embedding of the chain $\mathbf{Q}$. We shall show that $\mathbf{Q}$ is embeddable in A. Applying the previous 6.5.2, let I be an initial interval of A which is maximal with respect to inclusion, among those initial intervals having a scattered stratified augmentation. Let J be a final interval, similarly defined. Let E be the base of A; the difference $E - (|I| \cup |J|)$ is non-empty. Since if not, then take scattered stratified augmentations I^+ of I and J^+ of J. Take as new incomparability classes the classes of I^+ followed by, in order, the class of non-empty intersections of the classes of J^+ with $E - |I|$. This then yields a stratified augmentation of A whose principal chain is the sum of the principal chain of I^+ and a restriction of the principal chain of J^+, hence a scattered chain: this contradicts our assumption.

Let u be an element in the difference $E - (|I| \cup |J|)$. We shall prove that the restriction of A to those elements $< u$, on the one hand, and the restriction to those elements $> u$, on the other hand, have only augmentations in which $\mathbf{Q}$ is embeddable. Note first that there exist elements $< u$ which do not belong to $|I|$. For if not, then the union $I \cup \{u\}$ would be an initial interval of A and a proper extension of I, yet it has a scattered stratified augmentation, obtained by taking a scattered stratified augmentation of I, completed by a last class formed by the singleton of u: this contradicts maximality of I.

Suppose that the initial interval of those elements $< u$ has a scattered stratified augmentation, say U. Take again a scattered stratified augmentation I^+ of I. Then on the union $|I| \cup |U|$, take the incomparability classes of I^+, in their order, followed by, again in their order, the non-empty intersections of classes of U with $|U| - (|U| \cap |I|)$. We thus obtain a scattered stratified augmentation of $A/(|I| \cup |U|)$, contradicting the maximality of I. Analogous proof for the final interval J and the final interval of those elements $> u (\text{mod } A)$.

Now let u_0 be u. Since the initial interval $< u$ has only stratified augmentations in which $\mathbf{Q}$ is embeddable, there exists an element $u_1 < u_0$ such that each of the two intervals, those elements $< u_1$ and those elements strictly between u_1 and u_0, has only stratified augmentations in which $\mathbf{Q}$ is embeddable. Similarly, there exists an element $u_2 > u_0$ with the same property for the final interval $> u_2$ and the interval of those elements strictly between u_0 and u_2. Iterating this and using dependent choice, we obtain a denumerable set of elements u_i (i integer) giving a restriction of A isomorphic with $\mathbf{Q}$. •

6.6 Simple convergence topology

6.6.1 Topology on the power set $\mathcal{P}(\mathbf{N})$

Start with the set $\mathbf{N}$ of the non-negative integers. For each couple of finite subsets $F, G \subset \mathbf{N}$, let $U_{F,G}$ denote the set of those subsets of $\mathbf{N}$ which include F and are disjoint from G.

Note that $U_{F,G}$ is non-empty iff F and G are disjoint.

For $F = G = 0$ (empty set), we get $U_{0,0}$ = the entire power set $\mathcal{P}(\mathbf{N})$.

The intersection $U_{F,G} \cap U_{F',G'} = U_{F \cup F', G \cup G'}$.

Let us define an **open set** to be any union of preceding U sets, then a **closed set** to be the complement of an open set. We see that the intersection of any two open sets is still an open set; so that **we have a topology on** $\mathcal{P}(\mathbf{N})$, which is classically called the **simple convergence topology**.

Note that $U_{F,G}$ is the intersection of all $U_{\{x\},0}$ where $x \in F$ and all $U_{0,\{y\}}$ where $y \in G$ (recall that 0 denotes the empty set).

The complement of $U_{\{a\},0}$ is $U_{0,\{a\}}$ for any integer a.

The complement of a U set is a union of U sets, thus an open set; so that each U set is both open and closed, i.e. the complement of an open set; more briefly each U is a **clopen set**. The U's are called the **basic clopen sets**.

This topology is Hausdorff, i.e. satisfies the separation axiom: given two subsets A and B of $\mathbf{N}$, supposed to be distinct, we can always assume the existence of a non-empty finite set F included in A yet disjoint from B; then we have a finite set G, possibly empty, which is included in B and disjoint from A; so that $U_{F,G}$ and $U_{G,F}$ separate A from B.

6.6.2 Convergent sequence, adherent element

Consider an ω-sequence of subsets H_i of $\mathbf{N}$(i integer). We say that this sequence **converges** and that a subsest H of $\mathbf{N}$ is the **limit** of the H_i iff every open set containing H also contains H_i from some point on. Equivalently, for every integer x, either x belongs to H and then x belongs to H_i from some index on; or x belongs to the complement of H and then x belongs to the complement of H_i from some index on.

Following the classical definition, a subset H of $\mathbf{N}$ is said to be **adherent** to a set $\mathcal{H}$ of subsets of $\mathbf{N}$ iff every open set which contains H as an element, also intersects $\mathcal{H}$. Equivalently every basic clopen set which contains H intersects $\mathcal{H}$.

Also H is adherent to $\mathcal{H}$ iff there exists a convergent ω-sequence in $\mathcal{H}$ whose limit is H.

6.6.3 The topology is compact

In other words **if $\mathcal{P}(\mathbf{N})$ is covered by a union of open sets, then it is covered by a finite number of these sets.**

• Let us consider an ω-sequence of couples $(F(i), G(i))$(i integer) of finite subsets of $\mathbf{N}$, and let $U_i = U_{F(i),G(i)}$. Assume that for each i, the union $U_0 \cup U_1 \cup ... \cup U_i$ is strictly included in $\mathcal{P}(\mathbf{N})$. We will show that the union of all these U_i is distinct from $\mathcal{P}(\mathbf{N})$.

Take an arbitrary function h which, to each i, associates an element $h(i)$ belonging to $F(i) \cup G(i)$ (which can be assumed to be non-empty). Giving to $h(i)$ the sign (+) or (-) according to whether it belongs to $F(i)$ or $G(i)$, we can choose h to verify the following condition. For each i, there exists an element in the difference $\mathcal{P}(\mathbf{N}) - (U_0 \cup U_1 \cup ... \cup U_i)$, which contains as an element all those $h(0), h(1), ..., h(i)$ of sign (-) , and none of sign (+). It follows that the equality $h(i) = h(j)$ (with $i \neq j$) implies that $h(i)$ and $h(j)$ have the same sign. Finally the set of all $h(i)$ of sign (-) does not belong to the union of the U_i for all integers i. •

It follows from the compactness that **the clopen sets are exactly the finite unions of the basic clopen sets.**

6.6.4 Simple convergence topology on $\mathcal{J}(A)$ (set of initial intervals of A)

Let A be a poset; we shall extend to the set $\mathcal{J}(A)$ of initial intervals of A, the preceding topology, which can be considered as defined on $\mathcal{J}(\mathbf{N})$ where $\mathbf{N}$ reduces to the partial ordering of identity on integers.

Let F, G be two finite subsets of the base $|A|$. Let $U_{F,G}$ be the set of those initial intervals of A which include F and are disjoint from G.

$U_{F,G}$ is non-empty iff each element of F is $<$ or $|(\text{mod}\, A)$ with each element in G.

For F and G empty, we obtain the entire set $\mathcal{J}(A)$.

Again define an **open set** $(\text{mod}\, A)$ to be any union of preceding U sets, and define a **closed set** as the complement of an open set.

The intersection of any two open sets is still an open set. So that we have a topology on $\mathcal{J}(A)$, which we call the **simple convergence topology on** $\mathcal{J}(A)$.

The complement of a U set is a union of U sets, thus an open set; so that our U sets are **clopen sets: both open and closed**. We call these U the **basic clopen sets**.

This topoplogy is Hausdorff: repeat the proof of 6.6.1.

We leave it to the reader to verify, as in the previous subsection, that **this topology is compact**. So that clopen sets are exactly the finite unions of basic clopen sets.

Recall that an initial interval H is said to be **adherent to a set $\mathcal{H}$ of initial intervals** iff every open set (equivalently every basic clopen set) which contains H as an element, also intersects $\mathcal{H}$.

An ordinal-indexed sequence of initial intervals H_i is said to be **convergent** and to have the **limit** H iff every open set containing H also contains H_i from some point on. Equivalently, for every element x in the base, either x belongs to H, and then x belongs to H_i from some point on; or x belongs to the complement of H, and then x belongs to the complement of H_i from some point on.

Remark. If H is the limit of a convergent ordinal sequence in $\mathcal{H}$, then H is adherent to $\mathcal{H}$. However in a non-denumerable topological space, H can be adherent to $\mathcal{H}$ without being the limit of any convergent ordinal sequence: see 6.10.1 below.

6.6.5 Isolated element, topologically scattered set of intervals, perfect set

Let $\mathcal{H}$ be a set of initial intervals of A. An element X of $\mathcal{H}$ is said to be **isolated** $(\text{mod}\, \mathcal{H})$ iff there exists an open set, and consequently a basic clopen set U such that X is the only element in the intersection $\mathcal{H} \cap U$.

A set $\mathcal{H}$ of initial intervals is said to be **topologically scattered** iff every non-empty subset $\mathcal{H}'$ of $\mathcal{H}$ contains at least an isolated element $(\text{mod}\, \mathcal{H}')$.

A set $\mathcal{K}$ of initial intervals of A is said to be **perfect** iff it is non-empty and has no isolated element $(\text{mod}\, \mathcal{K})$. So that **a set is topologically scattered iff**

it does not include any perfect subset.

Note that a perfect set is infinite; moreover its adherence is perfect; moreover its intersection with an open set is either perfect or empty.

6.6.6 Dense set in the topology of $\mathcal{P}(\mathbf{N})$

Dense set, open set union of the U_i's

A set is called **dense** iff its intersection with every non-empty open set is non-empty.

$\mathcal{H}$ is dense iff every set of integers is adherent to $\mathcal{H}$; or iff the complement of $\mathcal{H}$ cannot include any (non-empty) open set.

The only dense closed set is the full set $\mathcal{P}(\mathbf{N})$.

The intersection of two dense open sets (consequently any finite intersection of dense open sets) is itself a dense open set.

Consider an ω-sequence of couples $(F(i), G(i))$ of disjoint finite subsets of $\mathbf{N}$(i integer). It defines an open set which is the countable union of all the $U_i = U_{F(i),G(i)}$.

(1) **The open set defined by an ω-sequence $(F(i), G(i))$ is dense iff, for every couple of finite disjoint subsets F, G of $\mathbf{N}$, there exists an index i with $F(i)$ disjoint from G and $G(i)$ disjoint from F.**

• Assume that the sequence of couples $(F(i), G(i))$ satisfies our condition. Then for each couple of disjoint finite sets F, G there exists an i with $F \cup F(i)$ disjoint from $G \cup G(i)$ so that the intersection $U_{F,G} \cap U_{F(i),G(i)} = U_{F \cup F(i), G \cup G(i)}$ is non-empty. Hence the set defined by our ω-sequence is dense.

Conversely, assume that the set defined by our ω-sequence is dense, and consider the open set $U_{F,G}$ with arbitrary disjoint finite sets F and G. By hypothesis there exists an index i such that $U_{F,G} \cap U_{F(i),G(i)}$ is non-empty: hence we have $F \cup F(i)$ disjoint from $G \cup G(i)$, hence $F(i)$ disjoint from G and $G(i)$ disjoint from F. •

Consider the particular case of an ω-sequence where, for every index i, either $F(i)$ is a singleton with $G(i)$ empty or $G(i)$ is a singleton with $F(i)$ empty. Then the preceding statement takes the simple following form: **the open set defined by the sequence $(F(i), G(i))$ is dense iff there exist either infinitely many different singletons $F(i)$ or infinitely many different singletons $G(i)$.**

(2) **The open sets V_h**

Going back to the general case of arbitrary finite $F(i)$ and $G(i)$, let h be a function which, to each i, associates an element $h(i)$ in $F(i) \cup G(i)$. Avoiding the useless full set $\mathcal{P}(\mathbf{N})$, we can always assume that if $h(i) = h(j) (i \neq j)$, then either $h(i)$ belongs both to $F(i)$ and $F(j)$, or both to $G(i)$ and $G(j)$. Then we associate the sign (+) or (-) to $h(i)$, according to whether it belongs to $F(i)$ or to $G(i)$.

Let V_h be the open set formed by the subsets X of $\mathbf{N}$ for which there exists an i such that $h(i)$ has sign (+) and $h(i) \in X$, or $h(i)$ has sign (-) and $h(i) \notin X$. Then the open set, union of the $U_{F(i),G(i)}$, is the intersection of the V_h for all functions h previously defined.

For every function h associating to each integer i an element $h(i)$ of $\mathbf{N}$, taking infinitely many values $h(i)$ with any sign, the open set V_h previously defined is dense. **Thus there exist infinitely many dense open sets.**

(3) **Baire's condition**

Every compact topology satisfies BAIRE's condition:

Every countable intersection of dense open sets is non-empty; even it is dense.

• Let us sketch a direct proof for the present topology. Every open set is a union of sets $U_{F,G}$ thus by the previous subsection, a countable intersection of open sets V_h.

It suffices to prove BAIRE's condition for an arbitrary ω-sequence of dense open sets V_h, each of them defined by a function h taking infinitely many values with a same sign. To do this, replace each V_h by reducing h to only those values with sign (+) or only those with sign (-), assumed to be infinite in number. Denote by E_h the infinite set of these values. The open set V_h with sign (+) is reduced to the set of subsets C of $\mathbf{N}$ such that $C \cap E_h$ is non-empty. Similarly the open set V_h with sign (-) is the set of subsets C of $\mathbf{N}$ such that $(\mathbf{N} - C) \cap E_h$ is non-empty. We saw in 2.3.3 proposition (2) (countable case) that there exists a suitable set C for all the E_h, with $C \cap E_h$ and $(\mathbf{N} - C) \cap E_h$ infinite.

Thus there exists a set C including any given finite set and excluding any given finite set (disjoint each from the other). Hence there exists a C belonging to any given open set and belonging to each V_h of our ω-sequence. •

6.7 Topologically scattered poset

A poset A is said to be **topologically scattered** iff the set $\mathcal{J}(A)$ of its initial intervals is topologically scattered (or equivalently has no perfect subset), in the sense of the previous subsection.

Lemma. Every topologically scattered poset is finitely free and order-scattered (i.e has no embedding of Q).

• By 6.5.1 conditions (1) and (3), if A either has an infinite antichain or admits an embedding of $\mathbf{Q}$, then the set $\mathcal{J}(A)$ of its initial intervals admits an embedding of $\mathbf{Q}$. Denote by K those initial intervals of A which form a chain isomorphic with $\mathbf{Q}$ under inclusion. Then call L any initial interval of A which is the union of a strictly increasing ω-sequence (under inclusion) of K's. The set of these L is non-empty (obvious) and has no isolated element. Indeed each L is the union, hence the topological limit of a strictly increasing ω-sequence of K's; hence of an analogous sequence of L's situated between them (take care that a K is not in general the union of all preceding K's). So that these L constitute a perfect set; consequently $\mathcal{J}(A)$ is not topologically scattered. •

6.7.1 Initial intervals of a topologically scattered poset

Let A be an infinite, topologically scattered poset. Then the set of initial intervals of A is equipotent with A; uses axiom of choice.

• Let E be the base of A. By hypothesis, the set $\mathcal{J}(A)$ of initial intervals is topologically scattered with respect to the simple convergence topology. Use as follows the CANTOR-BENDIXSON procedure (using choice axiom). First take an ordinal-indexed sequence of all isolated elements H_i of $\mathcal{J}(A)$. To each H_i let us associate a basic clopen set $U_{F(i),G(i)}$ where $F(i)$ and $G(i)$ are two finite subsets of E, such that H_i is the only element of $\mathcal{J}(A)$ which belongs to this basic clopen set. Removing from $\mathcal{J}(A)$ all these isolated elements, it then remains a topologically scattered subset denoted by $(\mathcal{J}(A))'$, the first derivative of $\mathcal{J}(A)$.

Iterating the procedure, let us denote by H_j all the isolated elements in this first derivative; j is an ordinal index, strictly greater than all the preceding indices i. Again we obtain for each j a basic clopen set $U_{F(j),G(j)}$ such that H_j is the only element in the first derivative, belonging to this clopen set. Iterating this to successive derivatives, which are defined either by removing all isolated elements, or in the case of a limit ordinal rank, by taking the intersection of preceding derivatives, we finally reach the empty set.

Now note that during the procedure, a given basic clopen set $U_{F,G}$ can occur only once. Indeed the first time that this clopen set is used in order to remove an element H_k of $\mathcal{J}(A)$, then this $U_{F,G}$ can contain some H_i with $i < k$, but cannot contain as an element any of the following H_j with $j > k$. Consequently the cardinality of $\mathcal{J}(A)$ cannot exceed the cardinality of the set of all couples (F, G) of finite subsets of E; hence cannot exceed $\operatorname{Card} E$. •

6.7.2 Topologically scattered chain (in simple convergence topology)

Particular case where our poset is a chain C; we remain in the simple convergence topology among initial intervals, as defined in 6.6.4 (nothing to do with the topology commonly defined on the base $|C|$ by median intervals).

If C is order-scattered (i.e. scattered in the sense of 6.1), **then C is topologically scattered. Hence both notions of scattered and topologically scattered chain are equivalent** (axiom of choice).

• By hypothesis C is scattered, thus by 6.1.4 the set $\mathcal{J}(C)$ of initial intervals (ordered by inclusion) is scattered. Suppose our conclusion is false, so that $\mathcal{J}(C)$ contains a perfect subset $\mathcal{H}$. Consider as equivalent two elements x, y in $|C|$ iff x and y belong to the same initial intervals which belong to $\mathcal{H}$. Then in each equivalence class choose one representative element (axiom of choice). These representative elements constitute a scattered chain D (a restriction of C). Consequently there exist two consecutive elements a followed by b (mod D) (see 6.1). By construction of D there exists one and only one initial interval $V \in \mathcal{H}$ with $a \in V$ and $b \notin V$. The basic clopen set of those initial intervals of C which contain a yet not b, topologically isolates V in $\mathcal{H}$. Hence $\mathcal{H}$ is not perfect. •

6.7.3 Poset of antichains; thickness of a finitely free poset

Let X be a poset; we define the **poset of antichains** denoted $\mathcal{A}(X)$, as being made of all all antichains of X, ordered by reverse inclusion; then:

$\mathcal{A}(X)$ has a largest element, namely **the empty antichain** denoted $\emptyset$, and (with axiom of choice) minimal elements, namely the maximal antichains of X with respect to inclusion.

$\mathcal{A}(X)$ is well-founded iff X is finitely free.

Given a finitely free poset X, we call **thickness** of X, denoted $\text{Thick}(X)$ the height of the empty set in $\mathcal{A}(X)$. Clearly the thickness is 0 iff X is the empty set; the thickness is 1 iff X is a non-empty chain.

Given a poset X with thickness α and an element $u \in |X|$, then the restriction of X to elements $\geq u$ has thickness $\leq \alpha$; idem for the restriction to elements $\leq u$. Equality often happens: share a non-empty chain C (thickness 1) into the initial interval $\leq u$ and the final interval $\geq u$ $(\text{mod}\, C)$.

Given a poset X and an element $u \in |X|$, the singleton $\{u\}$ constitutes an antichain of X; moreover **the height** $\text{Ht}(u)$ **in** X **is** $< \text{Ht}(X)$; **more precisely** $\text{Ht}(X) = \text{Sup}(\text{Ht}(u) + 1)$ for all elements $u \in |X|$.

Let X_u denote the poset restriction of X to those elements which are $|u(\text{mod}\, X)$. Then $\text{Thick}(X_u) \leq \text{Ht}(\{u\})$ in $\mathcal{A}(X)$, hence $< \text{Ht}\,\emptyset$ in $\mathcal{A}(X)$ hence $\text{Thick}(X_u) < \text{Thick}(X)$.

6.7.4 Equivalence between finitely free scattered poset and topologically scattered poset

Theorem. Every finitely free, order-scattered poset is topologically scattered (axiom of choice). Consequently both notions are equivalent. Due to [202] POUZET 1981 p.847. For a more general result, see [174] MISLOVE 1984.

The statement is trivial for the empty poset, i.e. the poset with thickness 0. Moreover it is already proved for chains, i.e. posets with thickness 1: see 6.7.2.

Inside the proof, we use the following:

Perfect set partition lemma. Given a finitely free poset A with thickness α, suppose the theorem is true for every poset with thickness $< \alpha$, and moreover that $\mathcal{J}(A)$ includes a perfect subset. Then there exists at least one element $u \in |A|$ such that the restriction A^+ of A to elements $\geq u$ $(\text{mod}\, A)$ admits a perfect subset of $\mathcal{J}(A^+)$ and similarly the restriction A^- of A to elements $\leq u$ admits a perfect subset of $\mathcal{J}(A^-)$.

• Firstly prove that, if the theorem is true for every thickness $< \alpha$, then the lemma implies the theorem for α. Indeed let A be a finitely free, order-scattered poset with thickness α. Either $\mathcal{J}(A)$ is topologically scattered and we are done. Or $\mathcal{J}(A)$ contains a perfect set, then by the previous partition lemma there exists a $u \in |A|$ such that both restrictions to elements $\geq u$ and to elements $\leq u$ give a perfect subset of each set $\mathcal{J}(A^+)$ and $\mathcal{J}(A^-)$. Since by the previous subsection A^+ and A^- have thickness $\leq \alpha$, then by successive applications of the lemma we get a restriction of A isomorphic with the chain $\mathbf{Q}$ of rationals; hence A is not order-scattered: contradiction. •.

• To prove the partition lemma: let α be the thickness of A, and suppose the theorem is true for every thickness $< \alpha$.

By hypothesis $\mathcal{J}(A)$ includes a perfect subset $\mathcal{U}$. Let J, K be two distinct initial intervals of A belonging to $\mathcal{U}$. Fix ideas by assuming the existence of an element $u \in |J| - |K|$. In the following we keep the notation $\mathcal{U}$ for the (again perfect) set of those initial intervals X such that $u \in X \in \mathcal{U}$.

Call $\mathcal{U}^+$ the set $\mathcal{U}$ where each element X (= initial interval of A) is replaced by $X \cap A^+$; so that $\mathcal{U}^+ \subseteq \mathcal{J}(A^+)$. Arguing ad absurdum, suppose that $\mathcal{J}(A^+)$ does not include any perfect subset. Consequently there exists an isolated element in $\mathcal{U}^+$ which can be considered as the intersection $A^+ \cap I_0$ where $I_0 \in \mathcal{U}$.

Let A^* be the restriction of A to elements incomparable with $u \pmod A$. By the previous subsection A^* has thickness $< \alpha$, and by hypothesis the theorem is true for such a thickness; so that the set $\mathcal{J}(A^*)$ (initial intervals of A^*) does not contain any perfect subset. Call $\mathcal{V}$ the set of intersections $X \cap A^*$ where $X \in \mathcal{U}$ and furthermore $X \cap A^+ = I_0 \cap A^+$. Then there exists an element which is isolated in $\mathcal{V}$ and which can be considered as the intersection $A^* \cap X_0$ where $X_0 \in \mathcal{U}$.

Since X_0 (initial interval of A), is entirely defined by its restrictions to A^+, to A^* and by the obvious fact that all elements $\leq u \pmod A$ belong to X_0, we see that X_0 is isolated in $\mathcal{U}$ by means of some open sets (or equivalently basic clopen sets) in our topology on $\mathcal{J}(A)$. Hence $\mathcal{U}$ is not perfect: contradiction.

Similar argument and contradiction with A^- instead of A^+. Now the supposed perfect set $\mathcal{U}$ is formed of initial intervals X such that $u \notin X \in \mathcal{U}$. •

Corollaries.

(1) **Let A be an infinite, finitely free and order-scattered poset; then the set $\mathcal{J}(A)$ of initial intervals is equipotent with A** (axiom of choice). Due to [17] BONNET 1975. Consequence of 6.7.1 by taking into account the equivalence between finitely free, order-scattered poset and topologically scattered poset. This statement generalizes 6.1.4 proposition (2).

(2) **Let A be a denumerable poset; then the following two conditions are equivalent:**

(i) **A is finitely free and order-scattered (= no embedding of $\mathbf{Q}$)**;

(ii) **the set of initial intervals of A is denumerable.** We use axiom of choice to deduce (ii) from (i).

• If A has a denumerable free subset or if $\mathbf{Q}$ is embeddable in A, then there are continuum many initial intervals: hence (ii) implies (i). The converse is a particular case of 6.7.1 by taking into account the equivalence theorem 6.7.4 (using axiom of choice). •

(3) Let A be an infinite partial ordering; then **the following two conditions are equivalent** (uses axiom of choice).

(i) **A is finitely free and order-scattered;**

(ii) **there exists a scattered chain with the same cardinality as A, in which every totally ordered augmentation of A is embeddable** (uses axiom of choice).

• The condition (ii) implies that every totally ordered augmentation of A is scattered, hence (ii) implies (i) by 6.5.1, conditions (1) and (2).

Conversely, if A satisfies our (i), then again by 6.5.1 conditions (1) and (3), the partial ordering $\mathcal{J}(A)$ (initial intervals of A) is order-scattered. Hence by 6.5.3 proposition (2), there exists a scattered, totally ordered augmentation B of $\mathcal{J}(A)$. On the one hand, by the previous corollary (1), $\mathcal{J}(A)$ and B have same cardinal as A. On the other hand, for every totally ordered augmentation X of A, we have that X is a restriction of $\mathcal{J}(X)$ which is itself a totaly ordered restriction of $\mathcal{J}(A)$, hence of B. •

6.8 Indivisible relation or chain

We say that a relation R with infinite base E is **indivisible** iff for every partition of E into two complementary subsets C and $D = E - C$, either $R \leq R/C$ or $R \leq R/D$. Instead of "indivisible", some authors say "indecomposable" or "strongly indecomposable" ([153] LAVER 1973 p.105).

For example the chain ω is indivisible.

Every indivisible chain is a fortiori indecomposable. However the chain $\mathbf{Z}.\omega$, where $\mathbf{Z}$ is the chain of positive and negative integers, is indecomposable yet not indivisible: if we partition $\mathbf{Z}$ into ω and ω^- (the chain of negative integers), then we get the restrictions ω^2 and $\omega^-.\omega$, neither of which admits an embedding of $\mathbf{Z}.\omega$.

Every regular aleph is indivisible. Indeed if α is a regular aleph and we partition its base into two complementary subsets, then one of these subsets must be cofinal in α, and thus yields a restriction isomorphic with α.

6.8.1 For an ordinal, indivisibility is identical with indecomposability

Consequently by 2.8.7, **every infinite aleph is indivisible**.

• It suffices to show that every indecomposable ordinal is indivisible, hence that every ordinal power of ω, say ω^u, is indivisible (see 1.3.6).

Proceed by induction on the exponent u. Suppose firstly that u is a successor ordinal: $u = v + 1$. Then ω^u is a sum along ω of indivisible ordinals equal to ω^v. Hence a partition of ω^u into two complementary subsets gives one of the subsets as a sum along ω of components ω^v.

Now suppose that u is a limit ordinal. Then u is the supremum of a sequence of ordinals $v(i)$. We can suppose that the index i runs through the regular aleph $\mathrm{Cof}\, u$. By the induction hypothesis, each power $\omega^{v(i)}$ is an indivisible initial interval of ω^u, the union of which is ω^u.

The partition of the base yields, for each i, at least one restriction isomorphic with $\omega^{v(i)}$: call this the i-majority restriction. Then at least one of the two subsets in our partition yields an i-majority restriction for a set of indices i which is cofinal in $\mathrm{Cof}\, u$. Thus we obtain a restriction of ω^u which is isomorphic with the union, or supremum of the $\omega^{v(i)}$, hence isomorphic with the given ordinal $\omega^u = w^{(\text{Sup}\, v(i))} = \text{Sup}(\omega^{v(i)})$. •

6.8.2 Decomposition of an indivisible chain

Theorem. Given an indivisible chain A with infinite base E, there exists a partition of E into two complementary subsets C and $D = E - C$ with $A \simeq A/C \simeq A/D$

Due to HAGENDORF in 1975; uses axiom of choice.

• An indivisible chain A is a fortiori indecomposable. By 6.3.4 (unique decomposition theorem) either A is equimorphic with $A.2$, and so the desired partition exists. Or A is strictly right or left indecomposable: to fix ideas suppose that it is right indecomposable. Set $I = \text{Cof}\, A$ which is a regular aleph. Decompose A into a sum along I of intervals A_i(i running through I; axiom of choice).

With the interval A_0, associate the complementary final interval, in which A is embeddable, and in this final interval take a proper initial interval $A"_0$ in which A_0 is embeddable. Call $A'_0 = A_0$, so that A'_0 and $A"_0$ are two copies of A_0. After $A"_0$ it remains a final interval equimorphic with A. Iterate this by embedding, in this final interval, two consecutive copies A'_1 and $A"_1$ of A_1, the interval A_1 being embeddable in each copy and the remaining final interval being equimorphic with A. This procedure can be iterated for every index i belonging to I. Indeed for each ordinal index i, the i-sequence of the sums $A'_j + A"_j (j < i)$ of copies cannot exhaust A; since otherwise either i would be cofinal in I (contradicting the regularity of I), or i would have an immediate predecessor $i - 1$ with A'_{i-1} or $A"_{i-1}$ a final interval of A, which is excluded.

Finally, the union of the A'_i and the union of the $A"_i$ for all $i \in I$, yield the desired partition. •

It is proved in [200] POUZET 1981 that, **given an indivisible relation R with denumerable base E, there exists a partition of E into two complementary subsets C and D with $R \simeq R/C \simeq R/D$.** A more intuitive proof, due to [246] THOMASSÉ 1995, is given in 11.6.4 below.

Problem: the case of an uncountable indivisible relation.

From [153] LAVER 1973:

Any indivisible poset is obtained from ordinals by inversions and lexicographical sums such that each term is embeddable in the following term (this is stronger than the hereditarily indecomposable condition of 7.4 and 7.5.3).

For example starting from ω and the opposite chain ω^- we get the indivisible chain $\omega.\omega^-$ yet we cannot get the sum $\mathbf{Z} = \omega^- + \omega$ nor the product $\omega.\mathbf{Z}$.

6.9 Rigid chain

We say that a chain A is **rigid** iff A cannot be embedded in any strict restriction. For example, any finite chain is rigid.

1 - A denumerable scattered chain is not rigid: take an infinite 0-neighborhood, which is isomorphic either with ω or its converse or with their sum $\mathbf{Z}$, then take away an element from this neighborhood.

Using the Hausdorff decomposition 6.2.1, one easily sees that a denumerable chain is never rigid. Moreover, between two elements in a rigid chain, either there exist only finitely many elements or more than denumerably many elements.

For the existence of an infinite rigid chain, see [20] BONNET, SHELAH 1985.

2 - Let A be an infinite rigid chain. Let A_u be the chain A after adding a new element u. Then A_u **is immediately greater than** A (with respect to embeddability).

• Firstly $A < A_u$. Indeed if $A_u \leq A$, then by removing u we get a strict restriction of A which would be isomorphic with A: contradiction.

Suppose $A_u \leq A$; then the chain $B = A_u$ minus the element u is isomorphic with a strict restriction of A, contrary to the rigidity of A.

Now suppose the existence of an intermediate chain I with $A_u > I > A$. We can take for I a strict restriction of A_u which necessarily contains u. Then take a strict restriction A' of I which be isomorphic with A. The element u does not belong to the base $|A'|$, since otherwise A' would be a strict restriction of $A = A_u$ minus u. Finally A' is a strict restriction of I minus u which is itself a strict restriction of A_u minus u. Consequently A' is a restriction of $A = A_u$ minus u: so that A is not rigid: contradiction. •

Problem. Existence of three chains A, B, I such that B is obtained from A by addition of a unique element, yet $A > I > B$ with respect to strict embeddability.

3 - Any two distinct initial intervals of a rigid chain A are never isomorphic. Idem for any two distinct final intervals. Consequently if we add an element u and another element v in two different cuts of A serparated by infinitely many elements, then the two obtained chains A_u and A_v are not isomorphic.

A chain A is said to be **weakly rigid** iff every automorphism f from A onto a restriction of A moves at most countably many elements $x (f(x) \neq x)$.

Consequently every finite or denumerable chain is weakly rigid (yet not rigid in the denumerable case).

6.10 Exercises

6.10.1 Every ultrafilter on the integers is adherent to the set of trivial ultrafilters; yet it is not the limit of any convergent sequence of them

Let $\mathbf{N}$ be the set of non-negative integers; let us define a topology on $\mathcal{PP}(\mathbf{N})$.

Start with the basic clopen sets $U_{A,B}$ where A, B are finite sequences of subsets of $\mathbf{N}$. More precisely $A = (A_1, ..., A_m)$ and $B = (B_1, ..., B_n)$ and $U_{A,B}$ is the set of those elements of $\mathcal{PP}(\mathbf{N})$ which contain as elements A_1 and ... and A_m yet do not contain B_1 nor ... nor B_n.

As in 6.6, an open set is defined as an arbitrary union of basic clopen sets; as precedently we have the compacity; consequently any clopen set is a finite union of basic clopen sets.

(1) Firstly prove that every ultrafilter on $\mathbf{N}$ is adherent to the set of all trivial ultrafilters.

Let H be an arbitrary ultrafilter on $\mathbf{N}$ (thus an element of $\mathcal{PP}(\mathbf{N})$) which contains as elements the subsets $A_1, ..., A_m$ of $\mathbf{N}$ yet does not contain B_1 nor ... nor B_n (so that H contains the complementary sets $\mathbf{N} - B_1$, ... $\mathbf{N} - B_n$). Take an integer h which belongs to the A's yet not to any B. Then the trivial ultrafilter on $\mathbf{N}$, which is formed of those sets which contain h, belongs to $U_{A,B}$ where $A = (A_1, ..., A_m)$ and $B = (B_1, ..., B_n)$. Hence H **is adherent to the set of all trivial ultrafilters**.

(2) Secondly, given an ordinal-indexed sequence of subsets $H_1, ..., H_i, ..$ of $\mathcal{P}(\mathbf{N})$, thus elements of $\mathcal{PP}(\mathbf{N})$ and another subset $H \subseteq \mathcal{P}(\mathbf{N})$, the sequence H_i is said to be **convergent towards the limit** H iff every open set (equivalently every basic clopen set) which contains H as an element, also contains H_i from some point on. Equivalently, for every subset X of $\mathbf{N}$, either $X \in H$ and then $X \in H_i$ from some point on, or $X \notin H$ and then $X \notin H_i$ from some point on.

Consequently given a convergent ordinal sequence whose terms are trivial ultrafilters on $\mathbf{N}$, we see that from some point on, the sequence reduces to the repetition of one trivial ultrafilter, which is the limit element. Indeed consider two cases, according to whether the cofinality of our ordinal sequence is ω or is strictly larger. In the latter case, for convergence it is necessary and sufficient that only one trivial ultrafilter is cofinally repetited: it is seen by associating to each trivial ultrafilter H_i the unique singleton which belongs to it.

In the case of cofinality ω, denote by h_i the integer associated with the trivial ultrafilter H_i. Then if h_i takes infinitely many values, let V be a subset of $\mathbf{N}$ which contains infinitely many such and excludes infinitely many such values. Then V neither can belong to H_i from some point on, nor can be a non-element of H_i from some point on.

Finally a non-trivial ultrafilter cannot be the limit of any ordinal sequence of trivial ultrafilters.

6.10.2 A generalized condition of indecomposability

Let us generalize to an arbitrary chain the indecomposability condition already given for an arbitrary ordinal in 5.6.

A chain U is indecomposable iff, for every chain $X <$ or $|U$ (with respect to embeddability) we have $X.2 <$ **or** $|U$. See [104] HAGENDORF 1977 th. 1.12, who proves the stronger result: if $U \neq 2$ it suffices that $X <$ or $|U$ yields $X.2 \leq$ or $|U$.

• Suppose that U is indecomposable and that $X.2 \geq U$: then $X \geq U$.

Conversely, suppose that U is decomposable, so that $U = V + W$ with $V < U$ and $W < U$, yet that U satisfy our conclusion. Then $V.2 \not\geq U = V + W$ so

$W \not\leq V$; similarly $V \not\leq W$. Moreover $(W+V).2 = W+V+W+V \geq U$ so $W+V \geq U = V+W$ and hence $V \leq W$ or $W \leq V$, so that either $V.2 \geq U$ or $W.2 \geq U$: contradiction. •

Note that the chain $\mathbf{Z}$ of the integers is decomposable, yet verifies the condition that every $X < \mathbf{Z}$ satisfies $X.2 <$ or $|\mathbf{Z}$.

The proposition 5.6.1 about ordinals does not extend to chains. Indeed $U = \omega^-.\omega$ is indecomposable and $\omega < U$ but $\omega.2|U$ and $\not< U$.

If a chain U is indecomposable, then every initial interval $X < U$ satisfies $X.2 < U$. However the retro- ordinal $(\omega+1)^-$ is decomposable and every non-empty initial interval $X < U$ is isomorphic with 1, hence $X.2 < U$. Similarly with $(\omega^2+\omega)^-$. Similarly with $\mathbf{Q}+\omega_1$.

Problem. Let U be a chain. If every chain $X < U$ yields $X.2 < U$, then is U an indecomposable chain. Affirmative answer for scattered U (HAGENDORF in 1982, published in ToR-86 p.219).

6.10.3 An intermediate chain

Given an infinite chain A, either $A.2$ is equimorphic with A, or there exists at least one chain which is strictly intermediate between A and $A.2$ with respect to embeddability (HAGENDORF in 1984; published in ToR-86 p.176).

• Decompose $A = B+L$ where L is the maximum final interval which is left indecomposable: see 6.3.3. If $L = A$ then either $A.2$ is equimorphic with A, or $1+A$ is strictly intermediate between A and $A.2$. If L is empty, then each initial interval of A is $< A$, so that $A+1$ is strictly intermediate. In the following we assume that $L \neq A$ and L non empty.

Consider the sum A_1+A_2 where A_1, A_2 both isomorphic with A. Decompose this sum, obtaining $C+L'$ where L' is the maximum left indecomposable interval. Either $L' > A_2$: then $A.2$ is equimorphic with A. Or $L' = L$ and $C = A+B < A.2$ by 6.3.3. Either C is strictly intermediate between A and $A.2$. Or C is equimorphic with A. Then either $L.2$ is equimorphic with L: then $A = B+L \simeq B+L.2 = A+L \simeq C+L = A.2$. Or finally L is strictly left indecomposable. Since $A = B+L \simeq B+L+B$ and $B < A$ (see previous reference), then $L+B$ is embeddable in L, so that $1+B \leq L$ and $B < L$. Thus $L+B+L$ is strictly intermediate between $A = B+L$ and $A.2 = B+L+B+L$. •

6.10.4 Restriction of a topologically scattered poset

Every restriction of a topoplogically scattered poset is itself topologically scattered. This becomes obvious if we use the equivalence with finitely free, order-scattered posets. Yet in the absence of the equivalence theorem 6.7.4, the above statement is not obvious.

Start from a topologically scattered poset A and its restriction B. Call f the identity mapping from B onto itself. Then consider $\overline{f^{-1}}$ as being a surjective mapping from $\mathcal{P}(|A|)$ onto $\mathcal{P}(|B|)$($\mathcal{P}$ = power set), then a continuous and surjective mapping from the set of initial intervals $\mathcal{J}(A)$ onto $\mathcal{J}(B)$ (again in our simple convergence topology).

Then use a result of [183] PELCZYNSKI, SEMADENI 1959 (see also [219] RUDIN 1957): compact topologically scattered spaces are preserved under continuous images.

Chapter 7

Barrier, forerunning, well quasi-ordering of scattered chains, better partial ordering

7.1 Barrier, barrier partition theorem, successive elements (Nash-Williams)

Let E be a denumerable set of natural numbers, and U a set of finite non-empty subsets of E, whose union is E. The set U is a **barrier** iff:

(1) the elements of U are mutually non-inclusive;

(2) for every infinite subset X of E, there exists a finite initial interval of X (initial with respect to the usual ordering of the integers) which belongs to U.

Examples. For E take the set of all natural numbers, this case being the most frequently encountered. Then the set of singletons of the integers is a barrier. Similarly, the set of pairs of integers is a barrier, as well as the set of p-element subsets of the integers, for any given positive integer p.

Another example of a barrier: the union of the set of pairs whose minimum is 0, and the set of 3-element subsets with minimum 1, and for each integer $i \geq 2$, the set of i-element subsets with minimum $i - 2$.

7.1.1 Restriction of a barrier

Let U be a barrier and P an infinite subset of the union $\cup U$. Then the set of those elements of U which are subsets of P is a barrier.

Moreover, every barrier included in U is thus obtained, i.e. by taking an infinite set of integers and restricting U to those elements which are

subsets of this infinite set.

• The elements of U which are subsets of P are mutually incomparable with respect to inclusion. Moreover, for every infinite subset X of P, there exists a finite initial interval of X which is an element of U.

For the second assertion, let V be a barrier included in U, and P be the union $\cup V$. For each element r in U which is a subset of P, take an infinite subset R of P which begins with r: there exists an element of V which is an initial interval of R, and this can only be r. •

If U is a barrier and P an infinite subset of $\cup U$, then the set of intersections of P with the elements of U does not necessarily form a barrier. For instance, starting with the barrier U of pairs of integers, let P be the set of strictly positive integers (we remove zero). Then we obtain both the singleton $\{1\}$ coming from the pair $\{0, 1\}$, and for example the pair $\{1, 2\}$ in which the singleton is included.

7.1.2 Removing an initial interval

Let U be a barrier and r a finite set of integers which is a proper initial interval of an element of U. Then the set of elements of the form $x - r$, where x is any element of U beginning with the initial interval r, is a barrier.

• Let V be the set of our difference sets $x - r$. The union of the elements of U beginning with the initial interval r is infinite; so $\cup V$ is infinite. Any two of these elements are incomparable with respect to inclusion: this subsists when removing the initial interval r. Finally, for every infinite subset P of the union $\cup V$, which necessarily begins with an integer $> \operatorname{Max} r$, there exists a finite initial interval y of P, such that the union $r \cup y$ belongs to U, so y belongs to V. •

7.1.3 Lexicographic rank of a barrier

Every barrier is lexicographically well-ordered.

This is a particular case of 3.2.1. In other words, the set of the elements of a barrier, when ordered lexicographically starting with the usual ordering of integers, forms a denumerable well-ordering.

Every barrier U thus has a lexicographic rank, in the sense of 3.2.1: the order type of the well-ordering of the elements of U.

For example, the barrier of the singletons has rank ω. The barrier of the p-element subsets (p = positive integer) has rank ω^p. The barrier formed of the pairs with minimum 0, and the 3-element subsets with minimum 1, and for each integer $i \geq 2$, the i-element subsets with minimum $i - 2$, has rank ω^ω.

For each denumerable indecomposable ordinal γ (thus a power of ω), there exists a barrier with rank γ ([191] POUZET 1972).

More generally, the lexicographic ranks of barriers are exactly all the ordinals ω^p(p = positive integer) and $(\omega^\alpha).p$, where α is a denumerable ordinal and p a positive integer ([3] ASSOUS 1974).

7.1.4 Barrier partition theorem

Let U be a barrier and E be the union of U. Partition the elements of U into two complementary sets U' and U". Then there exists an infinite subset H of E such that the elements of U which are subsets of H, all belong to U' or all belong to U". See [179] NASH-WILLIAMS 1968.

Note that these elements form a barrier by 7.1.1. In particular, we obtain RAMSEY's theorem by considering a positive integer p and taking for U the set of all p-element subsets of E.

• Given two distinct elements of U, one is never included in the other, hence one is never an initial interval of the other: the theorem now follows from NASH-WILLIAMS separation theorem (see 3.2.4). •

7.1.5 Successive element

Given two finite sets of natural numbers r and s, we say that r **precedes** s or that s **succeeds** r, or that s is a **successive element** of r, denoted by $r \lhd s$, iff s is obtained from r by adding on integers which are strictly greater than $\text{Max}(r)$ and then removing $\text{Min}(r)$.

For example, given two integers a, b, the singleton $\{a\}$ precedes the singleton $\{b\}$ iff $a < b$. Given three integers $a < b < c$, the pair $\{a, b\}$ precedes the pair $\{b, c\}$.

(1) **Let U be a barrier and r, s two elements of U; put $m = \text{Card}\, r$. If $\text{Max}\, r < \text{Min}\, s$, then there exists a sequence of $m+1$ successive elements of U going from r to s, say $r = r_0 \lhd r_1 \lhd r_2 \lhd \ldots \lhd r_m = s$.**

• Let I be an infinite set of integers which begins with the initial interval r followed by s and finally by integers larger than $\text{Max}\, s$, all belonging to the union of U. Start with $r_0 = r$ and $I_0 = I$. Then define $I_1 = I_0$ minus its minimum, then let r_1 be the initial interval of I_1 which belongs to U, so that $r_0 \lhd r_1$. Iterate until reaching $r_m = s$. •

(2) **Let U be a barrier and r, s be two arbitrary elements of U. Then there exists an element t of U and two finite sequences of successive elements of U, the first sequence going from r to t and the second going from s to t.**

• Take an arbitrary element $t \in U$ with $\text{Min}\, t$ strictly larger than both $\text{Max}\, r$ and $\text{Max}\, s$, then apply proposition (1) above. •

7.1.6 Construction of two sequences with successive terms

Lemma. Let U be a barrier, r be an element of U and $m = \text{Card}\, r$.

Given an integer k larger than or equal to m, there exists a sequence of $m+1$ successive elements of U beginning with r, say $r = r_0 \lhd r_1 \lhd \ldots \lhd r_m$ and a sequence of $k+1$ successive elements of U, both sequences having same first and same last terms, say $r = s_0 \lhd s_1 \lhd \ldots \lhd s_k = r_m$.

• First construct a sequence of $k+1$ successive terms beginning with r, say $r = s_0 \lhd s_1 \lhd \ldots \lhd s_k$ using an infinite set of integers which begins with the initial interval r, all the integers belonging to the union of U. Since k is larger than or equal to $m = \text{Card}\, r$, the minimum of s_k is an integer strictly larger than $\text{Max}\, r$. Using the previous subsection proposition (1), we construct the sequence of $m+1$ successive elements of U going from r to $r_m = s_k$. •

7.1.7 A strictly increasing function on a barrier

Lemma. Let h be a function from the set ω of non-negative integers into ω. Let U be a barrier and f a function from U into ω, such that for each couple of successive elements $x \lhd y$ we have $f(x) < f(y)$. Then there exist two successive elements $s \lhd t$ in U with $h(f(s)) \leq f(t)$ (POUZET 1977).

• Suppose on the contrary that $s \lhd t$ always implies that $f(t) \leq h(f(s)) - 1$. Start with an arbitrary element s of U, which shall be fixed for the following discussion, and let $m = \text{Card}\, s$. We shall prove that there exists an integer k such that, for every sequence of $m+1$ successive elements of U beginning with s, say $s = s_0 \lhd s_1 \lhd \ldots \lhd s_m$, the value $f(s_m)$ is bounded by k, say $f(s_m) < k$; in spite of the infinitely many possible choices for $s_1, \ldots, s_m$.

Indeed, the value $f(s_1)$ is bounded by the maximum $h(f(s)) - 1$, which we denote by k_1. The value $f(s_2) \leq h(f(s_1)) - 1$ is bounded by the maximum $k_2 = \text{Max}(h(x)) - 1$, where x runs through the interval $0, 1, 2, \ldots, k_1$. We continue thusly until $f(s_m) \leq h(f(s_{m-1})) - 1$, bounded by $k_m = \text{Max}(h(x)) - 1$, where x runs through the interval $0, 1, 2, \ldots, k_{m-1}$. We then take $k = k_m + 1$. We can always assume that the bound $k \geq m$. Then by the preceding subsection, there exist two sequences, say $s = s_0 \lhd s_1 \lhd \ldots \lhd s_m$ and $s = t_0 \lhd t_1 \lhd \ldots \lhd t_k = s_m$. The first sequence requires that $f(s_m) < k$, and the second, given our hypotheses, requires that $f(s_m) - f(s) \geq k$, hence that $f(s_m) \geq k$: contradiction. •

7.1.8 Square and power of a barrier

Let U be a barrier. To each couple of elements $s, t \in U$ such that $s \lhd t$, associate the union $s \cup t$. Note that this union uniquely determines s as that element of U which is an initial interval of the union. Similarly t is uniquely determined, as being the union with its minimum element removed. The unions in consideration are mutually non-inclusive. Moreover, every infinite subset of the union set of U has such a union $s \cup t$ as an initial interval. Thus these unions form a barrier, called the **square** of U and denoted by U^2.

For each positive integer h, the power U^h is defined by induction: $U^1 = U$ and $U^{h+1} = (U^h)^2$.

7.2 Barrier sequence, theorem on the minimal bad barrier sequence (Nash-Williams)

7.2.1 Good or bad barrier sequence

Given a barrier U and a poset A, we define a **barrier sequence**, more precisely a U-**barrier sequence in** A or with values in A, to be a function with domain U such that the range is included in the base $|A|$. We also more simply speak of a U-**sequence**.

A barrier sequence f of U in A is called **good** iff there exist two elements $s, t \in U$ with $s \triangleleft t$ and $fs \leq ft (\text{mod}\, A)$. The sequence f is called **bad** otherwise.

In the trivial case of a barrier of singletons, recall that for u, v non-negative integers, we have $u \triangleleft v$ iff $u < v$ in the usual ordering of integers. We have again the notion of **good** or **bad** ω-sequence, in the sense of 4.2.1. Consequently, if A is a well partial ordering, then the barrier sequences whose domain is the barrier of singletons are good: see 4.3.2 proposition (2).

This result does not extend to all barriers. Indeed, let A be the well partial ordering of RADO (see 4.4.2) and let U be the barrier of all pairs of non-negative integers.

Then **the following function from** U **into** A **is a bad barrier sequence**: for all integers x, y $(x < y)$ our function takes the pair $\{x, y\}$ into the couple of integers $(x, y - x + 1)$.

• Two successive pairs are of the form $\{x, y\}$ and $\{y, z\}$ with three integers $x < y < z$. They are taken respectively into the couples $(x, y - x + 1)$ and $(y, z - y + 1)$. These two couples have distinct first terms. Therefore the second couple can be greater (modulo the partial ordering of RADO) than the first only if $y \geq x + (y - x + 1) = y + 1$: contradiction. •

7.2.2 Restriction of a barrier sequence, perfect barrier sequence

Let A be a poset and U, V be barriers with V included in U. As with any function, we say that, given a U-sequence f, its **restriction** to V, denoted by f/V, is the V-sequence taking for each element of V the same value as f.

Let U be a barrier and A a poset. A barrier sequence f (which is a function from U into the base of A) is called **perfect** iff, for each couple of elements s, t in U, the condition $s \triangleleft t$ implies $fs \leq ft (\text{mod}\, A)$.

Theorem. Let U **be a barrier and** f **a function from** U **into a poset. Then there exists a barrier** V **included in** U**, such that the restriction** f/V **is either perfect or bad** ([178] NASH-WILLIAMS 1965 p. 705).

• Consider the square U^2 of the barrier U, which is formed of unions $r = s \cup t$ where s, t are successive elements in U. Recall that each element $r \in U^2$ determines s as the initial interval of r which belongs to U, as well as t, which is r with its minimum removed. Partition the elements r of U^2 into two classes, according to whether $fs \leq ft$ (modulo the partial ordering) or not. By the barrier partition theorem (7.1.4), there exists an infinite set E of integers in which the elements of U^2 are all in the same class; and by 7.1.1 they constitute a barrier V. According to whether the class in consideration corresponds to the condition $fs \leq ft$ or to its negation, the restriction f/V is either perfect or bad. •

7.2.3 Inf-restriction, minimal bad barrier sequence

Let U, V be barriers (V included in U); let A be a poset and f be a U-sequence with values in A. A V-sequence g with values in A is called an **inf-restriction** $(\text{mod}\, A)$ of f iff for each element x of V we have $gx \leq fx (\text{mod}\, A)$.

We say that a barrier sequence f with domain U and with values in A is **minimal bad** $(\text{mod}\, A)$ iff f is bad and every bad inf-restriction of f is a restriction. In other words, for each barrier V included in U, each bad V-sequence g such that $gx \leq fx$ satisfies $gx = fx$ (and not $gx < fx$) for every $x \in V$.

This generalizes the notion of a minimal bad ω-sequence. Indeed in the case where U is the trivial barrier of singletons, f becomes an ω-sequence and we find the minimal bad ω-sequence of 4.2.2.

7.2.4 Existence of a minimal bad barrier sequence

Theorem. Let A be a well-founded poset and f be a bad barrier sequence with values in A. Then there exists a barrier sequence which is an inf-restriction of f and is minimal bad $(\text{mod}\, A)$ ([178] NASH-WILLIAMS 1965', lemma 24 p.707; uses dependent choice). This generalizes the existence of the minimal bad sequence in 4.2.6.

• Let U denote the barrier domain of f, and let r_0 be an element in U whose maximum integer a is the least possible. Let U_0 be a barrier included in U and which contains the element r_0. Let f_0 be a bad function from U_0 into A which is a n inf-restriction of f and for which $f_0(r_0)$ is minimal $(\text{mod}\, A)$, in the sense that no other choice of U_0 and f_0 gives a value $f_0(r_0)$ strictly less than the above value.

Let r_1 be an element of U_0 which is distinct from r_0 and whose maximum integer is the least possible. Let U_1 be a barrier included in U_0 and which contains the elements r_0, r_1. Let f_1 be a bad function from U_1 into A, which is an inf-restriction of f_0 and for which $f_1(r_1)$ is minimal. Iterating this, we obtain, for each integer i, a finite set r_i of integers, a barrier U_i and a function f_i from U_i into A (dependent choice). Each U_i is the set of those elements of U which are subsets of the union $\cup U_i$, and each of these unions is included in the preceding one.

Note that $f_1(r_0) = f_0(r_0)$, and in general $f_j(r_i) = f_i(r_i)$ for all integers i and $j > i$. The maximum of r_i increases (however not always strictly) in i and tends to infinity, since there are only finitely many sets of integers having a given maximum. We shall prove that the set V of the r_i is a barrier.

Indeed, two distinct r_i are mutually non-inclusive, since they belong to U. Moreover, every element r in U which is included in the union of the r_i is an r_i: letting a denote the maximum of r, it suffices to see that there only exist finitely many sets of integers with maximum a, hence that r will be one of the r_i provided that it is simultaneously included in $\cup U_0$ and $\cup U_1$ and $\cup U_2$.. . Here this is indeed the case, since r is included in the union of the r_i. Finally the barrier V is the set of the r_i, as well as the intersection of the U_i, as well as the set of those elements of U which are included in the union of the r_i. Moreover the union $\cup V$ is equal to the union of the r_i, as well as the intersection of the $\cup U_i$.

Let g denote the barrier sequence with domain V, and which is defined by $g(r_i) = f_i(r_i)$ for $i = 0, 1, 2, ..$. Note that g is a bad inf-restriction of f. We shall prove that g is minimal bad. Suppose the contrary: there exists a barrier W included in V, and a bad W-sequence h which is an inf-restriction of g, and an integer i for which $r_i \in W$ with $h(r_i) < g(r_i)(\text{mod}\, A)$. Let n be the least i for which this holds. Call W^+ the sub-barrier of V defined by $\cup W^+ =$ the union of r_0, of r_1, ... , of r_{n-1} and of all elements of W: hence $W \subseteq W^+ \subseteq V$. Let h^+ be the barrier sequence with domain W^+ with $h^+(r_i) = h(r_i)$ for $r_i \in W$, and $h^+(r_i) = g(r_i)$ for $r_i \in W^+ - W$; so in particular $h^+(r_n) = h(r_n) < g(r_n)(\text{mod}\, A)$. Note that $h^+(r) \leq g(r)$ for every $r \in W^+$: hence our barrier sequence h^+ is an inf-restriction of g.

To obtain a contradiction, it suffices to show that h^+ is bad. Indeed for $i = 0, 1, ..., n-1$ we have $h(r_i) = g(r_i)$ and either $r_i \in W^+ - W$ or $r_i \in W$: in both cases we have $h^+(r_i) = g(r_i)$. Since $h^+(r_n) < g(r_n)(\text{mod}\, A)$, we obtain a contradiction with the minimality of the element $f_n(r_n) = g(r_n)$: see the iterative definition of $f_0, f_1, ..., f_n$ in our second alinea.

Thus suppose that h^+ is good and obtain a contradiction as follows. By our assumption, there exist two successive elements in the barrier W^+, say $s \lhd t$, with $h^+(s) \leq h^+(t)$. Then $s \in W$, for otherwise $h^+(s) = g(s) \leq h^+(t) \leq g(t)(\text{mod}\, A)$, so g would be good: contradiction.

From the fact that $s \in W$ we deduce that $h^+(s) = h(s)$. Moreover s is distinct from each $r_i (i = 0, 1, ..., n-1)$; for otherwise we would have the following conditions: $h^+(s) = h(s) = g(s) \leq h + (t) \leq g(t)$ thus g would be good.

Since the maximum of r_i increases in i, the maximum of s is greater than or equal to the maximum of r_n. Hence every element of t belongs to the union $\cup W$, either because it belongs to s, or because it is strictly greater than the elements of the $r_i (i = 0, 1, ..., n)$. Finally $t \in W$ so $h(s) = h^+(s) \leq h^+(t) = h(t)$, so h is good: contradiction. •

7.3 Forerunning between barriers and barrier sequences

7.3.1 Forerunner or successor barrier

Given a barrier U, remember that the union $\cup U$ is the infinite set of those integers which belong to elements of U.

We say that the barrier U is a **forerunner** of a barrier V or that U **foreruns** V, or again that V is a **successor** of U with respect to forerunning, iff the union $\cup V$

is included in $\cup U$ and each element in V has a (unique) initial interval belonging to U.

For example, each barrier included in U is a successor of U. The barrier of all pairs of integers is a successor of the barrier of all singletons. The barrier of those pairs which are subsets of an arbitrary infinite set of integers is a successor of the barrier of all singletons.

The relation of forerunning is reflexive, transitive and antisymmetric: it is a partial ordering on the set of barriers.

The notion of forerunning, for barriers and for barrier sequences, is introduced by [178] NASH-WILLIAMS 1965 p.714. The following results about forerunning are at least implicitely obtained by him; yet some of them are improved and clarified by [154] LAVER 1978 and [217] ROSENSTEIN 1982.

7.3.2 Completion of a successor barrier

Let U be a barrier and V a successor of U with respect to forerunning. Assume that V is not simply a sub-barrier of U. Let p be the least maximum integer of those elements of U which are subsets of $\cup V$ yet do not belong to V. Call W the set of those elements of U which are not subsets of $\cup V$ yet which are formed of integers $\leq p$ or of elements of $\cup V$.

Then $V \cup W$ is a barrier, and furthermore a successor of U.

Note that p can be replaced by any smaller integer.

In the particular case where V is a sub-barrier of U, take p to be an arbitrary integer. Then the preceding definition of W is preserved and our conclusion still holds with the same following proof.

• The sets V and W are obviously disjoint. Prove first that any elements $s \in V$ and $t \in W$ are incomparable with respect to inclusion. Indeed, there exists an initial interval r of s with $r \in U$. If $s \subseteq t$ then $r \subseteq s \subseteq t$ and, since r and t belong to U, we have $r = s = t$. Yet r is a subset of $\cup V$ and t is not: contradiction. Now if $t \subseteq s$, then t is a subset of $\cup V$: contradiction.

Now let X be an infinite set of integers in $\cup(V \cup W)$, thus a set of integers $\leq p$ or which belong to $\cup V$ We shall prove that there exists an initial interval of X which belongs either to V or to W. Since U is a barrier, there exists an initial interval s of X, which belongs to U. Then either s is not a subset of $\cup V$, and hence $s \in W$ and we are finished. Or s is a subset of $\cup V$. In this case, either $s \in W$ and we are finished. Or $s \notin W$ and then $(\operatorname{Max} s) \geq p$ by the definition of the integer p. Then the elements of X which are strictly greater than $\operatorname{Max} s$ all belong to $\cup V$, as well as the elements of s. Thus there exists an initial interval of X which belongs to V. •

7.3.3 Forerunner or successor barrier sequence

Let A be a poset and δ be any function, which to each element in the base $|A|$ associates an ordinal. Let U be a barrier and let f be a U-sequence with values in A. Similarly let V be a successor barrier of U and g be a V-sequence with values in A. We say that f is a **forerunner of** g (mod A, δ) or that g is a **successor of** f (mod A, δ) with respect to forerunning, iff for each element $t \in V$ and for the initial interval s of t which belongs to U, either we have that $s = t$ and then $gt = fs$; or we have $s \subset t$ and then $gt < fs (\text{mod}\, A)$ and furthermore $\delta gt < \delta fs$ (in the usual ordering of ordinals).

Forerunning for barrier sequences is reflexive, transitive and antisymmetric: it defines a partial ordering on the set of barrier sequences with values in A, for a given function δ.

In following applications, we take A to be the embedding relation on a set of chains, and δ the (non-strictly) increasing function which to each chain x associates the neighborhood rank of x (see 6.2.4). For this reason we will call δ a **ranking function**. However we only need, in the present theory, that δ associates to each element of A any ordinal value.

7.3.4 Completion of a successor barrier sequence

(1) Let A be a poset and δ be a ranking function on A, i.e. any function with ordinal values. Let U be a barrier and V a successor of U. Let f be a U-sequence with values in A, and g be a V-sequence with values in A. Assume that g is a successor of $f (\text{mod}\, A, \delta)$. Let p be the least maximum integer of those elements in U which are subsets of $\cup V$ yet do not belong to V. Call W the set of those elements in U which are not subsets of $\cup V$, yet are formed of integers $\leq p$ or belonging to $\cup V$.

Then $V \cup W$ is a barrier by 7.3.2; **the barrier sequence h, defined to be equal to f on W and to g on V, is a successor of** $f (\text{mod}\, A, \delta)$.

(2) **Moreover if f and g are bad, then h is bad** ([217] ROSENSTEIN 1982 ch.10).

As in 7.3.2, p can be replaced by any smaller integer. Moreover in the particular case where V is a sub-barrier of U, then p can be an arbitrary integer.

• (1) For every $s \in W$, we have $hs = fs$. For every $t \in V$ and for the initial interval s of t which belongs to U, we have $ht = gt$. Then in the case $s = t$ we have $gt = fs$ by hypothesis, hence again $ht = fs$. In the case $s \subset t$ by hypothesis we have $gt < fs$ (mod A) and $\delta gt < \delta fs$; hence we have $ht < fs$ (mod A) and $\delta ht < \delta fs$. •

• (2) Now suppose that f and g are bad, then prove that h is bad. For this consider two elements s' and $s"$ of $V \cup W$, with $s' \lhd s"$.

We distinguish four cases. If s' and $s"$ belong to V, then $hs' = gs'$ and $hs" = gs"$ and by hypothesis $gs' \not\leq gs" (\text{mod}\, A)$, hence $hs' \not\leq hs"$. Similarly argue if s' and $s"$ belong to W, by replacing g by f.

Suppose that $s' \in W$ and $s" \in V$, and let s be the initial interval of $s"$ which belongs to U, hence s is not included in s' (since $s' \in U$), hence $s' \lhd s$. Then $hs' = fs'$ and $hs" = gs" \leq fs (\text{mod}\, A)$. Since f is bad, we have that $fs' \not\leq fs (\text{mod}\, A)$: thus $hs' \not\leq hs" (\text{mod}\, A)$.

Finally suppose that $s' \in V$ and $s" \in W$, and let s be the initial interval of s' which belongs to U. Then either $s = s'$, in which case we have $hs' = gs' = fs$ and

$hs" = fs"$; and by hypothesis $fs \not\leq fs"(\bmod A)$, so that $hs' \not\leq hs"(\bmod A)$. Or $s \subset s'$, in which case $s \notin V$, yet s is a subset of $\cup V$, hence $(\operatorname{Max} s) \geq p$. In this case it follows that $s, s', s"$ are subsets of $\cup V$, hence $s" \notin W$: contradiction. •

7.3.5 Minimal bad barrier sequence with respect to forerunning

Let A be a poset, and δ be a ranking function with domain $|A|$. Say that a barrier sequence f with values in A is **minimal bad** $(\bmod A, \delta)$ with respect to forerunning, iff f is bad and if every bad successor of f reduces to a restriction of f.

In other words, denoting by U the barrier $\operatorname{Dom} f$, there does not exist any barrier V successor of U, non-included in U, with a V-sequence which is a bad successor of $f(\bmod A, \delta)$.

Theorem. Let A be a poset, and δ a ranking function with $\operatorname{Dom}\delta = |A|$. Let U be a barrier and f a bad U-sequence with values in A. Then there exists a minimal bad barrier sequence $(\bmod A, \delta)$ which is a successor of f ([154] LAVER 1978; uses dependent choice).

Note the analogy with the theorem on the minimal bad barrier sequence in 7.2.4. However here A is an arbitrary poset, not necessarily well-founded. Indeed we shall use the present statement in order to prove that certain posets are well-founded, and even are well partial orderings. In compensation, the present statement uses a ranking function associated with the poset A.

• Set $V_0 = U$ and $f_0 = f$. Suppose that there exists a barrier V_1, a successor of U which does not reduce to a sub-barrier of U, and a bad barrier sequence f_1 with domain V_1, a successor of $f(\bmod A, \delta)$. Choose V_1 and f_1 such that the least maximum integer p_0 of those elements of U which are subsets of $\cup V_1$ yet not elements of V_1 is the least possible. Using the preceding completion statement, we can assume that $\cup V_1$ contains all those integers $\leq p_0$ which belong to $\cup U$.

Iterate this (axiom of dependent choice). For each integer i, we have a barrier V_i and a bad V_i-sequence f_i, with f_{i+1} a successor of $f_i \bmod(A, \delta)$. If one of these f_i is minimal, we are finished, by using the transitivity of forerunning.

Suppose the contrary: for each integer index i, the barrier V_{i+1} is not a sub-barrier of V_i. Then to each i is associated an integer p_i, the least possible maximum of those elements of V_i which are subsets of the union set $\cup V_{i+1}$ yet not elements of V_{i+1}. By the transitivity of forerunning, we see that p_i increases in i, though not necessarily strictly. Indeed if we had $p_{i+1} < p_i$, then the barrier sequence f_{i+2}, or another giving p_{i+1} or giving an integer $\leq p_{i+1}$, would appear before f_{i+1}. However p_i tends to infinity, since there are only finitely many sets of integers with maximum p_i, and a given finite set can at most once be an element of V_i without being an element of V_{i+1}. By the preceding subsection, we can assume that each p_i belongs to the union set $\cup V_j$ for all $j \geq i$, hence for all integers j. Same remark for any element of $\cup V_i$ which is $\leq p_i$.

The union sets $\cup V_i$ form a decreasing ω-sequence with respect to inclusion. Let H be their intersection, which is infinite since every p_i belongs to H. For any infinite subsest X of H and any integer i, there exists a (unique) finite initial interval s_i of X which belongs to the barrier V_i. This s_i increases in i, although not necessarily strictly. Because of the decreasing values $\delta f_i(s_i)$, for X fixed, there exists an i from which point on s_i remains constant.

Call s this constant set, and note that, when X varies, these s form a barrier V with $\cup V = H$. Indeed by construction each infinite subset of H admits a certain s as an initial interval. Moreover given any two such finite sets s, t, there exists an i from which point on s and t are both elements of the barrier V_i: so that they are incomparable with respect to inclusion. Finally the barrier V is a successor of each V_i.

Define the V-sequence g as follows. For each $s \in V$, take an i from which point on $s \in V_i$: then we set $g(s) = f_i(s)$, this value being independent from the chosen i. The barrier sequence g so defined is bad, since the f_i are bad. Also g is a successor of each f_i, in particular a successor of $f_0 = f \ (\mathrm{mod}\, A, \delta)$.

It remains to see that g is minimal. Take an arbitrary barrier W successor of V, which is not simply a sub-barrier of V. Take a W-sequence h which is a bad successor of g $(\mathrm{mod}\, A, \delta)$. Then there exists an element t of W with an initial interval s belonging to V and distinct from t, so $s \subset t$. Consider the least i for which we have simultaneously $p_i > \mathrm{Max}\, s$ and s an element of V_i. By transitivity h is a bad successor of f_i. The set s belongs to V_i yet not to W, hence h should have been taken instead of f_{i+1}, because it leads to $\mathrm{Max}\, s$ (or to a lesser integer) instead of p_i: contradiction. •

7.4 Hereditarily indecomposable chain

Here is first an intuitive definition of an **hereditarily indecomposable chain**, or **h-indecomposable chain**.

The empty chain, every singleton chain, every chain isomorphic with a regular aleph, as well as the converse of such, are h-indecomposable.

If α is a regular infinite aleph and $A_i (i < \alpha)$ are h-indecomposable chains such that, for each i, the set of indices $j (i < j < \alpha)$ for which A_i is embeddable in A_j is cofinal in α, then the sum of the A_i along α or along its converse, are h-indecomposable.

Finally the only h-indecomposable chains are those which can be so constructed.

The preceding "definition" is easy to use, and in fact we shall usually take it. But it is incorrect from a logical point of view. It is a **definition scheme**, in the same sense as the axiom schemes of separation or substitution. Indeed the preceding alinea (the only h-indecomposable chains ..) could be explicited thusly: if an arbitrary condition C holds for every regular aleph and its converse, if C is preserved in taking any sum of chains A_i along a regular aleph or its converse, assuming that each A_i be embeddable in A_j for a cofinal set of indices j; then C holds for every h-indecomposable chain.

This type of procedure is admissible as an axiom scheme, which represents infinitely many axioms, each for a given condition C. Obviously this is no more admissible as a definition, which must have a finite length.

We propose the following finite (and admissible) definition, which is equivalent to the preceding scheme and so will permit us to use the scheme, in practice.

Let A be a chain and $\mathcal{J}_A$ a set of intervals of A. Suppose that $\mathcal{J}_A$ contains the empty interval and every singleton interval in A. Suppose that, given a regular infinite aleph α, if $\mathcal{J}_A$ contains the intervals $A_i (i < \alpha)$ which are mutually disjoint and situated in A without intermediate intervals, in the order of increasing i or of decreasing i, and such that for each i the set of indices j for which A_i is embeddable in A_j is cofinal in α, then $\mathcal{J}_A$ contains the union of A_i's (which is again an interval of A). Now we say that A is **h-indecomposable**, or **hereditarily indecomposable** iff the chain A itself belongs to any set $\mathcal{J}_A$ satisfying the preceding.

For example ω, ω^2, $\omega.\omega^-$, $\mathbf{Z}.\omega$ and their converses are h-indecomposable.

For every h-indecomposable chain, its converse is h-indecomposable.

Note that the regular aleph α, such that A is the sum of the A_i along α in the above construction, is the cofinality of A; the co-initiality in the case of a sum along the converse of α.

7.4.1 Preliminary lemmas

(1) **Every h-indecomposable chain is right or left indecomposable**, according to whether it is a sum along a regular aleph or along its converse.

(2) **An ordinal is h-indecomposable iff it is indecomposable** in the sense of 1.3.6. Proof by induction on the ordinal itself (start with a cofinal restriction of minimum length, i.e. isomorphic with the cofinality).

(3) **Every h-indecomposable chain is strongly scattered** (see definition in 6.2.3).

Moreover, in the decomposition of an h-indecomposable chain A into the intervals $A_i (i < \operatorname{Cof} A)$, **each A_i has a neighborhood rank strictly less than the rank of A.** Use for instance the definition scheme of the previous subsection, taking for C the condition " is strongly scattered ". Finally use 6.3.5.

7.4.2 First statement on hereditarily indecomposable chains

Lemma. Let A be a scattered chain. Suppose that the h-indecomposable restrictions of A form a well quasi-ordering with respect to embeddability. Then A is a finite sum of h-indecomposable chains ([151] LAVER 1968 and [152] 1971, uses dependent choice). For the well quasi-ordering, see 4.3.2.

• The chain A is strongly scattered (modulo the axiom of dependent choice), hence has a neighborhood rank. Suppose the statement is false, and let α be the least ordinal for which there exists an A of neighborhood rank α, whose h-indecomposable restrictions form a well quasi-ordering with respect to embeddability, yet A is not a finite sum of h-indecomposable chains.

We can always assume that A has a minimum or a maximum element. Indeed, take an element a of the base, and replace A by the initial or final interval with endpoint a: the neighborhood rank remains at most equal to α and the h-indecomposable restrictions still form a well quasi-ordering.

To fix ideas, suppose that A has a minimum element, and apply 6.2.5. Then either A is a finite sum of chains with neighborhood ranks strictly less than α. In which case, by hypothesis, each of these chains is a finite sum of h-indecomposable chains, so that A is as well. Or there exists a regular infinite aleph γ such that A is the sum of a γ-sequence of chains $A_i (i < \gamma)$ with neighborhood ranks $< \alpha$. In which case, by hypothesis, each A_i is a finite sum of h-indecomposable chains: from this point on, let the A_i designate these chains themselves.

We shall prove that there exists a final interval of γ which yields an h-indecomposable final interval on A: the induced interval being the sum of the A_i whose indices i belong to the final interval of γ. This will suffice, since by 6.2.5 proposition (2), the complementary initial interval is a finite sum of chains with neighborhood ranks $< \alpha$; hence by hypothesis, a finite sum of h-indecomposable chains. Thus we shall prove that from some index on, for each i, the set of $j (i < j < \gamma)$ for which A_i is embeddable in A_j is cofinal in γ.

Suppose the contrary. There exists an index $i(0)$ and an $i'(0) > i(0)$ such that for all $i \geq i'(0)$, the chain $A_{i(0)}$ is not embeddable in A_i. Then there exists an $i(1) > i'(0)$ and an $i'(1) > i(1)$ such that for all $i \geq i'(1)$, neither $A_{i(1)}$ nor, by the preceding, $A_{i(0)}$, is embeddable in A_i. Iterating this, we have an ω-sequence of chains which is bad with respect to embeddability. Hence the chains A_i do not constitute a well quasi-ordering; contradiction. •

7.4.3 Second statement

Lemma Let A be a scattered indecomposable chain. Suppose that the h-indecomposable restrictions of A form a well quasi-ordering with respect to embeddability. Then A is h-indecomposable ([151] LAVER 1968 and [152] 1971, uses axiom of choice).

• By 6.3.4 corollary (3), the chain A is strictly right or left indecomposable; to fix ideas, suppose it is strictly right indecomposable. Take a cofinal restriction of A which is isomorphic with the regular infinite aleph $\operatorname{Cof} A$ (axiom of choice). Let α denote this restriction, which we identify with the ordinal $\operatorname{Cof} A$. For each $i < \alpha$ let A_i be the interval of A between the element i (inclusive) and the element $i + 1$ (exclusive). Thus A is the sum of the A_i along α. Define as follows the intervals $B_i (i < \alpha)$, so that A is the sum along α of the B_i, and every B_i is embeddable in every B_j with $j > i$. Moreover we require that no B_i is a final interval of A, so that $B_i < A$ with respect to embeddability.

Let $B_0 = A_0$. Let u be an ordinal $< \alpha$, and suppose that we have defined the $B_i (i < u)$. Let A' be the sum of these B_i. Then A' is different from A, since otherwise either u would be cofinal in α, or u would have an immediate predecessor $u - 1$ with B_{u-1} a final interval of A, thus equimorphic with A, contradicting our hypotheses.

Let A" be the final interval of A, complement of A'. Since A is right indecomposable, A" is equimorphic with A. Set B_u to be a proper initial interval of A" (so $B_u < A$), in which B_{u-1} is embeddable (in the case that u is a successor ordinal), or in which the sum of the $B_i (i < u)$ is embeddable (in the case that u is a limit ordinal). Finally, enlarge this interval B_u, if necessary, so that the sum of the $B_i (i \leq u)$ includes the corresponding sum of the A_i: thus we are insured to exhaust A by the sum of all the $B_i (i < \alpha)$.

By the previous subsection, each B_i is a finite sum of h-indecomposable chains, which we again designate by B_i. For each couple (i, j) with $i < j < \alpha$, the chain B_i is indecomposable and is embeddable in a finite sum of the form $B_j + B_{j+1} + ..$, hence in a chain B of index $\geq j$. Thus the chain A is h-indecomposable. •

7.5 Embeddability theorem for scattered chains (Laver)

7.5.1 A bad successor barrier

Let $\mathcal{A}$ be a set of h-indecomposable chains (defined in the preceding section), which is quasi-ordered under embeddability. Assume that $\mathcal{A}$ is closed with respect to taking any h-indecomposable interval of a chain. Let δ be the ranking function which associates to each chain its neighborhood rank (6.2.4).

Lemma. Let U be a barrier and f a bad U-sequence with values in A. Then there exists a barrier V successor of U which is not a sub-barrier of U, and a bad V-sequence which is a successor of f $\operatorname{mod}(A, \delta)$ (uses axiom of choice).

• The chains in $\mathcal{A}$ which belong to the range of f are all infinite. For otherwise, either the empty chain or a chain consisting of a singleton would belong to $\operatorname{Rng} f$, and so the barrier sequence f would be good.

Partition the elements of U into two sets, according to whether the image under f is right or left indecomposable. At least one of these sets includes a barrier, which we again denote by U: see partition theorem 7.1.4. To fix ideas, suppose that all the chains are right indecomposable.

Compose the function f with the function γ which to each chain associates its cofinality: we obtain a barrier sequence $\gamma \circ f$ taking ordinal values, and more precisely values which are regular alephs. It follows that $\gamma \circ f$ has no bad restriction. Indeed take an ω-sequence of successive elements in U, indexed by integers i, say $s_0 \lhd s_1 \lhd \ldots \lhd s_i \lhd ..$; then the values $\gamma \circ f(s_i)$ cannot be strictly decreasing. By 7.2.2, there exists a barrier included in U and such that the restriction of $\gamma \circ f$ is perfect. Let us denote this barrier again by U.

To each h-indecomposable chain A in $\mathcal{A}$, associate a decomposition into a sum, along the ordinal $\gamma(A)$, of h-indecomposable intervals $A_i (i < \gamma(A))$; with the condition that each A_i has neighborhood rank strictly less than the rank of A, and that for each i, there are $\gamma(A)$ many $A_j (i < j < \gamma(A))$ in which A_i is embeddable:

see 7.4. This decomposition of each A into the A_i is chosen once for all and shall be called the **standard decomposition** (axiom of choice).

Since f is bad, for every s and s' in U with $s \lhd s'$, we have the non-embeddability $fs \not\leq fs'$. Consider the standard decomposition $fs = \Sigma A_i$ and $fs' = \Sigma A'_j$. Since $\gamma \circ f$ is perfect, we have $gfs \leq gfs'$, hence the index j runs through a regular aleph at least as large as that which i runs through. It follows that there exists at least one interval A_i which is not embeddable in any A'_j. The first of these A_i shall be associated with the union $t = s \cup s'$. The set of these unions is a barrier V. More precisely V is the square of a sub-barrier of U, hence a successor of U.

Now we define the barrier sequence g with domain V by $gt = A_i$. This g is a successor of $f(\bmod A, \delta)$: indeed for each t in V and for the initial interval s of t belonging to U, we have $s \subset t$ and $gt < fs$ under embeddability. Furthermore by 7.4.1 proposition (3), the interval A_i has a neighborhood rank strictly less than the rank of A: in other words $\delta gt < \delta fs$.

It remains to see that g is bad. Let t and t' be elements of V such that $t \lhd t'$. Then t is a union $s \cup s'$ with $s \lhd s'$; similarly t' is of the form $s' \cup s"$ with $s' \lhd s"$ and $s, s', s"$ elements of U. The chain gt is an interval A_i of the standard decomposition of $A = fs$; similarly gt' is an interval A'_j of the standard decomposition of $A' = fs'$; and finally A_i has been defined to be non-embeddable in any term of the standard decomposition of A': so $A_i \not\leq A'_j$, i.e. $gt \not\leq gt'$. •

7.5.2 Embeddability among h-indecomposable chains

Lemma. Every set of h-indecomposable chains forms a well quasi-ordering under embeddability (uses axiom of choice). It is even a better quasi-ordering, in the sense of 7.6.3 below.

• Suppose the contrary, that there exists an ω-sequence taking h-indecomposable chains as values, which is bad with respect to embeddability (4.3.2 proposition (2)). Consider this ω-sequence as a barrier sequence on singletons. By 7.3.5, there exists a minimal bad barrier sequence which is a successor of the above, modulo the partial ordering of embeddability and the ranking function which, to each chain, associates its neighborhood rank.

Yet in the previous subsection, using axiom of choice, we proved, on the contrary, that every bad barrier sequence of that kind has a bad successor barrier sequence which is not simply a restriction to a sub-barrier: contradiction. •

7.5.3 Finite decomposition of a scattered chain

Every scattered chain is a finite sum of h-indecomposable chains (uses axiom of choice).

In particular, **every indecomposable scattered chain is h-indecomposable**. In other words, for a scattered chain, the notion of indecomposability coincides with that of h-indecomposability.

• By the previous lemma, every set of h-indecomposable chains is well quasi-ordered under embeddability. Our proposition follows from 7.4.2 and 7.4.3. •

7.5.4 Embeddability theorem

Theorem. Every set of scattered chains forms a well quasi-ordering under embeddability. Due to [151] LAVER 1968; also [152] 1971; uses axiom of choice).

• The relation of embeddability for h-indecomposable chains, forms a well quasi-ordering, by the preceding 7.5.2 (axiom of choice). Now use HIGMAN's theorem on words, 4.5.2. The relation of embeddability between words, or finite sequences of h-indecomposable chains, is also a well quasi-ordering. The same holds for the embeddability among finite sums of h-indecomposable chains: indeed embeddability on sums is a reinforcement of embeddability on sequences. By 7.5.3, every scattered chain is a finite sum of h-indecomposable chains: the theorem is proved. •

In [152] LAVER 1971, this theorem is extended to countable unions of scattered chains. Unfortunately, the proof is too complicated to be presented here.

In [36] COROMINAS 1984, the theorem is extended to countable trees.

In [245] THOMASSÉ 1993, it is extended to those denumerable posets which do not embed N (the poset on four elements a, b, a', b' with $a < a', b < a', b < b'$, incomparability elsewhere).

7.5.5 Covering by a finite number of doublets

Given a scattered chain, the equivalence relation of covering by doublets of indecomposable chains (see 6.4.5) **has only a finite number of equivalence classes**. Each class is an indecomposable interval or a doublet (uses axiom of choice).

• Using the preceding 7.5.3, decompose our chain into a finite number of right or left indecomposable intervals. Replace any two contiguous such intervals by their union, provided this union is indecomposable. When it becomes impossible to effect these replacements, then the intervals thus obtained, or the unions of two contiguous intervals, constitute the covering by doublets. The uniqueness of this decomposition follows from propositions 6.4.1 and 6.4.2. •

For a non-scattered chain, the relation of covering by right or left indecomposable chains, is still an equivalence relation: see 6.4.3. Hence the union of both equivalence relations is again an equivalence relation. Recall that, for a non-scattered chain, there is no equivalence relation by doublets: see 6.4.5.

However **there can be infinitely many equivalence classes for this equivalence relation** (BONNET in 1972; published in [77] "Theory of Relations" 1986; uses axiom of choice).

• Start with A_0 = the chain of the reals. By 5.5.2 (DUSHNIK, MILLER), we have a strictly decreasing ω-sequence of chains A_i(i integer), where each A_i has cardinality of the continuum. Moreover we can require that $A_i \not\leq A_{i+1}.\omega$ for each i (same reference, proposition (3), uses axiom of choice). On the other hand, we have $\mathbf{Q} \leq A_i$ for each i ($\mathbf{Q}$ denotes the chain of rationals). Indeed A_i has at least ω_1 many elements, and neither the ordinal ω_1 nor its converse is embeddable in the reals, hence in A_i: use 5.7.2 in the particular case where $\alpha = 1$. Thus $A_i \not\leq \mathbf{Q} + A_{i+1} + \mathbf{Q} + \ldots + \mathbf{Q} + A_{i+h}$ for any two integers i and h.

Let $U = \omega_1^- + \omega_1$ and consider the sum of the ω-sequence $A_0 + U + A_1 + U + \ldots + U + A_i + U + \ldots$ We shall prove that each interval isomorphic with U is one of the desired equivalence classes; hence that there exist infinitely many equivalence classes.

Indeed, take two elements x and y in two consecutive components: for example x belongs to U and y belongs to A_i following the considered component U. We must join x to y by finitely many intermediate elements, such that any two consecutive elements be either right equivalent (i.e. covered by a same right indecomposable interval) or left equivalent. We can assume that x and y are themselves consecutive elements; then it suffices to see that they are neither right nor left equivalent.

Firstly, a non-final interval I which contains x and y is obviously decomposable into a finite sequence of disjoint sub-intervals in which I cannot be embedded. Secondly, a final interval is obviously not left indecomposable; nor is it right indecomposable; for otherwise, it would be necessary that A_i, for example, be embeddable in a sum of the form $U + A_{i+1} + U + \ldots + U + A_{i+h}$. But an interval of A_i which is a restriction of U is countable, since it is isomorphic with the union of a well-ordered set of reals and the converse of such a well-ordered set. So it must be that A_i is embeddable in $\mathbf{Q} + A_{i+1} + \mathbf{Q} + \ldots + \mathbf{Q} + A_{i+h}$, contradicting the previous discussion. •

7.6 Indecomposable sequence, better partial ordering

7.6.1 Tail of an ordinal sequence; indecomposable sequence

Given a poset A and an ordinal sequence u of length α which takes values in A, we call a **tail**, or **final interval** of u, any sequence obtained by taking a final interval β of the length α, then the restriction u/β, then reindexing this restriction, substituting for β the ordinal isomorphic with β.

Recall that the connected notion of **initial interval of a sequence** has already been introduced in 4.1.4. An ordinal sequence u with values in A is said to be **indecomposable** (mod A) iff u is embeddable (mod A) in every non-empty tail of u (embedding among sequences has been introduced in 4.1.2). This requires that the ordinal length of the sequence u be itself an indecomposable ordinal, hence a power of ω: see 1.3.6.

In the contrary case, the ordinal sequence is said to be **decomposable** (mod A).

Note that, with an indecomposable ordinal, ω for example, one can construct decomposable sequences of length ω: start with the free partial ordering based on two elements a, b taken to be incomparable, then take the ω-sequence $b, a, a, a, \ldots$. We can even construct ω-sequences none of whose tails are indecomposable: start with the identity poset on a denumerable base E and take an ω-sequence in E, without repetition.

7.6.2 α-better partial ordering

Let α be an indecomposable ordinal. A poset A is said to be an α-**better partial ordering** iff every α-sequence in A has a non-empty indecomposable tail. Obvious definition of a $(< \alpha)$ or $(\leq \alpha)$-**better partial ordering**.

Every well partial ordering is an ω-better partial ordering, and conversely.

The "converse" part uses dependent choice; however ZF suffices if the given partial ordering is denumerable.

• Let A be a well partial ordering, and u be an ω-sequence in A, with terms u_i(i integer). Consider those integers h such that there exist only finitely many integers i with $u_i \geq u_h (\operatorname{mod} A)$. Then there are only finitely many such h. For otherwise there would exist an ω-sequence $h(i)(i$ integer) with $u_{h(j)} <$ or $| u_{h(i)} (\operatorname{mod} A)$ for all i and $j > i$, hence a bad ω-sequence. The tail obtained by beginning with an index strictly greater than these values h is an indecomposable sequence.

Conversely, suppose that A is not a well partial ordering, and let u be a bad ω-sequence: see 4.3.2 proposition (2) (dependent choice). For such a sequence, no tail is embeddable in a proper tail of itself. •

7.6.3 Better partial ordering

A partial ordering is said to be a **better partial ordering** iff it is an α-better partial ordering for every ordinal α (more precisely for every indecomposable α). By the preceding, every better partial ordering is a well partial ordering.

The only obvious example of a better partial ordering is the ordering defined on a unique element. In this case every ordinal sequence repeats the element. If α is the length of the sequence, then the indecomposable tail is obtained by taking the last term of the Cantor normal development of α.

Every restriction, every partially ordered augmentation of an α-better partial ordering is an α-better partial ordering. Same statement with "better partial ordering".

A quasi-ordering is said to be an α-**better quasi-ordering** iff the partial ordering of the equivalence classes (each formed of elements simultaneously greater than and lesser than each other) is an α-better partial ordering. Equivalently iff every α-sequence has non-empty indecomposable tail.

A quasi-ordering is said to be a **better quasi-ordering** iff it is an α-better quasi-ordering for every α.

7.6.4 On Rado's well partial ordering

The well partial ordering of RADO, defined in 4.4.2, is an example of a well partial ordering which is not a better partial ordering.

More exactly **Rado's well partial ordering is not an ω^2-better partial ordering**.

• Recall that this partial ordering is defined on couples of non-negative integers. Take the lexicographical ω^2-sequence (0,0), (0,1), (0,2), ... , (1,0), (1,1), (1,2), ..

A tail U is necessarily an ω^2-sequence which contains, for a sufficiently large integer p, the couples $(p,0), (p,1), (p,2), ..$ followed by $(p+1,0)$. If we attempt to embed U in its tail which begins with $(p+1,0)$, then for each integer i the couple (p,i) is not embeddable before $(p+i,0)$. Hence to embed all these couples (p,i) where p is fixed and i varies, we exceed all the terms of U: it is then impossible to embed the term $(p+1,0)$. •

7.6.5 Connection between an α-better ordering A and sequences in A

(1) **Let α be an indecomposable ordinal. If A is a $(\leq \alpha)$-better partial ordering, then the set of ordinal-indexed sequences in A with lengths $< \alpha$ constitutes a well quasi-ordering with respect to embeddability.** Uses dependent choice; ZF suffices if A is countable.

In the case where $\alpha = \omega$, this is just HIGMAN's theorem on the well partial ordering of words (4.5.2) in view of the equivalence between well partial ordering and ω-better partial ordering, 7.6.2.

• Since A is a $(\leq \alpha)$-better partial ordering, every sequence in A with length less than or equal to α is a finite sum of indecomposable (mod A) sequences. Indeed, it suffices to operate regressively, starting with the indecomposable tail of the given sequence.

Now consider a set of indecomposable (mod A) sequences of lengths $< \alpha$. Let B be the quasi-ordering between these sequences, with respect to embeddability. We shall prove that B is a well quasi-ordering. Take an ω-sequence of these indecomposable sequences s_i(i integer). Then take their sum u, whose length is $\leq \alpha$, since α is a power of ω.

By hypothesis, there exists a tail of u which is indecomposable; let s_p (p integer) be its first term. In particular, this tail is embeddable in its proper tail which begins with s_{p+1}. Then s_p is embeddable in a finite sum of $s_i(i \geq p+1)$; hence by indecomposability, this s_p is embeddable in some $s_i(i \geq p+1)$. Thus the sequence of the s_i is good, and so B is a well quasi-ordering (dependent choice, see 4.3.2 proposition (2)).

By HIGMAN's theorem already mentioned, every set of words formed of indecomposable (mod A) sequences of lengths $< \alpha$ constitutes a well quasi-ordering. The same is true for the set of finite sums corresponding to these words, their quasi-ordering with respect to embeddability being a reinforcement of the quasi-ordering of the words: use 4.5.3 proposition (1). Since every sequence in A with length $< \alpha$ is such a sum, our proposition is proved. •

(2) **Let α be an indecomposable ordinal and A a partial ordering. If the set of α-sequences in A constitutes a well-founded quasi-ordering (with respect to embeddability), then A is an α-better ordering.**

• Take an α-sequence in A, and then the set of all non-empty tails of this sequence. By hypothesis one of these tails is minimal, thus minimum with respect to embeddability, i.e. indecomposable. •

(3) Let α be a denumerable indecomposable ordinal and A a partial ordering. **If the set of sequences in A with lengths $< \alpha$ constitutes a well quasi-ordering (under embeddability), then A is an α-better ordering.**

• Suppose that A is not an α-better ordering. Then there exists an α-sequence in A which, by countability of α, has a strictly decreasing ω-sequence of tails s_i(i integer); where it is understood that, for each i, we can choose s_{i+1} to begin with a given term of s_i situated as far in s_i as we desire.

Then there exists an initial interval t_0 of s_0 and an integer i from which point on $t_0 \not\leq s_i$. For otherwise, every proper initial interval of s_0 would be embeddable in some proper initial interval of s_i for every i; then by the preceding discussion we could embed all of s_0 in s_1 for example: contradiction. Let $h(0) = 0$ and let $h(1)$ be the least integer such that for all larger integers i we have $t_0 \not\leq s_i$. There exists an initial interval t_1 of $s_{h(1)}$ and an i, from which point on we have $t_1 \not\leq s_i$. Iterating this, we obtain a bad ω-sequence (with respect to embeddability) of intervals t_i of lengths $< \alpha$. Finally the set of sequences in A of lengths $< \alpha$, does not constitute a well quasi-ordering. •

From propositions (1) and (2), it follows that **a necessary and sufficient condition for A to be a better partial ordering is that every set of ordinal sequences in A forms a well quasi-ordering under embeddability** (necessity uses dependent choice).

7.6.6 Connection with barriers

Let α be a denumerable indecomposable ordinal, let A be a ($\leq \alpha$)-better partial ordering and U a barrier of lexicographic rank $\leq \alpha$. Then every U-sequence in A is good ([189] POUZET 1970; uses dependent choice; ZF suffices if A is countable); see also [203] POUZET 1985.

• By 7.6.2, A is a well partial ordering; so that if U reduces to the barrier of the singletons, then any U-sequence in A is good by 4.3.2 proposition (2). In the following, we remove from U all singletons and suppose $\alpha \geq \omega^2$.

For each integer i, let U_i be the infinite set of those elements of U which begin with i. These elements, when ordered lexicographically, form a sequence with length $< \alpha$. indeed they are less than any element of U which begins with $i + 1$. Let f be a U-sequence in A, and let f_i denote the restriction of f to the domain U_i. Consider f_i as an ordinal sequence with values in A, by replacing each element of U_i by its height modulo U_i in the lexicographic ordering. Thus f_i becomes a sequence in A with length $< \alpha$. Letting i vary, it follows from the preceding subsection, proposition (1) (dependent choice), that there exist two integers $i, j > i$ such that f_i is embeddable in f_j. In other words, there exists a function h from U_i into U_j, which is strictly increasing with respect to the lexicographic ordering, such that for each element s of U_i, we have $fs \leq fhs \bmod A$.

Suppose first that $s = \{i, j\}$ is an element of U. Then its image $t = hs$ begins with j, hence $s \lhd t$ and $fs \leq ft \pmod A$, so f is good. In the other case, we have elements in U beginning with $i, j, u (u > j)$, whose images under h are elements beginning with j, hence with $j, v(v > j)$. As u varies, there are at most finitely many values v used. Indeed we can never reach the images under h of elements beginning with $i, j+1$. Hence there exists a maximum v.

Moreover, there exists a maximum k for which all those elements beginning with i, j, k have as images, elements beginning with j followed by an integer $\geq k$. Indeed $k = j + 1$ satisfies this condition. And on the other hand, there are only finitely many k which verify this condition, since those k which do so, are less than or equal to the maximum possible v.

The elements of U beginning with i, j, k have images which begin with j, k. Indeed, if an element of U beginning with i, j, k had an image beginning with $j, k' \geq k+1$, then the elements beginning with $i, j, k+1$ would have images beginning with $j, k' \geq k+1$, contradicting the maximality of k.

Suppose now that $s = \{i, j, k\}$ is an element of U. Then its image $t = hs$ begins with j, k, hence $s \lhd t$ and $fs \leq ft (\bmod A)$, so that f is good. In the opposite case, we have elements of U beginning with $i, j, k, u > k$, whose images under h begin with $j, k, v > k$.

Iterating this, we obtain a strictly increasing ω-sequence $i(0) = i, i(1) = j, i(2) = k, ..$ such that, for each integer r, the elements of U which begin with $i(0), i(1), ..., i(r), ..$ have images under h which begin with $i(1), ..., i(r)$. By the definition of barrier, there exists an r for which $s = i(0), ..., i(r)$ is an element of U; its image $t = hs$ satisfies $s < t$ and $fs \leq ft (\bmod A)$, hence f is good. •

Corollary. if A is a better partial ordering (i.e. α-better for every denumerable ordinal α), then every barrier sequence with values in A is good.

In other words, A is a better partial ordering with respect to barriers, or $^{-}$better partial ordering in the sense of the next section.

7.7 Better partial ordering with respect to barriers; equivalence of both notions

We say that A is a **better partial ordering with respect to barriers**, or more simply a **$^{-}$better partial ordering** iff every barrier sequence with values in A is good (see definition in 7.2.1). This notion is historically anterior to the better partial ordering, and goes back to [177] NASH-WILLIAMS 1965. It is proved in [179] NASH-WILLIAMS 1968 that every $^{-}$better partial ordering is a better partial ordering, which we obtain in 7.7.11 below. Taking into account the converse statement in the previous subsection (corollary), we shall conclude that both notions are equivalent.

A quasi-ordering is said to be a **$^{-}$better quasi-ordering** iff the partial ordering of its equivalence classes, each formed by elements simultaneously greater and lesser, is a $^{-}$better partial ordering.

7.7.1 Connection with well partial orderings

Every restriction, every partially ordered augmentation of a $^-$better partial ordering is a $^-$better partial ordering.

Every $^-$better partial ordering is a well partial ordering (uses dependent choice; ZF suffices for a countable partial ordering).

• Since every barrier sequence is supposed to be good, in particular every ω-sequence (which can be identified with a barrier on singletons) is good. Then our statement follows by 4.3.2 proposition (2), using dependent choice. •

7.7.2 An elementary $^-$better partial ordering

The poset formed by two incomparable elements is a $^-$better partial ordering.

• Let U be a barrier and s an element of U. By 7.1.6, there exists a sequence of finite length h of elements t_i of $U(i = 1, ..., h)$; and another sequence of length $h + 1$ of elements $t'_i(i = 1, ..., h + 1)$ with $s \lhd t_1 \lhd t_2 \lhd ... \lhd t_h$ and $s \lhd t'_1 \lhd t'_2 \lhd ... \lhd t'_{h+1}$, both having the same last term $t_h = t'_{h+1}$. Then there necessarily exist two successive elements with the same image in the poset, since there are only two elements in it. •

7.7.3 Case of a well-ordering

Every well-ordering, and in particular every finite chain is a $^-$better partial ordering.

• Let U be a barrier. Take an ω-sequence of elements s_i of $U(i$ integer) such that $s_i \lhd s_{i+1}$ for each i. Then for each function f from U into the well-ordering A, there exists an i satisfying $f(s_i) \leq f(s_{i+1})(\mathrm{mod}\, A)$; so that f is good. •

7.7.4 $^-$Better ordering and partition; the finite case

(1) Let A be a poset. Partition its base into two disjoint subsets, thus obtaining the two restrictions B and C. **If B and C are $^-$better partial orderings, then so is A.**

• Suppose that our conclusion is false. Let U be a barrier and f be a bad U-sequence in A. Partition the elements of U into two classes, according to whether the image under f belongs to the base $|B|$ or to $|C|$. By the barrier partition theorem 7.1.4, there exists a sub-barrier V of U with f/V a bad barrier sequence from V into B or into C. Hence either B or C is not a $^-$better partial ordering. •

(1') In view of 7.6.3 and 7.7.11, the above (1) subsists if we delete the word "disjoint".

(2) **Every finite poset is a $^-$better partial ordering.**

(3) **If A is a $^-$better partial ordering, then so is every partially ordered extension of A to its base augmented by a finite number of elements.**

Statements (2) and (3) follow from (1). Indeed the ordering on a singleton is itself a better partial ordering (as noticed in 7.6.3), thus a $^-$better partial ordering by 7.6.6, corollary.

7.7.5 Direct product of $^-$better partial orderings

Lemma. If A and B are both $^-$better partial orderings, then the direct product $A \times B$ is a $^-$better partial ordering (direct product is defined in 4.8); see [178] NASH-WILLIAMS 1965' p.706.

• Let U be a barrier and f a function from U into $A \times B$. Call p the projection which, to each couple (a, b) in the product of the bases, associates the first term a, and call q the projection which associates the second term b. Since A is a $^-$better partial ordering, by 7.2.2, there exists a barrier V included in U, with $(p \circ f)/V$ perfect. Since B is a $^-$better partial ordering, there exist two elements s, t in V with $s < t$ and $qfs \leq qft(\text{mod}\, B)$. Moreover $pfs \leq pft(\text{mod}\, A)$, since $p \circ f$ is perfect. Thus $fs \leq ft$ modulo the product $A \times B$: hence f is good. •

7.7.6 A $^-$better stratified partial ordering

To each integer i, associate a finite set F_i of elements; so that, for fixed i, the elements in F_i are mutually incomparable, and so that each element in F_{i+1} is strictly greater than all the elements of F_i. We thus obtain a stratified poset, in the sense of 2.10.1. Moreover this poset is well-founded and F_i is the set of elements of height i (see 2.7.1).

This stratified poset is a $^-$better partial ordering.

Particular case of the next subsection. Assuming that the base is denumerable (which in general requires the countable axiom of choice on finite sets), we have the following easy proof.

• Start with the set of couples of integers, hence the direct product of ω with itself, which (by 7.7.3 plus the previous subsection) is a $^-$better partial ordering. For F_0 take a set of couples of integers of the form (x, y) with $x+y = a_0 = \text{Card}\, F_0$. Then for F_1 take a set of couples (x, y) with $x \geq a_0$ and $y \geq a_0$ and $x + y = a_0 + a_1$ where $a_1 = \text{Card}\, F_1$; and so forth. Our stratified poset is isomorphic with a restriction of the direct product $\omega \times \omega$: this is a $^-$better partial ordering. •

7.7.7 On proper initial intervals of a poset

Let A be a poset. If every proper initial interval of A is finite, then A is a $^-$better partial ordering (POUZET in 1977, published in ToR-86 p.228).

As in the preceding statement, this proposition gives well partial orderings whose elements have finite heights. Yet they are more varied: indeed it is no longer required that each element of height $i + 1$ be greater than every element of height i.

• Note first that every non-empty subset of the base has a minimal element; for otherwise, this would yield an infinite proper initial interval. Moreover every free set, or antichain, is finite; thus A is a well partial ordering. By hypothesis, each element of the base has a height which is an integer.

Moreover, if A is infinite, then by our hypothesis A is directed, thus A is an ideal. For otherwise, if a and b are two elements without any common upper bound, then either the set of non-upper bounds of a, or the set of non-upper bounds of b, is an infinite initial interval which is distinct from A: contradiction.

Let U be a barrier and g a function from U into A. We shall prove that g is good. Let h be the function which, to each element in the base $|A|$, associates its height $(\mathrm{mod}\, A)$, which is an integer. Let f be the composition $h \circ g$. Since the chain ω of the heights is a $^{-}$better partial ordering, there exists a barrier V included in U, such that the restriction f/V is perfect (see 7.2.2). Let V' be the subset of the squarred barrier V^2, formed of the unions $s \cup t$ of elements s, t in V such that $s \triangleleft t$ and $fs = ft$. By the barrier partition theorem 7.1.4, there exists an infinite set H of integers which either only contains elements of V', or only contains elements of $V^2 - V'$. Let W be the barrier V restricted to H; then W^2 is either included in V' or included in $V^2 - V'$.

In the first case, there exists an integer p such that $fs = p$ for every element s in W. Indeed, given two elements s, t of W, by 7.1.5 proposition (2) there exists a third element u of W with two finite sequences of successive elements of W, say $s \triangleleft s_1 \triangleleft \ldots \triangleleft u$ and $t \triangleleft t_1 \triangleleft \ldots \triangleleft u$. Then we have $fs = f(s_1) = \ldots = fu$ and $ft = f(t_1) = \ldots = fu$. For each element x of the base $|A|$ whose height $(\mathrm{mod}\, A)$ is p, let W_x be the subset of W formed of those s such that $gs = x$. The elements x are mutually incomparable $(\mathrm{mod}\, A)$, hence there are only finitely many such. Thus there exists an x with a barrier X included in W_x. For two elements s, t in X, which we can take as successive, we have $gs = gt$; hence g is good.

In the second case, recall that f is perfect. Thus for every s, t in W, the condition $s \triangleleft t$ implies $fs \leq ft$. But here $fs \neq ft$ so that $fs < ft$. Let k be the function from the set ω of the integers, into ω, which to each integer i associates the least j for which each element of height $\leq i$ is less than $(\mathrm{mod}\, A)$ every element of height j, thus also less than every element of height $\geq j$. This value $j = ki$ exists, since for each element a with height i, the set of non-upper bounds of a is a proper initial interval of A; hence it is finite and there are only finitely many heights of its elements.

By 7.1.7, there exist two elements s, t in W satisfying $s \triangleleft t$ and $kfs \leq ft$. Thus each element of the same height as gt is greater $(\mathrm{mod}\, A)$ than every element of the same height as gs; in particular $gs \leq gt (\mathrm{mod}\, A)$: hence g is good. •

7.7.8 Ideals of a well partial ordering

(1) **Let A be a well partial ordering which has finitely many infinite ideals. Then A is a $^{-}$better partial ordering.**

(2) **Let A be a poset which has only finitely many infinite initial intervals. Then A is a $^{-}$better partial ordering** (POUZET in 1977, published in ToR-86 p.230; uses dependent choice; ZF suffices if A is countable).

• (2) follows from (1), since the poset A under consideration is necessarily well-founded and finitely free.

Suppose first that A is a directed well partial ordering, with no other infinite ideal than itself. Then A has no infinite proper initial interval. Indeed, every infinite well partial ordering has as a restriction, at least one infinite ideal: see 4.7.1 (dependent choice); so by the preceding subsection, A is a $^{-}$better partial ordering.

In the general case, we argue by induction. Given a positive integer p, suppose the proposition holds for any well partial ordering with at most p infinite ideals, and let A be a well partial ordering with $p+1$ infinite ideals. Let I be an ideal of A, which is maximal with respect to inclusion. Partition the base $|A|$ into the union C of those ideals distinct from I, and the complement D of C. The restriction A/C is a well partial ordering having only p infinite ideals, and the restriction A/D is still a directed well partial ordering, hence an ideal, having no other infinite ideal than itself (provided D is infinite). Each is thus a $^{-}$better partial ordering; so by 7.7.4, A is itself a $^{-}$better partial ordering. •

7.7.9 $^{-}$Better partial ordering of words

Let A be a $^{-}$better partial ordering; then the set of all words (i.e. finite sequences) in A forms a $^{-}$better partial ordering under embeddability (uses dependent choice).

• Suppose on the contrary that there exists a barrier U and a bad U-sequence f taking as values words. The partial ordering of words is a well-founded partial ordering: see 4.1.3. So we can assume that f is minimal bad: see theorem 7.2.4 (dependent choice). Since f is bad, for every s in U, the word fs is non-empty.

Let g and h be barrier sequences with domain U, defined as follows. For each s in U, the value gs is the word composed of the first term of fs; the value hs is fs with the first term removed. By 7.2.3, there exists a sub-barrier V of U such that g/V is perfect (since A is a $^{-}$better partial ordering). On the other hand, the restriction h/V is good; indeed since f is minimal bad, in going from f to h, for each s in V, the word fs is replaced by hs which is strictly less than fs with respect to embeddability. Thus there exist two elements s, t in V with $s \lhd t$ and $hs \leq ht$ with respect to embeddability. Now, as $gs \leq gt (\operatorname{mod} A)$, we have $fs \leq ft$ with respect to embeddability; so f is good: contradiction. •

7.7.10 $^{-}$Better quasi-ordering of ordinal-indexed sequences

Generalize as follows the previous proposition.

Let A be a $^{-}$better partial ordering. Then any set of ordinal-indexed sequences with values in A forms a $^{-}$better quasi-ordering under embeddability ([179] NASH-WILLIAMS 1968; the following proof is due to MILNER in 1984; published in ToR-86 p. 231; uses dependent choice).

• Let B be a set of ordinal-indexed sequences in A; this B is quasi-ordered under embeddability (see 4.1.2). We can assume that B reduces to a partial ordering, by replacing sequences by their equivalence classes under embeddability. Take the ranking function δ which to each sequence u associates the length of u. Suppose that B is not a $^-$better partial ordering: there exists a barrier U and a bad U-sequence with values in B. By theorem 7.3.5 (dependent choice), there exists a minimal bad barrier sequence (mod B, δ), say f; we call again U the barrier domain of f.

Partition the elements s of U into three disjoint classes, according to the three following possibilities: either the sequence fs has length 1, or its length is a limit ordinal, or its length is a successor ordinal > 1. Using the barrier partition theorem 7.1.4, we can assume that the entire barrier U reduces to one of the three considered classes.

In the first case, for each s in U the sequence fs reduces to an element of A. Since A is a $^-$better partial ordering, the U-sequence f is necessarily good: contradiction.

Examine the second case where all lengths are limit ordinals. Take any two successive elements $s \lhd t$ in U, so htat $fs \not\leq ft$ under embeddability. Since the length of fs is a limit ordinal, there exists a proper initial interval of fs which is non-embeddable in ft: see 4.1.4. Consider the square barrier $V = U^2$ and to each element v of V associate its initial interval s which belongs to U and the final interval $t = v$ minus its minimum integer, so that $s \lhd t$ and $v = s \cup t$. Then to this v associate gv, the minimum proper initial interval of fs which is non-embeddable in ft.

Note that $\delta gv < \delta fs$ since gv has length strictly smaller than fs. Therefore f foreruns $g(\text{mod}\, B, \delta)$ yet g does not reduce to a restriction of f. Finally g is bad: take any two successive elements $v \lhd w$ in V and the corresponding initial intervals $s \lhd t$ in U; then gv is non-embeddable in ft thus non-embeddable in gw which is an initial interval of ft. This contradicts our hypothesis that f is minimal bad.

Examine the third case where all lengths are successor ordinals different from 1. To each element s of U associate the last term ls of the sequence fs, and also the sequence gs which is fs minus its last term. By 7.2.2 we can replace U by a sub-barrier again called U, such that the U-sequence l is perfect. Therefore the U-sequence g must be bad, since f is bad.

Take the square barrier $V = U^2$, and to each element v of V associate the initial interval s of v which belongs to U. Then put $hv = gs$. Note that $\delta hv = \delta gs < \delta fs$, therefore f foreruns $h(\text{mod}\, B, \delta)$ yet h is not a restriction of f. Finally h is bad; indeed with the same notations than in the second case, $hv = gs$ is non-embeddable in $hw = gt$. This contradicts our hypothesis that f is minimal bad. •

7.7.11 Equivalence between both notions

Every $^-$better partial ordering is a better partial ordering (uses dependent choice); therefore both notions coincide, by 7.6.6, corollary.

• Let A be a $^-$better partial ordering. By the previous subsection, every set of ordinal-indexed sequences in A forms a $^-$better quasi-ordering under embeddability, thus a well quasi-ordering by 7.7.1. Using 7.6.5 proposition (3), we see that A

is an α-better ordering for each α: in other words a better partial ordering. •

Consequently we can everywhere delete the symbol $(^-)$ before "better partial (or quasi-) ordering".

7.8 Exercises

7.8.1 Every finitely free poset has a cofinal restriction which is a better partial ordering

Due to POUZET in 1979, answering a conjecture due to GALVIN; published in ToR-86 p. 237; uses axiom of choice.

1 - Let A be a finitely free poset. Partition A into a finite union of ideals: see 4.7.2 (axiom of choice). For each ideal, take a cofinal restriction which is a well-founded poset, hence a directed well partial ordering: see 2.13.2. Then in view of 7.7.4 proposition (1'), it would suffice to prove our assertion for each directed well partial ordering.

2 - Let A be a directed well partial ordering. By 4.13.5, there exists a cofinal restriction of A which is isomorphic with the direct product of a finite number of regular alephs. This cofinal restriction is a better partial ordering by 7.7.3 and 7.7.5.

7.8.2 Each countable poset is an ω_1-better partial ordering

More generally, consider a poset A such that every bad sequence in A is countable, and show that A is an ω_1-better partial ordering (see ToR-86 p. 238).

7.8.3 A well partial ordering for which every barrier sequence on ω^2 is good, but which is not a better partial ordering

Let A be the partial ordering defined on all ordered triples of natural numbers x, y, z by the following condition: $(x, y, z) \leq (x', y', z')$ iff $x \leq x', y \leq y', z \leq z'$ and additionally either $x = x'$ or $x < x'$ and $y < x'$, or finally $x < x'$ and $z < y'$.

1 - Prove the transitivity of the above. Note that A is well-founded, there being only finitely many predecessors of (x, y, z).

Prove that A is finitely free, hence a well partial ordering. For this, note that the direct product ($x \leq x'$ and $y \leq y'$ and $z \leq z'$) is a well partial ordering. Suppose that there exists an ω-sequence of triples (x_i, y_i, z_i) (i natural number), which are mutually incomparable (mod A). Then extract an ω-sequence with x_i increasing, y_i increasing, z_i increasing. Since incomparability requires that x_i be strictly increasing in i, there exists an i for which $x_i > x_0$ and $x_i > y_0$: contradiction.

2 - Prove that A is not a better partial ordering. Indeed we have the bad barrier sequence with domain ω^3 (i.e. the set of all 3-element subsets of the set of natural

numbers), defined as follows: each 3-element subset $\{x, y, z\}$ with $x < y < z$ is taken into the ordered triple (x, y, z). To see this, let $s = \{x, y, z\}$ and take an integer $r > z$ and set $t = \{y, z, r\}$, this being the only possible manner to get $s \lhd t$. Then (x, y, z) and (y, z, r) are incomparable $(\mathrm{mod}\, A)$.

3 - Now let f be a barrier sequence with domain ω^2 (the set of unordered pairs of integers). For each pair s of integers, let $x(s), y(s), z(s)$ be the coordinates of $f(s)$. We shall prove that f is good.

Denote by B the direct product ($x \leq x'$ and $y \leq y'$ and $z \leq z'$), which is a better partial ordering by 7.7.3 and 7.7.5. Since A and B are both based on the same set of ordered triples, consider f as a barrier sequence in B and then replace f by a restriction which be perfect $(\mathrm{mod}\, B)$: see 7.2.2. After renumbering, this restriction of f still has domain ω^2, and now we again consider f as taking values in A.

Thus for any two pairs of integers, say s, t with $s \lhd t$, we have $x(s) \leq x(t)$ and similarly with y and z. Then either there exist s, t with $s \lhd t$ and $x(s) = x(t)$, in which case f is good and we are finished.

Or $s \lhd t$ implies necessarily that $x(s) < x(t)$. Then by RAMSEY's theorem, we can require either that for all integers $i < j < k$ we have $y(\{i, j\}) = y(\{i, k\})$, or that we have the strict inequality $y(\{i, j\}) < y(\{i, k\})$. In the first case, take a strictly increasing ω-sequence of integers $i_0 < i_1 < \ldots$, so that successive pairs give strictly increasing values for x. Then for h sufficiently large $y(\{i_0, i_h\}) < x(\{i_h, i_{h+1}\})$ and obviously $x(\{i_0, i_h\}) < x(\{i_h, i_{h+1}\})$: so that f is good.

In the second case, take again a strictly increasing ω-sequence of integers giving, for h sufficiently large, $z(\{i_0, i_1\}) < y(\{i_1, i_h\})$ and obviously $x(\{i_0, i_1\}) < x(\{i_1, i_h\})$, and so f is good.

4 - Generalize the preceding for the partial ordering on the set of k-tuples of integers, defined by the direct product $x_1 \leq x'_1$ and ... and $x_k \leq x'_k$, and additionally either $x_1 = x'_1$, or $x_1 < x'_1$ with either $x_2 < x'_1$ or $x_3 < x'_2$ or ... or $x_k < x'_{k-1}$.

Chapter 8

Faithful extension of a relation, bivalent tableau, faithful augmentation, Szpilrajn chain

8.1 Faithful extension between relations

8.1.1 Case of one relation with one embedding

Let R, S be two n-ary relations ($n \geq 1$); assume that S does not admit an embedding of R. Then there exists a strictly greater extension T of S which does not admit an embedding of R.

We call it a **faithful extension** of S modulo R.

Moreover **we can choose T to be an immediate extension of S**: see [104] HAGENDORF 1977.

• Suppose first that R and S are unary. Let a^+ be the cardinality of the set of elements giving the value (+) to R, and a^- the analogous cardinality for (-); similarly let b^+ and b^- be the analogous cardinalities for S. Since $R \not\leq S$, either $b^+ < a^+$ or $b^- < a^-$. Suppose the first case holds, the argument being analogous for the second case. It suffices to take an extension of S in which b+ is preserved and b^- is replaced by an immediately larger cardinal.

Suppose that R and S have arity $n \geq 2$. Add to the base E of S a set D^+ which is disjoint from E, and define the extension T^+ of S with base $E \cup D^+$, taking the value (+) for those n-tuples containing at least one term in D^+. Also choose D^+ with cardinal (aleph) sufficiently large to have $T^+ > S$. Do the same with the value (-), thus obtaining D^- and $T^- > S$. We claim that $R \not\leq T^+$ or $R \not\leq T^-$, which yields our conclusion.

Indeed suppose the contrary, and consider R as a restriction of T^+. The base of R is not a subset of E, since $R \not\leq S$, hence there exists an element u^+ such that R takes the value (+) for each n-tuple containing at least one occurrence of u^+. There exists an analogous element for the value (-): contradiction proving the faithful extension.

Moreover T^+ and T^- are immediate extensions of S, by 5.1.3. •

8.1.2 Many relations with the same non-embedding

(1) **Let R be a relation with arity ≥ 2; assume that $S_1 \not\geq R$ and $S_2 \not\geq R$ (with respect to embeddability). Then there exists a common extension of S_1 and S_2 which does not admit an embedding of R.**

The contrapositive is : if $R \leq X$ for every X which satisfies both $X \geq S_1$ and $X \geq S_2$, then $R \leq S_1$ or $R \leq S_2$.

• Take the case of a binary relation, and suppose that S_1 and S_2 have disjoint bases E_1 and E_2. Let S^+ be the common extension with base $E_1 \cup E_2$ taking the value (+) for those couples having one term in E_1 and the other in E_2. Analogously define S^- for the value (-). It suffices to see that either S^+ or S^- does not admit an embedding of R.

Suppose the contrary; then there exists a partition of the base $|R|$ into two non-empty disjoint subsets such that, for every element u in one subset and v in the other, we have $R(u,v) = R(v,u) = +$. Same conclusion with the value (-). Note that, given two partitions of the base, each with at least two non-empty disjoint sets, then there exist two elements u, v in the base which are separated both by the first partition and by the second. Thus there exist two elements u, v giving simultaneously $R(u,v) = +$ and - : contradiction. •

The proposition is obviously false for unary relations.

The statement immediately extends to any finite or infinite sequence:

(2) **Consider an arbitrary sequence of relations S_i(i ordinal) of common arity ≥ 2. Let R be a relation of the same arity. If $S_i \not\geq R$ for every i, then there exists a common extension of all those S_i which do not admit an embedding of R.**

The contrapositive is: **if $R \leq X$ for every X which is $\geq$ all the S_i, then there exists an i with $R \leq S_i$.**

• As precedently we may suppose that all the S_i have disjoint bases. Let R^+ denote the common extension of S_i on the union of the bases, which takes the value (+) for all those n-tuples (n = arity) containing at least two terms taken from two distinct bases. Analogously define the extension R^-. Terminate as in preceding (1). •

8.1.3 One relation with two or three non-embeddings

Let S have arity ≥ 2 and non-empty base; moreover $S \not\geq R_1, S \not\geq R_2$ and $S \not\geq R_3$. Then there exists a proper extension S^+ of S which respects the non-embeddabilities $S^+ \not\geq R_1, S^+ \not\geq R_2$ and $S^+ \not\geq R_3$.

For the arity 1 or for empty base, the proposition is obviously false, even with only R_1 and R_2.

• Consider the four following extensions of S. Let S_1 be obtained by adding a new element a and setting $S_1(a,x) = S_1(x,a) = +$ for every x in the base $|S|$. Let S_2 be similarly obtained with (-) instead of (+). Let S_3 be obtained with $S_3(a,x) = +$ and $S_3(x,a) = -$ for every x in $|S|$ and moreover $S_3(a,a) = +$. Finally let S_4 satisfy the same conditions, except that $S_4(a,a) = -$.

Suppose our conclusion is false. Then there exist two $S_i (i = 1,2,3,4)$ which admit an embedding of a same R. Hence in the base of this R there exists, say to fix ideas an element a_1 playing the role of a in S_1 and an a_3 playing the role of a in S_3.

Suppose firstly that the base $|R|$ has cardinality ≥ 2. Then a_1 and a_3 are distinct, since $R(x,a_1) = R(a_1,x) = +$ for every $x \neq a_1$ and $R(x,a_3) = -$ for every $x \neq a_3$. Moreover $R(a_1,a_3)$ takes simultaneously the value (+) and the value (-): contradiction. Analogous argument with S_1 and S_2, with S_1 and S_4, with S_2 and S_3, with S_2 and S_4, with S_3 and S_4.

Suppose now that the base $|R|$ has cardinality 1, and to fix ideas, suppose that R takes the value (-). Since $S \not\geq R$ and by hypothesis $|S|$ non-empty, necessarily S is reflexive. As previously define extensions S_1, S_2, S_3 which now are reflexive. Either $R_1 = R_2 = R_3 = R$ and then our conclusion holds. Or R_1 and possibly R_2 are distinct from R, thus have cardinalities ≥ 2. Again suppose our conclusion is false: then R_1 for instance is embeddable in at least two $S_i (i = 1,2,3)$, and the argument terminates as previously. •

8.2 Faithful extension between chains (Hagendorf, Jullien)

8.2.1 Existence of a faithful extension

Let A be an infinite chain, B a chain in which A is not embeddable. Then there exists a chain strictly greater than B in which A is not embeddable; uses axiom of choice; [103] HAGENDORF 1972; the case for scattered chains, i.e. chains without any embedding of $\mathbf{Q}$, was already proved by [130] JULLIEN 1969.

• If B is a finite chain, then it suffices to take $B + 1$.

Suppose that B is infinite; we distinguish two cases: $B < A$ and $B|A$ (with respect to embeddability).

In the case $B < A$, let B' and B" be two mutually incomparable chains, each immediately greater than B (see 5.6.3). Suppose the proposition is false. Then $A \leq B'$ and $A \leq B$". It is impossible to have both equimorphisms $A \simeq B'$ and $A \simeq B$", since $B'|B$". Hence $A < B'$ for instance, and so A is a strictly intermediate between B and B': contradiction.

It remains to examine the case where B is infinite and $B|A$. We argue ad absurdum by supposing that A is embeddable in every chain strictly greater than B. In an arbitrary manner decompose B into $C + C'$, where C is an initial interval and C' the complementary final interval. Let U be the least ordinal such that $C+U+C' > B$; similarly let V be the least retro-ordinal such that $C+V+C' > B$. By 5.6.1, the ordinal U and the converse of V are indecomposable ordinals.

By hypothesis $A \leq C+U+C'$. Consider an isomorphism of A onto a restriction of this sum: this decomposes A into three intervals D, U_A, D' with $A = D+UA+D'$ and $D \leq C, U_A \leq U, D' \leq C'$. If $U_A < U$ then $A \leq C + UA + C' \leq B$, contradicting the hypothesis that B is incomparable with A. Thus $U_A = U$ and then $A = D + U + D'$. Similarly we define the decomposition of A into intervals E, E' and an interval isomorphic to the retro-ordinal V such that $A = E+V+E'$ and $E \leq C, E' \leq C'$.

Envisage all possible relative positions of U and V, which are intervals of A. If the interval U is included in E, then $D+U \leq E \leq C$ so $D+U+D' \leq C+C' = B$: in other words $A \leq B$, contradicting the hypothesis. If there exists a final interval of U in E', then $U+D' \leq E' \leq C'$ so that $A = D+U+D' \leq C+C' = B$, contradicting the hypothesis. Thus we must suppose that the interval U, isomorphic to an indecomposable ordinal, is included in the interval V, isomorphic to a retro-ordinal. Hence $U = V = 1$, thus $D = E$ and $D' = E'$.

Let x be an arbitrary element of the base $|B|$. Let B_x be the initial interval of those elements $\leq x(\mathrm{mod}\, B)$ and B'_x the final interval $> x(\mathrm{mod}\, B)$. By the preceding, we have $B_x+1+B'_x > B$ hence $\geq A$. Moreover, by the axiom of choice, associate to each x an element fx in the base $|A|$, such that if A_x designates the initial interval of those elements $< fx(\mathrm{mod}\, A)$, then we have $A = A_x + 1 + A'_x$ with $A_x \leq B_x$ and $A'_x \leq B'_x$.

We shall prove that the function f which to each element x in $|B|$ associates fx in $|A|$, is strictly increasing. From this we will deduce that $B \leq A$, which contradicts the hypothesis of incomparability of A and B.

To do this, we argue ad absurdum by supposing that f is not strictly increasing: thus there exist two elements x, y in $|B|$ with $x < y(\mathrm{mod}\, B)$, and so with $B_x+1 \leq B_y$. Yet $fy \leq fx(\mathrm{mod}\, A)$ and so $A'_x \leq A'_y \leq B'_y$ thus $A = A_x + 1 + A'_x \leq B_x + 1 + B'_y \leq B_y + B'_y = B$, contradicting the hypothesis. •

Corollary 1. If two chains A, B have the same strictly greater chains (with respect to embeddability), then A and B are equimorphic.

Corollary 2. If A and B are infinite and if every chain $> B$ is $\geq A$ and conversely, then A and B are equimorphic.

8.2.2 A counterexample

The faithful extension no longer holds if we replace A by two chains A_1 and A_2 (JULLIEN 1969).

For example neither $A_1 = \omega + 1$ nor $A_2 = \mathbf{Z} = \omega^- + \omega$ is embeddable in ω. However, every chain which is strictly greater than ω admits an embedding of A_1 or of A_2.

The faithful extension no longer holds if we replace B by two chains B_1 and B_2.

For example $A = \omega + \omega^-$ is neither embeddable in $B_1 = \omega^-.\omega$ nor in its converse $B_2 = \omega.\omega^-$. Yet A is embeddable in every chain in which both B_1 and B_2 are embeddable: see 6.3.6, corollary.

Problem posed by SABBAGH in 1975. Existence of two chains A, B which are incomparable with respect to embeddability, and have a supremum chain C. More precisely, for every chain X we would have $X \geq C$ iff $X \geq A$ and $X \geq B$.

It has been proved by [104] HAGENDORF 1977 that this supremum chain does not exist for A or B scattered; the general case remains unsolved.

For the dual of the above statement, i.e. the existence of the infimum, we have the easy example $A = \omega + 1$ and $B = \mathbf{Z}$ with the infimum $C = \omega$.

8.3 Faithful infinite extension: Malitz' and Lopez' counterexamples

(1) The statement 5.10.3 can be rewritten as follows.

Let R be a finite relation. Let $A_1, ..., A_h$ be a finite set of finite relations with common arity; if there exist extensions of R with arbitrary large finite cardinalities, which are $\not\geq A_1$ and ... and $\not\geq A_h$, then there exists a denumerable extension of R which respects the same conditions.

(2) **Let $A_1, ..., A_h$ and $B_1, ..., B_k$ be two finite sets of finite relations with common arity, and let R satisfy $R \not\geq A_1$ and ... and $\not\geq A_h$ as well as $R \geq B_1$ and... and $\geq B_k$. Then there exists an integer u such that every R with cardinality at least equal to u satisfying the preceding conditions has a restriction R' respecting the same conditions, and such that R' has a denumerable extension still respecting the conditions.**

• Let v be the sum of the cardinalities of the relations B_1 through B_k. For each R satisfying the conditions, there exists a restriction R' of R with cardinality at most equal to v, which satisfies the conditions. Consider all these R', which are only finitely many, up to isomorphism. For each of them, either there exists a denumerable extension satisfying the conditions. Or there exists an integer $u(R')$ which is strictly greater than the cardinalities of all extensions of R' respecting the conditions. Then it suffices to set u to be the maximum of these $u(R')$. •

8.3.1 Malitz' counterexample

Can we require that $R' = R$; in other words, does there exist an integer u such that, if R has cardinality greater than or equal to u and satisfies the conditions, then there exists a denumerable extension of R satisfying them.

A negative answer is due to [166] MALITZ 1967, whose construction was already used in 5.10.4.

• Take the base of integers from 0 to $n-1$. Let I_n be the usual chain of these integers; let C_n be the consecutivity relation ($y = x+1$); let 0_n be the unary relation called the singleton of zero, i.e. the relation taking (+) for 0 and (-) elsewhere; and let U_n be the relation singleton of $n-1$. Finally let R_n be the quadrirelation $(I_n, C_n, 0_n, U_n)$. From $n = 7$ on, all the R_n have the same restrictions of cardinalities 1, 2 and 3 , up to isomorphism. Let $A_1, ..., A_h$ be those quadrirelations of the same arity and cardinalities 1, 2, 3 which are not embeddable in R_7, and hence in $R_n (n \geq 7)$. We see that every extension of an R_n to a new element added to its base admits an embedding of one of the $A_1, ..., A_h$.

An analogous but rather complicated counterexample is obtained for binary relations by [158] LOPEZ 1973. •

8.3.2 Lopez' counterexample

Given the finite relations $A_1, ..., A_h$ and $B_1, ..., B_k$, one can ask whether there exist two integers u, v such that, for every R with cardinality greater than or equal to u, there exist v elements of the base, such that the restriction of R to its base with these v elements removed respects the embedding inequalities in the B's and has an extension of arbitrary large cardinality respecting the non-embedding inequalities in the A's. Negative answer by LOPEZ 1973.

• For the base, take the set of points, or couples of integers called the abscissa and ordinate, and which vary from 0 to $n-1$. Let R_n be the multirelation on this base, which is composed of the following 4 unary relations and 6 binary relations. The unary relation 0_n takes the value (+) for the points with abscissa 0. The relation U_n takes (+) for the points with abscissa $n-1$. Similarly $0'_n$ and U'_n are defined by interchanging abscissas and ordinates. The stratified partial ordering I_n takes the value (+) for each couple of points $(i, x), (j, y)$ whose abscissas satisfy $i < j < n$, with arbitrary ordinates x, y; moreover I_n is reflexive. The equivalence relation E_n takes (+) for any two points with a same abscissa and arbitrary ordinates. The equivalence classes of this relation are thus the classes of elements which are pairwise incomparable modulo I_n. The binary relation C_n, which by abuse of notation we shall call a **consecutivity**, takes the value (+) for each couple of points $(i, x), (i+1, y)$ whose abscissas are consecutive. Finally, the stratified poset I'_n, the equivalence relation E'_n and the consecutivity C'_n are obtained from the preceding by interchanging abscissas and ordinates.

From $n = 7$ on, every R_n has the same restrictions $B_1, ..., B_k$ with cardinalities 1, 2, 3 (up to isomorphism). Let $A_1, ..., A_h$ be the other multirelations of the same arity and cardinalities 1, 2, 3. We see that every proper extension of $R_n (n \geq 7)$ admits an embedding of at least one of the A's. Indeed, add a new element t to the base $|R_n|$. Consider the case where either 0_n or U_n or $0'_n$ or U'_n takes the value (+) for t, and reduce this to the case of MALITZ' counterexample in the previous subsection. Now consider the case where all the preceding unary relations take the value (-) for t. Then either there exists an equivalence class of E_n to which t belongs: again reduce to MALITZ. Or t occurs between two consecutive equivalence classes of E_n. In this case, use the consecutivity C_n to see that the extension of R_n thus obtained admits an embedding of one of the A's .

Now suppose the existence of u and v satisfying our hypothesis; take $n > u$ and $> v$. Let S_n be a restriction of R_n in which the B's are embeddable, and which is obtained by removing v points. Then in each equivalence class of E_n, there remains at least one element of $|S_n|$; similarly for E'_n. Add a new element t to the base $|S_n|$, and attempt to require that the extension of S_n to its base with t added admit only embeddings of the B's and not of the A's. This leads us to situate t in the chain of the equivalence classes of E_n. By using C_n, one sees that t necessarily belongs to one of the equivalence classes: t cannot be situated between two consecutive classes. Thus we obtain an element in the base $|S_n|$, which is equivalent with t modulo E_n, and another element equivalent with t modulo E'_n. From this, we deduce that t is the unique element common to both equivalence classes. Thus we have again a restriction of R_n obtained by removing $v-1$ points: this is our extension of S_n.

Iterating this, we obtain R_n itself, and at the following step we obtain a proper extension of R_n, in which necessarily one of the A's is embeddable. •

8.4 Bivalent tableau (= bipartite graph): C. Rauzy

A **bivalent tableau** is the system formed by two sets: the set E of **columns** and the set F of **rows**; and a function which, to each element in the cartesian product $E \times F$, associates either the value (+) or the value (-). We may consider E and F as being two disjoint sets.

We may consider a bivalent tableau T as being a **bipartite graph** defined on the two disjoint sets E and F with the following interpretation: we draw an arrow from any element $x \in E$ to any element $y \in F$ iff $T(x, y) = +$. However specialists of bipartite graphs generally ignore the following notion of embeddability and extensivity problems.

Given a tableau T on $E \times F$, we have the immediate notions of the **restriction** of T to a subset E' of E and a subset F' of F.

Given two tableaux T on $E \times F$ and T' on $E' \times F'$, we define an **isomorphism** from T onto T' via two bijections: e from E onto E' and f from F onto F'.

Consequently we say that T is **embeddable in** T', denoted by $T \leq T'$, iff there exists an isomorphism from T onto a restriction of T'. In other words, an injection e from E into E' and an injection f from F into F', preserving the values: $T'(ex, fy) = T(x, y)$ for all $x \in E$ and $y \in F$.

The tableau T is said to be **extensive by the tableau** U (relatively to rows) iff either $T \leq U$, or there exists an U^+ obtained from U by adding a row, such that $T \not\leq U^+$. otherwise, if $T \not\leq U$ and yet $T \leq U^+$ for every U^+ obtained from U by adding a row, then we say that T is **inextensive by** U (relatively to rows). Note that extensivity generalizes embeddability.

We leave it to the reader to see that **the tableau T with two columns (below on the left), which is not embeddable in the tableau U with four columns (below on the right), is inextensive by** U (relatively to rows):

T			U			
+	+		+	-	+	+
-	-		+	-	+	-
+	-		+	-	-	-
+	-		-	+	-	+

Hint. First note that you cannot add different values to the third and fourth columns in U, without embedding T; if you add + + to these columns, then you must add - to the second column (because of the second and third columns); and then T is embedded in the second and fourth columns: contradiction. Now if you add - - to the third and fourth columns, then you must add + to the first column (because of the first and fourth columns); and then T is embedded in the first and third columns: contradiction.

We say that a tableau T is p-**extensive** (p integer) iff T is extensive by every tableau having p rows. In other words, for every tableau U having p rows and such that $T \not\leq U$, there exists U^+ (obtained from U by adding a row) which respects the non-embeddability $T \not\leq U^+$.

Finally T is said to be **extensive** (relatively to rows) iff T is extensive by every tableau; it is said to be **inextensive** otherwise.

Every inextensive tableau must have a row of (+), a row of (-), and at least two identical rows. This is the case for the above example T with two columns and four rows.

It is proved by [159] LOPEZ 1977 that **the above inextensive tableau T is however p-extensive for every integer $p \geq 5$.**

Problem. For each bivalent tableau which is finite, i.e. which has finitely many rows and columns, do there exist infinitely many integers p for which T is p-extensive. More strongly does there exist an integer p such that T is q-extensive for every integer $q \geq p$.

An important particular case, yet unsolved, of the preceding problem, concerns tableaux T with only two columns. Some results of [214] Claire RAUZY 1986:

(1) **Any tableau having an arbitrary number a of rows (+ +), an arbitrary number b of (- -), then c lines (+ -) and d lines (- +), with the condition that the absolute value $|d - c| = 0$ or 1, is p-extensive for every integer p;**

(2) **Any tableau with exactly one row (++), one row (−−) and n rows (+−), is p-extensive for $p \leq 2n - 1$ and for $p \geq 6n - 4$.**

In particular, we find an attenuation of LOPEZ' result: indeed the above tableau with one row (+ +), one row (- -) and two rows (+ -) corresponds to $n = 2$, so that we have the p-extensivity for $p \geq 6.2 - 4 = 8$ (instead of 5).

• Proof of (1) in the particular case where $d = c$. Define the tableau T as having a rows (+ +), b rows (- -), c rows (+ -) and c rows (- +). Consider an arbirary tableau U with p rows, such that $T \not\leq U$. A given column in U is said to be (+)-**major** iff the number of (+) is $> (p + a - b)/2$; otherwise, if the number of (+) is $\leq (p + a - b)/2$, then the column is said to be (+)-**minor**.

Let us decide to "aggravate" inequalities between columns, by adding the value (+) to each (+)-major column, and adding (-) to each (+)-minor column. To any couple of columns in U, let us associate the four following integers: x = number of rows (+ +); y = number of (- -); z = number of (+ -); t = number of (- +); so that $p = x + y + z + t$.

In order that T be embedded in U after its completion by an additive row, let us consider three cases. Firstly T can appear by adding a row (+ +) to a couple of (+)-major columns. Then we necessarily have $x = a-1, y \geq b, z \geq c, t \geq c$, so that the total number of values (+) in these two columns is $2x+z+t = p+x-y \leq p+a-1-b$: contradiction, since two (+)-major columns in U contain at least $p+a-b$ (and even $p+a-b+1$) values (+) .

Secondly T can appear by adding a row (- -) to a couple of (+)-minor columns. Then we have $x \geq a, y = b-1, z \geq c, t \geq c$, so that the total number of values (+) in these two columns is again $p+x-y \geq p+a-b+1 > p+a-b$: contradiction, since two (+)-minor columns in U contain at most $p+a-b$ values (+) .

Thirdly T appears by adding a row (- +) with (-) added to a first (+)-minor column and (+) added to a second (+)-major column. Then we must have $z < t$ (indeed $x+z \leq (p+a-b)/2 < x+t$) and $x \geq a, y \geq b, z \geq c, t = c-1$ so that $c \leq z < t = c-1$ hence $c < c-1$: contradiction. •

We leave it to the reader to adapt the preceding proof to the cases where $d = c+1$ or $c = d+1$.

• Proof of the first case of (2), i.e. for $p \leq 2n-1$. Define the tableau T as having one row (+ +), one row (- -) and n rows (+ -). Consider an arbitrary tableau U with $p \leq 2n-1$ rows, such that $T \not\leq U$. A given column in U is said to be (+)-**minor** iff the number of values (+) is $\leq n-1$; otherwise, if the number of (+) is $\geq n$, and consequently the number of (-) is $\leq p-n \leq n-1$, then the column is (+)-**major**. Let us decide to "attenuate" inequalities between columns by adding the value (+) to each (+)-minor column and (-) to each (+)-major column.

Given a couple of columns in U, either both are (+)-minor columns and consequently the row (+ +) is added in order that T be embedded in U after its completion. In this first case, there exists no row (+ +) in the given couple, yet there exists at least one row (- -) and at least either n rows (+ -) or n rows (- +). Thus we have at least n values (+) in one of our two (+)-minor columns: contradiction.

Or secondly both columns are (+)-major and consequently the row (- -) is added. Then there exists no row (- -) in our couple, yet there exists at least one row (+ +) and at least either n rows (+ -) or n rows (- +). Thus we have at least n values (-) in one of our (+)-major columns: contradiction.

Or thirdly we have in our couple one first (+)-minor column and one second (+)-major column; consequently the row (+ -) is added. Then there exist exactly $n-1$ rows (+ -) in our couple of columns, and at least one row (+ +) and one row (- -). So that our first (+)-minor column contains at least n values (+): contradiction. •

8.5 Poset of height 2 and associated tableaux: Hazim Sharif

The general problem of faithful extension among finite posets is trivial. Indeed, given two finite posets A, B with $A \not\leq B$ (with respect to embeddability), define B^* by adding to B an element incomparable to all other ones; then $B^>$ by adding an element strictly greater than all other ones. Then assuming that $A \leq$ both B^* and $B^>$, there exists in the base $|A|$ an element u which is incomparable to all other ones and an element v which is strictl greater to all other ones; so that $u|v$ and $v > u(\bmod A)$: contradiction proving that either A^* or $A^>$ is a faithful extension of A modulo B.

Things become much more interesting if, following an idea of Robert BONNET, we require that the faithful extension have **same height as the given partial ordering**. For instance consider two posets A, B with common height 2, such that $A \not\leq B$ and ask for the existence of a poset B^+ extension of B with height 2, which respects the non-embeddability $A \not\leq B^+$.

First of all if A has no element which is incomparable $(\bmod A)$ with each other element in $|A|$, then it suffices to add to the base $|B|$ an element u which is incomparable $(\bmod B)$ with each element of $|B|$. Denote by B^+ the obtained extension of B: then $A \not\leq B^+$.

In the following subsection, we choose a poset A having at least one element incomparable with all other ones.

8.5.1 The inextensive poset $(|X|..)$ by means of the bracelet

Start with the poset denoted by $|X|$ with two mutually incomparable elements a, b of height 0 plus two incomparable elements a', b' of height 1, and the four comparisons $a < a', a < b', b < a', b < b'$. Add to $|X|$ two elements mutually incomparable and incomparable with all other ones: we denote by $A = (|X|..)$ the obtained **inextensive poset** on six elements.

Now define on ten elements the so-called **bracelet** B with four **plates**, each isomorphic with $|X|$, plus two symmetrical **pendants**.

More precisely take four elements a, b, c, d which shall be minimal, or with height 0, then four corresponding elements a', b', c', d' which shall be maximal, or with height 1, disposed as the eight vertices of a cube with two opposite horizontal faces and four vertical ones. The two horizontal faces represent two antichains $\{a, b, c, d\}$ and $\{a', b', c', d'\}$ disposed so that the vertices a and a' have symmetrical positions in the cube; idem for b and b', c and c', d and d'.

Each of the four lateral faces such that $\{a, b, d', c'\}$ (with ac' and bd' vertical edges) is the base of a poset called a **plate** and isomorphic to $|X|$ with $a|b, c'|d', a < c', a < d', b < c', b < d'(\bmod B)$. Furthermore we choose two opposite vertices in the cube, say the minimal vertex d and the maximal vertex d' and we add two elements called the **pendants**: a minimal element e such that $e < d'$ and incomparable with all other elements; and a maximal element $e' > d$ and incomparable with all other elements.

Note that there exist, up to isomorphism, 63 extensions of B to its base augmented by an element which respects the height 2. Indeed on one hand we can add a minimal new element which is necessarily incomparable with a, b, c, d, e and may be either strictly less or incomparable with a', b', c', d', e': this gives $2^5 = 32$ possibilities.

On the other hand we can add a maximal element which is necessarily incomparable with a', b', c', d', e' and may be either strictly greater than or incomparable with a, b, c, d, e; yet it is necessary to have at least one comparison $> a$ or $> b$ or ... or $> e$, in order that our new element have height 1: so we have only 31 possibilities.

8.5.2 Non-extensivity statement

$A \not\leq B$ yet A is embeddable in each of the 63 posets with height 2 obtained from B by adding a new element (see [78] FRAÏSSÉ, HAZIM SHARIF 1993; the example is due to Hazim Sharif).

• The reader can easily check that $A = (|X|..)$ is not embeddable in B.

The poset B being symmetrical by the transformation of a into a', ..., e into e', it suffices to check the 32 first possibilities, for instance it suffices to add to B an arbitrary minimal element s, obtaining the extension B^+.

Let us check, by a unique argument, the eight cases where $s < b', s < e'$ and s is either strictly less than or is incomparable with a', c', d' (and obviously s is incomparable with the minimal elements a, b, c, d, e) . We see that A is isomorphic with the restriction of B^+ to the elements s, d, b', e' (which constitute an $|X|$) and b, e (incomparable with them).

Assuming that $s|a', s|b'$ and s either strictly less than or incomparable with c', d', e', we see that A is isomorphic with the restriction of B^+ to the elements c, d, a', b' (which constitute an $|X|$) and s, e (incomparable with them): this checks eight new cases.

Assuming that $s < a', s|b'$ and $s < e'$, then A is isomorphic with the restriction of B^+ to s, d, a', e', a, e: this checks four new cases. Assuming that $s < a', s < c'$ and $s|e'$, then A is isomorphic with the restriction of B^+ to a', c', b, s, e, e': this checks again four new cases.

Assuming that $s < a', s|b', s|c'$ and $s|e'$, then A is isomorphic with the restriction of B^+ to a, d, b', c', s, e: this checks two new cases. Assuming that $s|a', s < b', s|d'$ and $s|e'$, then A is isomorphic with the restriction of B^+ to b, c, a', d', s, e': this checks two new cases. Assuming that $s|a', s < b', s < d'$ and $s|e'$, then A is isomorphic with the restriction of B^+ to s, a, b', d', a', e': this checks again two new cases. Assuming that $s < a', s < b', s|c', s|d'$ and $s|e'$, then A is isomorphic with the restriction of B^+ to a, b, c', d', e', s: this checks one new case. Finally assuming $s < a', s < b', s|c', s < d'$ and $s|e'$ then A is isomorphic with the restriction of B^+ to c, s, a', d', c', e': this is the 32-nd and last case. •

8.5.3 Tableau associated with a finite poset of heigh 2

Following an idea of Maurice POUZET and Claire RAUZY, **to each finite poset A with height 2, we associate a non-empty tableau $T(A)$ having at least one value** $(+)$.

To each element in A with height 1 we associate a column; to each element with height 0, having at least one strictly greater element modulo A, we associate a row.

At the intersection of a row and a column, we put the value (+) or (-) according to whether the two corresponding elements are comparable or incomparable (mod A). Finally to each element in A which is incomparable to all other ones, we associate either a row of values (-), or a column of (-).

Conversely we see that **any finite non-empty tableau taking at least one value (+) is associated to one and only one poset of height 2 (up to isomorphism)**.

The different tableaux associated to a same poset are said to be **equivalent**: we pass from one to another by replacing a row of (-) by a column of (-) or conversely. To be precise the tableau with n colums associated to A is denoted $T_n(A)$.

For example to the chain on two elements is associated the unique tableau with one row, one column and value (+). To the poset denoted by (Λ.) with three minimal elements a, b, c and one element $a' > a, a' > b$ and $a'|c$ we associate T_1 with three rows and one column (+ + -) (write it vertically), and T_2 with two rows and two columns, which are (+ +) and (- -).

8.5.4 Correspondence lemmas

(1) **Let $T(A)$ and $T(B)$ be two tableaux respectively associated with the finite posets A and B (of height 2); assuming that $T(A) \leq T(B)$ (under the embeddability between tableaux), then $A \leq B$ (under the embeddability between relations)**.

• By hypothesis $T(A)$ is isomorphic with a non-empty restriction of $T(B)$; more precisely there exists at least one row and one column in $T(A)$ with the value (+) at their intersection. By hypothesis each row and each column of $T(A)$ bijectively corresponds to a vertex in A as well as a vertex in B: so that $A \leq B$. •

(2) **Given two finite posets A, B of height 2, then $A \leq B$ (with respect to embeddability between relations) iff for every tableau $T(B)$ associated with B there exists a tableau $T(A)$ associated with A, satisfying $T(A) \leq T(B)$ (with respect to embeddability between tableaux)**.

• If our condition is satisfied, then $A \leq B$ by statement (1).

Conversely suppose that $A \leq B$ and let $T(B)$ be any tableau associated with B. Each vertex in B bijectively corresponds to either one row or one column of $T(B)$. Consider A as a restriction of B: the preceding correspondence yields a non-empty tableau $T(A)$ associated with A and restriction of $T(B)$. •

(3) **It is possible that $A \leq B$ and yet there exists a tableau $T(A)$ which is not embeddable in any $T(B)$**.

• Let $A = (V.)$ and $B = W$, where V denotes the poset with one element of height 0 less than two elements of height 1; the point in $(V.)$ denotes an additional element which is incomparable to the three preceding ones. Analogous interpretation for W. Then $(V.) \leq W$. The only tableau associated with W is $T_3(W)$ with 2 rows and 3 colums, more precisely the columns (+ +), (+ -) and (- +). Now the tableau $T_2(V.)$ is formed of two identical columns (+ -) so that it is not embeddable in $T_3(W)$. The paradox comes from an element of height 1 in W which necessarily becomes a column in $T_3(W)$ yet becomes a row in $T_2(V.)$. •

8.5.5 Equivalence lemma

Let $T_1, ..., T_k$ be the whole finite sequence of equivalent tableaux associated to a given finite poset with height 2. Let X be a finite non-empty tableau with at least one value $(+)$. If $X \not\geq T_1$ and ... and $X \not\geq T_k$, then the same conditions are true for each tableau equivalent with X.

• Let Y be a tableau equivalent with X; to fix ideas we may assume that Y is obtained from X by suppression of a column of (-) then addition of a row of (-). Now if $Y \geq T_i$ for some $i (1 \leq i \leq k)$, then there exists in T_i at least one row of (-) and obviously another row with a $(+)$. Suppress this row of values (-) and add a column of values (-): then we obtain an equivalent tableau which is a $T_j (j \neq i)$, and we have $X \geq T_j$: contradiction. •

8.5.6 Inextensivity theorem

Let A, B two finite posets of height 2, and suppose that A is inextensive by B. Let $T_1(A), ..., T_k(A)$ be the whole sequence of tableaux associated with A; let $T(B)$ an arbitrary tableau associated with B; then:

(1) $T(B) \not\geq T_1(A)$ **and ... and** $T(B) \not\geq T_k(A)$;

(2) **let U be an extension obtained from $T(B)$ by addition of either an arbitrary row or an arbitrary column; then $U \geq T_1(A)$ or ... or $U \geq T_k(A)$.**

• (1) Since A is inextensive by B, we have that $A \not\leq B$ hence $T_1(A) \not\leq T(B)$ and ... and $T_k(A) \not\leq T(B)$ by 8.5.4 (1).

(2) Denote by B^* the poset with height 2 which corresponds to the tableau U. Then by the correspondence lemma (1) we have that $B < B^*$; hence $A \leq B^*$ by hypothesis of inextensivity. Finally by 8.5.4 (2) we have that $U = T(B*) \geq T_1(A)$ or ... or $U \geq T_k(A)$. •

8.5.7 Example of the tableaux associated with $(|X|..)$ and with the bracelet

Consider again our example of the poset $A = (|X|..)$ which admits the three following associated tableaux: $T_2(A)$ with 2 columns, $T_3(A)$ with 3 columns, $T_4(A)$ with 4 columns:

$$\begin{array}{cc} + & + \\ + & + \\ - & - \\ - & - \end{array} \qquad \begin{array}{ccc} + & + & - \\ + & + & - \\ - & - & - \end{array} \qquad \begin{array}{cccc} + & + & - & - \\ + & + & - & - \end{array}$$

Now consider the poset called **bracelet** (see 8.5.1), which admits no element incomparable to all other ones; so that there exists a unique associated tableau. Each element a, b, c, d, e has height 0 thus transforms into a row; each element a', b', c', d', e' has height 1 thus transforms into a column. The symmetry of the bracelet gives a diagonal symmetry of the following associated tableau:

	a'	b'	c'	d'	e'
a	-	+	+	+	-
b	+	-	+	+	-
c	+	+	-	+	-
d	+	+	+	-	+
e	-	-	-	+	-

The reader can easily check that neither T_2 nor T_3 nor T_4 admit an embedding in the above tableau. However if we add an arbitrary sixth column of five values, either T_2 or T_3 or T_4 is necessarily embedded. For example add a sixth column of values (+): then we see that T_4 is embedded in the 3rd and 5th rows. Add a column of values (-): then again T_4 is embedded in the two first rows. We can avoid an embedding of T_4, for instance by adding a sixth column with the following values: (+) in the three first rows, (-) in the two last rows. Yet in this case T_3 is embedded in the first, third and fifth rows.

8.6 Faithful augmentation, Szpilrajn chain, Jullien's theorem

We say that a chain C is **Szpilrajn** iff, for every poset A in which C is not embeddable, there exists, modulo the axiom of choice, a totally ordered augmentation of A in which C is not embeddable.

If the chain C is Szpilrajn, then so is every chain equimorphic with C, as well as with the converse C^-.

Among finite chains, only the empty chain and the singleton chain are Szpilrajn. Indeed, for each integer $p \geq 2$, the free poset on p elements, when augmented until becoming a chain, necessarily yields the chain of cardinality p.

8.6.1 Lemma

If a chain C does not admit an embedding of $C + 1$, then for each $p \geq 2$ the chain $C + p$ is not Szpilrajn.

For example $\omega + 2, \omega + 3$, .. are not Szpilrajn.

Indeed it suffices to start with the poset formed by the chain C followed by p mutually incomparable elements.

8.6.2 The chain ω of integers is Szpilrajn

• We shall see this, equivalently, for the converse ω^-. A poset A in which ω^- is not embeddable is well-founded (5.2.2, dependent choice). Then there exists a well-ordered augmentation of A: see 2.9.2 proposition (2) (axiom of choice; ZF suffices if A is countable). •

8.6.3 The chain Q of rationals is Szpilrajn

This is a form of the equivalence of conditions (1) and (2) of 6.5.3 (using axiom of choice).

8.6.4 Denumerably Szpilrajn chain

We say that a chain C is **denumerably Szpilrajn** iff C is denumerable and, for every denumerable poset A in which C is not embeddable, there exists a totally ordered augmentation of A in which C is not embeddable.

If a chain is denumerable and Szpilrajn, then it is denumerably Szpilrajn.

However, we shall see below that $\omega + 1$ is denumerably Szpilrajn without being Szpilrajn.

We see that $\omega + p$, with an integer $p \geq 2$, is not denumerably Szpilrajn (take again the chain ω followed by p incomparable elements).

8.6.5 The chain $\omega + \omega^-$ is not denumerably Szpilrajn

.

• Start with $A = \omega^-.\omega$ and $B = A^- = \omega.\omega^-$, assumed to be defined on two disjoint bases. Take the poset on $|A| \cup |B|$ with each element of A incomparable to each element of B. Then the chain $\omega + \omega^-$ is not embeddable in this poset. Yet it is embeddable in every totally ordered augmentation, by 6.3.6, corollary. •

8.6.6 A lemma on Szpilrajn chains

Let C, D be two chains. If $C + 1$ and $1 + D$ are Szpilrajn, then so is $C + 1 + D$.

Same statement for "denumerably Szpilrajn" ([130] JULLIEN 1969).

• Let A be a poset in which $C + 1 + D$ is not embeddable. Call H the initial interval of those elements x above which there is a chain in A isomorphic with $1 + D$; the element x being the minimum of this chain $1 + D$. Then $C + 1$ is not embeddable in H. Moreover, the chain $1 + D$ is not embeddable in the final interval H' complement of H.

Take a totally ordered augmentation of A formed by an initial interval which is an augmentation of H yet in which $C + 1$ is not embeddable; followed by a final interval which is an augmentation of H' yet in which $1 + D$ is not embeddable. Then $C + 1 + D$ is not embeddable in such a chain. •

In particular, **the chain $\mathbf{Z} = \omega^- + \omega$ of the positive and negative integers is Szpilrajn**, since $\omega^- = \omega^- + 1$ and $\omega = 1 + \omega$ are Szpilrajn.

8.6.7 Theorem. The ordinal $\omega + 1$ is denumerably Szpilrajn yet not Szpilrajn

See [130] JULLIEN 1969; also for the two corollaries); uses denumerable subset axiom.

• We shall prove that $\omega+1$ is denumerably Szpilrajn. Start with a denumerable poset A in which $\omega + 1$ is not embeddable. If A has a maximum element, then even ω is not embeddable in A, and so by the above 8.6.2 there exists a totally ordered augmentation of A in which ω is not embeddable.

If A has no maximum element, then take an ω-sequence of elements a_i(i integer) forming a cofinal set in A with the condition $a_j >$ or $| a_i (\mathrm{mod}\, A)$ for all i and $j > i$. Possibly we can have only a finite cofinal set of maximal elements, thus a finite sequence.

Associate to each i the restriction A_i of A to those elements which are $\leq a_i$ but $\not\leq a_0, \ldots, \not\leq a_{i-1}$. The bases $|A_i|$ are mutually disjoint and their union is the base $|A|$. None of the A_i admit an embedding of ω since otherwise, taking into account the maximum a_i, the ordinal $\omega + 1$ would be embeddable. By 8.6.2, there exists a totally ordered augmentation B_i of A_i in which ω is not embeddable. Then the sum of the B_i according to increasing i is a totally ordered augmentation of A, in which $\omega + 1$ is not embeddable.•

• Now we prove that $\omega+1$ is not Szpilrajn. Take an uncountably infinite set U. To each enumeration f, without repetition, of a denumerable subset of U and to each integer i, associate the couple (f, i). Take as base E the union of U and the set of these couples. Set $(f, i) < (f, j)$, for the same enumeration f, iff $i < j$ in the usual ordering of integers. Moreover for each couple, set $(f, i) < f(i)$: this second term being an element of U. Hence by transitivity we have $(f, i) < f(j)$ for all $j > i$. Apart from these cases, two distinct elements of E shall be incomparable. Call A the thus defined poset on E: we easily see that ω is embeddable in A, but not $\omega + 1$.

Let B be a totally ordered augmentation of A, based on E. There exists at least one element u of U having at least denumerably many elements $x < u(\mathrm{mod}\, B)$ with $x \in U$. In other words there exists at least a denumerable set which is not cofinal modulo B (denumerable subset axiom). Take an enumeration f of such elements. The poset A, hence the chain B as well, admits as a restriction the chain of the (f, i) where f is fixed and $i = 0, 1, 2, \ldots$: a chain of order type ω which has the element u as an upper bound $(\mathrm{mod}\, B)$. Thus $\omega + 1$ is embeddable in B. •

Corollary 1. If a denumerable chain C satisfies the strict inequality $C + 1 > C$ (with respect to embeddability), then $C + 1$ is not Szpilrajn.

• Either C has no maximum element. Then use the preceding argument: replace ω by C; replace each f by an injection of the base $|C|$ onto a denumerable subset of U; the inequality $i < j$ being made modulo C .

Or C has a maximum element. Denote by $C - 1$ the chain C after removing its maximum: $C - 1$ does not admit an embedding of C. Then it suffices to consider the poset obtained from $C - 1$ by adding two maximal and mutually incomparable elements. •

Corollary 2. The ordinal $\omega.2$, and more generally $\omega.p$ with p integer ≥ 2, is not Szpilrajn.

• Take an uncountably infinite set U which is the union of uncountably many disjoint denumerable subsets U_k. On each U_k take a chain isomorphic with ω. Furthermore any two elements of U which belong to two distinct U_k will be incomparable. Then it suffices to terminate as for $\omega + 1$. •

8.6.8 A list of Szpilrajn chains

It is proved by GALVIN and MAC KENZIE in 1969 (mentioned without proof in ToR-86 p.197) that ω **is the only denumerable Szpilrajn ordinal**.

The following is published in [19] BONNET, POUZET 1982.

The only denumerable Szpilrajn chains (up to equimorphism) are the following and their converses:

the chain $\mathbf{Q}$ of rationals; the scattered chains $P_1 = \omega, P_2 = \omega^-.\omega, P_3 = \omega.\omega^-.\omega$; in general $P_{i+1} = (P_i^-).\omega$ for each integer i, and more generally for each countable successor ordinal.

We set $P_\omega = \Sigma P_i (i \text{ integer})$. More generally, given any denumerable limit ordinal u, we set $P_u = \Sigma P_i$ for any ω-sequence of indices i forming a cofinal set in u.

Finally for each preceding P_i(i countable ordinal), the sum $(P_i)^- + P_i$ is szpilrajn. Note that $1 + P_i \simeq P_i$, so the latter sum is obviously Szpilrajn by 8.6.6 above.

For any two countable ordinals i and $j > i$, we have that $P_i^- + P_j$ is Szpilrajn by 8.6.6. However this is already understood, since the latter sum is equimorphic with P_j.

Chapter 9

Local isomorphism, free operator, chainable and monomorphic relations, relational or strong interval

9.1 Permutation, transposition, local isomorphism

9.1.1 Preliminaries

The notions of isomorphism, automorphism and embeddability are defined in chapters 1, 2 and 5; until now they were mainly used for posets and chains. We now apply these notions to arbitrary relations and multirelations. For convenience, the definitions and statements will be given for the case of relations. Unless otherwise indicated, they can be extended to multirelations: the role played by the arity of the relation, there being played by the maximum arity of the component relations in the multirelation.

Let E be a set, f a permutation of E, and F a finite subset of E. There exists a sequence $a_1, ..., a_h$ of elements of F, without repetition, such that for each $x \in F$, the image fx is obtained as well by composition of the successive transpositions $(a_1, fa_1), ..., (a_h, fa_h)$.

• Partition the elements of F into maximal partial orbits of the form $(u_1, u_2, ..., u_k)$ (where k is an integer) and $u_2 = fu_1, ..., u_k = fu_{k-1}$. Associate to each maximal partial orbit the sequence of transpositions $(u_k, fu_k), (u_{k-1}, u_k), ..., (u_1, u_2)$ if $fu_k \neq u_1$; and the sequence of transpositions $(u_{k-1}, u_k), ..., (u_1, u_2)$ if $fu_k = u_1$ (the case of a cycle, or total orbit). Then each element x of our partial (or total) orbit has image fx. It suffices to order in an arbitrary manner the set of orbits, hence the set of corresponding sequences of transpositions. •

9.1.2 Automorphism lemma

Let R be a relation with base E and f a permutation of E. If for each element $x \in E$, the transposition (x, fx) is an automorphism of R, then f is an automorphism of R ([157] LOPEZ 1969).

• Let n be the arity of R. If f modifies R, then there exists a sequence of n terms $x_i (i = 1, ..., n)$ with $R(x_1, ..., x_n) \neq R(fx_1, ..., fx_n)$. By the preceding proposition, there exists a sequence $y_1, ..., y_h$ formed of elements x_i such that the composition of the transpositions $(y_1, fy_1), ..., (y_h, fy_h)$ takes each x_i into fx_i and hence modifies R. Then at least one of the transpositions modifies R. •

9.1.3 Lemma of the altered restriction

Let R be an n-ary relation. If a transposition (a, b) modifies R, then there exists a subset F of the base $|R|$ with cardinality $\leq n + 1$, which contains a, b and such that (a, b) modifies R/F.

• Let $a_1, ..., a_n$ be such that $R(a_1, ..., a_n) \neq R(fa_1, ..., fa_n)$, where f denotes the transposition (a, b). Then necessarily a and b occur among the a_i and the $f(a_i)(i = 1, ..., n)$. Thus it suffices to take F to be the set of the a_i and the $f(a_i)$, which has cardinality less than or equal to $n+1$: indeed we have $a_i = f(a_i)$ except where one of these values is a and the other is b. •

9.1.4 Local isomorphism

Let R, R' be two relations of the same arity. A **local isomorphism** from R into R' is an isomorphism from a restriction of R onto a restriction of R'.

For example, if R, R' are two posets, then a local isomorphism from R into R' is a bijective mapping f from a subset of the base $|R|$ onto a subset of $|R'|$, with $fx < fy \pmod{R'}$ iff $x < y \pmod{R}$, for every x, y in $\mathrm{Dom}\, f$. In other words, f is order preserving, as well as its converse f^{-1}.

The definition of local isomorphism extends immediately to the case of two multirelations of the same arity. It follows from 1.7.6 that given two multirelations R, R' of the same arity, where $R = (R_1, ..., R_h)$ is the concatenation of the component relations $R_1, ..., R_h$ and $R' = (R'_1, ..., R'_h)$ is the concatenation of $R'_1, ..., R'_h$; then a bijective mapping f from a subset of $|R|$ onto a subset of $|R'|$ is a local isomorphism from R into R' iff f is simultaneously a local isomorphism from R_1 into R'_1 and ... and from R_h into R'_h.

For example, if R, R' are chains and S, S' are groups (thus ternary relations), then the concatenations RS and $R'S'$ are ordered groups provided that the known axioms for ordered groups are satisfied. Then a bijective mapping f from a subset of the base $|RS|$ onto a subset of the base $|R'S'|$ is a local isomorphism if, in addition to being order preserving between R and R', we have $(fx) \circ (fy) = fz$ modulo the group S' iff $x \circ y = z$ modulo S, for all $x, y, z \in \mathrm{Dom}\, f$.

Consider the empty function introduced in 1.7.6. Then extending the conventions of this paragraph, we say that, for every integer $n \geq 1$, the empty function is a local isomorphism from every n-ary relation into every other n-ary relation.

Moreover for all sets E, E' **the empty function is a local isomorphism from the 0-ary relation $(E,+)$ into $(E',+)$ and from $(E,-)$ into $(E',-)$, but not from $(E,+)$ into $(E',-)$** or conversely by exchanging (+) and (-).

Finally given two multirelations R, R' of the same arity, the empty function is a local isomorphism from R into R' either when all the component relations have positive arities, or when, for each index i corresponding to the 0-ary components R_i, R'_i, these latter have the same value (+) or (-).

9.1.5 Restriction, extension of a local isomorphism

If f is a local isomorphism from R into R', then f restricted to an arbitrary subset of its domain is still a local isomorphism from R into R'.

If f is a local isomorphism from R into R' , then the inverse function f^{-1} is a local isomorphism from R' into R. If additionally g is a local isomorphism from R' into R", then the composition $g \circ f$ is a local isomorphism from R into R".

In particular, if f is an isomorphism from R onto R', then every restriction of f is a local isomorphism. More particularly, the identity function with domain a subset of the base $|R|$ is a local isomorphism from R into R itself.

However, **a local isomorphism from R into R' is not in general extendible to an isomorphism from R onto R', even if R and R' are isomorphic or even identical.**

• If R is the chain of non-negative integers and R' the chain of negative integers, and f the function taking $0, 1, ..., p$ into $f(0) = -p-1, f(1) = -p, ..., f(p) = -1$; then f is not extendible (R and R' are not isomorphic).

Another example. If $R = R' =$ the chain of non-negative integers, then the function taking 0 into 1 is a local isomorphism; yet it is not even extendible to a range which, apart from 1, contains 0. •

Lemma. Let R, R' be two n-ary relations. A sufficient condition for a bijection f with domain $F \subseteq |R|$, to be a local isomorphism from R into R', is that for every subset X of F with cardinality $\leq n$, the function f restricted to X is a local isomorphism from R into R'.

If R, R' are two multirelations with the same arity, then the preceding proposition still holds, by setting n to be the maximum of the arity.

• Let $x_1, ..., x_n$ be a sequence of n elements in F. The set $X = \{x_1, ..., x_n\}$ is a subset of F and has cardinality $\leq n$. Hence by hypothesis we have $R'(fx_1, ..., fx_n) = R(x_1, ..., x_n)$ and consequently f is a local isomorphism. •

9.1.6 Local automorphism

A **local automorphism** of a relation R is a local isomorphism from R into R.

An automorphism f of R, hence also the restriction of f to an arbitrary subset of the base, is a local automorphism of R.

However in general, a local automorphism is not extendible to an automorphism: see our example (previous subsection) with the chain of integers and the local automorphism which takes 0 into 1.

The following statement is a particular case of the lemma in previous subsection:

Let R be an n-ary relation and let f be a bijection whose domain and range are subsets of the base $|R|$. **If every restriction of f to $\leq n$ elements is a local automorphism of R, then f itself is a local automorphism of R.**

9.1.7 Case of two relations with the same base

(1) **Let R be of arity m and S of arity n, both having the same base. If every local automorphism of R, defined on $\leq n$ elements, is a local automorphism of S, then every local automorphism of R is an automorphism of S.**

(2) With the same notations, **if R and S have the same local automorphisms on $\leq$ Max(m, n) elements, then they have same local automorphisms; and in particular they have same automorphisms.**

However, two distinct chains isomorphic with ω, both having the same base, have the same automorphism (the identity being their only automorphism), yet do not have the same local automorphisms.

9.2 Free interpretability

9.2.1 Generalities; intermediacy relation, cyclic relation

Let R, S be two multirelations with the same base. We say that S is **freely interpretable** in R iff every local automorphism of R is a local automorphism of S.

For example, let A be a poset. Then the relation of **strict partial ordering** $A'(x, y) = +$ iff $x < y \pmod A$ (in other words iff $A(x, y) = +$ and $x \neq y$) is freely interpretable in A; and A is freely interpretable in A'.

The relation of **intermediacy** or **betweenness** $B(x, y, z) = +$ iff z is between x and $y \pmod A$, is freely interpretable in A. We represent this intermediacy relation by the free formula: $(x \leq z \leq y) \vee (y \leq z \leq x)$ where our partial ordering is denoted $\leq$ and where, for instance, the first parenthesis is an obvious abbreviation for $(x \leq z \wedge z \leq y)$.

If one excepts particular cases (cardinality 1 or A reduced to the identity relation), A is not freely interpretable in B: for example take the usual ordering of the integers 1, 2, 3; then the permutation which reverses this ordering into 3, 2, 1 is a local automorphism for the betweenness relation B, but not so for A.

Another example. Let A be a chain; the **ternary cyclic relation** C associated with A is defined to satisfy $C(x, y, z) = +$ iff $(x \leq y \leq z) \vee (y \leq z \leq x) \vee (z \leq x \leq y) \pmod A$; we see that C is freely interpretable in A. If one excepts particular cases (cardinalities ≤ 2) the chain A is not freely interpretable in C: the circular permutation (1,2,3) is an automorphism for C yet not for the chain.

Free interpretability is reflexive and transitive, so that it defines a pre-ordering on every set of relations with common base. It is not antisymmetric; for example take an ordering $\leq$ and the corresponding strict ordering $<$ which are freely interpretable each by the other; or take an arbitrary relation R and its negation $\neg R$ (see 1.7.4).

A multirelation freely interpretable in R is still freely interpretable in any concatenation RS where S has same base as R.

A necessary and sufficient condition for S to be freely interpretable in R is that every component relation of S is freely interpretable in R.

Let R be a relation with base E. If any bijection whose domain and range are subsets of E is a local automorphism of R, then R is freely interpretable in every relation with base E. Take for example the identity relation on E, or its negation, or the n-ary relation taking always the value $(+)$.

Let R, S be two relations with common base, and let n be the arity of S (the maximum of this arity in the case of a multirelation). Then **a sufficient (and obviously necessary) condition for S to be freely interpretable in R is that every local automorphism of R whose domain has cardinality $\leq n$ is a local automorphism of S.**

This is, in other words, the proposition (1) of 9.1.7.

9.2.2 Dihedral quaternary relation associated with a chain

Given a positive integer n, we denote by T_n the **cyclic group** or **group of translations** generated by the cyclical permutation $(1, 2, 3, ..., n)$. Also J_n denotes the **reflection group** formed by the identity plus the reflection which transforms $1, 2, 3, ..., n-2, n-1, n$ respectively into $n, n-1, n-2, ..., 3, 2, 1$.

The **dihedral group** D_n is generated by the cyclic group T_n of translations and the reflection group J_n. Here we are interested by the particular case $n = 4$, hence by the dihedral group D_4.

Given a chain A, we have introduced the cyclic ternary relation C associated with A in the previous subsection. We additionally need the notion of the **dihedral quaternary relation** D **associated with** A, or equivalently associated with the ternary cycle C which is itself associated with A.

We can say that, starting from four elements x, y, z, t we have $D(x, y, z, t) = +$ either if at least two elements are equal, or if $x < y < z < t \pmod A$, or if the same is true for any one of the eight 4-tuples obtained from (x, y, z, t) by translations and reflections.

The following free logical formula expresses the dihedral relation by mean of the ternary cyclic relation:

$x = y \vee x = z \vee x = t \vee y = z \vee y = t \vee z = t \vee (C(x, y, z) \wedge C(z, t, x)) \vee (\neg C(x, y, z) \wedge \neg C(z, t, x))$;

by taking into account that, in the the cyclic relation C, the conjunction $(C(x, y, z) \wedge C(z, t, x))$ implies $C(y, z, t)$ and $C(t, x, y)$; idem for the negatiion $\neg C$.

Note that the dihedral relation associated with the ternary cycle C also means that either two variables are equal, or that x and z are situated in the two opposite intervals defined by y and t on C. Thus we can say that the dihedral relation is a kind of betweenness, or an intermediacy on the cyclic relation.

9.2.3 Free interpretability and concatenation

Let R, R' be two m-ary relations and S, S' be two n-ary relations; let E be the common base of R, S, and E' the common base of R', S'. **If every restriction of the concatenation $R'S'$ to $\leq n$ elements is embeddable in RS, and if S is freely interpretable in R, then S' is freely interpretable in R'.**

• Let f' be a local automorphism of R', such that $F' = \mathrm{Dom}\, f'$ has cardinality $\leq n$. By hypothesis, there exists an isomorphism g from the restriction $(R'S')/F'$ onto a restriction of RS, and there exists an isomorphism h from $(R'S')/\overline{f'}(F')$ onto a restriction of RS. Let $f = h \circ f' \circ (g^{-1})$: then f is a local automorphism of R and hence of S, the latter being freely interpretable in R. Thus f' is a local automorphism of S'. Hence every local automorphism of R' defined on at most n elements is a local automorphism of S'. By 9.2.1, S' is freely interpretable in R' (proof communicated by PAILLET). •

For example, let R be a chain and S the associated ternary cyclic relation. Let R' and S' be such that every restriction of $R'S'$ with cardinality ≤ 3 is embeddable in RS. Then R' is a chain and S' is the ternary cyclic relation associated with R'.

9.2.4 Free interpretability: restriction and extension

If S is freely interpretable in R, then for every subset X of the base, the restriction S/X is freely interpretable in R/X (trivial).

(1) **If S is freely interpretable in R and if R' is an extension of R, then there exists an extension of S which is freely interpretable in R'.**

• Let n be the arity of S, and $(x_1, ..., x_n)$ be an n-tuple in the base $|R'|$. If there exists a local isomorphism f from R' into R having domain $\{x_1, ..., x_n\}$, then put $S'(x_1, ..., x_n) = S(fx_1, ..., fx_n)$. This last value does not depend on the chosen isomorphism, because S is freely interpretable in R. Now if there does not exist such a local isomorphism from R' into R having domain $\{x_1, ..., x_n\}$, then put $S'(x_1, ..., x_n) = +$. Then S' is an extension of S and is freely interpretable in R'. •

(2) Let R, S have a common base, and n be the arity of S. **If, for each subset X of the base with cardinality $\leq 2n$, the restriction S/X is freely interpretable in R/X, then S is freely interpretable in R.**

• Suppose that S is not freely interpretable in R. Then there exists a local automorphism f of R which modifies S. Thus there exists an n-tuple $(a_1, ..., a_n)$ of elements in the base, with $S(a_1, ..., a_n) \neq S(fa_1, ..., fa_n)$. Then the set $X = \{a_1, ..., a_n, fa_1, ..., fa_n\}$ has cardinality at most equal to $2n$; and S/X is not freely interpretable in R/X. •

(3) **If R is a chain, then the integer $2n$ can be replaced by $n + 1$**: see below 9.5.1 proposition (2). However in general, for $n = 2$, the value 4 cannot be replaced by a smaller integer.

• Take a poset R constructed from two chains R_1, R_2 (each with cardinality ≥ 2), by setting each element of R_1 to be incomparable with each element of R_2. Take S to be the poset similarly obtained from the chains $S_1 = R_1$ and $S_2 = R_2^-$

= converse of R_2. Then S is not freely interpretable in R, yet S/X is freely interpretable in R/X for each subset X with cardinal 1, 2 or 3. •

9.3 Free operator, connection with free interpretability

Given two finite sequences of integers, say m and n, a **free operator** $\mathcal{P}$ associates to each m-ary multirelation R an n-ary multirelation $\mathcal{P}(R)$ having the same base. We have the additional condition that for any two m-ary relations R, R' **each local isomorphism from** R **into** R' **is also a local isomorphism from** $\mathcal{P}(R)$ **into** $\mathcal{P}(R')$. In other words, $\mathcal{P}$ preserves local isomorphism.

The sequences m, n are called the **arities of** $\mathcal{P}$, and we say that $\mathcal{P}$ is an (m, n)**-ary operator**. We say that any m-ary relation is **assignable** to $\mathcal{P}$.

Note that if F is a subset of the base $|R|$, then $\mathcal{P}(R/F) = \mathcal{P}(R)/F$.

Each free operator is completely determined by its values on multirelations whose base is a finite subset of the set ω of integers: the value in the general case follows immediately by using local isomorphism. As all the multirelations based on subsets of ω constitute a set, we can define an (m, n)-ary operator as a function which, to each m-ary relation based on a finite subset of ω, associates an n-ary relation on the same base, with the preceding condition about local isomorphisms. After defining $\mathcal{P}$ as indicated, we complete by defining the value taken by $\mathcal{P}$ for each m-ary relation: all this within the framework of the axioms of ZF.

Examples. For a given integer n, the **negation** $\neg$ is a free operator which takes each n-ary relation R into the relation $\neg R$ having the same base and always taking the opposite value.

The **conjonction** $\wedge$ is a free operator which takes each birelation R, S with arities $\leq n$ into the n-ary relation $R \wedge S$ which takes value (+) iff both R and S take value (+) (if R is m-ary with $m < n$ one takes into account only the m first elements in the n-sequence of arguments).

Similarly we have free operators for each logical connection, for instance the **disjonction** $\vee$, the **implication** $\Rightarrow$, the **bilateral implication** $\Leftrightarrow$, etc.

The **converse** or **symmetrizing operator** which takes each binary relation R into its converse S satisfying $S(x, y) = R(y, x)$ for all x, y in the base. Also the diagonalization operator which takes each binary relation R into the unary relation $S(x) = R(x, x)$.

An (m, n)-ary free operator $\mathcal{P}$ is determined by the couples $(R, \mathcal{P}(R))$ for those m-ary multirelations R having a base of cardinal at most equal to $\operatorname{Max} n$. Hence there are only finitely many free operators of given arities.

9.3.1 Free interpretability and free operator

A multirelation S **is freely interpretable in** R **iff there exists a free operator taking** R **into** S.

• Consider the case of two relations: R is m-ary and S is n-ary and S is freely interpretable in R. For each m-ary relation X and each n-tuple $(x_1, ..., x_n)$ of elements in the base $|X|$, either there exists an isomorphism f from $X/\{x_1, ..., x_n\}$ onto a restriction $R/\{fx_1, ..., fx_n\}$: then we set $(\mathcal{P}(X))(x_1, ..., x_n) = S(fx_1, ..., fx_n)$; this last value does not depend on the choosen isomorphism. Or no such isomorphism exists: then we set $(\mathcal{P}(X))(x_1, ..., x_n) = +$. •

9.3.2 Injective operator

A free operator $\mathcal{P}$ is said to be **injective** iff $R \neq R'$ implies that $\mathcal{P}(R) \neq \mathcal{P}(R')$ for all R, R' assignable to $\mathcal{P}$; or equivalently iff for all R, R' every local isomorphism from $\mathcal{P}(R)$ into $\mathcal{P}(R')$ is a local isomorphism from R into R'.

(1) **Every injective operator has an inverse**. More precisely, if $\mathcal{P}$ is injective, then there exists a free operator $\mathcal{Q}$ such that $\mathcal{QP}(R) = R$ for each R assignable to $\mathcal{P}$.

• Consider the case of relations, and let m, n be the arities of $\mathcal{P}$. Then given an n-ary relation Y and an m-tuple $(x_1, ..., x_m)$ of elements in the base $|Y|$, we define $(\mathcal{Q}(Y))(x_1, ..., x_m) = X(x_1, ..., x_m)$ if the restriction $Y/\{x_1, ..., x_m\}$ is the image under $\mathcal{P}$ of an m-ary relation X having the same base; or $(\mathcal{Q}(Y))(x_1, ..., x_m) = +$ if there is no such X. •

(2) Hence **if $\mathcal{P}$ is injective, then R and $\mathcal{P}(R)$ are each freely interpretable in the other**. A converse of this result shall be proved in 9.3.6 proposition (3) below.

9.3.3 Comparison between arities

We say that the arity n (of a multirelation) is **greater than the arity** m iff each term m_i of m can be associated with a term n_j of n, with $m_i \leq n_j$, in an injective manner: i.e. two distinct indices i, i' in m correspond to two distinct indices j, j' in n.

If the arity n is greater than the arity m, then for any integer p, there are more n-ary multirelations with base having cardinality p, than there are of m-ary multirelations with the same base.

However, even if the above condition is true for every integer p, this does not necessarily imply that the arity n is greater than m.

• For example, take $m = (1, 1)$ and $n = (0, 2)$. Then for a base of cardinality p, there are (2 to the power $2p$) many m-ary birelations, and (2 to the power $(1+p^2)$) many n-ary birelations, with $1 + p^2 \geq 2p$. Yet n is not greater than m. •

Problem. If the arity n is greater than m, then there obviously exists an (m, n)-ary injective operator. Indeed to each m_i-ary component R_i of the multirelation R, it suffices to associate the n_j-ary component $S_j (m_i \leq n_j)$ whose value only depends on the m_i first terms, that value being equal to that of R_i.

Conversely, if n is not greater than m, then we conjecture that there exists no injective free operator with arities (m, n).

For example **if $m = (1,1)$ and $n = (0,2)$, then there exists no injective free (m,n)-ary operator**.

• Start with a base of three elements a, b, c and the birelation $(R_1 R_2)$ where R_1 takes the value (+) only for a, and R_2 only for b. Then the only local automorphisms of this birelation are the identity on each subset of the base. On the other hand, any arbitrary binary relation must take the same value, for instance for (a,a) and (c,c), and thus admits a local automorphism other than the identity on a subset of the base. Obviously the concatenation of our binary relation with a 0-ary relation changes nothing in the previous discussion.

Finally, the reader who is tempted by the pseudo-solution which associates to $(R_1 R_2)$ the binary relation $S(x,y) = R_1(x) \wedge R_2(y)$, will note that injectivity is no longer satisfied when R_1 is always (-). •

9.3.4 Partial operator

Let m be a finite sequence of integers, and $\mathcal{A}$ a set of m-ary multirelations with finite bases, which is closed under restriction and isomorphism. To be rigorous in the frame of the axioms of ZF, we assume that the bases of our multirelations are finite subsets of the set of integers.

A (m,n)-ary **partial operator** with domain $\mathcal{A}$ is a function $\mathcal{P}$ which to each m-ary multirelation R belonging to $\mathcal{A}$, associates an n-ary multirelation $\mathcal{P}(R)$ with the same base, such that $\mathcal{P}$ preserves local isomorphisms.

A necessary and sufficient condition for a partial operator $\mathcal{P}$ to be injective, is again that every local isomorphism from $\mathcal{P}(R)$ into $\mathcal{P}(R')$ be a local isomorphism from R into R', for every R, R' belonging to $\operatorname{Dom}\mathcal{P}$.

9.3.5 Canonical extension

Let $\mathcal{P}$ be an (m,n)-ary partial operator, where the arity n is greater than the arity m, in the sense of 9.3.3. We shall define as follows the **operator** $\mathcal{P}^=$ extending $\mathcal{P}$, and whose domain contains all m-ary multirelations.

First, to each term m_i of m, associate in an injective manner a term n_j of n, such that $m_i \leq n_j$; in the following we denote m_i by m and n_j by n.

Given an m-ary relation R and elements $x_1, ..., x_n$ in the base $|R|$, either the restriction $R' = R/\{x_1, ..., x_n\}$ is an element of $\operatorname{Dom}\mathcal{P}$, and then we set $(\mathcal{P}^=(R))(x_1, ..., x_n) = (\mathcal{P}(R'))(x_1, ..., x_n)$; or not; then $(\mathcal{P}^=(R))(x_1, ..., x_m, ..., x_n) = R(x_1, ..., x_m)$.

The operator $\mathcal{P}^=$ thus defined is called the **canonical extension** of $\mathcal{P}$.

If $\mathcal{P}$ is already defined for all m-ary relations of cardinality $\leq n$ (replace by $\operatorname{Max} n$ for a multirelation), then the canonical extension is the unique free operator extending $\mathcal{P}$.

In general $\mathcal{P}$ can be injective without its canonical extension $\mathcal{P}^=$ being injective.

• Let $m = n = 1$ and let $\mathcal{P}$ associate to each unary relation taking always the value (+), the unary relation always (-); yet $\mathcal{P}$ is undefined for unary relations taking at least once the value (-). Then $\mathcal{P}^=$ takes every unary relation into the unary relation with same base, always (-); so that $\mathcal{P}^=$ is not injective. •

Let k be an integer; if a partial operator $\mathcal{P}$ is defined for every m-ary relation of cardinality $\leq k$ and only for such, and if $\mathcal{P}$ is injective, then the canonical extension $\mathcal{P}^=$ is injective.

Moreover in the case where $\mathcal{P}$ is an (m,m)-ary partial operator, then we have $(\mathcal{P}^=)^{-1} = (\mathcal{P}^{-1})^=$.

9.3.6 The injective extended operator

Theorem. Let m, n be two arities with n greater than m. Let $\mathcal{P}$ be an (m,n)-ary injective partial operator with domain $\mathcal{A}$ and range $\mathcal{B}$. Let k be the largest integer for which $\mathcal{A}$ contains every m-ary multirelation of cardinality $\leq k$. Then:

(1) **There exists a partial operator $\mathcal{P}^+$ extending $\mathcal{P}$, which is injective and defined for every m-ary multirelation of cardinality** $\leq k+1$ ([194] POUZET 1973).

(2) **Every injective partial (m,n)-ary operator, where n is greater than m, is extendible to an injective free operator**: go from k to $k+1$, etc. until reaching Max n.

(3) **Let R be an m-ary relation, S an n-ary relation with n greater than m. If R and S are each freely interpretable in the other, then there exists an injective free (m,n)-ary operator which takes R into S**: start with the injective partial operator which takes every restriction of R into the restriction of S having the same base.

(2) and (3) are immediate consequences of (1); moreover (3) is the converse of 9.3.2 proposition (2).

• Proof of (1). It suffices to consider the case of an m-ary relation of cardinality $k+1$ which does not belong to the domain $\mathcal{A}$, say R. Let $\mathcal{P}_k$ be the restriction of our partial operator $\mathcal{P}$ to the set of all relations with cardinalities $\leq k$.

Consider the sequence of relations $S_0, S_1, ..., S_h$ (h integer), where $S_0 = (\mathcal{P}_k^=)(R)$; then $S_1 = (\mathcal{P}_k^=)(\mathcal{P}^{-1})(S_0)$ provided that S_0 belongs to the range $\mathcal{B}$; then $S_2 = (\mathcal{P}_k^=)(\mathcal{P}^{-1})(S_1)$ provided that S_1 belongs to the range $\mathcal{B}$; and so on, until the first index h for which $S_h = (\mathcal{P}_k^=)(\mathcal{P}^{-1})(S_{h-1})$ does not belong to the range $\mathcal{B}$. Then we define $(\mathcal{P}^+)(R)$ to be S_h.

To see that this procedure always terminates by giving a relation which does not belong to the range $\mathcal{B}$, note that $S_i (i = 0, 1, ..)$ are mutually non-isomorphic, provided that they belong to $\mathcal{B}$. Indeed, if f were an isomorphism from S_i onto $S_j (1 \leq i < j \leq h)$, then f would be an isomorphism from S_{i-1} onto S_{j-1} (because the injectivity of $\mathcal{P}$ and $\mathcal{P}_k^=$), and so forth. Thus f would be an isomorphism from S_0 onto S_{j-i}, hence from R onto $\mathcal{P}^{-1}(S_{j-i-1})$. But S_{j-i-1} belongs to $\mathcal{B}$ hence R belongs to $\mathcal{A}$: contradiction.

Note that $\mathcal{P}^+$ obviously preserves local isomorphisms. And if we take a proper restriction of R which belongs to the domain $\mathcal{A}$, then $\mathcal{P}^+$ acts on this restriction as does $\mathcal{P}$.

Finally it remains to see that $\mathcal{P}^+$ is injective. Indeed, take two m-ary relations R, R' with cardinality $k+1$ and suppose that $\mathcal{P}^+(R) = \mathcal{P}^+(R')$: we must show that $R = R'$. Firstly note that R belongs to $\mathcal{A}$ iff $\mathcal{P}^+(R)$ belongs to $\mathcal{B}$, by the preceding construction. Either R and R' belong to $\mathcal{A}$, and then we have $\mathcal{P}^+(R) = \mathcal{P}(R)$ and similarly with R', so that $R = R'$ because the injectivity of $\mathcal{P}$.

Or neither R nor R' belongs to $\mathcal{A}$. Then we claim that, if we denote by h the number of successive transformations S_i associated with R, and by h' the number similarly associated with R', then $h = h'$. Indeed suppose that $h < h'$. Then we obtain that $R = \mathcal{P}^{-1}(S'_{h'-h-1})$; and since this $S'_{h'-h-1}$ belongs to $\mathcal{B}$, it follows that R belongs to $\mathcal{A}$: contradiction. Finally we obtain $S_i = S'_i$ for each i and hence $R = R'$. •

Problem due to [194] POUZET 1973. Can the above be generalized to the case of interpretability by logical formulas (first order predicate calculus with identity). That is, if R, S are relations of the same arity, each of which is interpretable in the other via a logical formula, then does there exist a logical formula which operates injectively and takes R into S.

9.4 Constant relation

A relation R with base E is said to be **constant** iff every permutation of E is an automorphism of R. For example the n-ary relation taking always the value (+), the n-ary relation always (-), the binary relation of identity (taking the value (+) when $x = y$ and (-) when $x \neq y$.

The definition extends to multirelations: we see that a multirelation is constant iff its components relations are all constant.

If R is constant, then every restriction of R is constant.

9.4.1 A characterization of constant relations

An n-ary relation R is constant iff $R(x_1, ..., x_n) = R(y_1, ..., y_n)$ for all sequences $x_1, ..., x_n$ and $y_1, ..., y_n$ such that the transformation which takes x_1 into y_1, ... , and x_n into y_n, is an injective function.

In other words, iff for each pair of indices $i, j (1 \leq i < j \leq n)$ we have that $x_i = x_j$ iff $y_i = y_j$.

It follows that R is constant iff:

(1) all restrictions of R having a same cardinality less than or equal to the arity, are isomorphic; and

(2) these restrictions are constant.

Neither of the above conditions (1) and (2) is alone sufficient; counterexamples:

For every unary relation, any restriction of cardinality 1 is constant.

For every reflexive and symmetric binary relation, any restriction of cardinality 1 or 2 is constant.

For every tournament (binary relation whose restrictions of cardinal 2 are chains), all restrictions with cardinal 1 are isomorphic; similarly with 2.

Given an n-ary constant relation R and an n-ary relation S: if every restriction of S with cardinality $\leq n$ is embeddable in R, then S is constant. Moreover if S has the same base as R, then $S = R$.

Given an n-ary constant relation R with base E, then for every superset E' of E there exists a constant extension of R to E'. If Card $E \geq n$, this extension is unique.

9.4.2 Other characterizations for constant relations

(1) **A relation R with base E is constant iff, for all elements a, b in E, the transposition (a, b) is an automorphism of R.**

• If R is constant, then our conclusion is obvious.

Conversely if R is not constant, then there exists a permutation f of E which modifies R. Using the automorphism lemma 9.1.2, there exists an element $a \in E$ such that the transposition (a, fa) modifies R. •

(2) **A relation R is constant iff every bijection between two subsets of the base is a local automorphism of R.**

• If our condition holds, then in particular every permutation of the base is an automorphism, hence R is constant.

Conversely if R is constant, then an arbitrary bijection f between two finite subsets of the base is extendible to a permutation of the base, which by hypothesis is an automorphism of R. Finally we obtain the case for a bijection with infinite domain (included in the base) by applying 9.1.6. •

(3) A necessary and sufficient condition for an n-ary relation R to be constant, is that **for each sequence $x_1, ..., x_n$ in $|R|$ and each index $i (1 \leq i \leq n)$ and for each element u in the base, distinct from $x_1, ..., x_n$, the bijection f with domain $\{x_1, ..., x_n\}$ which takes x_i into u and which preserves each $x_j \neq x_i (1 \leq j \leq n)$, satisfy $R(x_1, ..., x_n) = R(fx_1, ..., fx_n)$** (communicated by HODGES).

• If R is constant, then our conclusion follows from the preceding statement (2).

Conversely if our condition holds, then by transitivity, for each $m \leq n$, every bijection of an m-element set onto another is a local automorphism of R. Using 9.1.6, every bijection between any two subsets of the base is a local automorphism of R, and so by the preceding (2), the relation R is constant. •

(4) **An n-ary relation R is constant iff every restriction of R having cardinality $\leq n + 1$ is constant.**

• If R is constant, we already noticed that each restriction is constant.

Conversely, if each restriction to at most $n + 1$ elements is constant, then the condition in our preceding statement (3) is satisfied, since the set $\{x_1, ..., x_n, u\}$ has cardinality $\leq n + 1$. •

The proposition is false if we replace $n + 1$ by n: see 9.4.1: the example of any unary relation, or any reflexive and symmetric binary relation.

9.4.3 Calculation

For each integer n, let $H(n)$ be the number of partitions of the set $\{1, 2, ..., n\}$; in other words, the number of binary equivalence relations on this set. For $n = 0$ we adopt the convention that $H(0) = 1$, by considering that the empty binary relation is an equivalence relation.

Then $H(0) = H(1) = 1$ and we have the following recurrence relation:

$H(n+1) = C_0^n H(0) + C_1^n H(1) + ... + C_n^n H(n)$, where the $C_m^n = \frac{n!}{m!(n-m)!}$ for $m \leq n$ are the usual binomial coefficients.

• Given an equivalence relation on $\{1, 2, ..., n+1\}$, let A be the equivalence class of $n+1$, and let k be the cardinality of the complementary set $\{1, ..., n+1\} - A$: so $0 \leq k \leq n$. For each k the number of possible choices for A, or equivalently for the complement of A, is C_k^n. Hence the number of possible partitions of this complement is the product $C_k^n.H(k)$. •

Let E be a set with cardinality $\geq n$. Then there are (2 to the power $H(n)$) many constant n-ary relations with base E, where $H(n)$ is the previously defined function giving the number of partitions of $\{1, ..., n\}$.

• Let U be an equivalence relation with base $\{1, ..., n\}$. We say that an n-tuple $(x_1, ..., x_n)$ of elements in E is a (U, n)-tuple if for all $i, j (1 \leq i < j \leq n)$ we have $x_i = x_j$ iff i and j are equivalent modulo U. By 9.4.1, an n-ary relation R with base E is constant iff to each equivalence relation U based on $\{1, ..., n\}$ is associated a value $v(U)$ (equal to (+) or (-)), such that R takes the value $v(U)$ on all (U, n)-tuples of elements of E; the proposition follows. •

Some initial values of H: we have $H(2) = 2$; this yields the 4 constant binary relations: always (+), always (-), identity and its negation.

We have $H(3) = 5$: this yields 32 constant ternary relations; then $H(4) = 15, H(5) = 52, H(6) = 203$.

9.5 Chainable relation

Let A be a chain with base E. Then a relation R with base E is said to be **A-chainable** iff it is freely interpretable in A. In other words, iff every local automorphism of A is a local automorphism of R.

A relation R is said to be **chainable** iff there exists a chain A for which R is A-chainable.

For example the strict total ordering $<$ obtained from the chain or total ordering $\leq (\text{mod}\, A)$ by changing the value (+) into (-) along the diagonal, is A-chainable.

The ternary cyclic relation defined by giving (+) to the triple (x, y, z) iff $x \leq y \leq z$ or $y \leq z \leq x$ or $z \leq x \leq y (\text{mod}\, A)$, is A-chainable.

Similarly for the relation of intermediacy or betweenness $(\text{mod}\, A)$; similarly for the dihedral relation associated with A (see 9.2.2).

9.5.1 Lemmas on chainability

(1) Let R and R' be two n-ary, A-chainable relations on the same base. **If there exists an n-element set F for which $R/F = R'/F$, then $R = R'$.**

In other words, an n-ary, A-chainable relation is determined by its restriction to an n-element set. This follows from 1.7.1 and the uniqueness of the isomorphism from a finite chain onto another equipotent chain.

If R is A-chainable, then for every subset X of the base, the restriction R/X is (A/X)-chainable.

(2) **If for each subset X of the base of R with cardinality $\leq n+1 (n =$ arity of $R)$, the restriction R/X is (A/X)-chainable, then R is A-chainable.**

• Using 9.1.7 proposition (1), it suffices to see that for any two sequences of n elements $x_1 < x_2 < ... < x_n$ and $y_1 < y_2 < ... < y_n (\bmod A)$, the function taking x_i into y_i for $i = 1, ..., n$ is a local automorphism of R. For this, first define $z_i = \mathrm{Min}(x_i, y_i) (\bmod A)$ for $i = 1, ..., n$: thus $z_1 < z_2 < ... < z_n (\bmod A)$.

Now start with the sequence $x_1, x_2, ..., x_n$, then go to the sequence $z_1, x_2, ..., x_n$, so using the chainability of R/X by A/X with $X = \{z_1, x_1, ..., x_n\}$ whose cardinal is either n or $n+1$. Then go to the sequence $z_1, z_2, x_3, ..., x_n$, and so forth until $z_1, z_2, ..., z_n$, and then to $z_1, z_2, ..., z_{n-1}, y_n$ and so forth until we finally reach $y_1, y_2, ..., y_n$. •

(3) Given two multirelations R, S the concatenation RS is A-chainable iff R and S are both A-chainable.

An n-ary relation R is A-chainable iff, for each integer $i \leq n$, the restrictions of the concatenation AR with cardinality i are all isomorphic.

These two observations follow from the uniqueness of the isomorphism between two chains of a same finite cardinality.

9.5.2 Case of a constant relation

A constant relation with base E is A-chainable for every chain A having base E.

Indeed every bijection between two subsets of E, and in particular every local automorphism of an arbitrary chain on E, is a local automorphism of the given constant relation: see 9.4.2 proposition (2).

However, if a relation R with base E is A-chainable for every chain A on E, then R is not necessarily constant. For example, a chain A on two elements is both A-chainable and (A^-)-chainable, where A^- is the converse of A. We need a condition between the arity and the cardinal of the base: see proposition (1) below.

Another example: the ternary cyclic relation on three elements is A-chainable for any one of the six chains A based on these three elements. Recall that the ternary cyclic relations takes $(+)$ for triples (x, y, z) such that $x \leq y \leq z$ or $y \leq z \leq x$ or $z \leq x \leq y$ modulo the chain A.

(1) Let R be an n-ary relation with base E. **Assume that $\mathrm{Card}\, E \geq n+1$ and that either E is finite, or that E is infinite and orderable** (see 2.4.4).

If R is A-chainable for every chain A on E, then R is constant.

It even suffices that, for every $(n+1)$-element subset F of E and every chain A on F, the restriction R/F is A-chainable.

• It suffices to see that R satisfies the condition of 9.4.2 proposition (3). Indeed, take a sequence $x_1, ..., x_n$ in E, and let u be an element of E, distinct from the $x_i (i = 1, ..., n)$. Now fix i and let A be a chain on E for which x_i and u are consecutive, and such that those $x_j \neq x_i (1 \leq j \leq n)$ are either $< x_i$ or $> u (\text{mod}\, A)$. Then the bijection with domain $\{x_1, ..., x_n\}$, which takes x_i into u and preserves each $x_j \neq x_i$, is a local automorphism of the chain A, hence of R: thus the condition in 9.4.2 is satisfied (proof communicated by HODGES). •

(2) **If R is a chainable relation with base E, then for every superset E' of E there exists a chainable extension of R with base E'**; for $E' - E$ infinite, this uses the ordering axiom (2.4.4).

• Let A be a chain with base E, and R be an A-chainable relation. It suffices to take a totally ordered extension A' of A to E', and then to apply 9.2.4 proposition (1), to obtain an extension of R which be A'-chainable. •

9.5.3 Chainability and restriction

A necessary and sufficient condition for a relation R to be chainable is that every finite restriction of R be chainable (sufficiency uses the ultrafilter axiom; ZF suffices if R is countable).

• If R is A-chainable, then we already said in 9.5.1 that for every subset F of the base, the restriction R/F is (A/F)-chainable.

Conversely suppose that every finite restriction of R is chainable. Then associate, to each finite subset F of the base, the set U_F of chains X on F, for which R/F is X-chainable. By hypothesis U_F is non-empty for every finite F. Given F finite and a subset G of F, then each chain which is an element of U_F, when restricted to G, gives an element of U_G. By the coherence lemma 2.4.1, equivalent with the ultrafilter axiom (yet ZF suffices for a countable base), there exists a relation A based on $|R|$, such that for each F, the restriction A/F belongs to U_F, so that A/F is a chain. This A is a chain based on $|R|$. Each local automorphism of A having finite domain is a local automorphism of R: hence R is A-chainable: see end of 9.2.1. •

Corollary. Given a set of chainable relations R, which is directed under extension, then the common extension of these R on the union of their bases is chainable.

• This follows from the preceding proposition. Indeed let S be the common extension of relations R. Then for each finite subset F of the base $|S|$, the restriction S/F admits an R as an extension: consequently S/F is chainable. •

The preceding proposition will be strengthened in 13.3.3 (corollary): for each arity n there exists an integer $p(n)$ such that every n-ary relation is chainable provided that all its restrictions of cardinalities $\leq p(n)$ are chainable.

9.5.4 Existence of a chainable restriction

(1) **Given an arbitrary denumerable relation R, there exists a denumerable subset D of the base and a chain A with base D, which is isomorphic with ω, and such that the restriction R/D is A-chainable.**

(2) **Given an arity n and an integer p, there exists an integer $q \geq p$ such that every n-ary relation of cardinality $\geq q$ has a chainable restriction of cardinality p.**

• (1) Let n be the arity of R, which we can assume is ≥ 2. Let C be a chain with base $|R|$, which is isomorphic with ω. Consider the concatenated multirelation RC. We say that two n-element subsets X, X' of the base are **equivalent** iff the restrictions $(RC)/X$ and $(RC)/X'$ are isomorphic; similarly for i-element subsets ($i \leq n$). There are only finitely many equivalence classes for our equivalence relation.

Using RAMSEY's theorem, there exists a denumerable subset D of the base, in which all n-element subsets are equivalent, as well as all i-element subsets for each $i \leq n$. Let $A = C/D$, a chain isomorphic with ω. Then every local automorphism of A having domain of cardinality $\leq n$ is a local automorphism of R/D. By 9.2.1, the restriction R/D is freely interpretable in A. •

• (2) Analogous proof using the finitary version of RAMSEY's theorem. •

9.5.5 Chainability and embeddability

Given the ordinal ω, which we confuse with the chain of integers, each ω-chainable relation is minimal with respect to embeddability among denumerable relations. In other words, every denumerable relation which is embeddable in an ω-chainable relation R is isomorphic with R.

Let α be an aleph, i.e. an ordinal which is equipotent with no strictly smaller ordinal. Then each α-chainable relation is minimal with respect to embeddability among relations of cardinality α.

The proposition (1) in previous subsection asserts that for every denumerable relation R, there exists a denumerable minimal relation which is embeddable in R.

This result does not extend to the cardinality of the continuum. Indeed by 5.5.2 proposition (1) (DUSHNIK, MILLER), every chain with continuum cardinality and which is embeddable in the reals has a restriction of continuum cardinality which is strictly lesser, with respect to embeddability.

Problem. Does there exist a relation with continuum cardinality, which is minimal among relations with continuum cardinality, yet which is not chainable by mean of the continuum aleph (i.e. the smallest ordinal with continuum cardinality: the axiom of choice being used).

9.5.6 Calculation

For each couple of non-negative integers n and $r \leq n$, let S_r^n denote the **Stirling number**, or number of partitions of an n-element set into r non-empty classes.

In other words, the number of equivalence relations with cardinality n and having exactly r non-empty equivalence classes.

We have $S_0^0 = 1$: the equivalence relation with empty base is supposed to exist; obviously it has 0 non-empty equivalence classes. We have $S_0^n = 0$ for each $n \geq 1$, since an equivalence relation on n elements has at least one non-empty equivalence class.

We have $S_1^n = S_n^n = 1$ for each strictly positive integer n; moreover we have the recursion equality $S_r^n = S_{r-1}^{n-1} + r.S_r^{n-1}$ for every $r(1 \leq r < n)$.

• Suppose that the set $\{1, 2, ..., n\}$ is partitioned into r non-empty classes, and put aside the element n. Then an equivalence relation having r non-empty classes can be obtained, either by starting with an equivalence relation on $\{1, 2, ..., n-1\}$ having $r-1$ classes, to which we add the singleton $\{n\}$ as an r-th class: this yields S_{r-1}^{n-1} possibilities. Or by starting with an equivalence relation on $\{1, 2, ..., n-1\}$ having r classes, and then adding the integer n to one of these r classes: this yields $r.S_r^{n-1}$ possibilities. •

For example $S_2^3 = 3$, corresponding to the 3 possible partitions of $\{1, 2, 3\}$ into a singleton and its complement. Another example, $S_2^4 = 7$, corresponding to the 4 partitions of $\{1, 2, 3, 4\}$ into a singleton and its complement, plus the 3 partitions into 2 classes having 2 elements.

Let n be a non-negative integer and A be a chain with a base of cardinality $\geq n$. Then there are (2 to the power $V(n)$) many A-chainable n-ary relations, where $V(n) = 0!S_0^n + 1!S_1^n + 2!S_2^n + ... + n!S_n^n$. For positive n, we can remove the first term, since $S_0^n = 0$.

• Let F be a subset of the base, with cardinality n. We have already seen that an A-chainable n-ary relation is determined by its restrictions to F: see 9.5.1 proposition (1). Thus it suffices to calculate the number of n-ary relations based on F and (A/F)-chainable. Let us say that two n-tuples $x_1, ..., x_n$ and $y_1, ..., y_n$ in F are **equivalent** iff the function mapping x_i into $y_i (i = 1, ..., n)$ is a local automorphism of the chain A, thus iff this function is bijective and order-preserving (mod A). Then an n-ary relation R is (A/F)-chainable iff we have $R(x_1, ..., x_n) = R(y_1, ..., y_n)$ for all equivalent n-tuples.

Now there are exactly $V(n)$ equivalence classes. Indeed, each class is defined by partitioning the set of indices $1, 2, ..., n$ into a finite number r of non-empty classes, by letting $x_i = y_j (1 \leq i < j \leq n)$ iff i and j are in the same equivalence class (this yielding S_r^n possibilities); and finally by totally ordering the set of these r classes and letting $x_i < x_j$ iff the class of i is less than the class of j, modulo this total ordering of the classes. This yields $r! = 1.2.....r$ possibilities.

Finally each A-chainable relation is defined by giving the value (+) or the value (-) to each of the equivalence classes of the n-tuples in F: this yields (2 to the power $V(n)$) possibilities. •

Calculation of the first values of V. We have $V(0) = 1$, since $0! = 1$ and $S_0^0 = 1$: we obtain, on the base F, the two 0-ary relations $(F, +), (F, -)$.

We have $V(1) = 1$, giving the two unary relations always (+) and always (-).

We have $V(2) = 3$, giving 8 binary relations: always (+), the chain A, the converse chain, the identity relation, plus the 4 other relations obtained by inter-

changing (+) and (-); or equivalently by replacing (+) by (-) on the diagonal.
Then we have $V(3) = 13, V(4) = 75$.

9.6 Monomorphic relation

Let p be an integer; a relation R is said to be p-**monomorphic** iff the restrictions of R to any two p-element subsets of the base, are isomorphic.

Every relation is 0-monomorphic.

Every relation based on at most p elements is p-monomorphic.

Any reflexive binary relation is 1-monomorphic. Any binary tournament is 2-monomorphic; recall that a tournament is a reflexive, antisymmetric and comparable binary relation.

A relation R is said to be **$(\leq p)$-monomorphic** iff R is q-monomorphic for each integer $q \leq p$.

A relation is said to be **monomorphic** iff it is p-monomorphic for every integer p.

For example a chain, and more generally a chainable relation, is monomorphic.

There exist monomorphic non-chainable relations: for instance the binary cycle, or cycle of consecutivity, on three elements.

If R is p-monomorphic, then every restriction of R is p-monomorphic.

Similarly for $(\leq p)$-monomorphic and for monomorphic.

Every denumerable relation has a denumerable monomorphic restriction; since, by 9.5.4 proposition (1), it has a denumerable, chainable restriction.

For each integer p, there exists a unary relation which is p-monomorphic yet not $(p-1)$-monomorphic. Take a base of p elements, and a unary relation taking the value (+) on a single element, and the value (-) on the others.

For each integer p, there exists a relation which is p-monomorphic yet not $(p+1)$-monomorphic.

• Take a $(p+1)$-ary relation R satisfying $R(x_1, ..., x_{p+1}) = +$ whenever at least two of the elements $x_1, ..., x_{p+1}$ are equal. Thus all restrictions of cardinality p are isomorphic. To avoid that R be $(p+1)$-monomorphic, take two distinct $(p+1)$-element sets F and G; define R to take value (+) for every $(p+1)$-sequence of elements in F without repetition, and (-) for every such $(p+1)$-sequence in G. •

9.6.1 A characterization of p-monomorphy

A relation R is p-monomorphic iff each restriction of R with cardinality $p+1$ is p-monomorphic

Indeed we can go from one p-element subset of the base to another by adding an element then removing another element; or by a finite sequence of these two operations.

More generally, given $r \leq p$, a relation R is r-monomorphic iff each restriction of R with cardinality $p+1$ is r-monomorphic (for $r < p$ we must assume that Card $R \geq p+1$).

A relation is ($\leq p$)-monomorphic iff each restriction with cardinality $\leq p+1$ is ($\leq p$)-monomorphic.

9.6.2 Chainability and monomorphism

Every monomorphic relation having infinite base is chainable.

In other words, **for an infinite relation, monomorphism coincide with chainability**; uses the ultrafilter axiom and the denumerable subset axiom; ZF suffices for a denumerable relation.

• Let R be a monomorphic relation with infinite base E. Take a denumerable subset D of E (denumerable subset axiom). Then using 9.5.4 proposition (1), there exists a chain A on D such that R/D is A-chainable. Since R is monomorphic, to each finite subset F of E, associate an isomorphism f taking R/F onto a restriction $R/\overline{f}(F)$ such that $\overline{f}(F)$ is a finite subset of D. To this isomorphism f , associate the image of the chain $A/\overline{f}(F)$ under f^{-1}; then R/F is chainable by this image.

For each finite subset F of E, let U_F be the set of chains with base F and by which R/F is chainable: by the preceding, U_F is non-empty for each F. Moreover, for every finite subset F of E and every G included in F, each chain belonging to U_F, when restricted to G, gives an element of U_G. By the coherence lemma (2.4.1, equivalent to the ultrafilter axiom), there exists a chain C based on E, such that for every finite subset F, the restriction C/F belongs to U_F. Then for every F, the restriction R/F is (C/F)-chainable, and hence R is C-chainable by 9.5.1. •

9.6.3 Connection between p-monomorphies

Let R be a relation with base E and p an integer $\leq \operatorname{Card} E$. **If R is p-monomorphic, then R is also r-monomorphic for each integer** $r \leq \operatorname{Min}(p, (\operatorname{Card} E) - p)$ ([195] POUZET 1976).

In particular if E is infinite, or even with cardinality $\geq 2p-1$, then R is p-monomorphic iff R is ($\leq p$)-monomorphic.

• Let F be an arbitrary r-element subset of E. We say that an r-element subset of E has the color (+) iff the restriction of R to this set is isomorphic with R/F. Since R is p-monomorphic, every p-element subset includes the same number of r-element subsets having the color (+). On the other hand, we have $r+p \leq \operatorname{Card} E$; using 3.5.3 proposition (1), we see that all the r-element subsets of E have the color (+), and hence that R is r-monomorphic. •

9.6.4 Other connections between monomorphy and chainability

If a unary relation R is 1-monomorphic, then R is constant and hence chainable.

If R is a binary relation, then R can be (≤ 2)-monomorphic with infinite base, yet not be chainable. Indeed, take R to be reflexive, antisymmetric and comparable, i.e. a tournament with at least a cycle $R(a,b) = R(b,c) = R(c,a) =$ +.

If R is a binary (≤ 3)-monomorphic relation based on at least 4 elements, then R is chainable.

• To fix ideas, suppose that R is reflexive. Either each restriction of R to two elements is symmetric. Then since, by hypothesis, these restrictions are all isomorphic, it follows that R is a constant, thus a chainable relation.

Or the restrictions of R to two elements are chains, or oriented restrictions. Then the restrictions of R to 3 elements cannot be cyclic. Since otherwise, if we consider four elements, the four cyclic restrictions to 3 elements would be mutually incompatible. Thus all restrictions to 3 elements are chains, and then R is itself a chain. •

The two preceding cases, that of unary and binary relations, are but the first terms in a general process. We shall see in 13.3.3 that for each integer n, there exists an integer $p \geq n$ such that, if R is an n-ary $(\leq p)$-monomorphic relation with cardinality at least p, then R is chainable. More specifically for each n there exists a least integer $p \geq n$ and for this p there exists a least $q \geq p$ such that if R is a $(\leq p)$-monomorphic n-ary relation with cardinality $\geq q$, then R is chainable. By the previous subsection, we can suppose simply that R is p-monomorphic, instead of $(\leq p)$-monomorphic.

On the other hand, **for each integer $n \geq 2$ there exists an n-ary relation R with infinite base, such that R is $(\leq n)$-monomorphic yet not chainable.**

• To each n-element set associate a unique n-tuple, formed of the elements of this set without repetition, in such a way that these particular n-tuples do not form a chain, for which they would be the restrictions of cardinality n. Then let R take the value (+) for our particular n-tuples, and the value (-) for all other n-tuples. •

Problem. Does there exist a positive integer n_0 such that for each $n \geq n_0$ and each integer p there exists an n-ary relation which is p-monomorphic yet not $(p+1)$-monomorphic (from the preceding it follows that such relations would necessary be finite).

9.7 Tournament and monomorphy (Jean, Pouzet)

A binary relation A is called a **tournament** iff it is reflexive, antisymmetric and comparable: for all x, y either $A(x,y) = +$ (and thus $A(y,x) = -$ by antisymmetry) or $A(x,y) = -$ (and $A(y,x) = +$). In the first case we say that x **precedes** y or that y **follows** x $(\bmod A)$.

For each element u of the base E of A, the elements in $E - \{u\}$ are divided into two complementary sets, formed respectively of those elements which precede u and those which follow u.

Every restriction of a tournament is a tournament.

A tournament is a chain iff it is transitive.

The **binary cycle on three elements**, chosen to be reflexive, is a tournament called the **3-cycle**. Yet a binary reflexive cycle with cardinality ≥ 4 is not antisymmetric, thus is not a tournament.

A tournament is a chain iff it does not admit an embedding of the 3-cycle.

A binary relation is a tournament iff it admits no embedding of the binary relation with cardinality 1 and value (-) (ensures reflexivity); no embedding of the relation always (+) with cardinality 2 (ensures antisymmetry); finally no embedding of the identity relation with cardinality 2 (ensures comparability).

Given an element u in the base of a binary relation R, we shall say that a 3-cycle which is a restriction of R **passes through** u iff its base contains u.

9.7.1 The number of 3-cycles

Let A be a tournament and E its base, of finite cardinality $p \geq 5$.

If A is $(p-2)$-monomorphic, then:

(1) **For each element u of E, the number of 3-cycles passing through u is independent of u; for each pair of elements u, v of E, the number of 3-cycles passing through u and v is independent of u, v.**

(2) **Let (x, y) and (x', y') be any two arbitrary couples of elements in E which satisfy the conditions $x \neq y, x' \neq y', A(x, y) = A(x', y') = +$; and let f be an isomorphism from $A/(E - \{x, y\})$ onto $A/(E - \{x', y'\})$; then for each element z in $E - \{x, y\}$ and for its image $z' = fz$, the restrictions $A/\{x, y, z\}$ and $A/\{x', y', z'\}$ are either both 3-cycles, or both chains** (1977, published in ToR-86 p.261-262; see also [196] POUZET 1978).

• (1) Let u, v be two distinct elements in E. Since A is $(p-2)$-monomorphic, the number of 3-cycles which pass neither through u nor through v, is independent of u and v: let k_2 denote this number. Given an element u in E, the number of 3-cycles which do not pass through u is independent of u: apply the combinatorial lemma 3.5.1 proposition (1) with $p = 3$ and $p + q = (\text{Card}\, E) - 2$. Let k_1 denote this number.

Now let k denote the number of all the 3-cycles which are restrictions of A. For each element u, there exist $h_1 = k - k_1$ many 3-cycles which pass through u. Given u and v distinct, there exist k_1 many 3-cycles which do not pass through v, and among these, there exist k_2 many which do not pass through u either, hence $(k_1 - k_2)$ many which pass through u without passing through v. Finally there exist $h_2 = k - 2k_1 + k_2$ many 3-cycles which pass through u and through v. •

• (2) For elements x, y, z in E, let $h(x, y, z)$ denote the number, either 0 or 1, of the 3-cycles passing through x, y, z. Let $h(-x, y, z)$ denote the number of 3-cycles which pass through y and z without passing through x, etc.

We return to the two couples (x, y) and (x', y') and the elements z and fz in our statement. The number of 3-cycles which pass through z is equal to h_1 by the above (1). On the other hand, this number is equal to the following sum: $h(x, y, z) + h(-x, y, z) + h(x, -y, z) + h(-x, -y, z)$. The number of 3-cycles which pass through x and z is h_2 by the above (1), and is also equal to $h(x, y, z) + h(x, -y, z)$. Similarly the number of 3-cycles passing through y and z is h_2 and equals $h(x, y, z) + h(-x, y, z)$. Adding the second and the third equalities and then substracting the first, we obtain $h(x, y, z) = 2h_2 - h_1 + h(-x, -y, z)$. We have the same result when substituting x', y', z' for x, y, z. Now, since f is an isomorphism from $A/(E-\{x,y\})$ onto $A/(E-\{x',y'\})$, the numbers $h(-x,-y,z)$ and $h(-x',-y',z')$ are equal: so that $h(x,y,z)=h(x',y',z')$. •

9.7.2 Tournaments and 3-cycles

Let A be a tournament and E its base, which is finite and of cardinality $p \geq 5$. Suppose that A is (p-2)-monomorphic.

Then either A is a chain;

or else, **for any two couples (x,y) and (x',y') satisfying the conditions $x \neq y, x' \neq y', A(x,y)=A(x',y')=+$, every isomorphism from $A/(E-\{x,y\})$ onto $A/(E-\{x',y'\})$, when extended by taking x to x' and y to y', is an automorphism of A** ([124] JEAN 1969; the following proof is due to POUZET in 1977).

• Either all restrictions of A with cardinality 3 are chains: then A is a chain. Or there exists at least one 3-cycle. By the previous subsection, for each pair of distinct elements there passes at least one 3-cycle. Suppose, from this point on, that we are in this case.

Let (x,y) and (x',y') be two couples in our statement, and z an element in $E-\{x,y\}$, and z' the image of z under the isomorphism in our statement. Suppose that $A(x,z)=+$: we shall prove that $A(x',z')=+$ by arguing towards a contradiction in supposing the contrary, that $A(x',z')=-$ and hence that $A(z',x')=+$.

The restriction $A/\{x,y,z\}$ is not a cycle: by 9.7.1 proposition (2), the restriction $A/\{x',y',z'\}$ is not a cycle: hence $A(z',y')=+$. Let u be an element such that $A/\{x,y,u\}$ is a cycle: we know that such an element exists. Then we have $A(y,u)=A(u,x)=+$. We claim that $A(u,z)=+$. If not, then $A(z,u)=+$, so $A(z',u')=+$ as well, where u' is the image of u under the isomorphism in our hypothesis.

Then no 3-cycle passing through z' and u' would pass through x' nor through y' (recall that $A/\{x',y',u'\}$ is a cycle by 9.7.1, proposition (2). Yet $A/\{x,z,u\}$ would be a cycle; so that the number of 3-cycles passing through z and u would be strictly greater than the number of 3-cycles passing through z' and u', contradicting 9.7.1 proposition (1).

Thus our hypothesis $A(x,z)=A(z',x')=+$ allowed us to determine the values of the tournament A for all pairs included in $\{x,y,z,u\}$ and in $\{x',y',z',u'\}$. We terminate the proof by taking an element t for which $A/\{x,z,t\}$ is a cycle: we know that such an element exists. Thus t is distinct from y and from u. Moreover, no 3-cycle passing through z' and through the image t' of t (under the isomorphism in our hypothesis) can be completed by x' or by y'. Hence the number of cycles passing through z,t on the one hand, and through z',t' on the other hand, is different, again contradicting 9.7.1 proposition (1). •

Problem communicated by POUZET in 1978. Let A and A' be two tournaments having the same base E with finite cardinality $p \geq 5$. If for each subset X obtained by removing two elements from E, the restrictions A/X and A'/X are isomorphic, then are A and A' isomorphic.

9.7.3 When $(p-2)$-monomorphy induces $(p-1)$-monomorphy

Let A be a binary relation with finite cardinality $p \geq 5$. If A is $(p-2)$-monomorphic, then A is $(p-1)$-monomorphic.

Moreover, **if A is a non-totally ordered tournament, then for any two elements x, x' in the base, there exists an automorphism of A which takes x into x'** ([124] JEAN 1969).

• Since the relation A is $(p-2)$-monomorphic, A is also (≤ 2)-monomorphic by 9.6.3 (indeed $p \geq 5$ so that $2 < p-2$). Since A is 1-monomorphic, to fix our ideas, suppose that A is reflexive. Since A is 2-monomorphic, either its restrictions to 2-element sets are symmetric, in which case A is a constant relation, hence monomorphic. Or the restrictions to 2-element sets are oriented, in which case A is a tournament. By the previous subsection, either A is a chain, and hence monomorphic, or A is a non-totally ordered tournament.

In the latter case, let E be the base of A, and let x, x' be two elements of E. Either there exist two other elements y, y' with $A(x, y) = A(x', y') = +$; then by the previous subsection, there exists an automorphism of A taking x into x' and y into y'. Thus we have an isomorphism of the restrictions $A/(E - \{x\})$ and $A/(E - \{x'\})$; so that A is $(p-1)$-monomorphic.

Or there do not exist two such elements y, y'. Then A takes the value (+) for every couple with first term x, while taking the value (-) for every couple with first term x' (to fix ideas). Since E has cardinality strictly larger than 3, let t be an element distinct from x and x'. Then $A(x, x') = A(x, t) = +$; thus we have an automorphism of A which preserves x and takes t into x'. Similarly $A(x', x) = A(x', t) = -$; so we have an automorphism of A which takes x into t. By composition, we have an automorphism of A which takes x into x': thus A is $(p-1)$-monomorphic. •

In opposition to the preceding: for each integer $p \geq 4$, the binary cycle based on p elements is $(p-1)$ but not $(p-2)$-monomorphic.

Another example: for $p = 2q$, the partial ordering formed of q chains of cardinality 2, taken to be mutually incomparable.

Note that, for a binary relation with cardinality $p \geq 6$, if it is $(p-3)$-monomorphic, then it is (≤ 3)-monomorphic, hence chainable: see 9.6.3 and 9.6.4.

In general, for $r \geq 3$, every binary relation based on $p \geq r+3$ elements, which is $(p-r)$-monomorphic, is (≤ 3)-monomorphic, and thus chainable.

Problem due to JEAN in 1976, published in ToR-86 p.264).

Let $n \geq 2$ and $k \leq n$. Is every n-ary relation based on $p \geq 2n+1$ elements, which is $(p-k)$-monomorphic, necessarily $(p-k+1)$-monomorphic. Note that it is $(\leq k)$-monomorphic by 9.6.3.

For $n \geq 6$, is every n-ary relation based on $p \geq 2n+1$ elements, which is $(p-6)$-monomorphic, necessarily chainable (this is connected to JORDAN's hypothesis about permutation groups: see 12.3.3 below.

9.7.4 A connection with arithmetic

Let A be a tournament with cardinality p and which is $(p-2)$-monomorphic. If A is not a chain, then $p = 3$ modulo 4.

• Let E be the base of A, with cardinality p; and let u be an element of E. Let q denote the number of successors of $u (\text{mod}\, A)$; in other words, the number of x such that $A(u, x) = +$. If we replace u by another element u', then by the preceding discussion there exists an automorphism of A taking u into u', hence the number q of successors is preserved. The product $p.q$ is thus equal to the total number of couples of elements giving the value $(+)$ to A; hence equal to the number $p.(p-1)/2$ of pairs of elements. Thus $q = (p-1)/2$, which already shows that p is odd.

Given an element u in E, partition the set of elements distinct from u into the set F of $(p-1)/2$ many successors of u, and the set G of $(p-1)/2$ many predecessors of u. Let v and v' be two elements in F. By the preceding discussion, there exists an automorphism of A which preserves u and takes v into v'. Since it preserves u, this automorphism preserves F and G. Thus the number r of elements in F which are successors of v remains the same number when passing to v'. The product $r.(p-1)/2$ is thus equal to the total number of couples in F giving value $(+)$; hence equal to the number $(p-1)(p-3)/8$ of couples in F. Hence $r = (p-3)/4$, and thus $p-3$ is a multiple of 4. •

Anticipating the notions of 12.3.3, we say that a group of permutations is 2-**set-transitive** iff for any two (unordered) pairs of elements in the base, there exists a permutation of the group which takes the first pair into the second. The above proposition 9.7.2 then takes the following form: if A is a $(p-2)$-monomorphic, non-totally ordered tournament with cardinality p, then the group of automorphisms of A is 2-set-transitive (integer $p \geq 5$).

Groups which are 2-set-transitive have been studied in particular by [40] DEMBOWSKI 1968, p.96 note 2, under the name of 2-homogeneous groups. The term "set-transitive" is equally used and will be employed in this book to avoid confusion with homogeneous relations (12.1.1).

From the cited work, it follows that for every $(p-2)$-monomorphic, non-totally ordered tournament based on $p \geq 5$ elements, this p is an odd power of a prime congruent to 3 modulo 4.

Thus, after the binary cycle on 3 elements, we have a tournament on 7 elements which is 5-monomorphic, and hence 6, and 2, and 1-monomorphic, yet neither 3 nor 4-monomorphic. This tournament has chains of cardinality 3 and cycles of cardinality 3 as restrictions.

To construct it, start with an heptagon, or polygon with 7 vertices a, b, c, d, e, f, g. Cyclically orient the edges $ab, bc, ..., fg, ga$. Cyclically orient the starred heptagon $adgcfbe$ in the "opposite" direction $da, gd, ..., eb, ae$. In view of the rotational symmetry, it suffices to verify the isomorphism between the three sub-tournaments of cardinality 5: one is obtained by removing two consecutive vertices, a second by removing two vertices which are separated by one intermediate vertex, and a third by removing two vertices which are separated by two intermediate vertices.

9.8 Interval of a relation, strong interval of a poset

Let R be a relation with base E. A subset I of E is said to be an **interval** (mod R) or an R-**interval** iff every union of a local automorphism of R/I and of the identity on $E - I$ gives a local automorphism of R. See [76] FRAÏSSÉ 1984.

Note that, given the arity n of R, it is sufficient (and obviously necessary) that the union of any local automorphism of R/I defined on p elements ($p \leq n-1$) and of the identity on any $n-p$ elements of $E - I$ gives a local automorphism of R.

In particular, if R is a chain, we find an obvious characterization of the classical notion of an interval I by taking $n = 2$ and $p = 1$: for arbitrary elements $x, y \in I$ and an arbitrary $a \notin I$ either both $x, y < a$ or $x, y > a$ (mod R).

Given a relation R, the empty set, the whole base and the singleton of each element in the base are intervals (mod R).

Certain relations admit only the preceding intervals: for example **the consecutivity C associated with a chain A**: by definition $C(x, y) = +$ iff $x < y(\text{mod}\, A)$ and every element of the base is either $\leq x$ or $\geq y(\text{mod}\, A)$.

Case of a poset: strong interval.

As already noticed for a chain, in the case of a poset A, whose arity is 2, a subset I of the base is an A-**interval** or a **strong interval** iff for arbitrary two elements $x, y \in I$ and an arbitrary $a \notin I$ either both $x, y < a$ or $x, y > a$ or $x, y | a$.

9.8.1 Complement, intersection and filtering union of intervals

Given a relation R, let us call an R-**exterval** or simply an **exterval** the complement of an R-interval (with respect to the base).

A subset K of the base E is an exterval iff, for each local automorphism f of R/K and each subset G of $E - K$, either f is not extensible by the identity on G, or f is extensible by every local automorphism of $R/E - K$ with domain G.

Every intersection of R-intervals is an R-interval.

Given a set of R-intervals which is filtering under inclusion, then their union is an R-interval.

Start from the obvious fact that the union of two intervals in a chain is an interval provided that there exists a common element. This can be generalized as follows.

Let R be a relation, n its arity; let I, J be two R-intervals. **Assume that each restriction of $R/(I \cup J)$ with cardinality $\leq n-1$, admits an isomorphic relation among restrictions of $R/(I \cap J)$. Then the union $I \cup J$ is an R-interval.**

In the particular case of a poset, the arity is $n = 2$, so that $n - 1 = 1$ and furthermore all restrictions to one element are isomorphic. Consequently **in a poset, the union of two strong intervals is a strong interval provided that there exists a common element.**

9.8.2 Intervals of a restriction

Let R be a relation with base E. For each subset D of E and each R-interval I, the intersection $I \cap D$ is a (R/D)-interval.

Conversely **in the particular case where D is an R-interval, then the (R/D)-intervals are exactly the intersections with D of the R-intervals.**

This converse is not true for an arbitrary subset D of the base $|R|$.

Example of a poset. Take for R the boolean poset represented by a tridimensional cube with minimum u (the maximum being the opposite vertex). Take for D the set of the three vertices x, y, z which immediately follow u and set $F = \{x, y\}$ which is a (R/D)-interval, since the restriction R/D is a constant. Suppose there exists an R-interval G such that $F = G \cap D$. We see that G necessarily contains the element immediately after x, z which is incomparable with $y (\mathrm{mod}\, R)$. So that G contains z as an element, and $G \cap D \neq F$.

Another simpler example: start from the consecutivity relation on integers, where only the whole base, the empty set and singletons are intervals. Then take away all even integers: we get a constant relation, where all sets are intervals.

9.8.3 R-isomorphic and extendomorphic n-tuples

Given a set E, two h-tuples $a_1, ..., a_h$ and $b_1, ..., b_h$ in E are said to be **isomorphic** iff there exists a bijection which transforms a_i onto b_i for each $i = 1, ..., h$. In other words if two terms in the first h-tuple are equal, then the terms with same indices in the second h-tuple are also equal.

Given a relation R with base E and the previous h-tuples in E, they are said to be **R-isomorphic** iff they are isomorphic and moreover the bijection which transforms the first into the second h-tuple is a local automorphism of R. Note that the arity of R has a priori nothing to do with the length h of our h-tuples. The R-isomorphism is an equivalence, finer than the simple isomorphism.

Given an n-ary relation R with base E, a subset D of E and an integer h, we say that two h-tuples $a_1, ..., a_h$ and $b_1, ..., b_h$ in D are **$(R, E - D)$-extendomorphic** iff the transformation which takes a_i onto $b_i (i = 1, ..., h)$ and which saves each element of $E - D$ is a local automorphism of R. This defines, among n-tuples in D, an equivalence which is finer than the R-isomorphism (we find again the previous equivalence by taking $D = E$, in other words taking $E - D$ empty).

For instance if D is an R-interval, then for every h, any two h-tuples in D are $(R, E - D)$-extendomorphic provided that they are R-isomorphic.

If R is a chain, then already for $h = 1$, singletons are partitioned into finitely many classes of $(R, E - D)$-extendomorphism provided that D (and consequently $E - D$) reduces to a finite union of intervals.

Lemma of the complement. Given an n-ary relation R with base E and a subset D of E, suppose that for each $h (1 \leq h \leq n - 1)$, the h-tuples in D are partitioned into finitely many classes of $(R, E - D)$-extendomorphism. Then for every integer h, the h-tuples in $E - D$ are partitioned into finitely many classes of (R, D)-extendomorphism.

• For each $h(1 \leq h \leq n-1)$ and each class of $(R, E-D)$-extendomorphism among all the h-tuples in D, we take one representative h-tuple. By hypothesis the set of all elements of these representatives is a finite subset H of D.

Given two h-tuples in $E-D$, we put them in a same class iff they are (R, H)-extendomorphic. Since H is finite, there exist only finitely many such classes. To prove that any two such equivalent h-tuples are (R, D)-extendomorphic, it suffices to take an arbitrary subset F of D having cardinal $\leq n-1$ and to prove that our two h-tuples are (R, F)-extendomorphic.

Let G, G' be the two sets of values of our h-tuples, and g be a local automorphism of R which transforms the first into the second h-tuple and is extendible into a local automorphism of R by the identity on H. Let us prove that g is then extendible by the identity on F.

By definition of H, there exists a local automorphism f of R whose domain is F and whose range is an $F' \subseteq H$; moreover the union $f \cup I_G$ (identity on G) is again a local automorphism of R. Also $g \cup I_{F'}$ (identity on F') is by hypothesis a local automorphism of R. By composition $f \cup g$ is a local automorphism of R. Also $f^{-1} \cup I_{G'}$ (identity on G') is a local automorphism of R. By composition $g \cup I_F$ (identity on F) is a local automorphism of R: the lemma is proved. •

9.9 Hausdorff construction of finitely free, scattered posets (Abraham, Bonnet)

9.9.1 An ordinal sequence of strong intervals built from a finitely free poset

See [2] ABRAHAM, BONNET 1999 (p.64, section 3). Let us start from a finitely free poset A and an arbitrary element u_0 in its base. We construct as follows **a well-founded γ-sequence $u_\mu (\mu < \gamma$ ordinal) which is cofinal in A, with $u_\nu >$ or $|u_\mu \pmod A)$ for $\nu > \mu$.** Since by hypothesis A is finitely free, **the obtained cofinal restriction of A is a well partial ordering**.

The quoted authors imitate, in a more sophisticated manner, the construction of the well-founded cofinal sequence in theorem 2.7.2.

To u_0 we associate the set U_0 of elements $\leq u_0 \pmod A$. If u_0 is the maximum of A, then it constitutes our cofinal sequence.

If not, we take, if it exists, an arbitrary element $u_1 | u_0 (\mathrm{mod}\, A)$ to which we associate the set U_1 of elements $\leq u_1$ which do not already belong to U_0. In the contrary case where there exist, out of U_0, only elements $> u_0$, then we take any one of them as u_1 and we define U_1 as previously: so that all elements of U_1 are $> u_0$.

To be complete, even if there exist elements $|u_0$, we are authorized to take as u_1 an element $> u_0$, provided that each element $< u_1$ which does not belong to U_0 be itself $> u_0$. If we define U_1 as prevously, then each element of U_1 is $> u_0$.

In general suppose we have already obtained the $u_\mu (\mu < \alpha$ ordinal) and corresponding sets U_μ. If this α-sequence is cofinal, then we are done. If not, it remains some elements which do not belong to any U_μ; hence they are $>$ or $|u_\mu$ for each $\mu < \alpha$. We take as u_α an element which is minimal in this sense that each $u < u_\alpha \pmod A$ satisfies, for each μ, the same comparison which is true for u_α. In other words either $u > u_\mu$ or $u|u_\mu$ according to whether $u_\alpha > u_\mu$ or $u_\alpha|u_\mu$. Finally elements $u \leq u_\alpha$ yet not belonging to any $U_\mu(\mu < \alpha)$ constitute by definition the set U_α.

• Let us show that such a u_α always exists. Indeed suppose the contrary: there exists a strictlly decreasing ω-sequence $u(0) > u(1) > ... > u(i)(i$ integer) so that each time we go from $u(i)$ to $u(i+1)$ then there exists a least ordinal index $\mu(i)$ such that $u(i) > u_{\mu(i)}$ yet $u(i+1)|u_{\mu(i)} \pmod A$. Then necessarily for each integer i this is a different index $\mu(i)$ which is concerned. So that we get an ω-sequence without repetition of these indices. Since A restricted to the u_μ is a well partial ordering, then by 4.3.2 proposition (3), there exists an extracted ω-sequence which is strictly increasing $(\mathrm{mod}\, A)$. Consequently there exist two integers $i < j$ with $u_{\mu(i)} < u_{\mu(j)} \pmod A$, hence $u_{\mu(i)} < u_{\mu(j)} < u(j) \leq u(i+1) \pmod A$, hence $u(i+1) > u_{\mu(i)}$ which contradicts $u(i+1)|u_{\mu(i)}$.•

Finitely free weakening B of A and B-intervals (= strong intervals) associated with the previous well-founded cofinal sequence.

Let us define B on the same base as A and same restriction to each $U_\mu(\mu < \gamma$ ordinal). Moreover given two elements $x \in U_\mu$ and $y \in U_\nu(\nu \neq \mu)$, we set $x < y$ or $x > y$ or $x|y \pmod B$ according to whether $u_\mu < u_\nu$ or $u_\mu > u_\nu$ or $u_\mu|u_\nu \pmod A$.

We see that A is an augmentation of B (or B a weakening of A) and that each $U_\mu(\mu < \gamma)$ is a B-interval.

Finally B is finitely free. Indeed given an antichain F in B, each intersection $F_\mu = F \cap U_\mu$ is an antichain in A, hence is finite. Moreover if we had infinitely many indices μ with non-empty corresponding F_μ, then the corresponding u_μ would be mutually incomparable $(\mathrm{mod}\, A)$, so that A would not be finitely free.

9.9.2 The Hausdorff construction: theorem, definitions and preliminary lemmas

Theorem. If a condition C is true for well partial orderings and C is saved under (1) inversion; (2) lexicographic sum; (3) augmentation; then C is true for every finitely free, scattered poset. See the quoted article [2].

We need some definitions and preliminary lemmas.

1 - Let us define the **perpendicular sum** $X \perp Y$ as being the poset based on the union $|X| \cup |Y|$ (where the bases $|X|$ and $|Y|$ are assumed to be disjoint) with $x|y \pmod{X \perp Y}$ for $x \in |X|$ and $y \in |Y|$.

Given two posets X, Y with disjoint bases, recall the notation $\mathcal{A}(X)$ for the poset of antichains in X, ordered by reverse inclusion: see 6.7.3; then:

$\mathcal{A}(X \perp Y) = \mathcal{A}(X) \times \mathcal{A}(Y)$; consequently by 4.8.3, given two posets X, Y we have that $\mathrm{Thick}(X \perp Y) = \mathrm{Thick}(X) \oplus \mathrm{Thick}(Y)$, where $\oplus$ is the natural, or commutative sum of ordinals defined in 4.8.2, and Thick is defined in 6.7.3.

2 - We need the **Hessenberg based product** on ordinals, denoted $\odot$ and defined as follows:

$\alpha \odot 0 = 0$;

$\alpha \odot (\beta + 1) = (\alpha \odot \beta) \oplus \alpha$;

$\alpha \odot \beta = \mathrm{Sup}(\alpha \odot \gamma)(\gamma < \beta)$ for a limit ordinal β.

The reader can check the following inequalities between the usual ordinal product, the based product and the natural product $\otimes$ defined in 4.8.2: $\alpha.\beta \leq \alpha \odot \beta \leq \alpha \otimes \beta$.

Example given by the authors: $\alpha = \omega + 1$ and $\beta = \omega + 2$ give the following three distinct values $\alpha.\beta = \omega^2 + \omega.2 + 1$, $\alpha \odot \beta = \omega^2 + \omega.2 + 2$, $\alpha \otimes \beta = \omega^2 + \omega.3 + 2$.

They also mention the associativity $(\alpha \odot \beta) \odot \gamma = \alpha \odot (\beta \odot \gamma)$.

By lemma 1.11 of the quoted article:

Consider two finitely free posets X, Y and their lexicographical product (lexicographical sum of copies of X along Y); then this product is itself a finitely free poset whose thickness is $\mathrm{Thick}(X) \odot \mathrm{Thick}(Y)$

3 - We need the **Hessenberg based exponentiation** on ordinals, defined as follows:

$\alpha^0 = 1$ for $\alpha > 0$ (yet $0^\alpha = 0$);

$\alpha^{\beta+1} = \alpha^\beta \odot \alpha$;

$\alpha^\beta = \mathrm{Sup}(\alpha^\gamma)(\gamma < \beta)$ for a limit ordinal β.

9.9.3 The generalized Hausdorff hierarchy

For any ordinal ρ, the class $\mathcal{H}_0^\rho$(ρ ordinal) reduces, by definition, to the singleton of the poset 1 (defined on one element).

Then we denote by $\mathcal{H}_1^\rho$(ρ ordinal) the class of well partial orderings or inverses of such whose thickness is $\leq \rho$ (thickness is defined in 6.7.3). In particular $\mathcal{H}_1^1$ denotes the class of those chains which are either a well-ordering or the inverse of a well-ordering.

More generally if α is a limit ordinal, then we define: $\mathcal{H}_\alpha^\rho = \cup_{\beta<\alpha} \mathcal{H}_\beta^\rho$;

Given a successor ordinal $\alpha+1$, then we define: $\mathcal{H}_{\alpha+1}^\rho$ = class of lexicographical sums: $\Sigma_{i \in I} A_i$ where $A_i \in \mathcal{H}_\alpha^\rho$ and where the poset I is either a well partial ordering or its inverse, with thickness $\leq \rho$.

Finally define $\mathcal{H}^\rho = \cup_\alpha \mathcal{H}_\alpha^\rho$($\alpha$ ordinal); and $\mathcal{H} = \cup_\rho \mathcal{H}^\rho$($\rho$ ordinal).

Then we have the following (lemma 2.2 in the previous reference):

1 - **$\mathcal{H}^\rho$ is the least class which contains the well partial orderings with thickness $\leq \rho$ and is closed under lexicographical sums and inversions.** In particular $\mathcal{H}^1$ is the Hausdorff class of chains which contains ordinals and is closed under inversion and sum: see 6.2.7.

2 - **Each $\mathcal{H}_\alpha^\rho$ and $\mathcal{H}^\rho$ is closed under restrictions.**

3 - **Every poset in $\mathcal{H}^\rho$ is finitely free and scattered.**

4 - **The closure by augmentation of $\mathcal{H}^\rho$ is formed of finitely free, scattered posets; it is closed under restrictions, inversion, lexicographical sums and (obviously) augmentation.** Idem for the closure by augmentation of $\mathcal{H}$.

5 - **Any poset in $\mathcal{H}^\rho_\alpha$ has a thickness $\leq \rho^\alpha$ where our power notation means the Hessenberg exponentiation**, defined in the previous subsection.

9.9.4 Main lemmas

Given a poset A and an element $u \in |A|$, we denote $(\leq u)_A$ the restriction of A to all elements $\leq u \pmod A$; analogous notation $(\geq u)_A$.

Given a set $\mathcal{H}$ of posets, let us denote $\text{Aug}(\mathcal{H})$ **the set of all augmentations of posets in** $\mathcal{H}$.

First lemma. Let A be a finitely free poset with $\text{Thick}(A) \leq \rho$. **If for every x in the base $|A|$ we have that** $(\leq x)_A \in \text{Aug}(\mathcal{H}^\rho)$, **then** $A \in \text{Aug}(\mathcal{H}^\rho)$. Idem if we replace $(\leq x)_A$ by $(\geq x)_A$.

• Firstly take an ordinal α such that for every $x \in |A|$ we have $(\leq x)_A \in \text{Aug}(\mathcal{H}^\rho_\alpha)$; then we shall prove that $A \in \text{Aug}(\mathcal{H}^\rho_{\alpha+1})$. Construct a cofinal well-founded sequence $u_\mu(\mu < \alpha)$ in A with corresponding subsets U_μ, as in 9.9.1.

Consider the restriction $H = A/\{u_\mu\}(\mu < \alpha)$; this is a well partial ordering with $\text{Thick}(H) \leq \rho$; thus $H \in \mathcal{H}^\rho_1)$.

Take the lexicographical sum B of the U_μ's along H. Then A augments B; moreover:

(1) each U_μ is a restriction of $(\leq u_\mu)_A$, hence $U_\mu \in \text{Aug}(\mathcal{H}^\rho_\alpha)$, since $\mathcal{H}^\rho_\alpha$ is closed under restrictions; say that U_μ augments $V_\mu \in \mathcal{H}^\rho_\alpha$.

(2) the lexicographical sum $B = \Sigma_H V_\mu$ belongs to $\mathcal{H}^\rho_{\alpha+1}$.

Finally by 9.9.1 we see that A augments B hence $A \in \text{Aug}(\mathcal{H}^\rho)$. •

Second lemma. If A is a finitely free poset with $\text{Thick}(A) \leq \rho$ **but** $A \notin \text{Aug}(\mathcal{H}^\rho)$, **then there exists an element** $u \in |A|$ **such that**: $(\leq u)_A \notin \text{Aug}(\mathcal{H}^\rho)$ **and** $(\geq u)_A \notin \text{Aug}(\mathcal{H}^\rho)$.

• Suppose that A is a finitely free poset with $\text{Thick}(A) \leq \rho$ and let L be the set of elements u such that $(\leq u)_A \in \text{Aug}(\mathcal{H}^\rho)$ and G the set of u such that $(\geq u)_A \in \text{Aug}(\mathcal{H}^\rho)$.

Arguing ad absurdum, we assume that for every $u \in |A|$ either $(\leq u)_A \in \text{Aug}(\mathcal{H}^\rho)$ or $(\geq u)_A \in \text{Aug}(\mathcal{H}^\rho)$; in other words $A = L \cup G$. We must prove that $A \in \text{Aug}(\mathcal{H}^\rho)$.

L is an initial interval, and G a final interval of A. By the previous "first lemma" both L and G are in $\text{Aug}(\mathcal{H}^\rho)$. In other words there exist two posets $L^*, G^* \in \mathcal{H}^\rho)$ such that $L \in \text{Aug}(L^*)$ and $G \in \text{Aug}(G^*)$.

There are two possibilities: either for every x in L and y in G we have $x \leq y \pmod A$, or else there exist an x in L and a y in G with $x|y \pmod A$.

In the first case A is the sum of two posets which belong to $\text{Aug}(\mathcal{H}^\rho)$, and hence $A \in \text{Aug}(\mathcal{H}^\rho)$ as well, because $\mathcal{H}^\rho$ is closed under sums.

In the second case there are two incomparable elements in A, so that $\rho \geq 2$. Hence $\mathcal{H}^\rho_2$ contains the poset consisting of two incomparable members, and so $\mathcal{H}^\rho$ is closed under perpendicular sums. Since A is an augmentation of $L^* \perp (G^* - L^*)$ it follows that $A \in \text{Aug}(\mathcal{H}^\rho)$. •

Corollary. If A is a finitely free poset with $\text{Thick}(A) \leq \rho$ **but** $A \notin \text{Aug}(\mathcal{H}^\rho)$, **then** A embeds $\mathbf{Q}$. In other words:

If A is a finitely free, scattered poset with Thick$(A) \leq \rho$, **then** $A \in$ Aug$(\mathcal{H}^\rho)$. This proves the theorem.

• The previous lemma implies the existence of some element u in the base $|A|$ such that both restrictions $(\leq u)_A$ and $(\geq u)_A$ satisfy the hypothesis in the first previous statement. By iteration we get a restriction of A which is isomorphic with **Q**. •

9.9.5 The need for augmentations

The quoted authors give as an example the direct product A of the chain $\omega + \omega^-$ (where ω^- is the symmetric of ω) by the chain of two elements $a < b$. Let us denote by positive integers +1,+2,+3,... the elements of ω and by -1,-2,-3,... the elements of ω^-. Then our direct product has minimum $(a, +1)$ and maximum $(b, -1)$.

Firstly note that A is neither a well partial ordering nor the converse of a well partial ordering; so that we have to build, on the model of 9.9.1, a cofinal ordinal sequence u_μ whose corresponding sets U_μ give restrictions which are themselves well partial orderings or converse of such. It suffices for instance to start from the ω-sequence of couples $(a, +n)$, then take the entire set of couples $(a, -n)$ then similarly with couples having first term b. Finally we get two chains $\omega + \omega^-$, each being incomparable with the other one. We achieve by an obvious augmentation which restitutes the direct product A.

One sees that the only strong intervals of A are the empty set, each singleton and the following four sets: the entire base, the base minus the maximum, the base minus the minimum, and the base minus both minimum and maximum.

Consequently if A could be represented as a lexicographical sum $A = \Sigma_{i \in B} A_i$, then either B would be isomorphic with A, or else some A_i would be isomorphic with A.

9.10 Exercises

9.10.1 Extendible local automorphism

Given a relation R, we say that a local automorphism f is **extendible** iff there exists an automorphism of R extending f.

Consider the binary relation R defined as follows on a base of 10 elements $a, b, c, d, r, s, t, u, i, j$. Let $R(a, u) = R(a, r) = R(b, r) = R(b, s) = R(c, s) = R(c, t) = R(d, t) = R(d, u) = R(i, r) = R(j, s) = R(i, t) = R(j, u) = +$, while R takes the value (-) for all other couples.

The identity mapping on $\{a, b\}$ is obviously a local automorphism of R, and is obviously extendible to the identity automorphism on the entire base.

The mapping which preserves a and which takes i into j, is extendible to the automorphism which preserves a and c, and interchanges b and d, r and u, s and t, i and j.

Similarly the mapping which preserves b and which takes i into j, is extendible to the automorphism which preserves b and d, and interchanges a and c, r and s, t and u, i and j.

Let f denote the mapping which preserves a and b, and takes i into j. Then f is a local automorphism, since R is binary and the three restrictions of f to $\{a, b\}$, to $\{a, i\}$ and to $\{b, i\}$ are local automorphisms.

Prove that f is not extendible to the entire base. Note that r is the only element x such that $R(a, x) = R(b, x) = +$. Hence any automorphism extending f must preserve r. Taking i into j, we must have $R(i, r) = R(j, r)$, which is false.

Thus although all three restrictions of f to 2 elements are extendible to the entire base, f itself is not extendible.

9.10.2 Conditions for a relation to be constant

1 - Let A be the usual chain of the integers, and B be a chain on the same base, but which is isomorphic with the chain of rationals, hence dense and having neither a maximum nor a minimum element. Prove that, for each positive integer n and each permutation f of $\{1, 2, ..., n\}$, there exists a set F of n integers $u_1, ..., u_n$ such that, assuming that these integers are ordered by $u_1 < u_2 < ... < u_n \bmod A$, then they are reordered by $u_{f(1)} < u_{f(2)} < ... < u_{f(n)} \bmod B$.

2 - Let n be a positive integer, and R an n-ary relation based on the integers. Prove that if R is both A-chainable and B-chainable, then for every permutation f of $\{1, 2, ..., n\}$, there exists a set F of n integers such that, assuming that these elements of F are $u_1 < u_2 < ... < u_n \bmod(A/F)$, then the permutation f transposed by the bijection taking 1 into $u_1, ..., n$ into u_n, is an automorphism of R/F.

Hint. Take an n-element set G for which the restrictions A/G and B/G are identical, and then use the isomorphism of A/F onto A/G and of B/F onto B/G. From this, deduce that for each n-element set F, the restriction R/F is preserved by every permutation of F, thus R/F is a constant relation. Using 9.4.1, deduce that R itself is constant.

3 - Suppose that R is a denumerable n-ary relation satisfying the following: for every denumerable n-ary relation X, if every restriction to at most n elements is embeddable in R, then X is isomorphic with R. Then R is constant.

Hint. First prove that R is chainable; more strongly that, for every denumerable chain, there exists an isomorphic chain C with base $|R|$, such that R is C-chainable.

4 - Using the axiom of choice, extend the above result in (3) to the case of an arbitrary infinite relation.

Chapter 10

Age, rich relation, alpha-morphism, finitist and almost chainable relation, back-and-forth notions

10.1 Projection filter, older or younger relation, 1-extension, $(1,p)$-morphism, $(1,p)$-isomorphism

10.1.1 Generalities

Consider two sets E and E' and a function f from E into E'. Let m be an integer; consider two m-ary relations R with base E and R' with base E'. Then R is said to be the **inverse projection of R' under** f iff for every $x_1, ..., x_m$ in E we have $R'(fx_1, ..., fx_m) = R(x_1, ..., x_m)$.

In this case we write $R = f^{-1}(R')$.

If f is a bijection from E onto E', then we find the usual definition of an isomorphism from R onto R'.

Given sets E and E', consider the set of all functions from E into E'. A filter on this set shall be called a **projection filter** from E onto E'.

The relation R is called the **inverse projection of R' under the filter** $\mathcal{F}$, which we write by $R = \mathcal{F}^{-1}(R')$, iff for every $x_1, ..., x_m$ in E, the set of functions f satisfying $R'(fx_1, ..., fx_m) = R(x_1, ..., x_m)$ is an element of the filter $\mathcal{F}$. In other words, the preceding equality holds for almost every function mod $\mathcal{F}$.

If R' and f are given, then the inverse projection $\mathcal{F}^{-1}(R')$ exists and is unique. If R' and the filter $\mathcal{F}$ are given, then there exists at most one inverse projection.

If the inverse projection $\mathcal{F}^{-1}(R')$ exists and if $\mathcal{G}$ is a filter finer than $\mathcal{F}$, then $\mathcal{G}^{-1}(R') = \mathcal{F}^{-1}(R')$.

If $\mathcal{F}$ is an ultrafilter, then the inverse projection exists and is unique. Indeed either the set of functions f such that $R'(fx_1, ..., fx_m) = +$, or the analogous set with the value (-), is an element of the ultrafilter.

If the ultrafilter is trivial, morre precisely if it is formed of those sets of functions which contain as an element the function f, then the inverse projection under the trivial ultrafilter coincides with the inverse projection under f.

10.1.2 Injective filter

Let $\mathcal{F}$ be a projection filter from the set E onto E'. Two elements a, b of E are said to be **equivalent modulo** $\mathcal{F}$ iff $fa = fb$ for almost every $f(\text{mod}\,\mathcal{F})$. This condition is reflexive, transitive and symmetric.

A projection filter from E onto E' is said to be **injective** iff for any two distinct elements a, b of E, we have $fa \neq fb$ for almost every function f.

To each finite subset F of E, associate the set U_F of those functions from E into E' which are injective on F. When F varies, the sets U_F and their supersets constitute an injective filter, which is minimum in this sense that any given projection filter from E onto E' is injective iff it is finer than our injective filter.

If E' is infinite, then there always exists an injective filter from E onto E'.

Every filter which is finer than an injective filter, is injective.

In the case of a trivial ultrafilter, formed of those sets of functions which contain the function f, the ultrafilter is injective iff the function f is injective. In this case, necessarily $\text{Card}\, E' \geq \text{Card}\, E$.

10.1.3 Older relation, younger relation

Given two relations R, S of the same arity, we say that R is **younger** than S, or that S is **older** than R iff every finite restriction of R is embeddable in S. This comparison is reflexive and transitive, but not antisymmetric, even up to isomorphism. For example the chain ω of integers and its converse are each older than the other.

Let R be a relation with base E, and R' be a relation of the same arity, with base E'. Then there exists an injective projection filter $\mathcal{F}$ satisfying $R = \mathcal{F}^{-1}(R')$ iff R' is older than R.

• Let m denote the common arity of R and R'. Suppose that $R = \mathcal{F}^{-1}(R')$ and let F be a finite subset of E. The set of functions f which are injective on F is an element of $\mathcal{F}$. For each m-tuple $a_1, ..., a_m$ in F, the set of functions f satisfying $R'(fa_1, ..., fa_m) = R(a_1, ..., a_m)$ is again an element of $\mathcal{F}$. Since F is finite, the set of all m-tuples in F is finite, so that the intersection of the preceding sets of functions is an element of $\mathcal{F}$. And every function f which belongs to this set gives an isomorphism from R/F onto $R'/\overline{f}(F)$.

Conversely suppose that R' is older than R. For each finite subset F of E, there exists a function f from F into E', which is an isomorphism from R/F onto $R'/\overline{f}(F)$. Let U_F designate the set of those functions from E into E', whose restriction to F is such an isomorphism. Then these U_F and their supersets constitute a projection filter, obviously injective, under which R is the inverse image (or projection) of R'. •

10.1.4 Projection filter and concatenation

The preceding definitions and propositions immediately extend to the case of multirelations. Thus we have the following statement.

Let R, R' be two multirelations with respective bases E, E'.

If R' is older than R, then for every S' with base E', there exists an S with the same arity and with base E, such that the concatenation $R'S'$ is older than RS (uses the ultrafilter exiom; ZF suffices for E denumerable).

• Take an injective ultrafilter $\mathcal{F}$ such that $R = \mathcal{F}^{-1}(R')$, then set $S = \mathcal{F}^{-1}(S')$.

Another proof: for each finite subset F of E, there exists a relation S_F with base F, such that $(R/F, S_F)$ is embeddable in $R'S'$. Let U_F designate the set of these S_F for a given F, and apply the coherence lemma 2.4.1. In the denumerable case, the ultrafilter is unnecessary. •

10.1.5 Filter identical on a set

Let E, E' be two sets and D a subset of the intersection $E \cap E'$. A projection filter $\mathcal{F}$ from E onto E' is said to be D-**identical** iff there exists an element of $\mathcal{F}$, formed of functions from E into E', each of whose restriction to D is the identity. In other words, almost every function modulo $\mathcal{F}$, when restricted to D, is the identity.

If $R = \mathcal{F}^{-1}(R')$, where $\mathcal{F}$ is a D-identical filter, then R and R' have the same restriction to D.

If E' is infinite and E is a superset of E', then there exists a filter from E onto E' which is injective and E'-identical.

• For each finite subset F of E, take the set U_F of those functions from E into E' whose restriction to E' is the identity and which are injective on F. Then the supersets of these U_F constitute the desired filter. •

10.1.6 The 1-extension

Let R be a relation with infinite base E, and let E^+ be a superset of E. An extension R^+ of R to E^+ is said to be a **1-extension** of R iff for every finite subset F of E^+ there exists an isomorphism from R^+/F onto a restriction of R, which reduces to the identity on $F \cap E$.

For example if we add a last element to the chain ω of the integers, then we have a 1-extension. However if we add an element before the minimum element 0, then we obtain an extension isomorphic with ω, yet which is not a 1-extension.

Every 1-extension of R has the same finite restrictions as R, up to isomorphism. The converse is false: take again the chain of integers and its extension by adding an element before 0.

The notion of 1-extension is transitive.

If R^+ with base E^+ is a 1-extension of R with base E, then for every intermediate set $H(E \subseteq H \subseteq E^+)$ the restriction R^+/H is a 1-extension of R.

Let R be a relation with base E and R^+ be a relation with the same arity, which is based on a superset E^+ of E. Then R^+ is a 1-extension of R iff there exists a projection filter $\mathcal{F}$ from E^+ onto E, which is injective and E-identical, and which satisfies $R^+ = \mathcal{F}^{-1}(R)$.

Take up the proof of 10.1.3 with the functions from E^+ into E which reduce to the identity when restricted to E.

In connection with logic, note that R^+ is a 1-extension of R iff there exists an extension of R^+ which is an elementary extension of R (defined in 10.10.3 below); i.e. an extension which satisfies the same "elementary" logical formulas as R, where the free variables in the formulas are replaced by elements of the base $|R|$. See for example [30] CHANG, KEISLER 1973, prop. 5.2.2.

10.1.7 1-extension and concatenation; weak forms of the Lőwenheim-Skolem theorem

Let R be a multirelation with infinite base E, and let R^+ be a 1-extension of R, based on a superset E^+ of E. Then for every S with base E, there exists an extension S^+ of S to E^+ such that the concatenation R^+S^+ is a 1-extension of RS (uses the ultrafilter axiom). Take an ultrafilter $\mathcal{F}$ which is injective, E-identical and which verifies $R^+ = \mathcal{F}^{-1}(R)$; then set $S^+ = \mathcal{F}^{-1}(S)$.

In particular, for each infinite set E, every superset of E, considered as a multirelation reduced to its base, is a 1-extension of E.

Thus for every relation R with infinite base E and for every superset E^+ of E, there exists a 1-extension of R which is based on E^+. This is a weak form of the upward LÖWENHEIM-SKOLEM theorem.

Let R be a multirelation with infinite base E, and let D be an infinite subset of E. Then there exists a set D^+ equipotent with D, satisfying $D \subseteq D^+ \subseteq E$, such that R is a 1-extension of R/D^+ (uses axiom of choice; weak form of the downward LÖWENHEIM-SKOLEM theorem).

• Let us say that two finite subsets F, F' of E are equivalent iff they have the same intersection $G = F \cap D = F' \cap D$ and there exists an isomorphism from R/F onto R/F' which reduces to the identity on G. For a given finite subset G of D, there exist countably many equivalence classes. Now choose a representative F from each class (axiom of choice), then define D^* as the union of the F corresponding to all the G, so that D^* is equipotent with D. Iterate this, setting $D_1 = D^*$ then $D_2 = (D_1)^*$ etc.; finally $D^+ =$ union of D_i(i integer). •

10.1.8 A common extension

Given three relations R, R', S such that R and R' are both younger than S, then there exists an isomorphic copy of R' and a common extension

of this copy and R**, which is younger than** S (uses the ultrafilter axiom; ZF suffices if R and R' are countable).

• Let E, E' be the bases of R, R' which we assume to be disjoint; and let D be the base of S. To each couple (F, F') where F is a finite subset of E and F' a finite subset of E', associate the set $U_{F,F'}$ of all functions h from $E \cup E'$ into D, whose restrictions to F and to F' are respectively an isomorphism from R/F onto $S/\overline{h}(F)$ and an isomorphism from R'/F' onto $S/\overline{h}(F')$. By hypothesis, for all F and F', the set $U_{F,F'}$ is non-empty. Moreover, for any finite supersets $G \supseteq F$ and $G' \supseteq F'$, we have the inclusion $U_{G,G'} \subseteq U_{F,F'}$. Thus the supersets of the $U_{F,F'}$ constitute a filter on the set of functions from $E \cup E'$ into D. Take a finer ultrafilter (or use the coherence lemma 2.4.1).

Match an element a in E, with an element a' in E', iff $ha = ha'$ for almost every function h modulo the ultrafilter. Note that it is impossible to match up an element in E to more than one element in E', as well as to match up an element in E to another element in E, and similarly when exchanging E and E'. Indeed, the matching is an equivalence relation, and if a, b are distinct elements of E, then almost every function h takes $R/\{a, b\}$ isomorphically onto $S/\{ha, hb\}$. Thus $ha \neq hb$ for almost all h.

Let n be the common arity of the three relations. Take a sequence of n elements $a_1, ...a_n$ in $E \cup E'$ and define the relation R^+ by $R^+(a_1, ..., a_n) = +$ or - according to whether $S(ha_1, ..., ha_n) = +$ for almost every h or = - for almost every h, modulo the ultrafilter. This relation R^+ is a common extension of R and R'. Take an isomorphic copy R" of R', by keeping each element of E' which is matched to no element, and for each element of E' which is matched, by taking that element matched to it: then R" is isomorphic with R'. Finally, the desired common extension of R and R" is the restriction of R^+ to E augmented by those elements of E' which are matched to no other element. •

10.1.9 The 1-morphism and the 1-isomorphism

Let p be an integer. A local isomorphism f with domain F, from a relation R into a relation S, is said to be a $(1,p)$-**morphism** from R into S iff for every set $F^* = F$ augmented by at most p elements from the base $|R|$, there exists a local isomorphism from R into S which is an extension of f to the domain F^*.

The local isomorphism f is said to be a $(1,p)$-**isomorphism** from R into S iff f is a $(1,p)$-morphism and the inverse function f^{-1} is a $(1,p)$-morphism from S into R.

Every local isomorphism is a (1,0)-isomorphism.

Every $(1,p)$-morphism is a $(1,q)$-morphism for each $q \leq p$; similarly for a $(1,p)$-isomorphism.

We say that a local isomorphism f from R into S is a 1-**morphism** iff it is a $(1,p)$-morphism for every integer p. In other words if f is extendible to every finite superset of Dom f.

We say that f is a 1-**isomorphism** iff it is a $(1,p)$-isomorphism for every integer p. In other words f is extendible to every finite superset of Dom f and to every finite superset of Rng f. Or again iff f and f^{-1} are both 1-morphisms.

If S is an extension of R, then the identity on any subset of the base $|R|$ is a 1-morphism from R into S.

For example let R be the chain of the non-negative integers and S be the chain of the positive and negative integers, so that S is an extension of R. Then the identity on the singleton $\{0\}$ is a 1-morphism from R into S. Yet it is not even a (1,1)-isomorphism, since the negative integer (-1) is strictly less than 0 $(\text{mod}\, S)$ and there exists no element in $|R|$ which is strictly less than 0 $(\text{mod}\, R)$.

Let S be an extension of R. Then S is a 1-extension iff, for each finite subset F of $|R|$, the identity on F is a 1-morphism from S into R, hence also a 1-isomorphism.

The restriction of a $(1,p)$-morphism to a subset of its domain, is a $(1,p)$-morphism. The composition $g \circ f$ of two $(1,p)$-morphisms f and g, with $\text{Dom}\, g =$ Rng f, is a $(1,p)$-morphism. Same statements with $(1,p)$-isomorphisms, 1-morphisms, 1-isomorphisms.

10.1.10 $(1,p)$-equivalent relations

Let p be a non-negative integer; we say that two relations ane $(1,p)$-**equivalent** iff they have the same restrictions to $\leq p$ elements, up to isomorphism.

We have already defined $(1,p)$-isomorphism in 10.1.9. Then two relations R and S are $(1,p)$-equivalent iff the empty function is a $(1,p)$-isomorphism from R into S. Note that the conservation of $(1,p)$-isomorphism by composition ensures the transitivity of the $(1,p)$-equivalence.

10.2 Closed under embeddability, directed under embeddability, age

A set $\mathcal{R}$ of relations of the same arity is said to be **closed under embeddability** (up to isomorphism) iff for all relations X, Y, if X belongs to $\mathcal{R}$ and $Y \leq X$ (with respect to embeddability) then there exists a relation isomorphic with Y which belongs to $\mathcal{R}$.

A set $\mathcal{R}$ is said to be **directed under embeddability** iff for all relations X, Y in $\mathcal{R}$, there exists a relation Z in $\mathcal{R}$ satisfying $Z \geq X$ and $Z \geq Y$.

Given a relation R and an integer p, the set of restrictions of R with cardinality $\leq p$ is closed under embeddability.

For certain relations R, this set is also directed: for instance if R is a chain.

However if R is the consecutivity relation on the integers, then except for $p = 0$ and $p = 1$, this set is not directed.

Given a relation R and a positive integer p, the set of finite restrictions of R with cardinality $\geq p$ is directed but not closed under embeddability.

10.2.1 Age, representative relation

Let R be a relation. Then the set of finite restrictions of R is closed and directed under embeddability. Considered up to isomorphism, this set is called the **age** of R. We shall say that R **represents this age**, or that R is a **representative** of the age.

For the reader desiring a rigorous definition, the age of R is the set of isomorphic copies of the finite restrictions of R, which are based on finite subsets of the set ω of the integers.

Ages are countable sets: they can be compared under inclusion. For example the age of all finite chains, which is represented by an infinite chain, is included in the age of all finite posets, which itself is represented by those infinite posets in which every finite poset is embeddable.

Given two relations R, S of the same arity, R is younger than S, or S is older than R iff the age of R is included in the age of S.

The relations R and S have the same age iff each is older (and younger) than the other.

To each age $\mathcal{R}$ there corresponds the **negation age** of $\mathcal{R}$, obtained by replacing each element of $\mathcal{R}$ by its negation (i.e. the relation taking always the opposite value); or equivalently by replacing a representative by its negation.

For example, the negation of the age of all finite chains is the age of all strict finite chains: with $<$ instead of $\leq$.

An age may be identical to its negation: for example the set of all finite binary relations, or the set of all finite symmetric binary relations.

10.2.2 An age is an ideal

Given a set $\mathcal{R}$ of finite relations of the same arity, which is closed and directed under embeddability, there exists a countable relation whose age is $\mathcal{R}$ (up to isomorphism).

In other words, **an age is an ideal for the partial ordering of embeddability among finite relations**.

• Since the set $\mathcal{R}$ is countable (up to isomorphism), enumerate it by using integer indices i, thus obtaining a sequence of finite relations A_i. Let $B_0 = A_0$; then B_1 shall be the common extension, up to isomorphism, of B_0 and A_1, which belongs to R, hence to our sequence, with the least possible index; and so on. Then take the common extension of the B_i on the union of their bases. •

Given a set of ages, totally ordered under inclusion, the union and the intersection are both an age.

Given a set of ages for which inclusion is a directed partial ordering, then the union is an age.

However **in general the union or the intersection of two ages is not a age.**

• For the case of a union, take the age formed of all finite unary relations always (+), and the age of finite unary relations always (-).

For the case of an intersection, start with two disjoint, non-empty sets E, F. Take the binary relation A based on the union $E \cup F$ with $A(x, y) = +$ for x or y belonging to E, and $A(x, y) = -$ otherwise. Take the relation B with $B(x, y) = +$ for x and y belonging to E, and value (-) otherwise. Then the only common finite restrictions of A and B are the relations always (+) and the relations always (-): they do not constitute a directed set. •

A set $\mathcal{R}$ of finite relations of common arity, which is closed under embeddability, is an age iff $\mathcal{R}$ is not the union of two subsets distinct from $\mathcal{R}$ and closed under embeddability. This is a classical property of ideals: see 2.13.1.

Let A be a relation of arity ≥ 2, with finite base. Then **the set (up to isomorphism) of those finite relations which do not admit an embedding of A is an infinite age.** This follows from 8.1.2 (faithful common extension).

10.2.3 Existence of a representative

Given a relation R and an age $\mathcal{T}$ including the age of R, there exists an extension of R which is a representative of $\mathcal{T}$ (uses the ultrafilter axiom; ZF suffices for R countable).

• Take a relation T representative of the age $\mathcal{T}$, then a common extension of R and of an isomorphic copy of T, this extension being chosen to be younger than T: see 10.1.8. •

On the other hand, **starting with a relation R younger than T, there does not necessarily exist a restriction of T having the same age as R.**

• Take R to be the binary denumerable relation always (+). For each positive integer i, define S_i to be the binary relation of cardinality i, always taking the value (+), where the bases E_i of the S_i are taken to be disjoint. Let T be the common extension of the S_i, based on the union of the E_i; any two elements x, y from distinct E_i, giving the value (-). •

10.2.4 The number of countable representatives

In [165] MACPHERSON, POUZET, WOODROW 1992, it is proved (with axiom of choice) that the number of countable representatives (up to isomorphism) of any given age, is either 1 or ω or the continuum power.

10.2.5 Age and universal class

We already mentioned a connection between universal classes and ages (see 5.10.4). A directed universal class is an age. In the other sense an age with finitely many bounds is a universal class.

More generally any age is an intersection of denumerably many finitely bounded ages, hence of denumerably many universal classes.

10.3 Rich relation

A denumerable n-ary relation is said to be **rich** iff every denumerable n-ary relation is embeddable in it.

10.3.1 Construction of a rich relation

For every positive integer n, there exists a denumerable rich n-ary relation.

• This is obvious for $n = 1$: take a relation with the value (+) infinitely many times and the value (-) infinitely many times.

We shall prove the proposition for $n = 2$; the proof extends immediately to the case of $n \geq 2$.

As base, take the set of integers. We shall define our binary rich relation in several "stages". In stage 1, let E_1 be the set of the integers 0, 1 and let $R(0,0) = +$ and $R(1,1) = -$. In stage 2, let E_2 be the set of integers from 2 through 17, and define R as follows. From $x = 2$ to $x = 9$, attribute to R all the $2^3 = 8$ possible systems of values (+) and (-) for $R(0,x)$ and $R(x,0)$ and $R(x,x)$. Similarly from $x = 10$ to $x = 17$, attribute to R all the 8 possible systems of values (+) and (-) for $R(1,x)$ and $R(x,1)$ and $R(x,x)$.

In general, let i be a positive integer, and suppose that $E_1, ..., E_i$ are defined, and that R is already defined on their union, so that for any binary relation U based on $\{1, 2, ..., i\}$, there exists a sequence $u_1, ..., u_i$ where u_j belongs to $E_j (j = 1, ..., i)$; so that the transformation of $1, ..., i$ into $u_1, ..., u_i$ is an isomorphism from U onto the restriction $R/\{u_1, ..., u_i\}$.

Then we take a finite set E_{i+1} of integers x, all larger than those of the union $E_1 \cup ... \cup E_i$. And we partially define our relation in stage $i+1$, by attributing all possible systems of values $R(u,x)$ and $R(x,u)$ and $R(x,x)$, for each element u in $E_1 \cup ... \cup E_i$. We achieve to define R restricted to the new union $E_1 \cup ... \cup E_{i+1}$, by letting R take arbitrary values on couples not considered: for example (0,1) in stage 1, or (2,3) in stage 2.

Now we see that R is rich. Indeed start with an arbitrary denumerable binary relation X with base the positive integers. We substitute to each positive integer i an element u_i in E_i, so that the considered transformation be an isomorphism from X into a denumerable restriction of R. •

10.3.2 Two immediate properties of richness

(1) **For an arity $n \geq 2$, not all rich denumerable n-ary relations are isomorphic.**

• Start with a rich relation R. Add a new element a to its base, and let $R(a,x) = +$ for all x in the base $|R|$. Similarly add a new element b with the value (-) instead of (+). If the two extensions were isomorphic, then the element a would be taken into an a' by the isomorphism, and we would have $R(a',b) = +$ and = - simultaneously: contradiction. •

(2) **Let R be a rich n-ary relation with denumerable base E, and let F be a finite subset of E. Then the restriction $R/(E-F)$ is rich.**

• It suffices to prove the statement in the particular case where F reduces to the singleton of an element a of E. Take R' to be isomorphic with R and having base E' disjoint from E; then take a common extension S of R and R' with base $E \cup E'$, completed arbitrarily for n-tuples with some terms in E and some terms in E'. Then S is embeddable in the rich relation R, so that there exists a restriction of R isomorphic with S. Then whether a belongs to the image of E or of E', we have that R is embeddable in the restriction of R to its base minus a. •

10.3.3 Connection with indivisible relations

Recall that R with infinite base E is said to be **indivisible** iff for every partition of E into two complementary subsets C and $D = E - C$, then either $R \leq R/C$ or $R \leq R/D$: see 6.8.

A rich binary relation is not indivisible, since one can partition its base into the set of those x for which $R(x,x) = +$ and its complement: we obtain two restrictions into neither of which R is embeddable.

However, given a binary relation R, give the value $(+)$ to all couples (x,x) on the diagonal: the obtain the **reflexified relation** R .

Then **every reflexified rich binary relation is indivisible.**

• Let R be such a reflexified rich binary relation on the integers. Take an ω-sequence of isomorphic copies R_i of R, with mutually disjoint bases E_i(i integer). Take the common extension S of all the R_i, such that for x belonging to E_i and y to E_j, we have $S(x,y) = R(i,j)(i,j$ distinct integers). Then given a partition of the base $|S|$ into two complementary subsets, either some E_i is included in one of the subsets; or the other subset contains at least one element in each E_i. In both cases R is embeddable in one of the two restrictions of S. Since S is equimorphic with R, then R is indivisible. •

10.4 α-morphism, α-older relation, non-embeddability rank, non-richness rank

10.4.1 Definitions

Let R, S be two relations of the same arity; let f be a local isomorphism from R into S with domain F; and let α be an ordinal.

We say that f is an α-**morphism** from R into S in the following cases:

(1) every local isomorphism is a 0-morphism;

(2) if α is a successor ordinal, say $\alpha = \beta + 1$, then f is an α-morphism iff for every set $F^* = F$ augmented by a finite number of elements in $|R|$, there exists a β-morphism from R into S, extending f to the domain F^*;

(3) if α is a limit ordinal, then f is an α-morphism iff it is a β-morphism for every $\beta < \alpha$.

For $\alpha = 1$, we find again the notion of 1-morphism: see 10.1.9.

Every α-morphism is a β-morphism for each $\beta < \alpha$.

An α-morphism, when restricted to any subset of its domain, is still an α-morphism.

The composition of two α-morphisms is an α-morphism.

An isomorphism from R onto S is an α-morphism from R into any extension of S, for every ordinal α. Similarly for a local isomorphism from R into S which is extendible to an isomorphism from R onto S.

In particular, the identity on a subset of the base $|R|$ is an α-morphism from R into every extension of R, for every ordinal α.

We say that a relation S is α-**older** than R, or that R is α-**younger** than S, iff the empty function is an α-morphism from R into S.

The comparison "α-older" is reflexive and transitive. An α-older relation is β-older for every ordinal $\beta < \alpha$.

If R is embeddable in S, then S is α-older than R for every ordinal α. The notion of 1-older coincides with the notion of older, defined in 10.1.3; similarly for 1-younger.

10.4.2 Case of the ω_1-older relation

Lemma. Let R, S be two denumerable relations; if S is ω_1-older than R, then R is embeddable in S (uses countable axiom of choice).

• Let f be an ω_1-morphism from R into S, with finite domain F, and let G be an arbitrary finite superset of F (included in the base $|R|$). It suffices to prove that f is extendible to an ω_1-morphism g from R into S, with domain G. Indeed, starting with the empty set as F and iterating this, we end up with an isomorphism from the denumerable relation R onto a restriction of S (since it suffices that every finite restriction of f be an isomorphism: see 9.1.5).

Since $G = F$ augmented by a finite number of elements, and since for each countable ordinal α, the function f is an $(\alpha+1)$-morphism, there exists a bijection g, extending f to the domain G, which is an α-morphism from R into S.

There are only countably many bijections g with domain G and range included in the denumerable base $|S|$. While ω_1 many countable ordinals α: so there exists at least one g which is an α-morphism for every countable ordinal α, hence which is an ω_1-morphism from R into S. Note that this argument uses the fact that a countable union of countable ordinals is a countable ordinal; in other words, we use the countable axiom of choice, as already in 1.2.5. •

10.4.3 Non-embeddability rank

For R, S denumerable relations with same arity, either $R \leq S$ (and hence S is α-older than R for every ordinal α); or $R \not\leq S$, and by the preceding, there exists a greatest countable ordinal α for which S is α-older than R, and so a least ordinal $\alpha + 1$ from which point on S is not β-older than $R(\beta \geq \alpha + 1)$.

We call $\alpha + 1$ the **non-embeddability rank** of R in S. This rank remains unchanged if we replace either R or S by an equimorphic relation.

For example start with the chain **Q** of rationals and with the ordinal ω^α. By induction, we see that ω^α is α-older than **Q**. In other words, the non-embeddability rank of **Q** in ω^α is at least equal to $\alpha + 1$.

• More precisely a local isomorphism f from **Q** into an ordinal, with $\mathrm{Dom}\, f = \{a_1, ..., a_p\}(a_1 < ... < a_p \bmod \mathbf{Q})$ is an $(\alpha+1)$-morphism provided that each of the intervals $(0, fa_1), (fa_1, fa_2), ..., (fa_{p-1}, fa_p), (fa_p, \infty)$ in the ordinal, is at least equal to ω^α: this is the step from ω^α to $\omega^{(\alpha+1)}$. •

The lemma of the previous subsection generalizes the proposition 5.3.2 which states that **Q** is embeddable in a denumerable chain iff every countable ordinal is itself embeddable.

As a particular case of the non-embeddability rank, let R be a denumerable non-rich relation. Then we define the **non-richness rank** of R to be the least countable ordinal α for which R is not α-older than a denumerable rich relation. This rank does not depend on the chosen rich relation, since all rich relations of a given arity are equimorphic.

10.4.4 Problems of the non-embeddability rank

The non-embeddability rank leads to two kinds of problems, according to whether we fix R and consider α-older relations than R, or on the contrary if we fix S and consider α-younger relations than S.

First fix S (supposed to be denumerable) and note that in the usual cases, there exists a countable ordinal α such that every countable α-younger relation than S is embeddable in S .

For example, if S is the denumerable binary relation always $(+)$, or again the chain **Q** of rationals, then it suffices that a denumerable relation X be 1-younger than S, for X to be embeddable in S.

If S is rich, then every countable relation (which is 0-younger than S) is embeddable in S.

If S is the chain ω of non-negative integers, then a 1-younger relation is necessarily a chain. A 2-younger relation can have at most finitely many elements between any two given elements, or before any given element. So that it can only be either a finite chain, or a chain isomorphic with S, hence embeddable in S.

Problem. For each countable relation S, does there exist a countable ordinal α such that every countable α-younger relation than S is embeddable in S. In other words, if X is not embeddable in S, then is the non-embeddability rank of X in S uniformly bounded by a countable ordinal.

Now fix R and consider the α-older relations than R. We saw in the previous subsection that for R = chain of rationals and for every countable ordinal α, then ω^α is α-older than R.

10.4.5 Existence of ω_1 non-embeddability ranks

For each countable relation R of arity at least equal to 2, and for each countable ordinal α, there exists an $X \not\geq R$, which is countable and α-older than R.

In other words, the non-embeddability rank of R in an X which does not embed R, takes arbitrarily large countable ordinal values ([246] THOMASSÉ 1995 p.88-89).

A specific proof for rich relations, and more generally for inexhaustible relations, will be obtained in 10.6.1 below. This is the case, for instance, of the chain of non-negative integers, and of the denumerable binary relation always (+) .

• Proof for arity 2. To each countable ordinal α associate the following binary relation V_α. Start from the empty relation V_0, which is 0-older than any binary relation. Then for V_1 take the denumerable set of all finite binary relations U_i (up to isomorphism) with disjoint bases. We give the value (+) to each couple whose terms belong to two different bases $|U_i|$ and $|U_{i'}|(i' \neq i)$. Then V_1 is 1-older than every binary relation.

For V_2 take the denumerable set of all finite binary relations U_i (up to isomorphism) and to each U_i associate the denumerable set of all its possible finite extensions $U_{i,j}$ (up to isomorphism). Again we give value (+) to each couple whose terms belong to two different bases $|U_{i,j}|$ and $|U_{i,j'}|(j \neq j')$ (yet not to their intersection); or to each couple whose terms belong to two different $|U_{i,j}|$ and $|U_{i',j'}|$ with $i \neq i'$ whatever be j, j'. Then V_2 is 2-older than every binary relation.

In general to each countable ordinal α associate the denumerable set of (finite) decreasing sequences from any $\beta < \alpha$ until reaching 0. Then for each such decreasing sequence, take its first term β considered as an index, and to β associate all finite binary relations $U_i(i$ integer), considered up to isomorphism (so that there are denumerably many possibilities). Then to the second term $\gamma < \beta$ again considered as an index, associate all finite extensions $U_{i,j}$ of the preceding U_i (again denumerably many possibilities); and so on until reaching the index 0. Finally complete as precedently by value (+) for every non-valued couple. Then V_α is α-older than every binary relation.

Note that a denumerable binary relation always (-) (with indifferently values (+) or (-) on the diagonal) cannot be embedded in V_α.

Now call V_α^- the relation constructed as V_α by changing (+) into (-) for additional values. Obviously V_α^- is also α-older than every binary relation. Moreover, a denumerable binary relation always (+) (except possibly on its diagonal) cannot be embedded in V_α^-.

Consequently, no denumerable relation R can admit simultaneously an embedding in V_α and in V_α^-. Indeed by an immediate consequence of RAMSEY theorem, there exists in R either an embedding of a denumerable relation always (+) or an embedding of the relation always (-) (without consideration of the diagonal).

Finally given an arbitrary denumerable binary relation R, for each countable ordinal α we have that V_α and V_α^- are both α-older than R; yet either $V_\alpha \not\geq R$ or $V_\alpha^- \not\geq R$. Hence our statement is true by taking either $X = V_\alpha$ or $X = V_\alpha^-$. •

10.5 A relation rich for its age

10.5.1 Definition and examples

We say that a denumerable relation R is **rich for its age** iff every countable relation younger than R is embeddable in R.

By 10.2.3, **it suffices that every denumerable representative of the same age as R be embeddable in R.** In other words, the notion of "rich for its age" is identical to "maximal (with respect to embeddability) among the denumerable representatives of its age" .

First example, the rich denumerable binary relation is rich for its age (see 10.3.1).

Second example, the chain **Q** of rationals, or any denumerable chain in which **Q** is embeddable, or equivalently any chain equimorphic with **Q**, is rich for its age.

Third example. Start with the consecutivity relation on positive and negative integers, i.e. the binary relation taking the value (+) for those couples (x, y) where $y = x + 1$. Take denumerably many such consecutivity relations with mutually disjoint bases, and call these the **components**. Then complete by giving value (-) for couples of elements belonging to two distinct components. Then every denumerable younger relation is formed of analogous components which are either finite, or the consecutivity on positive integers, or the consecutivity on negative integers, or isomorphic with preceding components: such a relation is embeddable in our "rich" consecutivity relation.

10.5.2 A denumerable poset which is rich for its age

There exists a denumerable poset which is rich for its age, and in which all finite posets, thus all countable posets are embeddable.

• Recall the amalgamation lemma 1.7.3. Start with the ordering reduced to a singleton. Take the three possible posets of cardinality 2, extending our singleton ordering. These are obtained by adding a new element at the end, or at the beginning, or making it incomparable with the unique element of our given singleton. By the amalgamation lemma, there exists a common extension A_2 for our singleton ordering A_1 and for the three preceding extensions of A_1, which is itself a poset.

In general, suppose that we have a poset A_i(i integer) in which every increasing sequence of partial orderings with successive cardinalities $1, ..., i$ is embeddable. Take all possible partially extensions of A_i, whose base is augmented by a new element. There are only finitely many such extensions, and by the amalgamation lemma there exists a common extension A_{i+1} which is a finite poset. Finally the common extension of all the A_i(i integer) to the union of the bases is a denumerable poset in which every countable poset is embeddable. •

10.5.3 A denumerable tree which is rich for its age

There exists a denumerable tree which is rich for its age and in which all finite trees, thus all countable trees are embeddable.

• Our base will be the set of all finite non-empty sequences $(r_1, ..., r_p)$(p positive integer) of rationals. We say that such a sequence is less than another such sequence $(s_1, ..., s_q)$ iff $q \geq p$ and $r_1 = s_1, ..., r_{p-1} = s_{p-1}$, and finally $r_p \leq s_p$. The reader easily shows that this comparison is reflexive, transitive, antisymmetric, and that any two sequences which are less than a third one, are comparable: thus we have a denumerable tree.

Let us show that every countable tree is embeddable in it. Let R be a tree with denumerable base E. We can assume that E is the set of integers. Take a subset E' of E which contains the element 0, and such that the restriction $I = R/E'$ is a maximal chain: each element of $E - E'$ is incomparable $(\bmod R)$ to at least one element of E'.

Construct a totally ordered extension I^+ of I, by adding countably many cuts to the base of I. More precisely, for each element $x \in E - E'$, add the cut, or initial interval of I, formed of those elements which are less than $x(\bmod R)$. Then embed I^+ in the chain of the rationals. Each element of E', and in particular the integer 0, is taken into a rational; or in other words, into a sequence formed of a unique rational. As for each element $x \in E - E'$, it is taken into a finite sequence of rationals, the first of whose terms is the rational r which is the image of the cut of I previously associated with x. Let U_r denote the equivalence class of the elements $x \in E - E'$ to which r is thus associated.

In each equivalence class U_r, take the least integer $u(r)$, then take a subset U_r' of U_r which contains the element $u(r)$, and for which the restriction $I_r = R/U_r'$ is a maximal totally ordered restriction of the poset R/U_r. As before, construct an extension I_r^+ of I_r, by adding countably many cuts to the elements of the base $|I_r|$.

More precisely, for each element x of $U_r - U_r$, add the cut formed of those elements in the base $|U_r'|$ which are less than $x(\bmod R)$. Then embed I_r^+ in the chain of the rationals. Each element in U_r', and in particular the integer $u(r)$ is taken into a sequence of two rationals, the first of which is r and the second is its rational image. As for each element $x \in (U_r - U_r')$, it is taken into a finite sequence of at least three rationals, the first of which is r and the second is the image of the cut of I_r, belonging to I_r^+, previously associated with x.

Iterate the preceding construction by partitioning the elements of each class $U_r - U_r'$ into subclasses $U_{r,s}$ each associated with the couple of rationals r, s. Note that every element of the base E is taken into a finite sequence of rationals, since we successively take the least integer in E then in each class U_r, then in each class $U_{r,s}$ etc. , and then we choose a maximal chain going through this least integer. Also note that the relation of being less than, or greater than, or incomparable $(\bmod R)$, is preserved when replacing each element of the base E of R by the finite sequence of rationals which is associated with it. •

10.5.4 An age without any denumerable rich representative

There exists an age not having any denumerable rich representative. Proved by SPECKER in 1957, published in ToR-86 p.297.

• Start with all finite binary relations, each having its base partitioned into equivalence classes; on each class, construct a consecutivity relation by totally ordering the elements, and then giving the value (+) for couples of consecutive elements and the value (-) otherwise; hence in particular for couples whose terms belong to distinct classes. Now on the diagonal, hence for the couples (x, x) where x belongs to the base, assign an arbitrary value (+) or (-). A relation thus constructed shall be called a **modulated consecutivity relation**; they obviously constitute an age: closed and directed under embeddability.

Consider the following denumerable representatives of this age. Start with the consecutivity relation on the integers, modulated on the diagonal, so that every finite sequence of (+) and (-) is realized by at least one sequence of consecutive integers. To see that there exist continuum many such representatives, it suffices to bijectively associate each representative with an infinite set of integers. For example, if we assign arbitrary values (+) an (-) to those couples (x, x) for which the integer x is a square, then there remain arbitrarily large finite intervals between two consecutive squares, in order to ensure the existence of every finite sequence of (+) and (-).

Suppose that there exists a denumerable relation R which is rich for this age. Then each of the preceding relations must be embeddable in R. Thus R must have components which are obtained by modulation from either the consecutivity relation of the integers, or the consecutivity relation of the negative integers, or the consecutivity relation on the positive and negative integers. Since R is denumerable, there are countably many components; and in each component, there are countably many modulated consecutivity relations constructed on the integers (up to isomorphism). Indeed, for any such embedding, the images of 1, 2, 3, .. are determined as soon as the image of 0 is chosen. It follows that there are continuum many modulated consecutivity relations constructed from the integers, which are not embeddable in R: contradiction. •

Problem 1. Let R be a relation with denumerable base E, which is rich for its age, and F be a finite subset of E. Then if $R/(E-F)$ represents the same age as R, is it equimorphic with R.

Problem 2 concerning faithful extension. If R is not rich for its age, then does there exist an extension strictly greater than R under embeddability, which represents the same age as R and not rich for its age.

10.6 Inexhaustible relation, inexhaustible age

A relation R with infinite base E is said to be **inexhaustible** iff, for every finite subset F of E, the relation R is embeddable in its restriction $R/(E-F)$, hence is equimorphic to it.

By 10.3.2 proposition (2), every denumerable rich relation is inexhaustible.

Every indivisible relation is inexhaustible (see definition in 6.8).

For example the chain ω, the chain $\mathbf{Q}$, the denumerable binary relation always (+) are inexhaustible. More generally every indecomposable chain (by addition) is inexhaustible; yet the chain $\mathbf{Z}$ of positive and negative integers is inexhaustible and decomposable into ω and its converse ω^-.

Another example: the consecutivity relation on non-negative integers; yet the consecutivity relation on positive and negative integers is obviously not inexhaustible.

Lemma. Let F be a finite set. Consider an ω-sequence of denumerable relations R_i(i integer), all of the same arity ≥ 2, each of whose base includes F; such that for any two distinct integers i, j the intersection of the bases $|R_i| \cap |R_j| = F$; and finally with the compatibility condition $R_i/F = R_j/F$.

Let H be an inexhaustible relation of the same arity, with $H \not\leq R_i$ for each i. Then there exists a common extension of the R_i in which H is not embeddable.

• First there exists a common extension of the restrictions $R_i/(|R_i| - F)$, in which H is not embeddable: use 8.1.2 proposition (2). Complete this common extension arbitrarily, adding F to its base and preserving each R_i. Then since H is inexhaustible, the non-embeddability condition subsists. •

10.6.1 Inexhaustible relation and α-morphism

Let H be a denumerable inexhaustible relation of arity ≥ 2.

For each countable ordinal α and each finite subset F of the base $|H|$, there exists a countable relation R in which H is non-embeddable, and for which the identity on F is an α-morphism from H into R (uses countable axiom of choice).

• For $\alpha = 0$, take $R = H/F$.

Suppose that α is a successor ordinal $\alpha = \beta + 1$ and that our proposition holds for β. Then to each finite subset F_i of the base $|H|$, including F, associate a relation R_i in which H is non-embeddable, and for which the identity on F_i is a β-isomorphism from H into R_i. Keep the elements of F, but eventually change those in $|R_i| - F$ so that for i, j distinct, the set differences $|R_i| - F$ and $|R_j| - F$ are always disjoint. By the preceding proposition, there exists a common extension R of the R_i, in which H is non-embeddable, and where the identity on F is an α-morphism from H into R (take one R_i for each F_i by countable axiom of choice).

Now suppose that α is a non-zero limit ordinal, and let β_i(i integer) be an increasing ω-sequence of ordinals with supremum α. For each integer i, let R_i be a relation in which H is non-embeddable, and such that the identity on F is a β_i-morphism from H into R_i. Suppose that the difference sets $|R_i| - F$ are mutually disjoint, and take a common extension R of the R_i, in which H is non-embeddable. Then the identity on F is an α-morphism from H into R. •

The non-embeddability of H in R means either that R is properly embeddable in H, or that R and H are incomparable (with respect to embeddability).

In certain cases, for instance if H is the chain $\mathbf{Q}$ of rationals, the preceding proposition can be strengthened to $R < H$. Indeed we have seen in 10.4.3 that ω^α, which is $< \mathbf{Q}$, is α-older than $\mathbf{Q}$. And more generally that, for every finite subset F of the base $|\mathbf{Q}|$, there exists an α-morphism from $\mathbf{Q}$ into $\omega^{\alpha+1}$ with domain F.

However, the preceding proposition cannot always be so strengthened. Indeed for H, take the denumerable inexhaustible binary relation always $(+)$. Then the condition $R < H$ would imply that R is finite and always $(+)$; and such a relation R is not even 1-older than H.

10.6.2 Inexhaustible age

We say that an age $\mathcal{R}$ is **inexhaustible** iff there exists a relation representative of $\mathcal{R}$ which is inexhaustible.

If an age $\mathcal{R}$ is inexhaustible, then $\mathcal{R}$ verifies the following equivalent conditions.

(1) For any two finite relations A, B belonging to $\mathcal{R}$ and having disjoint bases, there exists a common extension of A and B, which belongs to $\mathcal{R}$.

(2) For every finite relation A belonging to $\mathcal{R}$ and every positive integer p, there exists an element B of $\mathcal{R}$ such that A is embeddable in the restriction of B to its base minus p arbitrary elements.

(3) Let R be a representative of the age $\mathcal{R}$, and E its base; then for every finite subset F of E, the restriction $R/(E-F)$ is again a representative of $\mathcal{R}$.

(4) For R, S representatives of $\mathcal{R}$ with disjoint bases, there exists a common extension of R and S which represents $\mathcal{R}$ (uses the ultrafilter axiom; ZF suffices for R, S denumerable).

• Starting with (1), take $p+1$ copies isomorphic with A and having disjoint bases, and a common extension B belonging to the age: we obtain (2).

The condition (2) immediately implies (3), and (3) immediately implies (1).

Note that it suffices that there exists a representative of $\mathcal{R}$ which satisfies (3) in order that every representative satisfy (3).

Finally (1) implies (4) by the coherence lemma 2.4.1, equivalent with the ultrafilter axiom. Apply this lemma to the unions $F \cup G$ where F is a finite subset of the base $|R|$ and G is a finite subset of $|S|$. Obviously (4) implies (1). •

10.6.3 Inexhaustible extension theorem

Every age which satisfies the condition (1) in the previous subsection, is inexhaustible.

Hence **each of the preceding conditions** (1) to (4) **is equivalent to inexhaustibility**.

More strongly **if $\mathcal{R}$ satisfies (1), then for every denumerable relation R representing $\mathcal{R}$, there exists a denumerable extension of R which is inexhaustible and represents $\mathcal{R}$** ([197] POUZET 1979 p.343).

• Suppose that the condition (1) holds. Given a set of relations R_i with mutually disjoint bases E_i, each a representative of $\mathcal{R}$ or of an age included in $\mathcal{R}$, there exists a common extension of the R_i, with base the union of the E_i, which represents $\mathcal{R}$ or an age included in $\mathcal{R}$. To see this, use the coherence lemma, as in the previous subsection for proving (4) from (1).

Now take an ω-sequence of isomorphic copies R_i of R (a representative of the age $\mathcal{R}$) , with disjoint bases. By the preceding, there exists a common extension S of the R_i, which represents $\mathcal{R}$. Let a^i_j denote the jth element of $E_i(i,j$ integers), where it is understood that there exists an isomorphism from R_i onto $R_{i'}$, which takes a^i_j into $a^{i'}_j$ (i and i' fixed, j varying).

Using RAMSEY's theorem, given an integer h, we obtain an ω-sequence extracted from the sequence of integers; such that after renumbering, the bijection which, for each integer i, takes a^i_0 into $a^{i+1}_0, ..., a^i_h$ into a^{i+1}_h, is a local automorphism of S.

By passing to the limit which respects each R_i, we obtain a relation S, still representing $\mathcal{R}$, for which the preceding holds for each integer h. Then S is isomorphic with its restriction obtained by removing the set E_0 from its base. Hence also by removing any finite union of the E_i; thus S is inexhaustible. •

10.6.4 Maximum inexhaustible age

Every age including an inexhaustible age, also includes a maximum inexhaustible age ([197] POUZET 1979 p.326).

More precisely let $\mathcal{R}$ be an age. The maximum inexhaustible age included in R is the set (denumerable up to isomorphism) of finite relations A for which any arbitrary finite set of isomorphic copies of A with disjoint bases, has a common extension in $\mathcal{R}$.

• It suffices to see that this set is directed. Let R be a denumerable relation representing $\mathcal{R}$. For each of the A considered, there exist infinitely many restrictions of R which are isomorphic with A, having disjoint bases. Thus given A and an analogous finite relation B, there exists a common extension of A and B, taken with disjoint bases, which has infinitely many isomorphic copies which are restrictions of R. •

10.7 A relation minimal for its age

A denumerable relation R is said to be **minimal for its age** iff there exists no relation representing the same age and strictly less than R under embeddability.

For example the chain ω of integers, and its converse, are the only minimal relations for the age of chains.

There exists an age having no minimal denumerable representative ([197] POUZET 1979 p. 318).

• Start with the set of all finite sequences and ω-sequences of 0 and 1, which is equipotent with the continuum.

First take the unary relation $\emptyset$ with value (+) for the empty sequence, and the value (-) for every non-empty sequence. Then take the unary relation U with value (+) for each non-empty finite sequence whose last term is 0, and with value (-) otherwise.

Take the binary relation R with value (+) for each couple formed of a finite sequence and of a consecutive sequence, the latter obtained by adding 0 or 1 as a last term. Additionally R takes the value (+) for each couple formed of a finite sequence s and an ω-sequence beginning by the initial interval s. Let R take the value (-) otherwise.

We see that the following are representatives of the same age. Each restriction, in particular each denumerable restriction of the trirelation $(R, U, \emptyset)$ whose base contains all finite sequences, and which satisfies the **density condition**, which asserts that for each finite sequence s belonging to the base, there exists at least one element in the base which is an ω-sequence beginning with s.

Moreover any two such denumerable restrictions are isomorphic iff they are identical. Indeed for each ω-sequence consider its successive finite initial intervals: identity is the only possible isomorphism.

Finally, if one of these denumerable dense restrictions of $(R, U, \emptyset)$ were minimal for its age, then by taking a proper dense restriction, we obtain a multirelation of the same age, but strictly less under embeddability: contradiction. •

10.8 Finitist relation

Given an infinite set E and a finite subset F of E, a relation R with base E is said to be F-**finitist** iff for every pair of elements $a, b \in E - F$, the transposition (a, b) is an automorphism of R. A relation is said to be **finitist** iff there exists an F for which it is F-finitist. To say that R is F-finitist is equivalent to saying that every permutation of the base E which is the identity on F, is an automorphism of R. This follows from the automorphism lemma 9.1.2.

Consequently, to say that R is F-finitist is equivalent to saying that R is freely interpretable in the sequence of special unary relations that we call **singleton relations**: each of which taking the value (+) on one element of F (free interpretability is defined in 9.2).

For example, a unary relation is finitist iff it takes the value (+) either on a finite subset of the base or on the complement of a finite subsset.

10.8.1 Kernel of a finitist relation

If R is F-finitist and G-finitist (F, G finite subsets of the base), then R is $(F \cap G)$-finitist.

• Take any two elements $a, b \in E - (F \cap G)$, then an element $c \in E - (F \cap G)$, and replace the transposition (a, b) by the composition $(a, c) \circ (b, c) \circ (a, c)$. •

Consequently, if R is finitist, then there exists a minimum finite set F (with respect to inclusion) for which R is F-finitist. We call this F the **kernel** of R.

A relation R is constant (see 9.4) iff R is finitist with empty kernel.

If u belongs to the kernel of R, yet not v, then (u, v) modifies R.

If two finitist n-ary relations with the same base, have the same kernel and the same restriction to the kernel plus n elements, then they are identical. Indeed they have the same restriction to each n-element set.

10.8.2 Characterization of finitist relations

(1) **If R is a relation with infinite base E, then R is finitist with a kernel of at most r elements iff there exist at most r elements x in E, each of which has at least $r+1$ elements y such that the transposition (x, y) modify R.**

• If R is finitist with a kernel of cardinality less than or equal to r, then our condition is obvious.

Conversely, suppose that our condition holds. Since E is infinite, let u be an element which is distinct from the x. Associate to u those v such that (u, v) modifies R: there are at most r many such v. Let F be the set of such v. For any two elements x, y in $E - F$, the transpositions (u, x) and (u, y), hence also (x, y) preserves R. Thus R is finitist and its kernel is included in F. •

(2) **Let R be a finitist n-ary relation with a kernel of cardinality r, and let S be another n-ary relation. If every restriction of S with cardinality $\leq (n+1)(r+1)^2$ is embeddable in R, then S is finitist and its kernel has cardinality less than or equal to r.**

• Suppose that S does not satisfy our conclusion. Then by the preceding proposition, there exist at least $r+1$ elements u, each of which has at least $r+1$ elements v such that (u, v) modifies S. There are at most $(r+1)^2$ many such transpositions; and for each transposition, there exists a set H of $n+1$ elements, including u and v, such that (u, v) modifies the restriction S/H (see 9.1.3). This property is preserved when taking an isomorphism on a restriction of R. Thus by the preceding proposition, R is not finitist, or if so, has a kernel with cardinality $> r$. •

10.8.3 Set of isomorphic copies of a finitist denumerable relation

Let R be a relation with denumerable base E.

If R is finitist with a non-empty kernel, then there exist denumerably many isomorphic copies of R with base E. If R has empty kernel, hence is constant, then all isomorphic copies of R with base E are identical to R.

If R is not finitist, then there exist continuum many isomorphic copies with base E.

• Suppose that R is not finitist. Then for each finite subset F of E, there exists a pair of elements x, y in $E - F$ such that the transposition (x, y) modifies R.

Let n be the arity; then there exists a $(n+1)$-element set F_0 and a pair of elements x_0, y_0 in F_0, such that the transposition (x_0, y_0) modifies the restriction R/F_0: see 9.1.3.

Call $R_0 = R$ and $R_{\{0\}}$ = image of R under the transposition (x_0, y_0). There exists a finite superset F_1 of F_0 and a pair of elements x_1, y_1 in $F_1 - F_0$, such that the transposition (x_1, y_1) modifies both R/F_1 and $R_{\{0\}}/F_1$. Yet (x_1, y_1) preserves R/F_0 and $R_{\{0\}}/F_0$, since neither x_1 nor y_1 belong to F_0.

Call $R_{\{1\}}$ = image of R under (x_1, y_1) and $R_{\{0,1\}}$ = image of R under the composition of (x_0, y_0) and (x_1, y_1). Then these 4 images of R are distinct, and even they yield 4 distinct restrictions to F_1.

Iterating this, we obtain, for each integer i, a finite set F_i including F_{i-1} and a pair of elements x_i, y_i in $F_i - F_{i-1}$. Finally for each countable set I of integer indices, we obtain an isomorphic copy R_I of R under the composition of all transpositions $(x_i, y_i)(i$ belonging to $I)$. The R_I are distinct for different sets I: this gives continuum many copies of R. •

10.8.4 Necessary condition to be finitist

If a relation R is finitist, then there exists an integer p such that every relation with the same base and which is $(1,p)$-equivalent with R, is isomorphic with R. The $(1,p)$-equivalence is defined in 10.1.10. The reciprocal statement is proved below, in 10.9.8

Note that the minimum value of p can be strictly greater than the cardinality of the kernel. For example if R is constant, hence with empty kernel, the minimum value of p is the arity of R, by 9.4.1.

• Let n be the arity and r the cardinality of the kernel of R. Consider a relation S with the same base E as R and which is $(1,p)$-equivalent with R, where $p = (n+1)(r+1)^2$. By 10.8.2 proposition (2), the relation S is finitist and its kernel has cardinality less than or equal to r. Now by interchanging R and S, we see, that the kernel of S has cardinality exactly r.

There exists an isomorphism f which embeds into S the restriction of R to an arbitrary p-element subset including the kernel. For a certain choice of this p-element subset, f takes the kernel of R into the kernel of S: take each element x of the kernel of R, then for each x take $r+1$ elements y such that (x, y) modifies R, then for each such couple (x, y) take a restriction of R to $n+1$ elements, including x and y, which is modified by the transposition (x, y). Then these conditions hold as well for S by the isomorphism f, since p is sufficiently large. So that the images fx constitute the kernel of S.

Finally extend f to a permutation of E, still denoted by f. Since p is at least equal to $r+n$, our permutation f takes R into a relation having the same kernel as S and the same restrictions as S to its kernel plus n elements: by 10.8.1, this image of R under f is S itself. •

10.9 Almost chainable relation

Given a set E, a finite subset F and a chain A with base $E - F$, a relation R with base E is said to be (F, A)-**chainable** iff each bijection which is the union of the identity on a subset of F and a local automorphism of A, is a local automorphism of R.

We say that R is F-**chainable** or is **almost chainable**.

In the case where F is empty, we find again the chainable relation.

10.9.1 Connection with free-interpretability

A relation R is F-chainable iff R is freely interpretable in the multirelation formed of the singleton unary relations of each element in F, and of any totally ordered extension of A for which F is either an initial interval, or a final interval, or the union of both.

Consequently **if R is F-chainable, then every relation freely interpretable in R is F-chainable**.

R is F-chainable iff there exists a chain A with the same base as R, such that the birelation RA is F-chainable.

10.9.2 Restrictions of an almost chainable relation

(1) **If R is F-chainable, then for each subset D of the base, the restriction R/D is $(F \cap D)$-chainable**.

In particular, if R with base E is F-chainable, then the restriction $R/(E-F)$ is chainable.

The converse is false. Indeed take the chain $\mathbf{Z}$ of positive and negative integers, modified by the condition $\mathbf{Z}(0,0) = -$ (instead of $+$). This modified $\mathbf{Z}$ is not almost chainable, since its profile is not bounded. .

(2) **Let R be a relation and F a finite subset of its base. If for each finite subset X of the base which includes F, the restriction R/X is F-chainable, then R is F-chainable** (uses the ultrafilter axiom).

• To each finite subset X including F, associate the set U_X of chains Y with base $X - F$ such that R/X is (F, Y)-chainable. By hypothesis U_X is non-empty for each X. If X' is included in X, then every chain belonging to U_X, when restricted to X', yields a chain belonging to $U_{X'}$. By 2.4.1 (equivalent to the ultrafilter axiom), there exists a chain A with base $E - F$, for which R is (F, A)-chainable. •

(3) **Let R be an infinite relation which is F-chainable, and R' the restriction of R to an infinite subset including F, and G be a subset of F. Now if R' is G-chainable, then R is G-chainable** (uses the ultrafilter axiom).

• Let E denote the base of R and E' the base of R'. For each finite subset X of E including F, there exists a bijection which is F-identical and takes X into a subset X' of E' and R/X into R/X'. Now R/X', thus R/X is G-chainable. Letting X vary, the previous proposition shows that R is G-chainable. •

10.9.3 Kernel of an almost chainable relation

Let R be an infinite relation and F, G be two finite subsets of the base $|R|$. If R is F-chainable and G-chainable, then R is $(F \cap G)$-chainable (uses the ultrafilter axiom).

Indeed let E denote the base $|R|$. The restriction $R/(E - G) \cup F$ is $(F \cap G)$-chainable by the previous subsection, proposition (1). Then R is $(F \cap G)$-chainable by the preceding proposition (3).

Consequently, if R is almost chainable, then there exists a minimum finite set F (with respect to inclusion) for which R is F-chainable. We call this F the **kernel** of R.

10.9.4 Existence of a denumerable almost chainable restriction

Let R be a relation with denumerable base E, and F be a finite subset of E. Then **there exists a denumerable subset D of E, including F and such that R/D is F-chainable**. Generalization of 9.5.4 proposition (1).

• Take a chain A with base E, which begins by the initial interval F. To obtain the denumerable set D apply RAMSEY's theorem, by considering two finite subsets X, X' of the base as equivalent iff both include F and there exists an isomorphism from $(R, A)/X$ onto $(R, A)/X'$ whose restriction to F is the identity. Let $n \geq 2$ be the arity of R. Using 9.2.4 proposition (2) and choosing sets X with cardinality $(\operatorname{Card} F) + 2n$, we insure the F-chainability of R/D by A/D concatenated with unary singletons built from all elements of F. •

10.9.5 Another proof of the profile increase theorem

Going back to this theorem, already proved in 3.6.1, the following remark due to POUZET, published in [74] vol.1 (1971) p.113 exercise 8-2 (see also english translation vol.1 p.108), avoids the use of the incidence matrix and the multicolor theorem.

(1) Given a denumerable relation R and an integer p, take a finite subset F of the base $|R|$, such that every restriction of R with cardinality p is embeddable in R/F. Then using the preceding proposition, take an F-chainable restriction of R: **the profile of this restriction is the same as the profile of R**, at least **up to the value** p.

Hence it suffices to prove the following:

(2) **The profile increases in the particular case where R is almost chainable**.

• Assume that R is F-chainable. To each isomorphism type of a restriction to p elements, associate a p-element subset G with R/G having this type, and the intersection $H = F \cap G$ having the least possible cardinality. Then associate to this G the isomorphism type of cardinality $p + 1$, obtained by taking G plus an element not belonging to F. Thus we define an injective function which, to each isomorphism type of cardinality p, associates an isomorphism type of cardinality $p+1$.

To see the injectivity: if the same isomorphism type of cardinality $p+1$ is obtained from two equipotent subsets H, H' of F, each augmented by elements not belonging to F, then every isomorphism of the first restriction onto the second restriction, takes H into H'. For otherwise H' would not be the intersection of F with least possible cardinality. •

10.9.6 Lemma

Let R be an almost chainable relation. Then every R' which is younger than R is almost chainable; moreover the cardinality of the kernel of R' is less than or equal to the cardinality of the kernel of R (uses the ultrafilter axiom).

• Let E, E' denote respectively the bases of R, R'. Let F be the kernel of R. By 10.9.1, there exists a chain A with base E, such that, denoting U_a the singleton unary relation associated with any element a of E, then R is freely interpretable in $(A, U_a, ...)$. By 10.1.4 (ultrafilter axiom), there exists a multirelation formed of a binary relation A' and of unary relations U'_a such that the concatenation $(R', A', U'_a, ...)$ is younger than $(R, A, U_a, ...)$.

The relation A' is a chain, since it is younger than A. Each U'_a is either a unary relation always (-) or the singleton unary relation of an element a' of E'. Let F' denote the set of these a'; hence F' is a subset of E' and $\text{Card}\, F' \leq \text{Card}\, F$. Moreover we can assume that F is an initial interval of A, and so F' is an initial interval of A'.

Finally R' is freely interpretable in $(A', U'_a, ...)$ by 9.2.1. Hence by 10.9.1 our relation R' is almost chainable with a kernel included in F'. •

10.9.7 Characterization of an almost chainable relation by mean of its profile

A necessary and sufficient condition for a denumerable relation R to be almost chainable is that the profile of R be bounded (the profile is defined in 3.6; sufficiency uses the ultrafilter axiom).

• Let R be almost chainable, with base E and kernel F. For each integer p, the isomorphism type of R when restricted to p elements, only depends on the intersection of F with the set of these p elements. Since the kernel is finite, the number of these isomorphism types is bounded by the number of restrictions of R/F.

Conversely, let R be a relation with denumerable base E, which is not almost chainable. To prove that the profile is unbounded, let h be an arbitrary integer; we shall construct h restrictions of R, all of the same finite cardinality and mutually non-isomorphic.

Start with a denumerable, chainable restriction R_0 of R (see 10.9.4 with F empty). The relation R is not younger than R_0: see the previous lemma, using the ultrafilter axiom. Hence there exists a finite subset F_0 of E, such that R/F_0 is not embeddable in R_0. Take a denumerable, F_0-chainable restriction R_1 of R. For every integer $p \geq \text{Card}\, F_0$, no restriction of R_0 with cardinality p is isomorphic with any restriction of R_1 to a p-element subset including F_0.

The relation R is neither younger than R_0 nor than R_1. Hence there exists a finite subset F_1 of E, such that R/F_1 is neither embeddable in R_0 nor in R_1. Take a denumerable, F_1-chainable restriction R_2 of R. For every integer $p \geq \text{Max}(\text{Card}\, F_0, \text{Card}\, F_1)$, a restriction of R_0 with cardinality p, and a restriction of R_1 to a p-element subset including F_0, and a restriction of R_2 to a p-element subset including F_1, are mutually non-isomorphic.

Iterating h times, we obtain h mutually non-isomorphic restrictions. •

10.9.8 Characterization of finitist relations by mean of $(1, p)$-equivalence

A necessary and sufficient condition for a relation R to be finitist is that there exists an integer p such that:

every relation on the same base, which is $(1, p)$-equivalent with R, is isomorphic with R.

See [70] FRAÏSSÉ 1954 paragraphs 20 and 21.

• The necessity is proved in 10.8.4.

Conversely, let E be the base of R, suppose that every relation with base E which is $(1, p)$-equivalent with R, be isomorphic with R. Here we assume E to be denumerable (the proof extends to any infinite base). Take a representative of each isomorphism type of a restriction of R to $\leq p$ elements, and let F be the union of the bases of these representatives. Since F is finite, by 10.9.4 there exists a denumerable restriction of R which extends R/F and is F-chainable. By hypothesis, this restriction is isomorphic with R: hence we can suppose that R is almost chainable.

Assume that R is (F, A)-chainable, where A is a chain which can arbitrarily be either isomorphic with ω (chain of integers) or with $\mathbf{Q}$ (chain of rationals). From this point on, the argument in 9.10.2, slightly modified to take account of the kernel F (or included in F), proves that for every pair of elements x, y in $E - F$, the transposition (x, y) preserves R. Hence R is F-finitist. •

10.10 Back-and-forth notions: isomorphism, operator, elementary extension

10.10.1 (k, p)-isomorphism and (k, p)-equivalence

See [71] FRAÏSSÉ 1956. Let us generalize as follows the $(1, p)$-isomorphism (see 10.1.9). Given a non-negative integer p and two m-ary relations R, S, any local isomorphism f from R into S is said to be a $(0, p)$-isomorphism. Given p and $k \geq 1$, suppose we already defined the $(k-1, p')$-isomorphism for any $p' \leq p$. Let

F be the domain and G the range of f. Then f is said to be a (k,p)-**isomorphism from** R **into** S iff for any set $\overline{F} = F$ plus $q \leq p$ elements in the base $|R|$, there exists an extension of f to the domain $\overline{F}$ which is a $(k-1, p-q)$-isomorphism from R into S, and conversely by changing f into f^{-1} and exchanging R, F and S, G.

The converse of a (k,p)-isomorphism, the composition of two (k,p)-isomorphisms, are (k,p)-isomorphisms.

Any isomorphism from R onto R', or any restriction of such an isomorphism, is a (k,p)-isomorphism from R onto R' for all values k, p.

If $k' \geq k, p' \geq p$, then every (k',p')-isomorphism is a (k,p)-isomorphism.

The notation becomes useless if $k > p$: indeed every (p,p)-isomorphism is a (k,p)-isomorphism for every $k > p$.

We say that R and S are (k,p)-**equivalent** iff the empty function is a (k,p)-isomorphism from R into S. By the preceding this is an equivalence relation in the usual sense (reflexive, symmetric and transitive relation). Moreover for $k' \geq k$ and $p' \geq p$, two (k',p')-equivalent relations are again (k,p)-equivalent.

Examples. Any two infinite chains are $(1,p)$-equivalent for each integer p. The chain ω of integers and the chain $\mathbf{Q}$ of rationals are not (2,3)-equivalent: take two consecutive integers $a, b = a+1$: there would exist two rationals $a' < b'$ such that the transformation of a', b' into a, b would be an (1,1)-isomorphism from $\mathbf{Q}$ into ω. Then it suffices to take a rational $c'(a' < c' < b')$ to see that the supposed (1,1)-isomorphism does not exist.

Two relations are **elementary equivalent** in the classical sense (going back to TARSKI's truth-value) iff they are (k,p)-equivalent for every k and p. It is well-known that they are not necessarily isomorphic. For instance the chain ω of integers is elementary equivalent with the sum $\omega + \mathbf{Z}$ where $\mathbf{Z}$ is the chain of negative and positive integers.

All previous notions are immediately extendible to multirelations.

10.10.2 (k,p)-operator

Let us generalize the free operator (9.3) by saying that a (m,n)-**ary operator** $\mathcal{P}$ associates to each m-ary multirelation R an n-ary relation $\mathcal{P}(R)$ on the same base (for commodity we restrict ourselves to the case where $\mathcal{P}(R)$ is a relation). Note that, contrary to the case of free operators, in general, given a subset F of the base $|R|$, we have that $\mathcal{P}(R/F) \neq \mathcal{P}(R)/F$.

Take the additional condition that **given two** m**-ary multirelations** R, R', **every** (k,p)**-isomorphism from** R **into** R' **is a local isomorphism from** $\mathcal{P}(R)$ **into** $\mathcal{P}(R')$. Then $\mathcal{P}$ is said to be a (k,p)-**operator**. We see that each (k,p)-operator is again a (k',p')-operator for any $k' \geq k, p' \geq p$.

Given a (k,p)-operator $\mathcal{P}$ and given $k' \geq k, p' \geq p$, then every (k',p')-isomorphism from R into R' is a $(k'-k, p'-p)$-isomorphism of $\mathcal{P}(R)$ into $\mathcal{P}(R')$.

Each logical formula Φ (in first order calculus with equality) represents a (k,p)-operator. Indeed let us choose a base E and replace each predicate in Φ by a relation of same arity, which is based on E. These relations constitute a multirelation R on E. Now let us replace the n-sequence of free variables $x_1, ..., x_n$ in Φ by an n-sequence $a_1, ..., a_n$ in E: then according to the classical TARSKI's truth-value, the formula Φ takes value + or -. We interprete this by saying that the formula transforms the multirelation R into the n-ary relation $S = \mathcal{P}(R)$ with base E, taking values $S(a_1, ..., a_n) = +$ or (-) according to whether the formula $\Phi(R, a_1, ..., a_n) = +$ or - for the given multirelation and the given elements $a_1, ..., a_n$. Conversely each (k,p)-operator can be represented by a formula: p is the number of quantifiers and k is the number of alternations of $\forall$ and $\exists$, when the formula is written in its prenex form.

In particular each closed formula (where each variable is submitted to a quantifier) represents one about finitely many classes of (k,p)-equivalence concerning multirelations of a given arity.

Yet the notion of (k,p)-operator is more subtle than the notion of elementary formula, if we consider the case of the empty base. For example the m-ary relation $\emptyset^m$ with empty base constitutes a class of (1,1)-equivalence. Indeed, given a non-empty base E and the empty base 0, there never exists a (1,1)-isomorphism from E into 0, nor from a unary relation R based on E into $\emptyset^1$.

In fact we can reduce the problem to the necessity of two existential quantifiers (and two analogous universal quantifiers) in the particular case where we go from a positive initial arity to the arity zero. Take the formula $\exists_x \rho(x)$ whose truth-value is obvious when the predicate ρ is replaced by a unary relation with non-empty base. We think that it can be extended in two different manners to the case of the unary relation $\emptyset^1$ with empty base. The most usual generalization consists to say that, since there does not exist any element in the empty base, then the result is (0,-). We denote $(\exists^-)^0_1$ this quantifier, with the subscript 1 for the initial arity and the superscript 0 for the final arity. We precise it as follows: it takes value (+) iff there exists at least one element in the base for which the initial unary relation is (+).

We also define the quantifier $(\exists^+)^0_1$ which works as the previous quantifier in the general case of a non-empty base. Yet it transforms the unary relation $\emptyset^1$ into (0,+). We precise it as follows: it takes value (-) iff there exists at least one element in the base and each element gives value (-) to the initial unary relation.

10.10.3 Elementary extension

Given a multirelation R with base E and an extension R' of R to a superset E' of E, we say that R' is an **elementary extension** of R iff, for any finite subset F of E, the identity on F is a (k,p)-isomorphism of R into R' for every k, p. Only for an infinite base E there exist an elementary extension to $E' \neq E$.

The classical translation is that, given any finite sequence $a_1, ..., a_n$ in E, and any logical formula Φ with predicates ρ and free variables (sometimes called constants) replaced by $a_1, ..., a_n$, then the truth-value $\Phi(R, a_1, ..., a_n)$ is unchanged if we replace R by R'.

10.10.4 The empty base and its paradoxes

See [75] FRAÏSSÉ 1982. Firstly let us consider the simple notion of **deduction** between two relations on a same base. We say that S deduces from R if each finite sequence of elements in the common base E which gives value (+) to R again gives value (+) to S; that is written $R \vdash S$. However in the case the empty base $E = 0$, then R and S do not take value (+) nor (-) except if they reduce to (0,+) or (0,-).

Consequently, if for instance we want to preserve reflexivity in all cases, then we have to extend our notion of deduction by saying that $R \vdash S$ iff no sequence can simultaneously give value (+) to R and (-) to S. For example if $\emptyset^1$ denotes the unary relation with empty base, then not only $\emptyset^1 \vdash (0,+)$ but also $\emptyset^1 \vdash (0,-)$ and conversely. On another side it is clear that $(0,-) \vdash (0,+)$ but the converse is false. Finally **the deduction is not always transitive** since we have $(0,+) \vdash \emptyset^1 \vdash (0,-)$ and not $(0,+) \vdash (0,-)$. Yet deduction remains transitive among relatons of a same arity.

Deduction immediately extends to operators: indeed **an operator $\mathcal{P}$ deduces $\mathcal{Q}$** iff, assuming they have same predicates and consequently work on same multirelations, we have $\mathcal{P}(R) \vdash \mathcal{Q}(R)$ for every multirelation R.

The previous considerations do not destroy the transitivity of deduction in a logical theory. Indeed formulas in a given theory can have a constant arity (for instance closed formulas which are 0-ary).

Let us mention another zerological paradox: when taking into account the case of empty base, **there exist a logical formula which cannot take the classical prenex form**. Let us start with some very particular universal classes. Take the class reduced to (0,+). It satisfies VAUGHT's criterion with $p = 1$ (see 5.10.1). This class has two bounds, which are (0,-) and $(U,+)$ where U is a singleton. Indeed the only strict restriction of $(U,+)$ is (0,+).

Now to define this class by a logical formula, we firstly have to transform any 0-ary relation into a unary relation with same base: let us denote $(-)^1_0$ the free operator which changes every 0-ary relation $(E,+)$ or $(E,-)$ into, say the unary relation on E which always takes value (-), and which changes (0,+) as well as (0,-) into the unary relation $\emptyset^1$. Before this free operator we put the quantifier $(\forall^+)^0_1$ which transforms into $(E,-)$ any unary relation having a non-empty base E and taking at least once the value (-). Moreover it transforms the unary relation $\emptyset^1$ into (0,+). The trouble is that the obtained formula transforms (0,+) and (0,-) as well into (0,+). So we have to eliminate the parasite solution (0,-), for instance by taking the conjunction of the previous formula with the free formula $(\rho)^0_0$ which transforms each 0-ary relation into itself. Finally (0,-) is transformed into itself, thus is eliminated. Our formula takes value (+) only for (0,+). Clearly it cannot be written under a prenex form, i.e. a unique free operator preceded by a quantifier.

10.11 Exercise

10.11.1 Finitist structure and Fraenkel-Mostowski model with negation of the axiom of choice

(1) Start with the set of non-negative integers, which we consider as urelements, i.e. as copies of the empty set, each having no element. On the set $\mathbf{N}$ of these integers, consider the set of finitist relations, in the sense of 10.8.

More generally, to each ordinal α, associate the set of **finitist structures** with **rank** α on $\mathbf{N}$, defined as follows by induction. Each element of $\mathbf{N}$, or **urelement**, is a structure with rank 0. Given a finite subset F of $\mathbf{N}$, a F-**finitist structure** A with rank $\alpha \geq 1$ is a set of finitist structures of ranks strictly less than α, with the condition that every permutation of $\mathbf{N} - F$ preserves A.

These finitist structures constitute a model of FRAENKEL-MOSTOWSKI's set theory: see [72] FRAÏSSÉ 1958. More precisely this model satisfies the axioms of ZF excepting extensionality, which must be weakened as follows: any two non-empty sets which have the same elements are identical.

One can easily transform the model so as to satisfy the full extensionality axiom, but while abandonning the foundation axiom. Indeed it suffices to consider each urelement as being equal to its own singleton. In this case, we add the empty set to the elements of $\mathbf{N}$, and we give to each of them the rank 0, the rank 1 being given only to those sets which contain several elements of $\mathbf{N}$ or one element of $\mathbf{N}$ plus the empty set.

(2) Show that **the axiom of choice is falsified in our model**. Define pairs, then couples of urelements as usually. Call **bi-couple** the pair $\{(x, y), (y, x)\}$ where x, y are two urelements. Note that each pair has rank 1, each couple has rank 2, each bi-couple has rank 3 and the set of all bi-couples has rank 4 and is F-finitist with F empty. Now if the axiom of choice was true, there would exist a choice set picking one and only one couple in each bi-couple. This would associate to each pair of urelements x, y one and only one of the couples (x, y) or (y, x). Obviously such a choice set is not finitist, hence it does not belong to our model.

(3) Construct the set of **words**, or finite sequences **without repetition** on the set $\mathbf{N}$ of the urelements; note that this set is F-finitist with F empty. Consider the function f which to each non-empty word u without repetition associates the word obtained from u by removing its last term. This function is a F-finitist structure, with F empty. $\operatorname{Dom} f$ is the set of non-empty words, $\operatorname{Rng} f$ is the set of all words (without repetition), including the empty sequence.

Then in the considered model $\operatorname{Dom} f$ **is strictly subpotent with** $\operatorname{Rng} f$ (example communicated by HODGES).

• A bijection from $\operatorname{Rng} f$ onto $\operatorname{Dom} f$ cannot be finitist. Indeed for such a bijection h, letting 0 denote the empty sequence, consider the ω-sequence of successive iterates $0, h(0), h^2(0) = h(h(0)), \ldots$. If this ω-sequence were F-finitist, with a finite subset F of $\mathbf{N}$, then all its terms would be non-repetitive sequences of elements of F. Since h is a bijection, these terms are all distinct, so that F should be infinite: contradiction. •

Problem. Is it possible to generalize to structures the automorphism lemma 9.1.2 and the lemma of altered restriction 9.1.3, so that we could define a finitist structure by using only transpositions instead of general permutations of $\mathbf{N} - F$. Consequently the intersection of two finite subsets F should be an F, so that each finitist structure should have a minimum F called its kernel.

Chapter 11

Relative isomorphism, saturated relation, existence criterion, solid or fragil family, interval-closure

11.1 Relative restriction, relative isomorphism, relage

11.1.1 A-relation, A-restriction, A-isomorphism and embeddability

Let A be a relation with finite base. An **A-relation** is a relation extension of A. An **A-restriction** of an A-relation R is a restriction of R to a superset of the base $|A|$. If A is not mentioned, then we shall say **relative restriction**.

We shall call an **A-isomorphism**, or more exactly an $|A|$-isomorphism from an A-relation R onto another A-relation S, any isomorphism from R onto S whose restriction to the base of A is the identity. Two extensions of A are said to be **A-isomorphic** iff such an isomorphism exists.

We say that an extension R of A is **A-embeddable** in another extension S of A iff there exists an A-isomorphism from R onto a restriction of S, again called an **A-embedding** of R into S . Here again we shall speak of **relative isomorphism** and of **relative embeddability**.

11.1.2 A-age or rel-age, A-older or younger relation

Given an extension R of A, the A-**age** of R will be the set of all finite A-restrictions of R, considered up to A-isomorphism. For the reader desiring a rigorous definition, the A-age of R is the set of those A-relations whose bases are finite sets including $|A|$ and included in the union $|A| \cup \omega$, where ω is the set of integers. It may be convenient to assume that the base $|A|$ is disjoint from ω.

We say that R **represents**, or is a **representative** of the considered A-age. If A is not mentioned, we shall speak of **rel-age**.

Each A-age is a set of finite A-relations, which is closed and directed under A-embeddability.

Conversely given a set $\mathcal{A}$ of finite A-relations, which is closed and directed under A-embeddability, there exists a countable A-relation which is a representative of $\mathcal{A}$. Same proof as in 10.2.2.

We say that an A-relation S is A-**older** than another A-relation R, or that R is A-**younger** than S, iff every finite A-restriction of R is A-embeddable in S. Or equivalently, the A-age represented by R is included in the A-age represented by S.

Given three A-relations R, R' and T where R and R' are both A-younger than T, there exists an A-isomorphic copy R" of R' and a common extension of R and R", which is A-younger than T (uses the ultrafilter axiom; ZF suffices if R and R' are countable; same proof as in 10.1.8).

Consequently, given an A-relation R and an A-age $\mathcal{T}$ including the A-age of R there exists an extension of R with the same cardinality, which is a representative of $\mathcal{T}$. Same conditions and same proof as in 10.2.3.

11.1.3 Induced rel-age, specification of a rel-age

Let A be a finite relation and B a finite extension of A. Starting with a B-age $\mathcal{B}$, the A-restrictions of the elements of $\mathcal{B}$, considered up to A-isomorphism, constitute an A-age which is said to be **induced** by $\mathcal{B}$.

In particular if A is empty, then the A-age reduces to the already known notion of an age: we shall speak of the **age induced** by a given B-age.

The notion of induced rel-age is reflexive, transitive and antisymmetric, and thus defines a partial ordering among rel-ages.

For every A-age $\mathcal{A}$ and every relation B belonging to $\mathcal{A}$, there exists at least one B-age which induces $\mathcal{A}$. We say that such a B-age **specifies** $\mathcal{A}$ or is a **specification** of $\mathcal{A}$.

For $\mathcal{A}$ and B given, in general there exist many B-ages specifications of $\mathcal{A}$.

• Start with the age of finite chains, so that A is empty, and let B be the chain on two elements $u < v$. There exists a specification, all of whose elements are chains beginning with u, v. There exists another specification for which we have at most p elements between u and v (p given integer). Finally there exists another specification with as many elements as one might wish between u and v; or with as many elements as one wish before u and between u and v and after v, etc. •

Consider two finite relations A and B extension of A, with an A-age $\mathcal{A}$ and a B-age $\mathcal{B}$. Then to say that $\mathcal{B}$ is a specification of $\mathcal{A}$ is equivalent to saying that every representative of $\mathcal{B}$ is a representative of $\mathcal{A}$.

If $\mathcal{B}$ is a specification of $\mathcal{A}$, then a representative of $\mathcal{A}$ which is an extension of B is not, in general, a representative of $\mathcal{B}$: consider again the case of chains with A empty and B defined on two elements.

It may happen that R be a representative of $\mathcal{A}$ and that no representative of $\mathcal{B}$ be isomorphic with R.

• Take for R the consecutivity relation on the positive and negative integers. Take A empty; let S be a consecutivity relation formed of two components each isomorphic with R. And for B, take the restriction of S to two elements, each in one of the components. Then every B-relation representing the B-age of S, has at least two components, hence cannot be isomorphic with R. •

11.1.4 Construction of a specification

Let A be a finite relation and B be a finite extension of A. Consider two A-ages $\mathcal{A}$ and $\mathcal{A}'$ including $\mathcal{A}$. Then for every B-age specification $\mathcal{B}$ of $\mathcal{A}$, there exists a B-age $\mathcal{B}'$ including $\mathcal{B}$ and which is a specification of $\mathcal{A}'$.

• Take a countable representative R of $\mathcal{A}'$ and a countable representative S of $\mathcal{B}$. This relation S is still a representative of $\mathcal{A}$. There exists a common extension T of R and S, up to A-isomorphism, which is a representative of $\mathcal{A}'$: see 11.1.2. This relation T is a B-relation, and is a representative of a certain B-age $\mathcal{B}'$ including $\mathcal{B}$. •

11.2 Maximal rel-age, maximalist relation

Given an age $\mathcal{R}$ and a finite relation A belonging to $\mathcal{R}$, an A-age specification of $\mathcal{R}$ is said to be **maximal** modulo $\mathcal{R}$ iff no other A-age specification of $\mathcal{R}$ strictly includes it.

Example. Take the age of all finite chains, and let A be the chain on two elements $u < v$. Then the only maximal A-age is obtained by authorizing an arbitrary finite number of elements before u, after v, and between u and v.

Another example. Start from the age represented by the consecutivity rrelation on the integers; and let A be the consecutivity relation on a fixed finite chain. Then there exists only one maximal A-age specification of the given age: namely the A-age of all the consecutivity relations of finite chains including A as an interval, and their A-restrictions.

Now taking the preceding age and taking for A the relation on two elements u, v with $A(u, v) = A(v, u) = -$, we have a maximal A-age by putting a unique element which is the successor of u and the predecessor of v, and as many elements as one wishes before u and after v.

Given a positive integer p, we have another maximal A-age, by putting exactly p elements between u and v, and as many as one wishes before u and after v.

Given an age $\mathcal{R}$ and a finite relation A belonging to $\mathcal{R}$, for each A-age specification of $\mathcal{R}$ there exists at least one maximal A-age which includes it (and is still a specification of $\mathcal{R}$).

We could apply the maximal ideal axiom. However this axiom, equivalent to the axiom of choice, is unnecessary: ZF suffices, since the considered elements are finite relations which can be based on the set of integers.

11.2.1 Extension and maximal rel-age

Given two finite relations A and B extension of A, and a maximal B-age $\mathcal{B}$, then the A-age induced by $\mathcal{B}$ is maximal.

• Call $\mathcal{A}$ the induced A-age, and suppose that $\mathcal{B}$ is maximal yet $\mathcal{A}$ is not maximal. Let C be an A-relation which does not belong to $\mathcal{A}$, and let $\mathcal{A}'$ be an A-age including $\mathcal{A}$ and containing the element C. Let $\mathcal{B}'$ be a B-age specification of $\mathcal{A}'$ and including $\mathcal{B}$: see 11.1.4. Since $\mathcal{B}$ is maximal, then $\mathcal{B}'$ is identical to $\mathcal{B}$ and hence C is an A-restriction of an element of $\mathcal{B}$; hence C is an element of $\mathcal{A}$: contradiction. •

On the other hand, **a specification of a maximal rel-age is not necessarily maximal.**

• Take two elements a, b and let A be the chain of the singleton of a, and B the chain $a < b$. The B-age formed of the finite chains, extensions of B in which a and b remain consecutive, is a specification of the A-age of all finite chains containing a. The latter is obviously maximal; yet the B-age under consideration is not maximal, since one can add elements between a and b. •

However, **given two finite relations A and B extension of A, then every maximal A-age which contains B has at least one maximal B-age as a specification.** This follows from the previous subsection and from 11.1.3.

11.2.2 Incompatibility lemma for rel-ages

Let $\mathcal{R}$ be an age and A a finite relation belonging to $\mathcal{R}$. Given a maximal A-age $\mathcal{A}$ and a finite A-relation B which does not belong to $\mathcal{A}$, there exists a relation B' belonging to $\mathcal{A}$, such that no A-age specification of $\mathcal{R}$ contains both B and B'.

• No A-age, specification of $\mathcal{R}$, contains both B and all the elements of $\mathcal{A}$. Suppose that our conclusion is false. Take a sequence of relations B'_i(i integer) belonging to $\mathcal{A}$, in which every element of $\mathcal{A}$ be embeddable. For each i, let C_i be a common extension of B and B'_i, which belongs to the age $\mathcal{R}$, and is obviously an extension of A. We can always suppose that B'_{i+1} is an extension of B'_i for each i. Moreover, for i fixed, we can suppose that for all $j > i$ we have the same restriction $D_i = C_j/(|B| \cup |B'_i|)$, by taking a suitable infinite sequence extracted from the j. Then the D_i constitute an A-age, specification of $\mathcal{R}$, which contains B and includes $\mathcal{A}$: contradiction. •

11.2.3 Maximalist subset

Given a relation R, a finite subset F of its base is said to be **maximalist** modulo R iff the (R/F)-age represented by R is maximal.

Equivalently, for each relation S of the same age as R, every 1-morphism from R into S with domain F is a 1-isomorphism (see 10.1.9).

Equivalently, for each extension S of R with the same age, and for every restriction B of S to a finite superset of F, there exists an isomorphism from B onto a restriction of R, which is the identity on F (see 11.1.2, generalizing 10.1.8).

Example. Consider the chain $\mathbf{Q} + \omega$ where $\mathbf{Q}$ is the chain of rationals. Then any finite subset of $\mathbf{Q}$ is maximalist, modulo the age of all finite chains. This is not the case for the pair of integers 0,1 in the final interval ω.

Every subset of a maximalist finite set is maximalist. This follows from 11.1.2.

11.2.4 Maximalist relation

A relation R is said to be **maximalist** iff every finite subset of its base is maximalist (mod R).

For example, the chain $\mathbf{Q}$ of the rationals is maximalist.

As a second example, start with the consecutivity relation on positive and negative integers, called a **component**; then take the consecutivity relation with p components (p integer) or with countably many components, already given as being rich for its age in 10.5.1, third example. For each positive integer p, the consecutivity relation with p components is maximalist. Similarly the relation formed of countably many components. All are of the same age.

A third example. Start with the consecutivity relation on the non-negative integers, say C. To each unary relation U on these integers, associate the relation C_U which differs from C on the diagonal, with $C_U(x, x) = U(x)$ for each integer x. Take a denumerable set of C_U with disjoint bases, which are mutually linked by the value (-), and which satisfy the following densisy condition. For each finite sequence s of values (+) and (-), take a component C_U with U beginning by s. We now define the birelation formed of the preceding binary relation, and a unary relation 0 which takes the value (+) for the minimum of each component, and (-) otherwise.

Then our denumerable birelations are all of the same age. There are continuum many such. Moreover **they are all maximalist**.

• Every 1-morphism from such a birelation R into another, say R', associates to each element of a component C_U of R, an element of a component C'_U of R', isomorphic with C_U, thus corresponding to the same unary relation U. The corresponding elements have a same rank in C_U and in C'_U; thus this is a 1-isomorphism. •

11.2.5 Maximalist extension theorem

For every denumerable relation R, there exists an extension of R which is denumerable, maximalist and of the same age.

More strongly, modulo the axiom of choice: **for every infinite relation R, there exists a maximalist extension of R with the same cardinality and the same age.**

• Let F be a finite subset of the base $|R|$. it suffices to prove that, by adding countably many elements, then we can obtain an extension S of R such that the (R/F)-age represented by S is maximal: the rest of the proof will follow by iteration.

Let $\mathcal{A}$ designate the (R/F)-age represented by R. Take a maximal (R/F)-age, say $\mathcal{B}$ including $\mathcal{A}$: see 11.2. Then take S to be an (R/F)-extension of R which is a representative of $\mathcal{B}$: see 11.1.2. •

• In the case of an uncountable base, the preceding proof is modified on two points. First by an ordinal indexing of all finite subsets (using the axiom of choice). Secondly by the construction of a common extension which preserves the age, or the rel-age (using the coherence lemma). •

11.2.6 A characterization of maximalist relations

A necessary and sufficient condition for a relation R to be maximalist, is that every extension of R with the same age be a 1-extension (the sufficiency uses the axiom of choice; ZF suffices if R is denumerable).

This proposition identifies the maximalist relations as those relations which are **existentially closed**: a notion used in the model-completeness criterion for logical theories: see [215] ROBINSON 1963.

• Let R be a maximalist relation, R' an extension of R with the same age, and F a finite subset of the base $|R|$. Since the (R/F)-age represented by R is maximal, it is identical to that represented by R'. Thus the identity on F is a 1-isomorphism from R into R'. In other words R' is a 1-extension of R.

Conversely, suppose that R is not maximalist. Thus there exists a finite subset F which is not maximalist $(\bmod R)$. By the preceding proposition (axiom of choice), there exists an extension R' of R having the same age and which is maximalist. Thus the identity on F is not a 1-morphism from R' into R, hence R' is not a 1-extension of R. •

11.3 Saturated subset, saturated relation

11.3.1 Saturated subset

Let R be a relation and F a finite subset of the base. We say that F is **saturated** modulo R iff for any relation S representing the same age as R, and any 1-morphism f from R into S with domain F, and any finite set G satisfying

$\overline{f}(F) \subseteq G \subset |S|$, there exists a 1-morphism from S into R with domain G, which extends f^{-1} (the 1-morphism is defined in 10.1.9).

For example, given the chain $\mathbf{Q}$ of the rationals and the chain ω of the integers, take $R = \mathbf{Q} + \omega$: then every finite subset of the initial interval $\mathbf{Q}$ is saturated $(\bmod R)$, but no subset of the final interval ω is saturated, except the empty set and the singletons. This example has already been given in 11.2.3 about maximalists subsets.

Another example. For the ordinal ω^2, the empty set is saturated.

• For any chain A and any finite subset $G = \{a_1, ..., a_p\}$ of its base, with $a_1 < ... < a_p (\bmod A)$; for a function g from A into ω^2 to be a 1-morphism, it suffices to put $g(a_1)$ into the second component ω of ω^2, and in general $g(a_i)(i = 1, ..., p)$ into the $(i+1)$st component, leaving infinitely many elements in each of the intervals defined by the images $g(a_i)$. •

11.3.2 Every saturated finite subset is maximalist

The maximalist subset is defined in 11.2.3.

In other words, **if F is saturated $(\bmod R)$, then the (R/F)-age represented by R is maximal**. Equivalently, for every S representing the same age as R, every 1-morphism from R into S with domain F, is also a 1-isomorphism.

The converse is false.

• Take R to be the consecutivity relation on the positive and negative integers, and take F to be a finite interval of this consecutivity relation. Then the (R/F)-age represented by R is maximal. However, take S to be the relation obtained from two components, each isomorphic with R, with the value (-) for couples whose elements belong to distinct components. Then a 1-morphism f from R into S with domain F is also a 1-isomorphism. But by taking G to be a finite superset of $\overline{f}(F)$ with elements in both of the two components, we see that no local isomorphism from S into R with domain G and which extends f^{-1}, is a 1-morphism. •

11.3.3 An extension with the same age

Let R be a relation, F be a saturated finite subset of the base; let G be a subset of F and g be a 1-morphism from R into a relation S representing the same age, with $G = \operatorname{Dom} g$. Then there exists an extension S^* of S of the same age, and a 1-morphism f from R into S^*, extending g to the domain F (this proposition and its following consequence are communicated by HODGES; uses the ultrafilter axiom).

• Let $H = \operatorname{Rng} g$, hence g takes R/G into S/H. Since g is a 1-morphism, the (S/H)-age represented by S is included in the image under g of the (R/G)-age represented by R. By 11.1.2 (ultrafilter axiom), there exists an extension S^* of S such that the (S/H)-age represented by S^* is exactly the image under g of the (R/G)-age represented by R.

We can even require that there exists an isomorphism from R onto a certain restriction of S^*, which on G is identical to g. So that this isomorphism, when restricted to F, is a 1-isomorphism from R into S^*. •

Consequently **every subset of a saturated finite subset is saturated** (uses ultrafilter axiom).

• Take up the preceding notations, where F is a saturated finite subset of the base $|R|$, and G is a subset of F. Let g be a 1-morphism from R into S with $G = \mathrm{Dom}\, g$. Using the preceding extension S^* of S and the 1-morphism f, we see that, for every finite superset K of $\overline{f}(F)$, included in the base $|S|$, there exists a 1-morphism from S^* into R with domain K, hence a fortiori a 1-morphism from S into R, which extends f^{-1}, hence extends g^{-1}. Thus G is saturated. •

11.3.4 Saturated relation

A relation R is said to be **saturated** iff every finite subset of its base is saturated. By the preceding, it suffices that each finite subset of the base be included in a finite saturated subset.

For example, the chain of rationals is saturated.

The rich relation defined in 10.3.1, is saturated.

The consecutivity relation with denumerably many components constructed from the chain of positive and negative integers.

The poset in 10.5.2 and the tree in 10.5.3, which are rich for their ages, are saturated.

Every saturated relation is maximalist by 11.3.2.

A non-saturated maximalist relation is obtained from 11.1.4 with the consecutivity relation on $\mathbf{Z}$ (positive and negative integers); only the analogous relation with denumerably many such components, is saturated.

In the terminology of Abraham ROBINSON, a saturated relation is called an existentially universal model of a universal theory. See for example [115] HIRSCHFELD, WHEELER 1975 p.31; or [233] SIMMONS 1976 p.384.

11.3.5 Saturation and isomorphy

Any two denumerable saturated relations of the same age are isomorphic.

More precisely, **let R, R' be denumerable saturated relations of the same age. Then every 1-morphism from R into R' with finite domain, is extendible to an isomorphism from R onto R'.**

• Let f be a 1-morphism from R into R', with finite domain F and range $F' = \overline{f}(F)$. For every finite superset G' of F' which is included in the base $|R'|$, the inverse function f^{-1} is extendible to a 1-morphism from R' into R with domain G'. By iterating this alternatively from R into R' and back, we obtain an ω-sequence of local isomorphisms, one extending the other. Taking care to include each element of the bases $|R|$ and $|R'|$ in the domains of local isomorphisms, we end up with an isomorphism from R onto R'. •

11.3.6 Saturation and richness

Every denumerable saturated relation is rich for its age.

• Let R be a denumerable saturated relation, and S be a denumerable relation of the same age. Thus the empty function is a 1-isomorphism from R into S. Since R is saturated, for any finite subset G of the base $|S|$, there exists a 1-morphism g from S into R with domain G.

Let $F = \overline{g}(G)$ and $f = g^{-1}$. Since the relation R is maximalist (see 11.2.4), the (R/F)-age $\mathcal{A}$ represented by R is maximal. Since g is a 1-morphism, the (S/G)-age represented by S is included in the rel-age image of $\mathcal{A}$ under f. By 11.2.1, there exists an extension T of S which is denumerable, which represents the same age as R and S, and which represents the rel-age image of $\mathcal{A}$ under f. In other words, f is a 1-isomorphism from R into T.

Iterating this, and since R is saturated, we have that given an arbitrary finite superset G_1 of G included in the base $|S|$, we obtain a 1-morphism g_1 from T into R, hence from S into R, which is an extension of g to the domain G_1. Then we obtain an extension T_1 of S which is denumerable and of the same age, and for which $f_1 = (g_1)^{-1}$ is a 1-isomorphism from R into T_1.

By including every element of the base $|S|$ in the domains of the successive g_i, we have that the union of all $g_i(i$ integer) is an isomorphism from S onto a restriction of R. •

Remark. With the third example of 11.2.4, we have an age without any rich representative. To see this, take up SPECKER's argument in 10.5.4. By the preceding proposition, every relation which is not rich for its age, is not saturated. So that with our third example of 11.2.4, we obtain an age without any saturated representative, but with continuum many maximalist representatives; remark due to HODGES in 1979, published in ToR-86 p.301.

11.4 Existence criterion for a rich relation (Pouzet, Vaught)

Preliminary lemma. Let R be a relation and F a finite maximalist subset of the base. Let $\mathcal{A}$ denote the (R/F)-age represented by R. **Then F is saturated iff, for any finite subset G of the base, which includes F, and for any (R/G)-age $\mathcal{B}$ which is a specification of $\mathcal{A}$, there exists a subset G' of the base, which is the image of G under an F-isomorphism from R/G onto R/G', such that the (R/G')-age represented by R includes the image of $\mathcal{B}$.**

This is simply the translation, in terms of rel-age, of the definition of a saturated subset of the base $|R|$: see 11.3.1. We shall use it in the proof of the following criterion.

Theorem. Given an age $\mathcal{R}$, there exists a denumerable rich relation for the age $\mathcal{R}$ iff for each finite relation A belonging to $\mathcal{R}$, there are countably many maximal A-ages specifications of $\mathcal{R}$.

Moreover in this case, for each denumerable relation R representing $\mathcal{R}$, there exists a denumerable saturated extension of R which represents $\mathcal{R}$ ([191] POUZET 1972, generalizing [249] VAUGHT 1961).

More strongly, **for every infinite representative R of $\mathcal{R}$, there exists a saturated extension of R, of the same cardinality, which represents $\mathcal{R}$** (uses axiom of choice if R is uncountable).

• Suppose that there exists a denumerable rich relation R representing the age $\mathcal{R}$. Then for each finite restriction A of R and each maximal A-age which is a specification $\mathcal{A}$ of $\mathcal{R}$, there exists an A-relation U representing $\mathcal{A}$ (see 11.2). Since R is rich, there exists a restriction isomorphic with U, hence a finite restriction A' of R which is isomorphic with A, such that the A'-age represented by R includes $\mathcal{A}$, hence is an isomorphic image of $\mathcal{A}$. Finally each maximal rel-age which is a specification of $\mathcal{R}$ is definable, up to isomorphism, by a finite restriction of R. Thus there are countably many such rel-ages.

Conversely, suppose that for each element A of the age $\mathcal{R}$, there are countably many maximal A-ages which are specifications of $\mathcal{R}$. Start with a denumerable representative R of $\mathcal{R}$, which we can always assume to be maximalist, by the maximalist extension theorem 11.2.5. Let F and G including F, be two arbitrary finite subsets of the base $|R|$. Take G into G' via an F-identical isomorphism h making $G'-F$ disjoint from the base $|R|$. Let $\mathcal{A}$ denote the (R/F)-age represented by R (it is a maximal rel-age), and let $\mathcal{B}$ be an arbitrary maximal $(\overline{h}(R/G))$-age which specifies $\mathcal{A}$. Then let T be a denumerable representative of $\mathcal{B}$, hence a representative of the (R/F)-age $\mathcal{A}$.

By 11.1.2 there exists an A-isomorphic copy T' of T and a common extension R_1 of R and T', such that if we denote by G_1 the image of G' when passing from T to T', then the (R_1/G_1)-age represented by R_1 includes the image of $\mathcal{B}$. Furthermore we can choose R_1 to be maximalist.

Pass from R to R_1, then iterate this, by using all possible quadruples $(F, G, \mathcal{A}, \mathcal{B})$. In view of the countability of the set of maximal rel-ages, in the limit, we obtain a denumerable extension of R, each of whose finite subsets is saturated. •

In the case where R is uncountable, the preceding proof requires the ultrafilter axiom to take a common extension by 11.1.2, and more strongly the axiom of choice to well-order the base, and then to well-order the set of finite subsets.

Corollary. Let R be a relation. If there exists a saturated finite subset of the base, then the age of R satisfies the preceding criterion, and there exists a saturated relation of the same age as R (uses the ultrafilter axiom).

• The empty set is saturated by 11.3.3 (ultrafilter axiom). Let F be empty in the preliminary lemma. Then for any finite subset G of the base, and any maximal (R/G)-age $\mathcal{B}$ which specifies the age of R, there exists a subset G' which is the image of G under a local automorphism of R, such that the (R/G')-age represented by R is isomorphic with $\mathcal{B}$. It follows that there are countably many maximal rel-ages arising from an arbitrary finite restriction of R: the criterion is satisfied. •

11.4.1 A classification of ages

Lemma. Given an age $\mathcal{R}$, if there exists no denumerable rich relation for $\mathcal{R}$, then there exists a strictly increasing ω_1-sequence (under embeddability) of denumerable relations representing $\mathcal{R}$, such that these relations cannot be all embedded in any single denumerable relation representing $\mathcal{R}$ (uses the axiom of choice and the continuum hypothesis).

• First take an ω_1-sequence of all relations $R_i(i$ countable ordinal) representing $\mathcal{R}$ and based on the integers. Let $S_0 = R_0$; let S_1 be a common extension of S_0 and R_1; then S_2 be a common extension of S_1 and R_2, etc.; where the S are denumerable and represent $\mathcal{R}$. For each countable limit ordinal α, let S_α be a denumerable common extension of the $S_i(i < \alpha)$. The $S_i(i$ countable ordinal) can be chosen to be strictly increasing, since there does not exist any rich relation. •

Classification. Suppose that there exists a denumerable rich relation with age $\mathcal{R}$; then (under the continuum hypothesis):

(1) **Either all denumerable relations which represent $\mathcal{R}$ are equimorphic.**

This is the case for denumerable binary relations always (+) for instance, which are all isomorphic.

This is the case for equivalence relations with denumerably many classes of cardinality 2, and possibly classes of cardinality 1.

(2) **Or there exists a strictly increasing ω-sequence (under embedability) of denumerable non-rich relations representing $\mathcal{R}$, such that every denumerable relation representing $\mathcal{R}$, in which all these relations are embeddable, is rich for $\mathcal{R}$.**

This is the case for relations formed of finitely many components, each isomorphic with the consecutivity relation on positive and negative integers.

(3) **Or there exists an ω_1-sequence satisfying the preceding conditions.**

This is the case for chains, with the sequence of the denumerable ordinals.

(4) **Or there exists a relation which represents $\mathcal{R}$ and which is immediately less than the denumerable rich relations for $\mathcal{R}$.**

Problem. The impossibility of case (4); this is, in other words, the problem 2 in 10.5.4.

Cases (2) and (3) are compatible (POUZET in 1999). Take the age of the binary denumerable "composite" relation formed by the chain **Q** of rationals and infinitely many consecutivity relations on **Z**, mutually connected by value (-). To get an ω-sequence satisfying (2), replace the infinitely many consecutivities by finitely many ones. To get an ω_1-sequence satisfying (3), replace **Q** by denumerable ordinals.

11.5 Solid or fragil family or relation (Thomassé)

The following notions and proofs (in the two following sections) are due to [246] THOMASSÉ 1997 (from a doctoral diss. going back to 1995).

Given a denumerable set E, by a **family on** E we mean a denumerable set of finite subsets of E. By a **Δ-family on** E, we mean an infinite (thus denumerable) set of disjoint finite subsets of E.

By a **realization** of the Δ-family $\mathcal{E}$ on E, we mean the union of any infinite subset of $\mathcal{E}$ (so that the realization is a denumerable subset of E).

11.5.1 Solid or fragil family

Let E be a denumerable set and $\mathcal{E}$ be a family on E. Then one and only one of the following (i) and (ii) is true.

(i) There exists a finite subset F of E and a Δ-family $\mathcal{F}$ on $E-F$ such that, for every element X of $\mathcal{F}$, the union $F \cup X$ belongs to $\mathcal{E}$.

(ii) There exists a Δ-family $\mathcal{F}$ on E such that every element of $\mathcal{F}$ intersects all elements of $\mathcal{E}$ except a finite number of them.

The family $\mathcal{E}$ is said to be **solid** or **fragil** according to whether (i) or (ii) is true. Moreover F is said to be a **center** of $\mathcal{E}$ iff F is minimal (under inclusion) among those finite subsets of E which satisfy (i).

• Obviously (i) and (ii) cannot be simultaneously satisfied.

Suppose that (ii) is falsified. If there exist denumerably many mutually disjoint elements in $\mathcal{E}$, then we are in the case (i) with F empty. Thus there exists an integer k such that we have $k-1$ yet not k mutually disjoint elements in $\mathcal{E}$. Let F_0 be the union of $k-1$ mutually disjoint elements of $\mathcal{E}$: then every element of $\mathcal{E}$ intersects F_0.

Denote by $\mathcal{E}_0$ the family $\mathcal{E}$ whose elements are replaced by their intersection with $E - F_0$. If $\mathcal{E}_0$ includes a Δ-family, then $\mathcal{E}$ satisfies (i). Assume the contrary: then construct F_1 and $\mathcal{E}_1$ from F_0 and $\mathcal{E}_0$ in the same manner than F_0 and $\mathcal{E}_0$ from $\mathcal{E}$. All elements of $\mathcal{E}$, except a finite number of them, intersect F_1.

Iterating as long as $\mathcal{E}_0, \mathcal{E}_1, ...$ do not contain any Δ-family, we obtain the sequences F_i and $\mathcal{E}_i (i$ integer). Since (ii) is false, these sequences are necessarily finite: we reach a last step defined by F_n and $\mathcal{E}_n$, the latter set including a Δ-family $\mathcal{F}^*$. Let $F^* = F_0 \cup F_1 \cup ... \cup F_n$. For every element X of $\mathcal{F}^*$ there exists an $F_X \subseteq F^*$ such that $F_X \cup X$ belongs to $\mathcal{E}$. Finally there exists a subset F of F^* and a sub-Δ-family $\mathcal{F}$ of $\mathcal{F}^*$ which satisfy (i). •

11.5.2 Some lemmas concerning solid and fragil families

(1) **Let E be a denumerable set and $\mathcal{E}$ be a solid family on E. Partition $\mathcal{E}$ into finitely many (not necessarily disjoint) subsets $\mathcal{E}_i (i \leq n)$. Then at least one $\mathcal{E}_i$ is a solid family and each center of $\mathcal{E}$ is a center of a certain solid family $\mathcal{E}_i$.**

• Let F be a center of $\mathcal{E}$, associated with a Δ-family $\mathcal{F}$ on E. At least one $\mathcal{E}_i$ has infinitely many elements of the form $F \cup X$ with $X \in \mathcal{F}$. Then this $\mathcal{E}_i$ is solid and has a center included in F. Yet F being a center of $\mathcal{E}$, is also a center of $\mathcal{E}_i$. •

(2) **Let $\mathcal{E}$ be a solid family on E, and C be a finite subset of E. Then replacing each element of $\mathcal{E}$ by its intersection with $E - C$, we obtain a solid family $\mathcal{E}'$ on $E - C$. Moreover each center of $\mathcal{E}'$ is the intersection with $E - C$ of a center of $\mathcal{E}$.**

• If each element of $\mathcal{E}$ includes C, then the statement is obvious. Let D be a subset of C; we denote by $\mathcal{E}_D$ the subset of $\mathcal{E}$ formed of those elements which include D and are disjoint from $C - D$. Let $\mathcal{E}'_D$ be obtained by replacing each element of $\mathcal{E}_D$ by its intersection with $E - D$ (or equivalently with $E - C$). Let $\mathcal{E}'$ be the union of these $\mathcal{E}'_D$: by (1) one of the $\mathcal{E}_D$ is solid. Consequently $\mathcal{E}'_D$ and $\mathcal{E}'$ are solid. Let F be a center of $\mathcal{E}'$. Again by (1), there exists a D such that F is a center of a certain $\mathcal{E}'_D$. Choosing D minimal under inclusion, the union $F \cup D$ is a center of $\mathcal{E}_D$. Thus $F \cup D$ includes a center F' of $\mathcal{E}$. Yet $F' - D$ is a center of $\mathcal{E}'$; hence $F' = F \cup D$. •

(3) Let $\mathcal{E}$ be a family on E. Let r be a function with domain $\mathcal{E}$ which takes integer values. For each X in $\mathcal{E}$ the integer $r(X)$ will be said the **rank** of X.

Then one and only one of the following conditions is true.

(i) **There exists a Δ-family $\mathcal{F}$ and a finite subset F of E such that we have elements X in $\mathcal{F}$ with arbitrary large values for the rank $r(F \cup X)$.**

(ii) **There exists a Δ-family $\mathcal{F}$ on E such that every element of $\mathcal{F}$ intersects each element of $\mathcal{E}$ whose rank is larger than a certain integer value.**

• Firstly it is obvious that (i) and (ii) cannot be simultaneously true.

Now suppose that (ii) is false. Either there exists a finite subset F_0 of E which intersects all elements of $\mathcal{E}$ whose rank is larger than some given integer value. Then in $\mathcal{E}$ we replace each element by its intersection with $E - F_0$, so obtaining $\mathcal{E}_0$. We construct F_1 from $\mathcal{E}_0$ in the same manner than F_0 from $\mathcal{E}$; and so on.

Since (ii) is falsified, we cannot have an infinite sequence $F_0, F_1, F_2, \ldots$. We necessarily reach a last step defined by F_n. Let $F^* = F_0 \cup F_1 \cup \ldots \cup F_n$ For every F' disjoint from F^* and every rank i, there exists a finite set H which is disjoint from F' and has an arbitrary large rank.

Let us denote by $\mathcal{E}_i$ the set of elements of $\mathcal{E}$ whose rank is $\geq i$. Clearly each $\mathcal{E}_i$ is solid. Moreover, by replacing each element of $\mathcal{E}_i$ by its intersection with $E - F^*$, we obtain a solid family with empty center. By the preceding (2), each $\mathcal{E}_i$ has a center included in F^*. Thus there exists a finite subset F of E which is the center of infinitely many $\mathcal{E}_i$. For each of them F is associated to a Δ-family $\mathcal{F}_i$. Then we obtain the Δ-family $\mathcal{F}$ by choosing an element in each Δ-family $\mathcal{F}_i$. •

11.5.3 Solid or fragil relation

A relation R with denumerable base E is said to be **solid** iff there exists a finite subset F of E and a Δ-family $\mathcal{F}$ on E such that, for every realization A of $\mathcal{F}$, we have that $R \leq R/(A \cup F)$.

A relation R with denumerable base E is said to be **fragil** iff there exists a Δ-family $\mathcal{F}$ on E such that, for every realization A of $\mathcal{F}$, we have that $R \not\leq R/(E-A)$.

A relation cannot be simultaneously solid and fragil: take two disjoint Δ-families, one of which being solid and the other fragil.

It is proved by [246] THOMASSÉ 1995 p.64-67 that **any denumerable relation is either solid or fragil**.

Examples. The chain $\mathbf{Q}$ of rationals is solid: take F empty; for the Δ-family $\mathcal{F}$ take the singleton $\{0\}$; then the pair $\{-1,+1\}$; then the set $\{-2,-1/2,+1/2,+2\}$; and so on by insertion of an element before, after and between all the already chosen rationals.

The consecutivity relation on integers is fragil: for the Δ-family, take infinitely many singletons.

11.6 Solid or fragil morphism (Thomassé)

Let R (with base E) and R' (with base E') be two binary denumerable relations; let f be a local isomorphism from R into R', with finite domain F thus finite range F'. Let α be a countable ordinal; then f is said to be an α-**solid morphism** from R into R' iff there exists a finite subset G' of E' with $G' \supseteq F'$ and a Δ-family $\mathcal{F}'$ on E', such that for every realization Y' of $\mathcal{F}'$ we have that f is an α-morphism from R into $R'/(G' \cup Y')$.

Each minimal G' (under inclusion) which satisfies the preceding is said to be a **kernel** of the α-morphism f.

Every α-solid morphism from R into R' is a fortiori an α-morphism from R into R'. Yet the converse can be false. Even the empty function (= empty set) is not necessarily an α-solid morphism from R into R itself: take an example in the preceding section with a fragil relation R.

Note that a local isomorphism is always a 0-solid morphism.

The local isomorphism f is said to be an α-**fragil morphism** from R into R' iff there exists a Δ-family $\mathcal{F}'$ on the base E' such that for every realization Y' of $\mathcal{F}'$ we have that f is not an α-morphism from R into $R'/(E'-Y')$.

11.6.1 Solid and fragil are incompatible

Given two relations R, R', a local isomorphism f from R into R' cannot be simultaneously an α-solid and an α-fragil morphism.

• Given two Δ-families $\mathcal{F}'$ and $\mathcal{F}$" on the base E' and a finite subset G' of E', we can choose a realization Y' of $\mathcal{F}'$ and a realization Y" of $\mathcal{F}$" with $G' \cup Y'$ disjoint from Y", so that $G' \cup Y' \subseteq E' - Y$". Then by our definitions f would be an α-morphism from R into $R'/(G' \cup Y')$ without being an α-morphism from R into $R'/(E' - Y")$: contradiction. •

11.6.2 Either solid or fragil

Given two relations R, R' and an ordinal α, every α-morphism of R into R' is either an α-solid morphism or an α-fragil morphism.

• Start with $\alpha = 0$: we already know that each 0-morphism is a 0-solid morphism.

Let $\alpha = \beta + 1$ and suppose that our statement is true for β. Let f be an α-morphism from R to R'. Let us denote by $F_0, F_1, ..., F_i, ..(i$ integer) an increasing ω-sequence of finite subsets of $E = |R|$ whose union is E and starting from $F_0 = \operatorname{Dom} f$. If there exists an integer i such that every extension of f to F_i is a β-fragil morphism, then f is an α-fragil morphism.

Denote $\mathcal{K}_i$ the set of kernels of all extensions of f to the domain F_i which are β-morphisms. Every element of $\mathcal{K}_i$ includes an element of $\mathcal{K}_j$ for each $j \leq i$. Since each element of $\mathcal{K}_i$ includes an image of F_i, there does not exist any kernel which would belong to all $\mathcal{K}_i$'s.

Let $\mathcal{E} = \mathcal{K}_0 \cup \mathcal{K}_1 \cup ... \cup \mathcal{K}_i \cup ...$ For each element K of $\mathcal{E}$ we denote $r(K)$ and call **rank** of K the maximum integer i for which $K \in \mathcal{K}_i$. By 11.5.2 proposition (3) we have the two following possibilities.

(i) There exists a Δ-family $\mathcal{F}'$ on $E' = |R'|$ and a finite subset F' of E' with $F' \supseteq \operatorname{Rng} f$, such that for every i we have an element G' of $\mathcal{F}'$ such that $F' \cup G'$ contains a kernel of an extension of f to the domain F_i which is a β-morphism. Then by a diagonal argument f is an α-solid morphism with kernel F'.

(ii) There exists a Δ-family $\mathcal{F}'$ such that for each element G' of $\mathcal{F}'$ we have a certain F_i such that f has no extension to the domain F_i which be a β-solid morphism from R into $R'/(E' - G')$. In other words all extensions are fragil, so that by a diagonal argument f is an α-fragil morphism.

Now let α be a limit ordinal, and assume that our statement is true for every $\gamma < \alpha$. Firstly note that if f is a γ-fragil morphism for a certain $\gamma < \alpha$, then f is an α-fragil morphism.

Secondly suppose that f is γ-solid for every $\gamma < \alpha$: then consider an increasing ω-sequence α_i which is cofinal in α. To each α_i, we associate $\mathcal{K}_i$ which is the set of kernels of f considered as an α_i-morphism. If this set is infinite, then f is α-fragil: otherwise each of its centers would be a kernel. Note that if K belongs to $\mathcal{K}_i$, then it includes an element of $\mathcal{K}_j$ for every $j < i$. If there exists a kernel K which belongs to all $\mathcal{K}_i$, then f is an α-solid morphism with kernel K.

If not, let $\mathcal{E} = \mathcal{K}_0 \cup \mathcal{K}_1 \cup ... \cup \mathcal{K}_i \cup ...$ and for each $K \in \mathcal{E}$ let us denote by $r(K)$ the maximum integer i for which $K \in \mathcal{K}_i$. By 11.5.2 proposition (3), since we assumed there is no K which belongs to all the $\mathcal{K}_i$, then we are in the case (ii) of the lemma: hence f is an α-fragil morphism. •

11.6.3 Couple of α-disjoint morphisms

Let f and g be two local isomorphisms from R into R', with a common finite domain $\operatorname{Dom} f = \operatorname{Dom} g$. Define as follows the condition that f and g form a **couple of α-disjoint morphisms from R into R'.**

Firstly f and g are said to be a couple of 0-disjoint morphisms iff $\operatorname{Rng} f$ and $\operatorname{Rng} g$ are two disjoint subsets of the base $|R'|$. Suppose f and g are two $(\alpha+1)$-morphisms from R into R'. Then they are said to be $(\alpha+1)$-disjoint iff for every finite subset F of the base $E = |R|$ with $F \supset \operatorname{Dom} f$, there exist two extensions of f and g with common domain F, which form a couple of α-disjoint morphisms from R into R'. Finally if α is a limit ordinal, then f and g are said to form a couple of α-disjoint morphisms iff they are β-disjoint morphisms for every $\beta < \alpha$.

Given a couple of α-disjoint morphisms with common finite domain F, their restriction to an arbitrary subset of F form a couple of α-disjoint morphisms. In such a case, the empty isomorphism constitutes, with itself, a couple of α-disjoint morphisms.

Lemma. Let R, R' be two denumerable relations. Assume that there exists a couple of ω_1-disjoint morphisms from R into R'. Then there exist two disjoint restrictions of R' which are isomorphic copies of R.

• By hypothesis the empty set constitutes with itself a couple of ω_1-disjoint morphisms from R into R'. Let $a_0, a_1, ..., a_i, ..$ be an enumeration of the base E of R. For each countable ordinal α, there exists a couple of distinct elements b_0, c_0 in the base E' of R', such that the transformation of a_0 into b_0 and the transformation of a_0 into c_0 constitute a couple of α-disjoint morphisms from R into R'.

However since E and E' are denumerable, there exist only denumerably many couples (b_0, c_0), so that there exists at least one such couple which is α-disjoint for every countable ordinal α, thus which is ω_1-disjoint. By iteration we get two isomorphisms from R into two disjoint restrictions of R'. •

11.6.4 A proof of the indivisibility theorem

See 6.8.2.

• Let R be an indivisible denumerable relation. For every partition of the base $E = |R|$ into two (not necessarily disjoint) subsets A, B, we have that R has an embedding either in R/A or in R/B. A fortiori for an arbitrary countable ordinal α the empty function is either an α-morphism from R into R/A or into R/B.

Assume that the empty function is an α-fragil morphism from R into R itself. Then there exists a Δ-family $\mathcal{F}$ such that for any two realizations A^* and B^* of $\mathcal{F}$ the empty function is neither an α-morphism from R into $R/(E - A^*)$ nor into $R/(E - B^*)$: contradiction by taking B^* disjoint from A^*, so that the union $(E - A^*) \cup (E - B^*) = E$.

Consequently for every countable ordinal α, the empty function cannot be an α-fragil morphism from R into R; thus it is an α-solid morphism by 11.6.2. In other words for each countable ordinal α, there exists a finite subset F of E and a Δ-family $\mathcal{F}$ on E such that, for every realization A of $\mathcal{F}$, the empty function is an α-morphism from R into $R/(F \cup A)$.

Either there exists a countable ordinal α with an associated finite set F and a Δ-family $\mathcal{F}$ such that for every realization A we have $R \leq R/(F \cup A)$. Then taking two disjoint realizations A and A', we have simultaneously $R \leq R/(F \cup A)$ and

$R \leq R/(F \cup A')$. By indivisibility of R, since F is finite, we have also $R \leq R/A'$ and our theorem is proved, since $F \cup A$ and A' are disjoint sets.

Or for every countable ordinal α and every corresponding F and $\mathcal{F}$, there exists a realization A such that $R \not\leq R/(F \cup A)$. Then by indivisibility $R \leq R/(E-(F \cup A))$. Hence the empty function is simultaneously an α-morphism from R into $R/(F \cup A)$ (by definition of the α-solid morphism), and an α-morphism from R into the disjoint restriction $R/(E-(F \cup A))$. In other words the empty function constitutes with itself a couple of ω_1-disjoint morphisms from R into R. Thus by the previous subsection there exists two disjoint restrictions which are isomorphic copies of R. •

11.7 Interval-filter and interval-closure

11.7.1 Interval-filter

Our main purpose is to extend to any relation the classical closure starting from rational numbers and leading to real numbers.

Let R be a relation with base E. Following [76] FRAÏSSÉ 1984, let us define a R-**filter** or **interval-filter** as being a non-empty set $\mathcal{F}$ of non-empty R-intervals which satisfy the two following conditions:

(i) every R-interval which includes an element of $\mathcal{F}$ is an element of $\mathcal{F}$;

(ii) the intersection of any two elements of $\mathcal{F}$ is an element of $\mathcal{F}$ (this is a R-interval: see 9.8.1).

As for usual filters, a R-filter is said to be **finer** than another one if it includes it. An R-**ultrafilter** is an R-filter which is maximal with respect to inclusion. An R-ultrafilter is said to be **trivial** when it is formed of all R-intervals which contain a same element.

Given a non-trivial R-ultrafilter $\mathcal{U}$, each element of $\mathcal{U}$ is an infinite subset of the base $|R|$ and the intersection of all elements of $\mathcal{U}$ is empty.

Given a non-empty R-interval D and an R-ultrafilter $\mathcal{U}$, either D intersects each element of $\mathcal{U}$ and then D belongs to $\mathcal{U}$. Or there exists at least one element of $\mathcal{U}$ which is disjoint from D. Consequently given two different R-ultrafilters $\mathcal{U}$ and $\mathcal{U}'$, there exist at least one element $D \in \mathcal{U}$ and one element $D' \in \mathcal{U}'$ with $D \cap D'$ empty.

In particular, given an arbitrary finite subset F of the base E and a non-trivial R-ultrafilter $\mathcal{U}$, there exists at least one element $D \in \mathcal{U}$ with $D \cap F$ empty.

If a finite union of R-intervals $D_i (i = 1, ..., n$ integer) is an element of an R-ultrafilter $\mathcal{U}$, then at least one of the D_i is an element of $\mathcal{U}$.

Every R-filter admits a finer R-ultrafilter; this is proved by using the maximal chain axiom (see 2.2.4). When R reduces to its base E (or to a relation taking always the value $(+)$, or more generally a relation which admits every permutation of E as an automorphism), then the R-filters and R-ultrafilters reduce to usual filters and ultrafilters on E.

11.7.2 Interval-closure

Let R be a relation with base E. To the non-trivial R-ultrafilter $\mathcal{U}$ we associate an arbitrary set U, finite or infinite, disjoint from E. **Then consider a relation S with base U, having the same arity as R, and satisfying the following request**:

for each finite subset X of U and each element $D \in \mathcal{U}$, there exists at least one isomorphism from S/X onto a restriction of R/D.

We say that such an S **agrees with** $\mathcal{U}$. The existence of a relation based on U and which agrees with $\mathcal{U}$ is obvious for any finite U. For an infinite U, it suffices to use the coherence lemma 2.4.1.

If S agrees with $\mathcal{U}$, then the same is true for any restriction of S.

The interval-closure $R^+(\mathcal{U}, S)$.

Given an arbitrary n-ary relation R based on E, an arbitrary non-trivial R-ultrafilter $\mathcal{U}$, an arbitrary set U disjoint from the base $|R|$ and an n-ary relation S based on U and which agrees with $\mathcal{U}$, then **there exists one and only one common extension of R, S with base $E \cup U$, called the interval-closure $R^+(\mathcal{U}, S)$ and defined as follows**:

Given an arbitrary n-sequence of elements $a_i (i = 1, ..., n)$ in $E \cup U$, separate those a_i which belong to E and call them the **fixed elements**. Take an R-interval D such that $D \in \mathcal{U}$ and D does not contain any fixed element. Then bijectively transform the a_i in U into the a'_i in D by a local isomorphism from S into R/D, and finally set $R^+(a_1, ..., a_n) = R(a'_1, ..., a'_n)$ where we identify $a'_i = a_i$ for any fixed element. **The so defined value does not depend from the choice of D in $\mathcal{U}$.**

Notice that the base $U = |S|$ **is an interval of the closure** $R^+(\mathcal{U}, S)$.

A simple example is given by taking R to be the usual ordering ω of non-negative integers. Then there exists one and only one non-trivial ω-ultrafilter: the well-known FRECHET filter generated by all intervals going from an arbitrary integer until ∞. We can take for U an arbitrary set and for S any chain based on U: in the closure, S becomes posterior to all integers.

11.7.3 Compatibility of several interval-closures

Start from an n-ary relation R with base E and consider two different non-trivial R-ultrafilters $\mathcal{U}, \mathcal{V}$ to which we associate two disjoint sets U, V (also disjoint from E). Consider independently two n-ary relations: firstly S based on U and which agrees with $\mathcal{U}$; secondly T based on V and which agrees with $\mathcal{V}$.

Then **the two interval-closures $R^+(\mathcal{U}, S)$ and $R^+(\mathcal{V}, T)$ are compatible, in this sense that there exists a common extension of R, S, T based on the union $E \cup U \cup V$, which is, more strongly, a common extension of both closures $R^+(\mathcal{U}, S)$ and $R^+(\mathcal{V}, T)$. It will be denoted $R^+(\mathcal{U}, S, \mathcal{V}, T)$ and unambiguously defined as follows**:

Given an arbitrary n-sequence $a_1, ..., a_n$ in $E \cup U \cup V$, as precedently separate those a_i which belong to E and again call them fixed elements. Then consider those

$a_i \in U$ and change them into the a'_i as precedently via a local isomorphism from S into R/H where $H \in \mathcal{U}$ and H does not contain any fixed element. Analogous strategy for those $a_i \in V$; again we denote a'_i their images, which belong to an element K in $\mathcal{V}$ where K does not contain any fixed element. The only caution is to choose $H \in \mathcal{U}$ and $K \in \mathcal{V}$ mutually disjoint, which is always possible since $\mathcal{V} \neq \mathcal{U}$. As previously identify $a'_i = a_i$ for any fixed element and set $R^+(a_1, ..., a_n) = R(a'_1, ..., a'_n)$. The so defined value does not depend from the choice of H in $\mathcal{U}$ and K in $\mathcal{V}$.

We leave it to the reader to extend the previous compatibility, firstly to any finite set of R-ultrafilters $\mathcal{U}, \mathcal{V}, \mathcal{W}...$, so defining $R^+(\mathcal{U}, S_U, \mathcal{V}, S_V, \mathcal{W}, S_W, ...)$ where $U, V, W, ...$ are arbitrary disjoint sets (and disjoint from E), where $S_U, S_V, S_W, ...$ are respectively based on $U, V, W, ...$ and respectively agree with $\mathcal{U}, \mathcal{V}, \mathcal{W}...$.

Finally the extension to any set, finite or infinite, of R-ultrafilters is easily proved, since the arity of R is finite.

Example. Start from the chain $\mathbf{Z}$ of negative and positive integers. Then we have two and only two $\mathbf{Z}$-ultrafilters. The right ultrafilter is formed of all intervals going from any integer to ∞, and the left one is formed of intervals from $-\infty$ to any integer. For S we take any chain: in the closure it will go to the right of $\mathbf{Z}$. Similarly we take any chain for T: it will go to the left of $\mathbf{Z}$. Finally in the closure each element of T will precede each element of S.

Example of a closure defined from infinitely many ultrafilters. The chain of reals as a closure of the chain $\mathbf{Q}$ of rationals. In this case non-trivial interval-ultrafilters fall into four possible cases. For each rational u, the right ultrafilter generated by intervals $]u, x]$ where $x > u$; the analogous left ultrafilter. Also the ultrafilter situated just before or just after an irrational number. To get the chain $\mathbf{R}$ of reals, It suffices to take U empty in all cases except, for instance, the right side of each irrational number where we take for U a singleton.

11.7.4 Elementary extensive interval-closure

Notion due to [118] ILLE 1990 p.219. With previous notations, $R^+(\mathcal{U}, S)$ is said to be **elementary extensive** iff for each element $D \in \mathcal{U}$ the restriction $R^+/(D \cup U)$ is an elementary extension of R/D (elementary extension is defined in 10.10.3).

If $R^+(\mathcal{U}, S)$ is elementary extensive, then taking for D the whole base E (which is an R-interval), we see that R^+ is an elementary extension of R.

The reciprocal proposition is false. Indeed start from the chain $\mathbf{Q}$ of rationals, consider the irrational number $\sqrt{2}$ and the $\mathbf{Q}$-ultrafilter $\mathcal{U}$ formed of all intervals D situated at the right of $\sqrt{2}$. Then take for S the chain on the unique element $\sqrt{2}$. On one hand $\mathbf{Q}$ plus the additional element $\sqrt{2}$ is an elementary extension of $\mathbf{Q}$. On the other hand the chain $\mathbf{Q}/D \cup \{\sqrt{2}\}$ having the first element $\sqrt{2}$ is not elementary equivalent with $\mathbf{Q}/D$: a fortiori it is not an elementary extension.

11.8 Non-classical ultraproduct and ultrapower

The reader is supposed to be familiar with the classical ultraproduct and ultrapower. Here we need a generalization going back to [73] FRAÏSSÉ 1966 and anticipated by [137] KOCHEN 1958.

Let I be a set and $\mathcal{F}$ an ultrafilter on I. Consider a family of relations R_i of the same arity n, where i ranges over I. A class $\mathcal{A}$ is said to be **induced** by $\mathcal{F}$ iff the set of indices i such that $R_i \in \mathcal{A}$ is an element of $\mathcal{F}$. Another way of saying this is that $R_i \in \mathcal{A}$ for **almost every** $i(\mathrm{mod}\,\mathcal{F})$.

For any class $\mathcal{A}$ of n-ary relations, either $\mathcal{A}$ or its complement is induced.

Any finite intersection of induced classes is an induced class (and therefore non-empty).

By the well-known compactness theorem, the intersection of all (k,p)-equivalence classes induced by $\mathcal{F}$ is not empty. It is **the elementary equivalence class induced by** $\mathcal{F}$.

Let E_i denote the base of R_i. We define a **function** to be any set f of couples $(i, f(i)(i \in I)$ such that $f(i) \in E_i$ (exactly one element for each i). Two functions f, g are said to be **equivalent** $\mathrm{mod}\,\mathcal{F}$ iff $f(i) = g(i)$ for almost every i.

Let B be **an arbitrary set of pairwise non-equivalent functions**. The **ultraproduct of the** R_i **over** $B(\mathrm{mod}\,\mathcal{F})$ is the n-ary relation R^+ with base B defined as follows. For each n-tuple of functions $f_1, ..., f_n$ in B we set $R^+(f_1, ..., f_n) =$ + or (-) according to whether $R(f_1(i), ..., f_n(i)) =$ + or (-) for almost every i.

The classical ultraproduct, where a representative function is taken in each class of equivalence modulo the ultrafilter, will be called the **complete ultraproduct**.

Suppose that all the R_i coincide, say with an n-ary relation R with base A, and let B contain all the constant functions $f(i) = a$ for all i, where $a \in A$. If we identify each $a \in A$ with the corresponding constant function, then B becomes a superset of A and the ultraproduct over B is an extension of R to B, called an **ultrapower** of R; in the classical case, we will call it a **complete ultrapower**.

Example: the chain ω of non-negative integers and its ultrapower $\omega + \mathbf{Z}$.

Take for I the set of non-negative integers and for $\mathcal{F}$ a non-trivial ultrafilter, which contains as an element the complement of each finite set of integers. Take all R_i identical with R. take for $\mathcal{F}$ the set of constant functions plus the set of functions $f(i) = i + c$ where c is a positive, null or negative integer. When $c < 0$ we set $f(i) = 0$ for $i < -c$ and $f(i) = i + c$ for $i \geq -c$.

The ultrapower is a chain of type $\omega + \mathbf{Z} = \omega + \omega^- + \omega$. Indeed given two functions $f_1, f_2 \in \mathcal{F}$, we have $S(f_1, f_2) =$ + or (-) according to whether $f_1(i) \leq f_2(i)$ or $> f_2(i)$ from some rank i. Consequently constant functions on one hand, functions $i+c$ (where c constant) on the other hand, are ordered by the value of this constant. Moreover, modulo S, each constant function is less than any function $i + c$.

11.8.1 Normal ultraproduct

Consider a family of n-ary relations $R_i (i \in I)$, an ultrafilter $\mathcal{F}$ on I and a set B of functions which are pairwise non-equivalent mod $\mathcal{F}$. The corresponding (reduced) ultraproduct is said to be **normal** iff the following condition holds.

For any finite set C of functions $f \in B$, any finite set D of functions g (not necessarily in B) and any two natural numbers k, p, there exist $U \in \mathcal{F}$ and, for each $g \in D$, a function $g' \in B$, such that for each $i \in U$ the mapping which saves each $f(i)(f \in C)$, and which takes each $g(i)$ onto $g'(i)(g \in D)$, is a (k,p)-automorphism of R_i. We call g' the **substitute** of g.

An obvious example of a normal ultraproduct is the classical, or complete ultraproduct.

A useful example of a normal ultrapower is the ultrapower reduced to constant functions.

Another example: the ultrapower $\omega + \mathbf{Z}$ previously defined from ω.

11.8.2 Isomorphism class induced by an ultraproduct

Let h be a non-negative integer; consider an h-tuple of functions $f_1, ..., f_h$, where h has a priori nothing to do with the arity n of our relations R_i. We put i, j in a same (k,p)-**isomorphism class** iff the transformation which takes resp. $f_1(i), ..., f_h(i)$ onto $f_1(j), ..., f_h(j)$ is a (k,p)-isomorphism from R_i into R_j. There exist finitely many such disjoint classes, so that only one of them is an element U of the ultrafilter $\mathcal{F}$. We call it the (k,p)-**isomorphism class induced by the ultraproduct**. Moreover we say that **the previous element U is (k,p)-connected with the h-tuple $f_1, ..., f_h$**.

In the case where $h = 0$, we find again the previous (k,p)-equivalence class induced by $\mathcal{F}$.

Lemma. If the ultraproduct is normal, and if U is (k,p)-connected with the h-tuple $f_1, ..., f_h$, then for any $i \in U$ the transformation which takes resp. $f_1, ..., f_h$ into $f_1(i), ..., f_h(i)$ is a (k,p)-isomorphism from the ultraproduct into R_i.

Taking $h = 0$: if the ultraproduct is normal, then **the (k,p)-equivalence class of the ultraproduct is identical to the (k,p)-equivalence class induced by the ultraproduct**.

In particular, if an ultrapower is normal, then the transformation which saves any finite set of constant functions is a (k,p)-isomorphism from the given relation R into the ultrapower.

Consequently a normal ultrapower of R is an elementary extension of R.

11.9 Three theorems on interval-closures: Ille

11.9.1 First theorem

Let R be a relation with an infinite base and $\mathcal{U}$ a non-trivial R-ultrafilter. Then there exists a relation S with non-empty base, such that $R^+(\mathcal{U}, S)$ is elementary extensive.

See [118] ILLE 1990 p.223.

The first theorem is proved as follows.

Let R be an n-ary relation with base E and let D be an R-interval. By 9.8.3, among n-tuples in $E - D$ there exist finitely many equivalence classes of (R, D)-extendomorphism. To each class C_i let us associate an n-ary relation $S_i (i = 1, ..., t$ integer) with base $E - D$, taking value $(+)$ exactly for those n-tuples in $E - D$ which belong to C_i.

Lemma. Given arbitrary integers k, p, if f is a (k, p)-automorphism of the concatenation $(R/E - D)S_1...S_t$, then the extension of f by the identity on D is a (k, p)-automorphism of R.

• Case of $k = 0$. For any n-tuple $x_1, ..., x_n$ in $E-D$, if f is a local automorphism of $(R/E - D)S_1...S_t$, then $x_1, ..., x_n$ and $f(x_1), ..., f(x_n)$ are in a same class of (R, D)-extendomorphism. Consequently the extension of f by the identity on D is a local automorphism of R.

By induction on k, the proof immediately extends to any two k, p. •

11.9.2 Construction of a normal ultrapower

Let R be an n-ary relation with base E and $\mathcal{U}$ be a non-trivial R-ultrafilter. We construct as follows an ultrapower R^+ of R by taking certain functions whose domain is E and whose range is $\subseteq E$. Our ultrapower is defined by an ultrafilter $\overline{\mathcal{U}}$ on E which is taken to be finer than $\mathcal{U}$.

Let $E(\mathcal{U})$ be the set of those functions f (from E into E) such that for every element $D \in \mathcal{U}$, we have $f^{-1}(D) \in \overline{\mathcal{U}}$. An obvious example of such an f is the identity from E onto E.

We identify each element $a \in E$ with the constant function $f_a(x) = a$ for every $x \in E$. Then E and $E(\mathcal{U})$ are disjoint. Indeed since $\mathcal{U}$ is non-trivial, for any $a \in E$ there exists at least one element $D \in \mathcal{U}$ such that $a \in E - D$, so that $f_a^{-1}(D)$ is empty.

Let us denote R^+ the ultrapower of R along $\overline{\mathcal{U}}$ which is based on the union of the two disjoint sets E and $E(\mathcal{U})$.

Given any finite subset H of E and any two integers k, p, the identity on H is a (k, p)-isomorphism of R into the ultrapower R^+. Consequently R^+ is an elementary extension of R (see 10.10.3).

• It suffices to prove that the ultrapower R^+ is normal in the sense of 11.8.1. Consider a sequence $a_1, ..., a_m (m$ integer) in E and a sequence $f_1, ..., f_r (r$ integer) formed of functions $f \in E(\mathcal{U})$ so that for every $D \in \mathcal{U}$ we have that $f^{-1}(D) \in \overline{\mathcal{U}}$. Also consider a sequence $g_1, ..., g_s (s$ integer) formed of functions with domain E

and range $\subseteq E$. We suppose these g belong neither to E (i.e. are not constants on E) nor to $E(\mathcal{U})$. Consequently there exists at least one element $X \in \overline{\mathcal{U}}$ and an element $D \in \mathcal{U}$ such that $a_1, ..., a_m$ belong to $E - D$ and $f_1(X), ..., f_r(X) \subseteq D$ and finally $g_1(X), ..., g_s(X) \subseteq E - D$.

Since $E - D$ is the complement of an R-interval D, by 9.8.3 for each integer, in particular for the arity n of R, there exist in $E - D$ finitely many equivalence classes C_i of (R, D)-extendomorphic n-tuples. To each class C_i we associate an n-ary relation S_i based on $E - D$, which takes value $(+)$ exactly for those n-tuples which belong to C_i. Let t be the number of the C_i or $S_i (i = 1, ..., t)$. Let us denote $F(X, E - D)$ the set of all functions whose domain is X and whose range is $\subseteq E - D$. Also denote $\overline{\mathcal{U}}/X$ the ultrafilter reduced to subsets of X. Also let $R(X, E - D)$ be the complete ultrapower of $R/E - D$ defined on $F(X, E - D)$.

There exist n-ary relations $S'_1, ..., S'_t$ based on $F(X, E - D)$, such that the concatenation $(R(X, E - D))S'_1, ..., S'_t$ is an ultrapower of $(R/E - D)S_1, ..., S_t$. Given any two integers k, p, there exist elements $b_1, ..., b_s$ in $E - D$ such that the transformation which takes b_i into $g_i/X (i = 1, ..., s)$ and which saves $a_1, ..., a_m$ is a (k, p)-isomorphism of $(R/E - D)S_1...S_t$ into $(R(X, E - D))S'_1...S'_t$.

Given any two integers k, p, there exists at least one subset X' of X which belongs to the restricted ultrafilter and such that for any $x \in X'$, the transformation which takes g_i/X into $g_i(x)(i = 1, ..., s)$ and which saves $a_1, ..., a_m$ is a (k, p)-isomorphism from the concatenation $(R(X, E - D))S'_1...S'_t$ into $(R/E - D)S_1...S_t$. By composition we get a (k, p)-automorphism of $(R/E - D)S_1...S_t$ which saves $a_1, ..., a_m$ and which takes b_i into $g_i(x)(i = 1, ..., s)$.

Finally using the previous lemma, let us extend the previous transformation by the identity on D. We get a (k, p)-automorphism of R which saves $a_1, ..., a_m$, which saves $f_1(x), ..., f_r(x)$ (since $x \in X$ and each $f_i(X) \subseteq D$), and which finally takes b_i into $g_i(x)(i = 1, ..., s)$. This is the normality condition (see 11.8.1) where the constant function b_i is the substitute of $g_i (i = 1, ..., s)$. •

11.9.3 An elementary extensivity

Let us take an arbitrary element $D \in \mathcal{U}$. With previous notations, we shall prove that $R^+/D \cup E(\mathcal{U})$ **is an elementary extension of** R/D.

• Start from the restriction R/D; we denote $\mathcal{U}/D$ the (R/D)-ultrafilter reduced to those elements of $\mathcal{U}$ which are subsets of D. They are (R/D)-intervals by 9.8.2. In the same way we denote $\overline{\mathcal{U}}/D$ the set of elements of $\overline{\mathcal{U}}$ which are included in D.

As previously we denote $F(D, D)$ the set of functions with domain D and range $\subseteq D$. Similarly to our previous notation $E(\mathcal{U})$ we denote $E(\mathcal{U}/D)$ the set of those functions $f \in F(D, D)$ such that for every $D' \in \mathcal{U}/D$ we have $f^{-1}(D') \in \overline{\mathcal{U}}/D$. Let us denote R^+_D the ultrapower of R/D which is based on $D \cup E(\mathcal{U}/D)$. By our previous statement, it is an elementary extension of R/D.

Now consider the transformation which saves each element of D and associates to each f in $E(\mathcal{U})$ the unique element of $E(\mathcal{U}/\mathcal{D})$ which coincides with f on $D \cap f^{-1}(D)$. Since for every f in $E(\mathcal{U})$, the intersection $D \cap f^{-1}(D)$ belongs to $\overline{\mathcal{U}}/D$, this transformation is an isomorphism from $R^+/(D \cup E(\mathcal{U}))$ onto R^+_D. •

11.9.4 Identification

The previous ultrapower R^+ **is the interval-closure** $R^+(\mathcal{U},T)$ **in the sense of** 11.7.2, **where** $\mathcal{U}$ **is the previously considered** R**-ultrafilter and** T **is the restriction** $R^+/E(\mathcal{U})$.

• Firstly $R^+/E(\mathcal{U})$ agrees with $\mathcal{U}$ in the sense of 11.7.2. Indeed given any finite subset H of $E(\mathcal{U})$ and any element $D \in \mathcal{U}$, by our previous statement there exists at least one local isomorphism from $R^+/E(\mathcal{U})$ into R/D with domain H.

The interval-closure of R, obtained from $\mathcal{U}$ and $R^+/E(\mathcal{U})$ which agrees with $\mathcal{U}$, is identical with R^+. Indeed if $a_1, ..., a_m$ belong to E and $f_1, ..., f_r$ belong to $E(\mathcal{U})$, then there exists at least one element $X \in \overline{\mathcal{U}}$ and one $D \in \mathcal{U}$ such that $a_1, ..., a_m$ belong to $E - D$ and for every $x \in X$ we have that $f_1(x), ..., f_r(x)$ belong to D, and finally the transformation which takes f_i into $f_i(x)$ for $i = 1, ..., r$ and which saves $a_1, ..., a_m$ is a local isomorphism from R^+ into R. •

The first theorem is now proved.

11.9.5 Second theorem

Given R**, the non-trivial** R**-ultrafilter** $\mathcal{U}$ **and any relation** S **which agrees with** $R, \mathcal{U}$**, then there exists an extension** $\overline{S}$ **of** S **such that the interval-closure** $R^+(\mathcal{U}, \overline{S})$ **is elementary extensive.**

See ILLE p.226-227.

• Let E be the base of R. Let E' be the base of the given interval-closure $R^+(\mathcal{U}, S)$. We suppose that the base $|S|$ is non-empty, so that $E' \supset E$, the difference $E' - E$ being the base $|S|$.

Let I be the set of all couples $i = (F, D)$ where F is a finite subset of E' and D is an element of $\mathcal{U}$. More precisely we denote F_i, D_i the components of i. We set $i' \leq i$ iff $F_{i'} \supseteq F_i$ and $D_{i'} \subseteq D_i$. Given two non-empty sets of indices closed under $\leq$, their intersection is non-empty and again closed under $\leq$. Consequently the supersets of previous sets constitute a filter on I. Let $\mathcal{V}$ be an ultrafilter on I, which contains as elements all these non-empty sets closed under $\leq$.

This $\mathcal{V}$ may play the same role as $\overline{\mathcal{U}}$ in the proof of 11.9.2. More precisely let $E(\mathcal{U})$ be the set of those functions f from I into E, such that for every $D \in \mathcal{U}$ we have $f^{-1}(D) \in \mathcal{V}$. This $E(\mathcal{U})$ is disjoint from E. We construct the normal ultrapower R^+ on the union $E \cup E(\mathcal{U})$. We see that R^+ is identical to the interval-closure $R^+(\mathcal{U}, R^+/E(\mathcal{U}))$ which is elementary extensive.

For each $i \in I$ we consider a local isomorphism t_i from $R^+(\mathcal{U}, S)$ into R/D_i with domain F_i, which saves each element of $F_i \cap E$. Then to any element $a \in E'$ is associated the function f_a from I into E, such that for every $i \leq (\{a\}, E)$ we have $f_a(i) = t_i(a)$. For any two elements $a \neq b$ in E' we have $f_a \neq f_b$, since for $i \leq (\{a, b\}, E)$ we have $f_a(i) \neq f_b(i)$.

For every sequence $a(1), ..., a(n)$ in E', where n is the arity of R, there exists an $X \in \mathcal{V}$ such that any $i \in X$ satisfies $R^+(f_{a(1)}, ..., f_{a(n)}) = R(f_{a(1)}(i), ..., f_{a(n)}(i))$. If $i \leq (\{a(1), ..., a(n)\}, E)$ then $R^+(\mathcal{U}, S)(a(1), ..., a(n)) = R(t_i(a(1)), ..., t_i(a(n)))$. Consequently for any element i in X which satisfies $i \leq (\{a(1), ..., a(n)\}, E)$, we have $R^+(\mathcal{U}, S)(a(1), ..., a(n)) = R^+(f_{a(1)}, ..., f_{a(n)})$. Hence by identifying each $a \in E'$ with f_a, we obtain the extension $\overline{S} = R^+/E(\mathcal{U})$ of S. •

11.9.6 Third theorem

Let R be a relation with base E. Let $\mathcal{U}_k (k \in K)$ be a (finite or infinite) set of non-trivial R-ultrafilters. Let S_k be a corresponding set of relations with disjoint bases (also disjoint from E) such that S_k agrees with $\mathcal{U}_k$ for each $k \in K$. Assume that $R^+(\mathcal{U}_k, S_k)$ is elementary extensive for each k. Then the total interval-closure $R^+(\mathcal{U}_k, S_k)(k \in K)$ is an elementary extension of R.

More strongly if D_k is any element of $\mathcal{U}_k$, then $R^+/(\cup_k D_k \cup |S_k|)(k \in K)$ is an elementary extension of $R/(\cup_k D_k)(k \in K)$, where R^+ denotes $R^+(\mathcal{U}_k, S_k)(k \in K)$.

See ILLE p.225.

Our conclusion does not subsist if, even reducing I to two R-ultrafilters, each giving an interval-closure, we only assume that each of these two closures is an elementary extension of R. For instance consider the chain $\mathbf{Q}$ of rationals. Denote $\mathcal{U}$ the interval-ultrafilter situated at the right of $\sqrt{2}$ and $\mathcal{V}$ the symmetrical interval-ultrafilter at the left of $\sqrt{2}$. In both cases, take for S the chain reduced to the singleton $\{\sqrt{2}\}$.

Then the first and the second closures are formed of rationals plus $\sqrt{2}$: both are elementary extensions of $\mathbf{Q}$. Yet the union is isomorphic with $\mathbf{Q} + 2 + \mathbf{Q}$: it is not elementary equivalent with $\mathbf{Q}$.

To prove the third theorem, the reader will be helped by the following two lemmas and their corollaries.

First lemma. Let R be a relation with base E. Let $\mathcal{U}_i (i \in I)$ be a **finite** family of non-trivial R-ultrafilters. Let R^+ be the interval-closure of R, obtained by associating to each $\mathcal{U}_i$ a relation S_i based on E_i. To each $i \in I$ we associate an element D_i of $\mathcal{U}_i$, all these D_i being mutually disjoint.

Then given any two integers k, p and any finite subset F of $E - \cup D_i (i \in I)$, if for each i in I we have that f_i is a (k,p)-isomorphism from $R^+/(D_i \cup E_i)$ into R/D_i (resp. from R/D_i into $R^+/(D_i \cup E_i)$), then the extension of all f_i which saves each element of F is a (k,p)-isomorphism from R^+ into R (resp. from R into R^+).

Proof by induction on k.

Corollary. If for each i in I, the restriction $R^+/(D_i \cup E_i)$ is an elementary extension of R/D_i, then R^+ is an elementary extension of R.

Second lemma. Let R be a relation with base E. Let $\mathcal{U}_i (i \in I)$ be a **finite** family of non-trivial R-ultrafilters. Let R^+ be the interval-closure of R, obtained by associating to each $\mathcal{U}_i$ a relation S_i based on E_i.

Then for each R-interval D, if $D \notin \cup \mathcal{U}_i (i \in I)$, then D is an R^+-interval.

Corollary. If $\mathcal{U}$ is a non-trivial R-ultrafilter which does not belong to the finite family $\mathcal{U}_i (i \in I)$, then the set of R^+-intervals which contain at least one element of $\mathcal{U}$ is a R^+-ultrafilter which contains the element $\mathcal{U} - \cup \mathcal{U}_i (i \in I)$.

Proof of the third theorem. Given $k \in K$, the base of S_k is denoted by E_k. For each finite subset I of K, let us denote $R^I = R^+/(E \cup (\cup_i E_i))(i \in I)$ which is the interval-closure of R obtained by associating to each $\mathcal{U}_i (i \in I)$ the relation S_i. Let $I \subseteq J$ be two finite subsets of K. By our second corollary for each $j \in J - I$ the set $\mathcal{V}_j$ of R^I-intervals which contain at least one element of $\mathcal{U}_j$ is a R^I-ultrafilter. Then R^J is the interval-closure of R^I obtained by associating to each $\mathcal{V}_j (j \in J - I)$ the relation S_j.

Now to any $j \in J$ we associate an element D_j of $\mathcal{U}_j$, all these D_j being mutually disjoint. For every $j \in J - I$ we have that D_j is an element of $\mathcal{V}_j$. Since $R^+(\mathcal{U}_j, S_j)$ is elementary extensive, then $R^J/(D_j \cup E_j) = R^+(\mathcal{U}_j, S_j)/(D_j \cup E_j)$ is an elementary extension of $R/D_j = R^I/D_j$. By the first corollary R^J is an elementary extension of R^I. Finally R^+, which is the common extension of the $R^I(I$ finite subset of $K)$ is an elementary extension of each R^I, and in particular of $R = R^\emptyset$.

More strongly for each $k \in K$ we consider an element D_k of $\mathcal{U}_k$ and we denote by A the restriction $R/(\cup_k D_k)(k \in K)$. The set $\mathcal{W}_k$ of A-intervals X such that there is an element C_k of $\mathcal{U}_k$ with $C_k \subseteq D_k \cap X$ is an A-ultrafilter. Then the interval-closure $A^+(\mathcal{W}_k, S_k)$ of A obtained by associating to $\mathcal{W}_k$ the relation S_k is elementary extensive. Since $R^+/((\cup_k D_k) \cup (\cup_k E_k))(k \in K)$ is the total interval-closure of $A^+(\mathcal{W}_k, S_k)(k \in K)$, then by our previous statement it is an elementary extension of A.

Chapter 12

Homogeneous relation, orbit, connection with permutation groups

12.1 Homogeneous relation

12.1.1 Generalities

Let p be an integer; a relation R is said to be p-**homogeneous** iff every local automorphism of R defined on p elements is extendible to an automorphism of R.

Every relation is 0-homogeneous, since the empty function is extendible to the identity mapping on the entire base.

Obvious definition of a $(\leq p)$-**homogeneous** relation.

A relation R is said to be **homogeneous** iff R is p-homogeneous for every integer p.

Example. A binary cycle on at least 5 elements is 1-homogeneous but not 2-homogeneous. Indeed, call a, b, c, d four consecutive elements; then the mapping which takes a, c into a, d is a local automorphism and is not extendible. The chain on 2 elements is not 1-homogeneous, but it is 2-homogeneous, since the only local automorphism on 2 elements is the identity transformation;

The following binary multirelation (with binary and unary component relations) is 2-homogeneous yet not 3-homogeneous.

• Start with the binary relation R already defined in 9.10.1 on the 10 elements $a, b, c, d, r, s, t, u, i, j$.

In order to insure the 2-homogeneity: for each couple, say (a, s) for instance, add a binary relation A such that $A(x, y) = +$ exactly for those couples (x, y) which are an image of (a, s) under one of the three following permutations:

(1) the mapping which preserves a and c and interchanges $(b, d), (r, u), (s, t), (i, j)$;

(2) the mapping which preserves b and d and interchanges $(a, c), (r, s), (t, u), (i, j)$;

(3) the composition of mappings (1) and (2) which preserves i and j and interchanges $(a,c),(b,d),(r,t),(s,u)$.

The images of (a,s) are $(a,t),(c,r),(c,u)$ and obviously (a,s) itself, the identity permutation being added to (1), (2), (3).

The obtained multirelation is not 3-homogeneous: as already mentioned in 9.10.1, the local automorphism which preserves a and b and interchanges (i,j) is not extendible to an automorphism on the entire base. •

The previous example is not 1-homogeneous. For example map a into b. Yet it becomes 1-homogeneous if we add, for instance, a unary relation taking value (+) for a and c, and another taking (+) for i and j.

Example of homogeneous relations. The relation always (+) (or always (-)). The chain **Q** of rationals.

Another homogeneous relation. The equivalence relation with finitely many, or with denumerably many classes having cardinality 2. In contrast with the preceding examples, here homogeneity does not subsist when we remove an element from the base.

Problem. For each arity n, does there exist a threshold $s(n)$ above which every n-ary relation which is $(\leq s(n))$-homogeneous, is homogeneous.

12.1.2 A characterization of homogeneous relations

Let E be a denumerable set. A relation R with base E is homogeneous iff, for any finite subset F of E, for any local automorphism f of R with domain F, and for any element $u \in E - F$, there exists a local automorphism of R which extends f to the domain $F \cup \{u\}$. See [69] FRAÏSSÉ 1954.

• If R is homogeneous, then it obviously satisfies our condition.

Conversely, suppose that the condition holds. Enumerate the elements of E as a_i(i integer) . Start with an arbitrary local automorphism f_0 of R with finite domain. Add a_0 to the domain (if it does not yet belong to Dom f_0), thus obtaining a local automorphism f'_0 extending f_0. Similarly add a_0 to Rng f'_0 (if necessary), thus obtaining a local automorphism f_1 extending f'_0, thus extending f_0. Iterate this, going from f_i to f_{i+1}(i integer) by adding a_i to the domain and then to the range. Finally the common extension of these f_i is an automorphism of R. •

This proposition does not extend, in general, to the case of an uncountable relation. Indeed consider the chain **Q** + **R** where **Q** is the chain of rationals and **R** is isomorphic with the chain of reals. Then the condition in our preceding statement is satisfied. Yet mapping an element of the initial interval **Q** into an element of the final interval **R**, we cannot extend this local automorphism to the entire base.

12.1.3 Connection with richness, age, embedding

(1)**For each integer n there exists a denumerable n-ary relation which is rich and homogeneous.**

• The rich denumerable relation R defined in 10.3.1 satisfies the following condition. Each finite restriction of R is arbitrarily extendible to its base plus an additional element, which belongs to the base $|R|$. So that R satisfies the condition of the previous subsection. •

(2) **Any two denumerable homogeneous relations with same age, are isomorphic.**

(3) **Let R be a denumerable homogeneous relation. Then every denumerable relation which is younger than R, is embeddable in R.**

• (2) Let R and R' be denumerable homogeneous relations with respective bases E and E', and the same age. Enumerate the elements of E as a_i, the elements of E' as a'_i(i integer). Embed the restriction $R/\{a_0\}$ into R', thus obtaining a local isomorphism f_0 from R into R' with domain $\{a_0\}$. By the previous subsection, f_0 is extendible to a local isomorphism g_0 from R into R', whose domain contains a_0 and whose range contains a'_0. Iterating this for each i, we obtain a local isomorphism f_i whose domain contains the elements $a_0, ..., a_i$ and which is extendible to a local isomorphism g_i whose range contains $a'_0, ..., a'_i$. The union of the f_i (or equivalently the g_i) is an isomorphism from R onto R'. •

Similar proof for (3) by going only from R' towards R.

12.2 Amalgamable set, amalgamable age

A set $\mathcal{R}$ of finite relations of the same arity is said to be **amalgamable** iff, for any relations A, B, C belonging to $\mathcal{R}$, for any isomorphism f from A onto a restriction of B and any isomorphism g from A onto a restriction of C, there exists a relation D belonging to $\mathcal{R}$, an isomorphism f' from B onto a restriction of D and an isomorphism g' from C onto a restriction of D, such that for each element x of the base $|A|$ we have $(f' \circ f)x = (g' \circ g)x$.

Obviously the relations in the considered set are defined up to isomorphism.

A set $\mathcal{R}$ of finite relations is said to be **strongly amalgamable** iff, for any relations B, C belonging to $\mathcal{R}$ and C' isomorphic with C, with a common restriction to the intersection of the bases $|B|$ and $|C'|$, then there exists a relation D in $\mathcal{R}$ (up to isomorphism) which is a common extension of B and C'.

For example, the age of all finite chains, and also the age of all finite posets are strongly amalgamable: see 1.7.3 (amalgamation lemma).

The age of those finite unary relations which take the value $(+)$ for at most one element, is amalgamable yet not strongly amalgamable.

The age of all finite trees is not amalgamable; the example given in 2.11.6 contradicts ordinary as well as strong amalgamability.

12.2.1 Amalgamation theorem

Given an age $\mathcal{R}$, there exists a countable homogeneous representative of $\mathcal{R}$ iff $\mathcal{R}$ is amalgamable.

• Let R be a countable homogeneous relation. Let A, B be two finite restrictions of R, where B is an extension of A, and let g be an isomorphism from A into a third finite restriction C of R. Then by hypothesis g is extendible to an automorphism of R. Let f' denote the restriction of this automorphism to the base $|B|$. Let f be the identity on $|A|$, let g' be the identity on $|C|$ and finally let D be the restriction of R to $|C| \cup \overline{(f')}(|B|)$. We thus have the condition of amalgamation.

Conversely, let $\mathcal{R}$ be an amalgamable age. Let B be a finite relation belonging to $\mathcal{R}$, and f a local automorphism of B. Let A denote the restriction of B to $\text{Dom} f$. We shall construct an extension D of B, still belonging to R, which is a local automorphism extending f to the domain $|B|$. For this, let $C = B$, and denote by g the identity on $|A| = \text{Dom} f$. Now using the amalgamability condition, obtain an extension D of B with an isomorphism g' from B onto a restriction of D, such that for every $x \in \text{Dom} f$, we have $g'x = (g' \circ g)x = (f' \circ f)x = fx$: so that g' extends f.

Start with an ω-sequence of finite relations A_i(i integer), the set of whose restrictions gives the amalgamable age $\mathcal{R}$. Let $B_0 = A_0$. Using the preceding, replace B_0 by its extension B_1 belonging to $\mathcal{R}$, such that every local automorphism of B_0 (there are only finitely many) has an extension which is a local automorphism of B_1 and whose domain is B_0. Moreover we require that A_1 be embeddable in B_1, which is possible since every age is directed.

Iterating this, we obtain an ω-sequence of elements B_i of $\mathcal{R}$, such that each B_{i+1} extends B_i, and A_i is embeddable in B_i for each integer i.

Take the common extension R of the B_i to the union of bases. Then every local automorphism of R with finite domain is extendible, with alternatively a domain and a range containing every element in every base $|B_i|$: the homogeneity is proved. •

12.2.2 Connection with the criterion

Given an amalgamable age, the criterion of 11.4 is satisfied, and the unique denumerable homogeneous representative of our age (up to isomorphism) is a saturated relation.

• Let R be a denumerable homogeneous relation, S a relation representing the same age, and f a local isomorphism from R into S, with finite domain F. Take an arbitrary finite superset G of $\overline{f}(F)$. There exists an isomorphism h from S/G onto a restriction of R. Since R is homogeneous, the local automorphism $h \circ f$ is extendible to an automorphism k of R. Then the composition $k^{-1} \circ h$ is an isomorphism from S/G onto a restriction of R, and extends f^{-1}. Replacing G by any finite superset (included in the base $|S|$), the same argument shows that $k^{-1} \circ h$ is a 1-morphism from S into R. Hence R is saturated (see the definitions in 11.3.4. •

It is proved by [108] HENSON 1972 that there exist continuum many mutually non-isomorphic denumerable homogeneous relations. Hence continuum many amalgamable ages. For other results on homogeneous relations, see [127] JONSSON 1965 and [23] CALAIS 1967.

12.3 Relational system, orbit, transitive group, adherence of a permutation group

12.3.1 Relational system, homogeneous system

Given a set E, a **relational system** with base E, or based on E, is any ordinal sequence whose terms are relations R_i(i ordinal) based on E. Letting n_i denote the arity of R_i, the ordinal sequence of integers n_i is said to be the **arity** of the system, and each R_i is said to be a **component** of the system.

By taking a finite sequence of relations based on E, we find again the multirelation based on E.

The notions of restriction of a system to a subset of the base, extension to a superset of the base, isomorphism, automorphism, local isomorphism or automorphism, all extend immediately to the case of relational systems.

The notion of homogeneous relation also extends to relational systems: a system R is said to be p-**homogeneous** iff every local automorphism of R, defined on p elements, is extendible to an automorphism of R. A system is said to be **homogeneous** iff it is p-homogeneous for every integer p.

However we have an important difference which prohibits certain generalizations. Indeed the number of relational systems of a given arity and of a given finite base is in general infinite. In particular RAMSEY's theorem, used when partitioning the p-element subsets of the base into a finite number of classes (colors) corresponding to different isomorphism types, can no longer be used systematically. The same remark holds for the coherence lemma.

A relational system R is said to be p-**homogeneous** iff every local automorphism of R, defined on p elements, is extendible to an automorphism of R.

A relational system is said to be **homogeneous** iff it is p-homogeneous for every integer p. These definitions generalize those of 12.1.1.

12.3.2 Orbit of an n-tuple or of an n-element set

Let E be a set, and H a set of permutations of E, not necessarily a group. Given an integer n and an n-tuple of elements $a_1, ..., a_n$ in E, we call the **orbit** of the n-tuple (modulo H) the set of n-tuple images of $a_1, ..., a_n$ under any permutation belonging to H.

Given an n-element set F, we call the **orbit** of F (modulo H) the set of n-element set images of F under any permutation belonging to H.

If H is a group, then the existence of a permutation belonging to H, which takes one n-tuple into another, is an equivalence relation between n-tuples. Similarly for n-element sets.

12.3.3 n-transitive group, n-set-transitive group

A group of permutations of E is said to be n-**transitive** iff for any two n-tuples there exists a permutation belonging to the group and taking one n-tuple into the other. In other words, iff there exists only one orbit of n-tuples.

For example the symmetric group formed of all permutations of E.

Another example. If the base E has finite cardinality at least equal to $n+2$, then **the alternating group formed of all even permutations, is n-transitive**.

An n-transitive group is m-transitive for any $m \leq n$.

Recall **Jordan's hypothesis, 1893: for $n \geq 6$, every n-transitive group (on finitely many elements) is either symmetric or alternating**. Affirmative solution; see [25] CAMERON 1981 p.9.

A group of permutations on E is said to be **n-set-transitive** iff any two n-element sets are transformable one into the other by a certain permutation of the group. In other words, iff there exists only one orbit of n-element sets.

An interesting example of a 2-set-transitive group has been considered in 9.7.

12.3.4 Adherent permutation, group closed under adherence

Let E be a set, n an integer and H a set of permutations of E. A permutation f of E is said to be **n-adherent** to H iff for any elements $a_1, ..., a_n$ in E, there exists a permutation of H taking a_1 into fa_1, ... , and a_n into fa_n.

Every permutation which is n-adherent to H is m-adherent for any $m \leq n$.

A set H of permutations of E is said to be **closed under n-adherence** iff every permutation of E which is n-adherent to H belongs to H.

If H is closed under n-adherence, then H is closed under m-adherence for each $m \geq n$.

A permutation is said to be **adherent** to H iff it is n-adherent for every n.

A set H of permutations is said to be **closed under adherence** iff every adherent permutation belongs to H.

(1) **Let R be a relational system with base E, formed of components all of arity $\leq n$. Then the group of automorphisms of R is closed under n-adherence.**

(2) **For every relational system, the group of automorphisms is closed under adherence.** Consequences of 9.1.6.

12.3.5 Three lemmas about adherence

(1) **Let G be a group of permutations of E, which is closed under n-adherence. Then there exists a ($\leq n$)-homogeneous relational system, formed of n-ary components, whose group of automorphisms is G** (uses axiom of choice when E is uncountable).

(2) **Let G be a group of permutations of E, which is closed under adherence. Then there exists a homogeneous relational system whose group of automorphisms is G** (same remark).

(3) Assume that E has finite cardinality h. Then for every group G of permutations of E, there exists a homogeneous multirelation of maximum arity h whose group of automorphisms is G.

• (1) To each n-tuple of elements of E, associate the orbit, i.e. the class of n-tuples which can be obtained from it by taking its image under any permutation in G. Then to each orbit, associate the n-ary relation based on E, which takes the value (+) for those n-tuples in the orbit, and (-) otherwise. Using the axiom of choice, take an ordinal sequence R of these relations (each bijectively associated to an orbit), which form a system. Every permutation belonging to G is an automorphism of R. Conversely, every automorphism g of R takes each n-tuple in E into an n-tuple belonging to the same orbit. Thus g belongs to G, since G is closed under n-adherence.

It remains to see that R is $(\leq n)$-homogeneous. Given a local automorphism f of R, defined on a domain of cardinality $\leq n$, take an arbitrary n-tuple containing all the elements $a_1, ..., a_n$ of F, with possible repetitions. This n-tuple and its image $fa_1, ..., fa_n$ belong to the same orbit of G. Thus there exists a permutation of E which extends f and belongs to G, hence which is an automorphism of R. •

• (2) Analogous proof, but where n takes all integer values and G is closed under adherence, instead of n-adherence. •

• (3) Particular case where G is closed under h-adherence, and with finitely many orbits. •

By the preceding propositions, **to every relational system, there corresponds a homogeneous system on the same base, having the same automorphisms, and without augmenting the maximum arity of the components**.

Note however that, starting from a simple relation, we can end up at a homogeneous relational system with infinitely many components. For example start with the chain of non-negative integers, whose only automorphism is the identity. Then we end up with the sequence of singleton unary relations of all the integers.

12.3.6 Homogeneity and set-transitivity

(1) **Let p be an integer, and R be a p-monomorphic, p-homogeneous relational system. Then the group of automorphisms of R is p-set-transitive**.

Indeed, for any two p-element subsets a, b of the base, there exists an isomorphism from R/a onto R/b, which is extendible to an automorphism of R.

(2) Let E be a set and G be a p-set-transitive group of permutations of E. **Then the closure G^* of G under adherence is again p-set-transitive, and conversely**.

Moreover there exists a homogeneous relational system whose automorphism group is G^*, and such a relational system is always p-monomorphic.

This follows from the previous subsection. Axiom of choice needed for E uncountable.

12.3.7 Counterexamples

(1) Example of **a group of permutations which is closed under** $(n+1)$-**adherence but not under** n**-adherence** (due to POUZET in 1979; published in ToR-86 p.320).

• Let E be a set of cardinality $n+2$, and f an odd permutation of E. Let G be a group of even permutations generated by $f \circ t$, where t is an arbitrary transposition interchanging two elements of E. Let g be a permutation $(n+1)$-adherent to G. Then g belongs to G . Indeed, there exists a transposition t such that g and $f \circ t$ are identical on $n+1$ elements of E, and hence are still identical on the $(n+2)$-nd and last element of E.

Since f is odd, it does not belong to G: it suffices to see that f is n-adherent to G. Indeed, for any $x_1, ..., x_n$ belonging to E, let y, z be two distinct elements which are also distinct from the x. Let t be the transposition (y, z). Then $(f \circ t)x_1 = fx_1, ..., (f \circ t)x_n = fx_n$, which proves the n-adherence of f. •

This example extends to the case where E is infinite, by taking $n+2$ elements in E and repeating the preceding construction.

(2) **Another example**. Let E be the set of all $(n+1)$-tuples of rationals, which we shall call the rational vector space of dimension $n+1(n$ integer $\geq 2)$. Let G be the group of linear permutations of E with positive determinants.

• The group G is not closed under$n$-adherence. Indeed, let f be a linear permutation of E with negative determinant, hence f does not belong to G. Take n arbitrary points $a_1, ..., a_n$ in E, then compose f with the symmetry with respect to a hyperplane passing through $a_1, ..., a_n$. We obtain a linear permutation with positive determinant, hence belonging to G, which takes a_1 into $fa_1, ..., a_n$ into fa_n.

On the other hand, G is closed under $(n+1)$-adherence. Indeed, let g be a permutation $(n+1)$-adherent to G. Take $n+1$ points $x_1, ..., x_{n+1}$ belonging to E, which are linearly independent. The images $gx_1, ..., gx_{n+1}$ are by hypothesis the images of the x under a linear permutation with positive determinant; hence they are linearly independent.

Let u be any point in E. Let $u_1, ..., u_{n+1}$ be the coordinates of u relative to the points x, so that we have $u = u_1.x_1 + ... + u_{n+1}.x_{n+1}$. Now define $v = u_1.x_1 + ... + u_n.x_n$ and $w = u_{n+1}.x_{n+1}$, so that $u = v + w$. The points $x_1, ..., x_n, v$ are linearly dependent. Since their images under g are by hypothesis of $(n+1)$-adherence identical to their images under a certain linear permutation belonging to G, the dependence relation is preserved: $gv = u_1.(gx_1) + ... + u_n.(gx_n)$. Similarly w and x_{n+1} are linearly dependent, so that we have $gw = u_{n+1}.(gx_{n+1})$; similarly $gu = gv + gw = u_1.(gx_1) + ... + u_{n+1}.(gx_{n+1})$.

Letting u vary, we see that g is a linear permutation which takes each x_i into $gx_i (i = 1, ..., n+1)$. Hence g has positive determinant, so that g belongs to G. •

In the preceding proof, the hypothesis $n \geq 2$ is used to go from $u = v + w$ to $gu = gv + gw$; for this, the 3-adherence of g is required.

On the other hand, take $n = 1$ thus $n+1 = 2$, so that E is the set of the rational points, or vectors in the plane, and G is the group of linear permutations of E with positive determinants. Then the permutation g which takes each vector $(0, v)$ into $(0, 2v)$ and which preserves each (u, v) when $u \neq 0$ (with u, v rationals), is 2-adherent to G; yet this g does not belong to G.

(3) Example, due to the same author, of **a group of permutations closed under adherence, yet not under n-adherence for any n.**

For each integer n, take a set E_n with cardinality n, where the E_n are mutually disjoint, and let E be their union. Take G to be the group of those permutations of E whose restriction to E_n, for each n, is an even permutation of E_n. Then a permutation of E which is adherent to G is necessarily an element of G. However, given an arbitrary integer n, a permutation f is n-adherent to G provided that its restriction to E_i is an even permutation of E_i for each $i \leq n+1$, and an arbitrary permutation of E_i for each $i \geq n+2$.

12.4 Increasing number of orbits: Livingstone, Wagner

12.4.1 Three statements

(1) **Theorem. Let p, q be integers, E a finite set with cardinality at least equal to $2p+q$ and G a group of permutations of E. Then the number of orbits (modulo G) of the $(p+q)$-element subsets of E is greater than or equal to the number of orbits of the p-element subsets** ([156] LIVINGSTONE, WAGNER 1965).

• Associate to the group G a homogeneous multirelation R whose automorphism group is G: see 12.3.5 proposition (3). Then two p-element sets a, b belong to the same orbit $(\bmod G)$ iff the restrictions R/a and R/b are isomorphic. Indeed by homogeneity, every isomorphism of one restriction onto another is extendible to an automorphism of R. Same result for the $(p+q)$-element sets. Thus our proposition follows from the fact that the number of isomorphism types of restrictions to $p+q$ elements is greater than or equal to the number of isomorphism types of the restrictions to p elements: see 3.6.1, profile increase theorem. •

(2) **Let E be a denumerable set, and G be a group of permutations of E. To each integer p associate the countable number of orbits of p-element subsets of E. Then this number increases with p** ([195] POUZET 1976).

• Consider the closure G^* of G under adherence, and note that for each p the orbits of the p-element sets $(\bmod G)$ are the same as the orbits of the p-element sets $(\bmod G^*)$. Take a relational system R whose automorphism group is G^*: see 12.3.5 proposition (2). The proof terminates as before, using the profile increase theorem 3.6.1. However we must note that this theorem extends to the case of a relational system with denumerably many components. Indeed, the multicolor theorem in 3.4.3 includes the case of infinitely many colors, hence here of infinitely many isomorphism types for certain values of p. •

(3) In particular let E have countable cardinality at least equal to $2p+q$, and

let G be a group of permutations of E. If G is $(p+q)$-set-transitive, then G is p-set-transitive (see the definition in 12.3.3).

• Indeed p-set-transitivity means that all the p-element subsets of E belong to the same orbit.

Another proof is obtained from 9.6.3, in view of the inequality: $p \leq \text{Min}(p+q, (\text{Card}\, E) - p - q)$; by using also 12.3.5 proposition (2). •

12.4.2 Orbits and groups of permutation

(1) Let p, q be two integers, and E be a set of cardinality $\geq 2p+q$; let G be a group of permutations of E. **Then every permutation of E which preserves the orbits of $(p+q)$-element sets $(\text{mod}\, G)$ also preserves the orbits of p-element sets.**

• Let G^* be the closure of G under adherence, and let R be a homogeneous relational system whose automorphism group is G^*: see 12.3.5 proposition (2). Let f be a permutation of E which takes each $(p+q)$-element set into another in the same orbit $(\text{mod}\, G)$, hence again in the same orbit $(\text{mod}\, G^*)$. Then f takes each restriction of R to p elements into an isomorphic restriction: see 3.6.2. Since R is homogeneous, for any p-element set a, there exists an automorphism of R, hence an element of G^*, hence an element of G, which takes R/a into the isomorphic restriction $R/\overline{f}(a)$. •

(2) Let E be a set of cardinality $\geq 2p+q$, and let G, H be two groups of permutations of E. **If every orbit of $(p+q)$-element sets $(\text{mod}\, G)$ is included in an orbit of $(p+q)$-element sets $(\text{mod}\, H)$, then every orbit of p-element sets $(\text{mod}\, G)$ is included in an orbit of p-element sets** $(\text{mod}\, H)$.

This result was conjectured by [10] BERCOV, HOBBY 1970, and a weaker version was proved by them; the present result is due to [195] POUZET 1976.

• Let a, b be two p-element sets belonging to the same orbit $(\text{mod}\, G)$. There exists a permutation g belonging to G, such that $\overline{g}(a) = b$. This permutation g preserves all the orbits $(\text{mod}\, G)$, and in particular the orbits of the $(p+q)$-element sets $(\text{mod}\, G)$, hence by hypothesis g preserves the orbits of the $(p+q)$-element sets $(\text{mod}\, H)$. By the previous proposition g preserves the orbits of the p-element sets $(\text{mod}\, H)$. Hence a and b belong to the same orbit $(\text{mod}\, H)$. •

12.5 Extensive subset, pseudo-homogeneous relation

Given a relation R with base E, we say that a finite subset F of E is **extensive** $\text{mod}\, R$ iff each local automorphism of R with domain F is extendible to an automorphism of R.

If F is extensive and G is a subset of E with R/G isomorphic to R/F, then G is extensive.

Note that R is homogeneous iff every finite subset of its base is extensive $(\text{mod}\, R)$.

A relation R with base E is said to be **pseudo-homogeneous** iff each finite subset of E is included in an extensive finite subset.

The relation of consecutivity on the positive and negative integers, i.e. on Z, is pseudo-homogeneous yet not homogeneous.

• Complete each finite set F by the least interval (mod $\mathbf{Z}$) including F: we obtain a finite extensive superset of F. In general F itself is not extensive: take $F = \{0, 2\}$ and the local automorphism which preserves 0 and takes 2 into 3. •

The consecutivity relation associated with Z is not rich for its age: the saturated, and thus rich denumerable relation representing this age, is obtained by taking denumerably many components each isomorphic with our consecutivity relation: see 10.5.1, third example.

The saturated denumerable tree which is described in 10.5.3 **is pseudo-homogeneous yet not homogeneous.**

• Repeating our argument of 2.11.6, take three incomparable elements a, b, c with an element $d < a$ and $d < b$, yet $d|c$. Take three other incomparable elements a', b', c' with an element $e' < b'$ and $e' < c'$ yet $e'|a'$. Then the local automorphism which takes a, b, c into a', b', c' is not extendible to an automorphism. Indeed the image of d would be $< b'$, thus comparable with e' and finally less than c', unless that e' be less than a'.

To see that our relation is pseudo-homogeneous, complete each finite set F to G in a manner that any two elements of F have a common predecessor in G. •

12.5.1 A characterization of pseudo-homogeneous relations

(1) **Every extensive subset is maximalist. Consequently every pseudo-homogeneous relation is maximalist** (see 11.2.3 and 11.2.4).

The converse is false: take a consecutivity relation formed of two, or several components, each isomorphic with the consecutivity relation on $\mathbf{Z}$: this is a maximalist non-pseudo-homogeneous relation.

(2) In the case of a pseudo-homogeneous relation, the proposition 12.1.2 becomes:

Let E be a denumerable set; then a relation R based on E is pseudo-homogeneous iff there exists a set of finite subsets F of E such that every finite subset is included in an F; furthermore, for any local automorphism f of R having an F as domain and for any finite subset G of E including this F there exists a local automorphism extending f to G.

12.5.2 A condition for isomorphism

The statement in 12.1.3 proposition (2) can be extended:

Any two denumerable pseudo-homogeneous relations with the same age, are isomorphic.

• Let R and R' be two denumerable pseudo-homogeneous relations of the same age. Start with a local isomorphism f from R into R', whose domain F is an extensive finite subset (mod R). Let $F' = \overline{f}(F)$ be the range, and note that for any finite subset G' of the base $|R'|$ including F', the inverse function f^{-1} is extendible to a local isomorphism from R' into R with domain G'. Indeed, by hypothesis there exists a local isomorphism g from R' into R with domain G'. Then the composition $g \circ f$ is a local automorphism of R with extensive domain F, hence $g \circ f$ is extendible to an automorphism h of R. Finally $h^{-1} \circ g$ is the desired local isomorphism from R' into R with domain G'.

Now it suffices to choose G' to be extensive (mod R'). Then by iterating this, we extend alternatively the local isomorphism from R into R', then the local isomorphism from R' into R, by taking successively each element of $|R|$ in the domain of one, and each element of $|R'|$ in the domain of the other. •

12.5.3 An age represented by a pseudo-homogeneous relation

The statement in 12.1.3 proposition (3) cannot be extended to pseudo-homogeneous relations: see the example already given of the consecutivity relation on **Z**.

There even exists an age which is represented by a pseudo-homogeneous relation, yet by no homogeneous relation and by no rich relation.

• Let C be the consecutivity relation on the non-negative integers. Denote by 0 the singleton unary relation taking the value (+) for the integer 0 only. Let A be a unary relation such that, for every finite sequence of (+) and (-), there exists a sequence of consecutive integers giving to A this sequence of values.

Then the trirelation $(C, 0, A)$ is pseudo-homogeneous, since each interval beginning with zero and going up to an arbitrary integer, is extensive with the identity as the unique local automorphism. However, no denumerable relation is rich for this age, because of the existence of continuum many possible components (constructed for instance, by starting from the consecutivity relation on **Z**) which can be added, without strengthening the represented age •

Example communicated by POUZET and published in ToR-86 p.338; note the analogy with SPECKER's argument in 10.5.4.

12.6 Pseudo-amalgamable age, pseudo-amalgamation theorem

Lemma. Let R be a pseudo-homogeneous relation. Then the set of the restrictions of R to extensive finite subsets (mod R) is amalgamable (see 12.2).

• Let A, B, C be three restrictions of R to extensive finite subsets of the base. Let f be an isomorphism from A onto a restriction of B, and g be an isomorphism from A onto a restriction of C. The images of the base $|A|$ under f and under g are extensive sets. Thus there exists an automorphism f' of R, extending f^{-1}; and similarly an automorphism g' of R, extending g^{-1}. Restrict f' to the domain $|B|$, and g' to the domain $|C|$. Then let D be the finite restriction of R to the union $\overline{f'}(|B|) \cup \overline{g'}(|C|)$. The amalgamation property is satisfied, since for each x in the base $|A|$ we have $(f' \circ f)x = x$ and $(g' \circ g)x = x$. •

We leave it to the reader to prove that every amalgamable set of finite restrictions of R, which generates the age of R under embeddability, is (up to isomorphism) a subset of the set of restrictions of R to extensive finite sets.

An age $\mathcal{R}$ is said to be **pseudo-amalgamable** iff there exists a subset of $\mathcal{R}$ which is amalgamable, and furthermore generates $\mathcal{R}$ under embeddability; hence this subset is directed.

Pseudo-amalgamation theorem. Given an age $\mathcal{R}$, there exists a countable pseudo-homogeneous representative of $\mathcal{R}$ iff $\mathcal{R}$ is pseudo-amalgamable ([23] CALAIS 1967, to whom the notion of pseudo-amalgamable age is due).

• Suppose that there exists a pseudo-homogeneous relation R which represents $\mathcal{R}$. Then the set of restrictions of R to extensive finite subsets of the base, is amalgamable, by the preceding lemma. And by definition of the pseudo-homogeneous relations, this set covers every finite subset of the base (under inclusion). Thus our age is pseudo-amalgamable.

Conversely, suppose that the age $\mathcal{R}$ is pseudo-amalgamable. Let $A_i(i$ integer) be a finite, or an ω-sequence of finite relations, which under embeddability gives $\mathcal{R}$, and which furthermore forms an amalgamable set.

We shall construct two sequences of finite relations B_i and $C_i(i$ integer), each of which is isomorphic with an $A_j(j$ integer); such that for each i we have the embeddability $A_i \leq B_i$, and B_i is a restriction of C_i, and C_i is a restriction of B_{i+1}. Moreover, for any two restrictions of B_i which are isomorphic to a same A_j, we require that every isomorphism from one onto the other, be extendible to an isomorphism from B_i onto another restriction of C_i. With this construction, the common extension of the B_i (or the C_i) to the union of their bases, satisfies the condition of 12.5.1 proposition (2), hence is pseudo-homogeneous and represents the given age.

For this sketched construction, start with $B_0 = A_0$. Suppose that B_i is already obtained, and let $f_{i,0}, \ldots, f_{i,p}$ be all the local automorphisms of B_i between restrictions isomorphic to a same $A_j(j$ variable). By amalgamation, by starting from the bijection $f_{i,0}$ and from the identity on the base $|A_j|$ which both transform A_j into a restriction of B_i, we obtain an extension $D_{i,0}$ of B_i, which belongs to our age $\mathcal{R}$, and such that $f_{i,0}$ is extendible to an isomorphism from B_i onto another restriction of $D_{i,0}$.

Iterate this to obtain the sequence, with finite length p, of successive extensions $D_{i,1}, \ldots, D_{i,p}$ corresponding to $f_{i,1}, \ldots, f_{i,p}$. Then take for C_i an extension of $D_{i,p}$ which is isomorphic to an A_j. Finally to be sure that the entire age is represented, take for B_{i+1} a common extension of C_i and A_{i+1}, which furthermore is chosen to be an A_j. •

12.7 Prehomogeneous relation

12.7.1 Relatively extensive subset

Given a relation R and two finite subsets of its base, say F and G including F, we say that F is **G-extensive** modulo R iff for every local automorphism f of R with domain G, the restriction f/F is extendible to an automorphism of R.

In the case where $G = F$, we find again the extensive finite set, in the sense of 12.5.

If F is G-extensive and if F', G' are the images of F, G under the same local automorphism of R, then F' is G'-extensive.

If F is G-extensive, then every subset of F is G-extensive. Furthermore F or each subset of F is G'-extensive for every finite superset G' of G.

A relation R is said to be **prehomogeneous** iff each finite subset F of its base admits a finite superset G for which F is G-extensive. Notion due to [181] PABION 1972. See also [191] POUZET 1972, C.R. p.813), and [206] POUZET, ROUX 1996.

Every homogeneous or pseudo-homogeneous relation is prehomogeneous.

12.7.2 Relative extensivity and maximalism

If F admits a finite superset G for which F is G-extensive, then F is maximalist.

• Take any extension S of R which represents the same age. Given an arbitrary finite subset H of the base $|S|$, including G, take an isomorphism h from S/H onto the restriction $R/\overline{h}(H)$. By hypothesis h/F is extendible to an automorphism of R. By composition, there exists an F-isomorphism from S/H onto a restriction of R: we see that F satisfies the third definition of a maximalist subset, 11.2.3. •

Consequently **every prehomogeneous relation is maximalist**.

The converse is false. Take up the counterexample in 12.5.1 propositiion (1); i.e. a consecutivity relation formed of several components, each isomorphic with the consecutivity on $\mathbf{Z}$.

12.7.3 Isomorphic prehomogeneous relations

Any two denumerable prehomogeneous relations with the same age are isomorphic. This generalizes 12.1.3 and 12.5.2.

• Let R, R' be two denumerable prehomogeneous relations with bases E, E', representing the same age. Let F be a G-extensive finite subset (mod R) and let f be a local isomorphism from R into R', with domain G. Let G' be the range of f. Take a finite superset H' of G' for which G' is H'-extensive (mod R'). By the same argument as in 12.5.2, we see that the inverse function $(f/F)^{-1}$ is extendible to a local isomorphism g' from R' into R with domain H. Similarly the inverse function $(g'/G')^{-1}$ is extendible to a local isomorphism from R into R', and so forth, going alternatively from R into R' and from R' into R.

To terminate the proof and obtain a common extension of our local isomorphisms which is an isomorphism from R onto R', it suffices to note that F can contain an arbitrary element of E then that G' can contain an arbitrary element of E'. •

12.7.4 Relation R_A associated with the chain A

Given a chain A, let R_A be the ternary relation freely interpretable in A, which is defined by $R_A(x, y, z) = +$ iff $x < z$ and $y < z (\text{mod}\, A)$ and $x \neq y$.

(1) **Let $\mathbf{Q}$ be the chain of the rationals; then $R_{\mathbf{Q}}$ is prehomogeneous but not pseudo-homogeneous.**

• Given a finite subset F of the base, let G be the set F augmented by one element which is strictly less $(\text{mod}\, \mathbf{Q})$ than the minimum element of F. Every local automorphism f of $R_{\mathbf{Q}}$ with domain G preserves the ordering of the elements $(\text{mod}\, \mathbf{Q})$, except when concerning the first two elements of G, where the ordering can be inverted. Hence the ordering of the elements of F is preserved; so that f/F is extendible to an automorphism of $\mathbf{Q}$, hence of $R_{\mathbf{Q}}$. Thus F is G-extensive $(\text{mod}\, R_{\mathbf{Q}})$; yet F is not extensive. •

(2) **Every denumerable younger relation than $R_{\mathbf{Q}}$ is of the form R_A, where A is a denumerable chain.**

• To each finite subset F of the base, associate the set of those chains C based on F and such that R_C be the restriction to F of the given relation. Then apply the coherence lemma 2.4.1. •

Corollary. The relation $R_{\mathbf{Q}}$ is rich for its age.

Hence **there exists an age having a rich representative, yet having no denumerable pseudo-homogeneous representative.**

Indeed a denumerable pseudo-homogeneous representative of the age under consideration would a fortiori be prehomogeneous, hence isomorphic with $R_{\mathbf{Q}}$.

Compare with the counterexamples in 12.5.3.

12.8 Isolated rel-age

12.8.1 Isolating pair, isolated rel-age

Let $\mathcal{R}$ be an age, and A, B be two finite relations belonging to $\mathcal{R}$, with B an extension of A. The couple (A, B) is said to be **isolating** (modulo $\mathcal{R}$) iff there exists one and only one maximal A-age specification of $\mathcal{R}$ which contains the element B. We say as well that this maximal A-age is **isolated by** B.

For example take $\mathcal{R}$ to be the age represented by the consecutivity relation on the non-negative integers. If A is the consecutivity relation of a finite chain, then the couple (A, A) is isolating, and the corresponding isolated maximal A-age is formed of all the consecutivity relations of finite chains including A as an interval, and their A-restrictions.

Now denote by A the common extension of the consecutivity relations of two finite chains, which we shall call the two **components**, with the value (-) for those couples formed of one element in each component.

Then there exist infinitely many maximal A-ages, obtained by deciding to put the first component before or after the second, with a non-zero finite number of intermediate elements. This can be done with the aid of a finite consecutivity relation denoted by B, which is an extension of A. Thus such a couple (A, B) is isolating.

Now with the same A, define the A-age obtained by never taking the union of the two components with finitely many intermediate elements. Then the maximal A-age thus defined is isolated by no couple of relations.

12.8.2 Two lemmas

(1) **If (A, B) is isolating (modulo $\mathcal{R}$), then so is (A, B') for every B' which is a finite extension of B and belongs to the age $\mathcal{R}$.**

Hence B' also belongs to the A-age specification of $\mathcal{R}$ and isolated by B.

If (A, B) is isolating, then so is (A', B) for each restriction A' of A.

• Let $\mathcal{A}$ be the unique maximal A-age to which the relation B belongs. Let $\mathcal{A}'$ be the A'-age induced by $\mathcal{A}$ (see 11.1.3). So $\mathcal{A}'$ contains B and is itself maximal, by 11.2.1. Suppose that there exists a maximal A'-age $\mathcal{B}'$ distinct from $\mathcal{A}'$ and containing the element B. Then there exists an A-age specification $\mathcal{B}$ of $\mathcal{B}'$, which contains B and is maximal. By hypothesis $\mathcal{B}$ is identical with $\mathcal{A}$, and hence $\mathcal{B}'$ (induced by $\mathcal{B}$) is identical with $\mathcal{A}'$ (induced by $\mathcal{A}$) : contradiction. •

(2) Let R be a relation, $\mathcal{R}$ the age represented by R, and F, G including F, two finite subsets of the base. If F is G-extensive (mod R), then **there exists a unique maximal (R/F)-age specification of $\mathcal{R}$, which contains the element R/G.**

Moreover the (R/F)-age thus isolated by R/G is represented by R.

• Suppose that F is G-extensive, yet that there exist two maximal (R/F)-ages specifications of $\mathcal{R}$, which contain the element R/G. In the first one, take an extension U' of R/G, and in the second, take an extension $U"$, such that no (R/F)-age specification of $\mathcal{R}$ contains both U' and $U"$: see 11.2.2.

Now take two finite subsets $H', H"$ of the base, with R/H' the image of U' under an isomorphism denoted by f', and $R/H"$ the image of $U"$ under $f"$. Set $F' = \overline{f'}(F)$ and $G' = \overline{f'}(G)$, and similarly $F" = \overline{f"}(F)$ and $G" = \overline{f"}(G)$. By the hypothesis of G-extensiveness, there exists an automorphism h' of R which takes F' into F with $h'/F' = (f'/F)^{-1}$ and another automorphism $h"$ which takes $F"$ into F with $h"/F" = (f"/F)^{-1}$. Then the (R/F)-age represented by R contains the element $R/\overline{h'}(H')$, which is the image of U' under an F-isomorphism; as well as the element $R/\overline{h"}(H")$, which is the image of $U"$: contradiction. Thus we have proved the uniqueness of the maximal (R/F)-age containing R/G.

We shall now prove our second conclusion. Let $\mathcal{A}$ denote the unique maximal (R/F)-age which contains the element R/G. For each extension U of R/G belonging to $\mathcal{A}$, there exists an isomorphism from U onto a restriction of R. Since F is G-extensive, by composition with an automorphism of R, we can require that our previous isomorphism be an F-isomorphism. Thus $\mathcal{A}$ is included in the (R/F)-age represented by R. Since it is maximal, it is the (R/F)-age of R. •

12.8.3 Connection with maximalist relations

Let R be a relation and $\mathcal{R}$ the age represented by R.

Suppose that R is maximalist, and let F and G including F be two finite subsets of the base, such that the couple $(R/F, R/G)$ is isolating (modulo $\mathcal{R}$). Then for every local automorphism h of R with domain G, the restriction h/F is a 1-isomorphism from R into R itself.

• Let $F' = \overline{h}(F)$ and $G' = \overline{h}(G)$, and take the image under h^{-1}/F' of the (R/F')-age represented by R. Since R is maximalist, this (R/F')-age is maximal. In taking its image under h^{-1}/F', we obtain a maximal (R/F)-age which contains the element R/G (up to F-isomorphism). Since the couple $(R/F, R/G)$ is isolating, we obtain precisely the (R/F)-age represented by R. Thus h^{-1}/F', hence also the inverse function h/F, is a 1-isomorphism from R into itself. •

12.8.4 Isolating pair and extensive subset

Let R be a relation with denumerable base E, and let $\mathcal{R}$ be the age represented by R.

Suppose that (1) **R is maximalist**, and

(2) **for each finite subset F of E, there exists a finite subset G including F, such that the couple $(R/F, R/G)$ is isolating** $(\mathrm{mod}\,\mathcal{R})$;

then R is prehomogeneous.

More precisely, for each of the preceding couples $(F, G$, the subset F is G-extensive $(\mathrm{mod}\,R)$.

• Let f be a local automorphism of R with domain G. We shall construct an automorphism of R extending f/F as follows. Set $F' = \overline{f}(F)$ and $G' = \overline{f}(G)$, and let F_1 be an arbitrary finite superset of F and G_1 a finite set including F_1, such that $(R/F_1, R/G_1)$ is isolating. By the preceding proposition, f/F is a 1-isomorphism from R into R. Thus there exists a function f_1 extending f/F to the domain G_1, which is a local automorphism of R.

Iterating this, by taking each element of the base, alternatively in the domain and in the range, we obtain an ω-sequence of extensive local automorphisms, whose union is an automorphism of R extending f. •

12.8.5 Connection with specifications

Let $\mathcal{R}$ be an age. Then either, for each finite relation A belonging to $\mathcal{R}$ there exists a finite extension B of A belonging to $\mathcal{R}$, such that (A, B) is isolating $(\mathrm{mod}\,\mathcal{R})$; or there exist continuum many maximal rel-ages specifications of $\mathcal{R}$.

• Let A be an element of $\mathcal{R}$, having no extension which together with A forms an isolating pair. Take two distinct maximal A-ages specifications of $\mathcal{R}$, and by 11.2.2 take a relation A_0 belonging to the first and a relation A_1 belonging to the second, such that there exists no A-age specification of $\mathcal{R}$ which contains both A_0 and A_1.

By hypothesis, among the relations belonging to $\mathcal{R}$ and extending A_0, there exists none which together with A forms an isolating pair. Thus there exist two distinct maximal A-ages specifications of $\mathcal{R}$, each of which contains the element A_0. Using again 11.2.2, take an element $A_{0,0}$ belonging to the first and $A_{0,1}$ belonging to the second, such that there exists no A-age specification of $\mathcal{R}$ which contains both $A_{0,0}$ and $A_{0,1}$. Moreover, since each A-age is directed, we can always require that $A_{0,0}$ and $A_{0,1}$ be extensions of A_0.

Dichotomously, in a similar fashion, obtain $A_{1,0}$ and $A_{1,1}$ starting with A_1. We obtain continuum many A-ages which we can always take to be maximal. •

12.9 Existence criterion for a prehomogeneous relation

(1) **Given an age $\mathcal{R}$, there exists a denumerable prehomogeneous representative of $\mathcal{R}$ iff for each finite relation A belonging to $\mathcal{R}$, there exists a finite extension B of A belonging to $\mathcal{R}$, such that (A, B) is isolating** (mod $\mathcal{R}$).

(2) Consequently **if there exists a denumerable representative of $\mathcal{R}$ which is rich for $\mathcal{R}$, then there exists a denumerable prehomogeneous representative of $\mathcal{R}$** ([191] POUZET 1972).

• By the existence criterion for a rich relation (see 11.4), if there exists a relation rich for its age $\mathcal{R}$, then there are countably many maximal rel-ages specifications of $\mathcal{R}$. Hence by the preceding proposition, for each element A of $\mathcal{R}$, there exists an extension B of A belonging to $\mathcal{R}$, such that the pair (A, B) is isolating. Thus our conclusion (2) follows from (1).

To prove (1), note first that, if there exists a prehomogeneous representative R of our age, hence if for every finite subset F of the base, there exists a finite G including F and for which F is G-extensive, then by 12.8.2 proposition (2), our criterion is satisfied, since the couple $(R/F, R/G)$ is isolating.

Conversely, assume the criterion in statement (1). Start with an ω-sequence of finite relations A_i(i integer) belonging to our age $\mathcal{R}$, each element of this age being embeddable in an A_i.

Set $B_0 = A_0$. Let C_0 be an element of $\mathcal{R}$ such that (B_0, C_0) is isolating. Let B_1 be an element in $\mathcal{R}$ which is a common extension of C_0 and A_1. Then let C_1 be such that (B_1, C_1) is isolating, and so forth. The common extension R of the B_i (or equivalently the C_i) is a representative of our age, and satisfies the second hypothesis of 12.8.4.

It remains to show that R is maximalist, which is the first hypothesis of 12.8.4. To that end, it suffices to see that each finite subset of the base $|R|$ is included in a maximalist finite subset. Indeed modify the preceding construction as follows. When we obtain C_0, take a finite extension D_0 belonging to our age, which is chosen so as to make the B_0-age represented by R maximal. Thus choose D_0 to admit an embedding of every extension of B_0 to one additional element which belongs to the unique maximal B_0-age isolated by C_0. When we obtain C_1, take a finite extension D_1 belonging to our age, which admits an embedding of every extension of B_0 to two additional elements, which belongs to the unique maximal B_0-age isolated by C_0, and which belongs to the unique maximal B_1-age isolated by C_1; and so forth. •

Recall that there exists an age having a prehomogeneous, and even a pseudo-homogeneous representative, yet having no rich representative: see 12.5.3.

12.10 Exercise

12.10.1 Permutation group generated by the finite local automorphisms of a relation

Let R be a relation, or a relational system, with base E. Consider the groups G_R of permutations of E, which are closed under adherence, and such that every local automorphisms of R with finite domain is extendible to an element of G_R. Note that the symmetric group (group of all permutations) is a G_R for every system R.

The intersection of all these G_R is a group, and shall be called the **group generated by the finite local automorphisms of** R.

1 - Note that the group thus generated is closed under adherence, and that every automorphism of R belongs to it.

2 - In the case of a homogeneous relational system R, the group generated is identical to the group of automorphisms of R.

Problem 1. Is the group thus generated always a G_R. In other words, is every finite local automorphism of R extendible to a permutation belonging to all the G_R.

Problem 2. Take R to be the chain of non-negative integers. What is the group generated. Is it identical to the symmetric group, or is it reduced to the identity.

Chapter 13

Bound of a relation, well relation, compatibility and chainability theorems, indicative group

13.1 Bound of a relation, bound of an age

13.1.1 Definitions and generalities

Given a relation R, a **bound** of R is any finite relation A with same arity, such that A is not embeddable in R, but every proper restriction of A is embeddable in R. Note that a bound has necessarily a non-empty base.

If we consider the partial ordering of embeddability between finite relations (considered up to isomorphism) and the initial interval formed of all finite restrictions of R (i.e. the age of R in the sense of 10.2.1), then we find the bounds of this initial interval, in the sense of 4.10. So that we shall speak of **a bound of the age represented by** R.

An age is completely defined by the set of its bounds. Indeed embedding defines a well-founded quasi-ordering among finite relations; see 4.10.1 proposition (2).

If R has finite cardinality p, then any bound of R has cardinality $\leq p+1$.

Examples. For a reflexive binary relation, the binary relation of cardinality 1 which takes the value (-), is a bound.

For a reflexive, symmetric relation, we have the preceding bound plus the chain of cardinality 2.

For a binary relation always (+) of cardinality ≥ 1, we have the two preceding bounds plus the identity relation of cardinality 2. For an infinite relation always

(+), we only have these three bounds. For a relation always (+) of finite cardinality p, we additionally have as a bound the relation always (+) of cardinality $p+1$.

For a reflexive, antisymmetric relation, we have as a bound the relation of cardinality 1 taking the value (-), the relation always (+) and the identity relation of cardinality 2.

For a reflexive, antisymmetric and comparable relation (for any x, y the relation takes the value (+) either for (x, y) or for (y, x)), we have the two preceding bounds plus the identity relation of cardinality 2.

For a chain of cardinality ≥ 2, we have the three preceding bounds plus the reflexive binary cycle of cardinality 3. If the chain is infinite, then these four bounds are the only possible. If the chain has finite cardinality p, then in addition we have as a bound the chain of cardinality $p+1$.

13.1.2 Finitely bounded relation or age

A finite relation, a chain, a relation always (+) or always (-) , an identity relation, all have finitely many bounds (up to isomorphism). We speak of a **finitely bounded relation**, or its **finitely bounded age**.

If every finite relation with the same arity is embeddable in R, then R has no bound. Example: a rich relation.

On the other hand, the consecutivity relation C on the integers has infinitely many bounds. Indeed for each integer $p \geq 3$, the binary cycle with cardinality p is a bound of C.

13.1.3 Lemmas

(1) Let R, S be two relations; if $R \leq S$ then no bound of S is embeddable in R.

(2) For R finite or S finite, we have $R \leq S$ iff no bound of S is embeddable in R; in other words, iff either S has no bound or every bound of S is non-embeddable in R.

(3) Any two finite relations having the same bounds are isomorphic.

(4) If R is finite and $R < S$, then there exists a bound of R which is embeddable in S; hence there exists a bound of R which is not a bound of S.

(5) For R finite or S finite, if the set of bounds of S is properly included in the set of bounds of R, then $R < S$.

• (1) is obvious.

(2) For R finite, (2) follows from 4.10.1 proposition (1), since embeddability between finite relations is a well-founded partial ordering.

However let us give a direct proof. If $R \leq S$, then no bound of S is embeddable in R.

Conversely if $R \not\leq S$, then either R is a bound of S; or there exists a restriction R_1 of R to its base minus one element, such that $R_1 \not\leq S$. Iterate this: after a finite number h of steps we obtain a restriction R_h of R which is a bound of S. Now suppose that R is infinite and S finite, so that $R \not\leq S$. Replace R by a restriction R' whose cardinality is finite but strictly greater than the cardinality of S. Then $R' \not\leq S$ hence R' and consequently R admits an embedding of a bound of S.

(3) follows from (2).

(4) Since S is not embeddable in R, hence S admits an embedding of at least a bound of R.

(5) follows from (2) and (3); note that for R finite and S infinite, there necessarily exists a bound of R which is embeddable in S. •

13.1.4 A counterexample

If R and S are both infinite, then statement (3) does not necessarily hold.

• Take R to be the chain of the non-negative integers. Take a sequence of finite relations $A_i (i$ integer) with disjoint bases, and such that every finite binary relation is isomorphic with an A_i. Take S to be the common extension of the A_i, which takes the value (+) for every couple whose terms belong to the bases of two distinct A_i. Then S has no bound; and yet $R \not\leq S$.

Another example. Let R and S be two infinite chains mutually incomparable with respect to embeddability: for instance take S = chain of non-negative integers and R = chain of negative integers. Recall that all infinite chains have the same four bounds already given above: relation (-) on 1 element, relation always (+) on 2 elements, identity relation on 2 elements and reflexive binary cycle on 3 elements. Thus no bound of S is embeddable in R. •

13.1.5 Three equivalent conditions on embeddability and bounds

Let R, S be two relations of the same arity; then the following three conditions are equivalent.

(1) Every finite restriction of R is embeddable in S; in other words R is younger than S in the sense of 10.1.3.

(2) No bound of S is embeddable in R.

(3) Every bound of S admits an embedding of a bound of R.

• Assume condition (1) and let A be a bound of S. If $A \leq R$, then also $A \leq S$, hence A is not a bound of S.

Conversely, if there exists a finite restriction A of R which is non-embeddable in S, then there exists a restriction of A, hence of R, which is a bound of S. Thus (1) and (2) are equivalent.

Assume condition (2), and let A be a bound of S; hence A is non-embeddable in R. Thus there exists a restriction of A which is a bound of R.

Conversely, if there exists a bound A of S which is embeddable in R, then A admits no embedding of any bound of R. Thus (2) and (3) are equivalent. •

In particular, if R and S are finite, then the embeddability $R \leq S$ is equivalent to the condition that no bound of S is embeddable in R (this is already in 13.1.3 proposition (1)), or again equivalent to the condition that every bound of S admits an embedding of a bound of R.

Another consequence. Let R, S be two relations of the same arity; then R and S have the same age iff they have the same bounds; also iff no bound of R is embeddable in S and conversely.

13.1.6 Complementary lemmas

(1) Complete as follows 13.1.3 proposition (4).

If R and S are infinite and $R < S$, then the set of bounds of R cannot be properly included in the set of bounds of S.

In other words, either there exists a bound of R which is not a bound of S, or R and S have same bounds.

• Let A be a bound of S which is not a bound of R. By the previous subsection propositions (1) and (3), there exists a restriction B of A which is a bound of R. Necessarily B is a proper restriction of A, so that B is embeddable in S and hence B cannot be a bound of S. •

(2) Given two relations R, S, if every finite restriction of R is embeddable in S, then every bound of R is a bound of S or is embeddable in S. Indeed, let A be a bound of R. Assume that A is non-embeddable in S; yet every proper restriction of A is embeddable in R, hence in S. Thus A is a bound of S.

(3) The converse is false. Take R to be an infinite chain, and S to be the chain of cardinality 3. Then every bound of R is one of the four relations with cardinalities 1, 2, 3 considered in 13.1.1. Hence every bound of R is a bound of S. Yet the chain of cardinality 4 is a bound of S and not a bound of R.

Another example with R and S finite. Take R to be the binary reflexive cycle of cardinality 3. There exist four bounds of R, up to isomorphism. These are: the binary relation of cardinality 1 and value (-); the binary relation always (+) of cardinality 2; the identity relation of cardinality 2; finally the chain of cardinality 3. Take S to be the common extension of these four bounds, taken with disjoint bases, S taking the value (+) for every couple whose terms belong to the bases of two distinct bounds. Then each bound of R is embeddable in S, yet R is non-embeddable in S.

13.2 Well relation

We say that a relation (or a multirelation) R is **well** iff the set of its finite restrictions, when partially ordered under embeddability, forms a well partial ordering. In other words, if any set of finite restrictions of R, mutually incomparable under embeddability, is finite.

For example, every chain is well.

Every tree is well, by KRUSKAL's theorem (5.4).

The consecutivity relation on the non-negative integers is well. Indeed, each finite restriction can be represented by a finite sequence of positive integers, each integer i representing a component of i consecutive integers. Then the embeddability between two finite sequences of integers implies the embeddability between the two corresponding finite restrictions. Now it suffices to recall that embeddability between finite sequences, or words of integers, is a well partial ordering by HIGMAN's theorem 4.5.2.

13.2.1 p-well multirelation

Given a non-negative integer p, we say that a multirelation R is **weakly p-well** iff upon concatenating R with any p unary relations on the same base, we obtain a well multirelation. R is **p-well** iff concatenating independently each finite restriction of R with any p unary relations on the same base, then we get a well partial ordering under embeddability. Both notions are due to [193] POUZET 1972, who asked if they are equivalent (an open problem until now).

If R is p-well, then every multirelation interpretable in R is p-well.

• Let S be freely interpretable in R, and let F, F' be two finite subsets of the base. If S/F concatenated with p unary relations A based on F, and S/F' concatenated with p unary relations A' on F', are incomparable with respect to embeddability, then the same is true for R/F concatenated with the A and R/F' concatenated with the A'. •

The consecutivity relation on the non-negative integers is well, but is not 1-well.

• For each integer i, take a sequence of $i + 2$ consecutive elements, and define a unary relation to take the value (-) for the first and the last element, and the value (+) between. •

13.2.2 Case of chainable relations

Every chain, and consequently every chainable multirelation is p-well for each integer p. Due to [193] POUZET 1972.

• It suffices to see that the multirelations $(A, B_1, ..., B_p)$, where A is a finite chain and the B are unary relations, form, up to isomorphism, a set which is well partially ordered under embeddability.

To each multirelation, associate the finite cardinality h of its base. Then associate a word of length h, obtained by replacing each $i = 1, ..., h$ by the sequence of values $B_1(x), ..., B_p(x)$, where x designates the ith element of the base, ordered modulo A. We say that two of these sequences are considered to be incomparable iff they are distinct. Thus we have a well partial ordering of identity with 2^p elements, or sequences, mutually incomparable. By HIGMAN's theorem (4.5.2 countable case, thus provable in ZF), the set of words formed of the preceding elements, constitutes a well partial ordering, under the usual embeddability of words. Consequently we obtain a well partial ordering, under embeddability, of our finite multirelations. •

13.2.3 Theorems on finite number of bounds

(1) **Let R be a multirelation; denote by m the maximum of its components'arities. If R is $2m$-well, then R has finitely many bounds** (see [193] POUZET 1972). Consequently:

(2) **Every chainable multirelation has finitely many bounds** ([85] FRASNAY 1965).

• Let U be a bound of R and E be the base of U. Take an arbitrary element $a_1 \in E$, and let $E_1 = E - \{a_1\}$. Consider those subsets F of E_1 for which there exists an $S \leq R$ with base E, which coincides with U on E_1 and on $F \cup \{a_1\}$. Note that when the arity $m = 1$, then no subset F, even empty, satisfies this condition. For otherwise S would be identical with U, hence $U \leq R$.

If there exists no such F, then associate to U the following multirelation: $(S, V_1, ..., V_m, A_1, ..., A_m)$ where $S \leq R$ and S coincides with U on E_1; where the V and the A are unary, V_1 taking the value (+) on E_1 and only on E_1, and A_1 being the singleton unary relation of a_1; the other V and A taking always the value (-).

Now suppose that there exist subsets F satisfying our preceding condition. Take such an F of maximum cardinality, say E_2. This E_2 is a proper subset of E_1; for if $E_2 = E_1$, then we would have $U \leq R$. Let a_2 be an arbitrary element of $E_1 - E_2$. Consider those subsets F of E_2 for which there exists an $S \leq R$ with base E, which coincides with U on E_1 and on $E_2 \cup \{a_1\}$ and on $F \cup \{a_1, a_2\}$. Notice that when the arity $m = 2$, then no such F, even empty, exists. For otherwise this would yield that S coincides with U on the union $E_2 \cup \{a_1, a_2\}$, contradicting the maximality of E_2.

If there exists no such F, then associate to the bound U the following multirelation: $(S, V_1, ..., V_m, A_1, ..., A_m)$ where $S \leq R$ and S coincides with U on E_1 and on $E_2 \cup \{a_1\}$; where V_1 takes the value (+) only on E_1, and V_2 takes (+) only on $E_2 \cup \{a_1\}$; the relation A_1 is the singleton of a_1 and A_2 is the singleton of a_2, and finally the other V and A are always (-).

Iterating this, we obtain, at most, a strictly descending sequence $E_1 \supset E_2 \supset ... \supset E_m$ and elements $a_1 \in E - E_1$ and $a_2 \in E_1 - E_2$ and ... and $a_m \in E_{m-1} - E_m$. Thus there exists an $S \leq R$ with base E, which coincides with U on E_1, on $E_2 \cup \{a_1\}$, on $E_3 \cup \{a_1, a_2\}$, ... , on $E_m \cup \{a_1, ..., a_{m-1}\}$. Yet no multirelation S of maximum arity m can satisfy the preceding conditions, and moreover coincide with U on $\{a_1, ..., a_m\}$. For otherwise S would coincide with U on $E_m \cup \{a_1, ..., a_m\}$, contradicting the maximality of E_m.

Associate to the bound U the multirelation $(S, V_1, ..., V_m, A_1, ..., A_m)$ where $S \leq R$ and S coincides with U on E_1, on $E_2 \cup \{a_1\}$, ... , on $E_m \cup \{a_1, ..., a_{m-1}\}$, the relation V_1 taking the value (+) only on E_1, and V_2 only on $E_2 \cup \{a_1\}$, ... , and V_m only on $E_m \cup \{a_1, ..., a_{m-1}\}$; and finally A_1 being the singleton relation of a_1, ... , and A_m the singleton relation of a_m. In each case we see that there exists an $S \leq R$ coinciding with U when V_1 or V_2 or ... or V_m takes the value (+). Yet it is impossible that in addition, S coincides with U on the set of the elements giving the value (+) to A_1 or A_2 or ... or A_m.

Suppose now that R is $2m$-well. Then the multirelations (S, V, A) previously associated to the bounds, form a well partial ordering under embeddability. Suppose that there exist infinitely many bounds. Then because of this well partial ordering and by RAMSEY's theorem, there exists an ω-sequence of multirelations (S, V, A) associated to the bounds, and such that each multirelation admits an embedding of each preceding one. More precisely, we can assume that each (S, V, A) is an extension of the preceding. Then there are at most m elements for which A_1 or ... or A_m takes the value $(+)$. So there exists an ω-sequence extracted from the preceding one, for which all the bounds U have the same restriction to these elements.

Let U be one of these bounds, E its base, and let U' be another bound whose base properly includes E. Then $U'/E \leq R$. Let $(S, V_1, ..., V_m, A_1, ..., A_m)$ designate the multirelation associated with U, and similarly $(S', V_1', ..., V_m', A_1', ..., A_m')$ the multirelation associated with U': the second multirelation extends the first one. The restriction U'/E coincides with S', hence with S and U, on the set where V_1 takes the value $(+)$, on the set where V_2 takes $(+)$, .. , on the set where V_m takes $(+)$. Moreover, U'/E coincides with U on the set of those elements giving the value $(+)$ to A_1 or A_2 or ... or A_m: contradiction. •

13.3 Chains compatible for a permutation group; group-compatibility and chainability theorems (Frasnay)

13.3.1 Compatibility modulo a group

Let m be a positive integer. A group G of permutations on the set of integers $1, ..., m$ is said to be an m-**ary group**, or a group with **arity** m.

Two chains A, B are said to be G-compatible, or compatible mod G iff for every set of m elements in the intersection of the bases, say $a_1 < a_2 < ... < a_m \bmod A$, the permutation u which reorders these elements according to $a_{u(1)} < a_{u(2)} < ... < a_{u(m)} \bmod B$, belongs to the group G.

We see that G-compatibility is reflexive and symmetric. If the chains A, B have the same base, then G-compatibility is transitive, hence is an equivalence relation.

In the general case, note that two chains are G-compatible when the intersection of the bases has cardinality strictly less than m (arity of G). Hence there is no transitivity, since for example if A and A' have the same base and if the base $|B|$ is disjoint from $|A| = |A'|$, then A and B on the one hand, and A' and B on the other hand, are G-compatible for every G, which is obviously not the case for A and A', assumed to be distinct.

If $m = 2$ and G reduces to the identity on the set $\{1, 2\}$, then G-compatibility means that the restrictions of both chains to the intersection of their bases is the same. Then we find again the notion of compatibility in the sense of 1.7.2; and there exists a common extension which is a chain based on the union of the two bases.

If m is an arbitrary integer, and G is the m-ary symmetric group, or group of all permutations of $\{1, ..., m\}$, then all chains are G-compatible.

Two chains are G-compatible iff their restrictions to the intersection of the bases are G-compatible.

Let A, B be two chains with the same base E, and let G be an m-ary group; let n be such that $m \leq n \leq \operatorname{Card} E$. **Then A and B are G-compatible iff, for each n-element subset X of E, the restrictions A/X and B/X are G-compatible.**

13.3.2 Group-compatibility theorem (recollement)

Let G be a group and m its arity. There exists an integer $n \geq m$ satisfying the following.

Given a (finite or infinite) set E of cardinality $\geq n$; and for each n-element subset F of E, given a chain A_F with base F, where the A_F are mutually G-compatible; then there exists a chain based on E, which is G-compatible with all the A_F ([85] FRASNAY 1965; our proof uses ultrafilter axiom; ZF suffices if E is countable).

• Suppose first that E is finite. Let A be a chain on E which we identify with an initial interval of the sequence ω of integers. Define the m-ary relation R on these integers, by setting $R(x_1, ..., x_m) = +$ iff either at least two of the x are identical, or the x are all distinct and the permutation which reorders the sequence $x_1, ..., x_m$ in increasing order (mod A) belongs to the group G.

Clearly every local automorphism of the chain A is a local automorphism of R. In other words R is freely interpretable in A, hence R is A-chainable. Using 13.2.2, we see that R is p-well for each integer p. Then R has finitely many bounds by 13.2.3.

Let $n \geq m$ be strictly greater than the cardinalities of bounds of R. We assume that the finite set E has cardinality $\geq n$. For each n-element subset F of E, let A_F be a chain on F, where the A_F are mutually G-compatible. Let S be the m-ary relation based on E, such that $S(x_1, ..., x_m) = +$ iff either at least two x are identical, or, denoting F any n-element subset of E containing $x_1, ..., x_m$ (now supposed to be distinct), then the permutation of $\{x_1, ..., x_m\}$ which reorders the sequence $(x_1, ..., x_m)$ according to the chain A_F belongs to G. Because of the G-compatibility of the A_F for any two F, the value of S does not depend upon the chosen n-element set F.

For each F the restriction S/F is embeddable in R: it suffices to take an isomorphism from A_F onto a restriction of A. The cardinality of each bound of R is $< n$, so no bound of R is embeddable in S: otherwise, such a bound would be embeddable in a restriction S/F, hence in R: contradiction.

It follows that S is embeddable in R. For otherwise some restriction of S would be a bound of R. Let f be an isomorphism from S onto a restriction of R. The inverse function f^{-1} takes a restriction of the chain A into a chain B, with the result that $S(x_1, ..., x_m) = +$ iff either two x are identical, or all the x are distinct and the permutation which reorders these x according to the chain B, belongs to

G. It follows that B is G-compatible with A_F for each n-element subset F of E. The proof is now achieved for the case that E is finite.

Now suppose E infinite. For each finite subset D of E, there exists a non-empty set U_D of chains based on D, each being G-compatible with the given chains, still denoted by A_F. For a subset D' of D, every chain belonging to U_D, when restricted to D', gives an element of $U_{D'}$. By the coherence lemma 2.4.1 equivalent to the ultrafilter axiom, there exists a chain based on E, whose restriction to each finite set D belongs to U_D. This chain is G-compatible with all the A_F. •

13.3.3 Chainability theorem

Let m be a positive integer. There exists an integer $p \geq m$ such that every m-ary relation with cardinality $\geq p$, which is $(\leq p)$-monomorphic, is chainable ([85] FRASNAY 1965; uses ultrafilter axiom; ZF suffices for a countable relation).

• Consider all groups of arity m, and let n be the maximum of the integers associated to these groups in the preceding group-compatibility theorem. By 9.5.4 proposition (2), there exists an integer $p \geq n$ such that every m-ary relation with cardinality $\geq p$ has a chainable restriction with cardinality n.

Let R be a $(\leq p)$-monomorphic m-ary relation with base E of cardinality at least equal to p. Then all the restrictions of R with cardinality n are isomorphic, hence they are all chainable. To each n-element subset F of E, associate a chain A_F based on F, such that the restriction R/F is A_F-chainable. Moreover, for any two n-element subsets F, F' of E, take $A_{F'}$ to be the image of A_F under one of the isomorphisms of R/F onto R/F'.

Let H be an m-element subset of the base E, and let F, F' be two n-element subsets, each of which includes H. The permutation of H which takes A_F/H into $A_{F'}/H$ is an automorphism of R/H. Indeed, take the image H' of H under the isomorphism from A_F onto $A_{F'}$; then take H' into H by preserving the order of elements (mod $A_{F'}$) and using chainability by $A_{F'}$.

Designate each element of H by its rank modulo A_F/H. Then the group of automorphisms of R/H becomes an m-ary permutation group. The preceding isomorphisms show that G depends neither on H nor on the choice of the n-element set F including H. By the preceding, A_F and $A_{F'}$ are G-compatible for any two n-element sets F and F'.

Now apply the group-compatibility theorem. There exists a chain A based on E, which is G-compatible with every A_F. We shall prove that R is A-chainable. Let H, H' be two m-element subsets of E. We shall first prove that the isomorphism from A/H onto A/H' takes R/H into R/H'. We can assume that $n \geq 2m$, hence that there exists an n-element subset F of E including both H and H'. The desired isomorphism can be obtained by composing three isomorphisms: from A/H onto A_F/H, from A_F/H onto A_F/H', then from A_F/H' onto A/H'. The first and the third isomorphisms belong to G, once each element is designated by its rank (mod A_F/H or A_F/H'). These are respectively an automorphism of R/H and an automorphism of R/H'. The second is an isomorphism from R/H onto R/H' by the definition of A_F.

Now let K, K' be two r-element subsets of E, with $r < m$. Suppose first that there exists an m-element subset H of E including both K and K'. Take an n-element set F including H. Transform K and K' by the isomorphism of A/H onto A_F/H (which belongs to G and hence is an automorphism of R/H). We see that the isomorphism from A/K onto A/K' takes R/K into R/K'. If no such m-element set H exists, then we take an m-element set H including K and another H' including K', and by the isomorphism from A/H onto A/H' we are in the preceding case. •

Corollary. Let m be an integer, p the integer $\geq m$ defined by the preceding theorem, and R an m-ary relation of cardinality strictly greater than p.

A sufficient (and necessary) condition for R to be chainable is that each restriction of R to $p+1$ elements is chainable, or even is $(\leq p)$-monomorphic (uses the ultrafilter axiom; ZF suffices if R is countable). This follows from the preceding theorem and from 9.6.1.

13.3.4 Non-trivial universal classes (Jean)

Given an integer m, **the class of all m-ary chainable relations is universal; similarly for m-ary monomorphic relations** ([123] JEAN 1967).

• Given an integer p, the class of p-monomorphic relations can easily be defined by a prenex formula with p universal quantifiers. The case of $(\leq p)$-monomorphic relations follows. Now by the chainability theorem above, for each arity m there exists an integer p such that every $(\leq p)$-monomorphic m-ary relation is chainable, the converse being obvious; our statement follows. Same proof with "monomorphic" instead of "chainable". •

For another similar example, see below, end of 13.10.3

13.4 Dilated group

13.4.1 Three equivalent conditions

Let $\mathcal{A}$ be the set of chains with the same finite base E, and let $p = \operatorname{Card} E$; then the following three conditions are equivalent.

(1) Given three chains A, B, C in $\mathcal{A}$, the image of C under the permutation which takes A into B, belongs to $\mathcal{A}$.

(2) There exists a p-ary group G such that $\mathcal{A}$ is formed of a chain and its images under the permutations belonging to G.

(3) There exists an integer $q \leq p$ and a q-ary group H such that $\mathcal{A}$ is formed of a chain and all chains on E which are H-compatible with it.

Indeed (1) and (2) are obviously equivalent; (2) is a particular case of (3); finally (3) easily implies (1).

13.4.2 Definition

Let G be a m-ary group, and E be an m'-element set with $m' \geq m$. Let A be a chain based on E. Consider the set of all chains on E which are G-compatible with A. By the preceding proposition these chains are the images of A under a certain m'-ary group G': this G' is called the m'-ary **dilated group** of G and denoted $G^{m'}$.

Let G be an m-ary group, H a subgroup of G, and let $m' \geq m$. Then the dilated group $H^{m'}$ is a subgroup of the dilated group $G^{m'}$.

Given three integers $m \leq m' \leq m"$ and an m-ary group G, then $(G^{m'})^{m"} = G^{m"}$.

13.4.3 Two distinct groups can have the same dilated group

• Let $m = 4$, and let G be the group formed of the identity on $\{1,2,3,4\}$. Let G' be formed of the identity plus the transposition (2,4). Then $G'^5 = G^5 =$ the group formed of the identity on $\{1,2,3,4,5\}$. Indeed start with the chain 1 2 3 4 5. A chain which is G'-compatible with it, orders the integers 1, 2, 4, 5 according to 1 2 4 5 or according to 1 5 4 2. Similarly, we have one of the chains 1 3 4 5 or 1 5 4 3, and one of the chains 2 3 4 5 or 2 5 4 3. The chain 1 5 4 2 requires that 5 be before 4, hence implies 1 5 4 3 and 2 5 4 3. Thus 5 must be both before and after 2: contradiction. Thus we have 1 2 4 5, which implies 1 3 4 5 and 2 3 4 5. •

13.4.4 Union, intersection, symmetry and dilated groups

(1) **Let G, H be two groups of the same arity m, and let $n \geq m$. Then $(G \cap H)^n = G^n \cap H^n$ and $(G \cup H)^n \supseteq G^n \cup H^n$; this inclusion is not always an equality.**

• The group $(G \cap H)^n$ is obviously included in G^n and in H^n.

On the other hand, if two chains are simultaneously G-compatible and H-compatible, then they are $(G \cap H)$-compatible.

The group $(G \cup H)^n$ obviously includes G^n and H^n.

Here is an example of two groups G, H with $G \cup H)^4 \supset G^4 \cup H^4$.

Let G be the group of cyclic permutations on $\{1,2,3\}$ and H be the group generated by the transposition (1,3) . Then $G \cup H$ is the symmetric group on $\{1,2,3\}$; hence $(G \cup H)^4$ is the symmetric group on $\{1,2,3,4\}$. On the other hand, G^4 is the group generated by the cycle (1,2,3,4); and H^4 is the group generated by (1,4)(2,3). So that the group generated by their union is the dihedral group on $\{1,2,3,4\}$ formed of 8 permutations generated by the considered cycle and the reflection (or inversion) (1,4)(2,3). (see 9.2.2). •

(2) **Let G, H be two m-ary groups. If H^{m+1} is strictly included in G^{m+1}, then $(G \cap H) \neq G$ and H is not necessarily included in G.**

• Suppose that $G \cap H = G$. Then G is included in H, so G^{m+1} included in H^{m+1}. To have H non-included in G, take G to be the group generated by the cycle (1,2,3,4), and take H to be the group on $\{1,2,3,4\}$ formed of the identity plus the transposition (2,4). Then H^5 reduces to the identity on $\{1,2,3,4,5\}$ (see 13.4.3) and G^5 is generated by the cycle (1,2,3,4,5). •

(3) **If G has a symmetric dilated group, then G is itself symmetric.**

• Let m be the arity of G, and let E be a set with cardinality $n \geq m$. By hypothesis G^n is symmetric, hence all chains on E are mutually G-compatible. Taking their restrictions to a m-element subset D of E, we obtain all chains based on D: hence G is symmetric. •

13.5 Group generated by a set of chains; bichain; contracted group

Let $\mathcal{A}$ be a set of chains with the same base E, and let m be an integer $\leq \operatorname{Card} E$. The least m-ary group G such that the chains in $\mathcal{A}$ are all G-compatible is said to be **the m-ary group generated by $\mathcal{A}$**. This is also the group generated, in the usual sense, by all permutations u of $\{1, ..., m\}$ obtained by taking any two chains in $\mathcal{A}$ and then any m elements in E, say $a_1 < a_2 < ... < a_m$ modulo one of the chains, and $a_{u(1)} < ... < a_{u(m)}$ modulo the other chain.

If $\mathcal{A}$ is just a set of two chains we call it a **bichain**, and G is **the m-ary group generated by this bichain**.

Start with an n-ary group G. Take a chain A of cardinality n. Then take $\mathcal{A}$ to be the set formed of A and all its images under the permutations belonging to G. Then for $m \leq n$ the m-ary group generated by $\mathcal{A}$ is said to be the **contracted group** of G to arity m, and denoted by G_m.

If H is a subgroup of G, then the contracted group H_m is a subgroup of G_m.

Two distinct groups can have the same contracted group. Indeed, there exists only one possible unary contracted group, and two possible binary (or 2-ary) contracted groups: the group formed of the identity, and the symmetric group formed of the identity and the transposition (1,2). More generally, the proposition follows from the increase with m of the number of m-ary groups.

Let G be an m-ary group, and $m" \leq m' \leq m$. Then **the twice contracted group $(G_{m'})_{m"}$ is an extension of $G_{m"}$.**

Problem. Are these two groups always identical.

More strongly, given a set $\mathcal{A}$ of chains all with the same base, and another set $\mathcal{B}$ of chains with the same base, if $\mathcal{A}$ and $\mathcal{B}$ generate the same m-ary group, then do they generate the same m'-ary group for $m' < m$.

13.5.1 The groups $(G^n)_m$ and $(G_m)^n$

Start with an m-ary group G, take $n \geq m$ and the dilated group G^n, and then the contracted group $(G^n)_m$. **Then $(G^n)_m$ is a subgroup of G, which can be distinct from G.**

Indeed start with G of arity 4, formed of the identity and the transposition (2,4). Then G^5 reduces to the identity (see 13.4.3); hence $(G^5)_4$ as well.

Start with an n-ary group G, take $m \leq n$ and the contracted goup G_m, and then the dilated group $(G_m)^n$. **Then $(G_m)^n$ is an extension of G, which can be distinct from G.**

Indeed start with the 5-ary group G on $\{1,2,3,4,5\}$ which preserves 4 and 5 and circularly permutes 1,2,3. Then G_4 preserves 4 and symmetrically permutes 1,2,3. So that $(G_4)^5$ preserves 4 and 5 and symmetrically permutes 1,2,3.

13.5.2 Union, intersection of contracted groups

Let G, H ne two m-ary groups and $n \leq m$. Then $(G \cup H)_n = G_n \cup H_n$ and $(G \cap H)_n \subseteq G_n \cap H_n$; this inclusion can be proper.

• By the preceding, the first group includes G_n and H_n, hence includes their union. Similarly the third group is included in the intersection.

Moreover, if a chain A is taken into a chain B by a permutation belonging to $G \cup H$, then there exists a finite sequence of chains going from A to B, and such that the passage from each one to the next is effected by a permutation either in G or in H. Hence the restriction of A to a given n-element set D, is taken into the restriction of B to D, by a finite sequence of permutations in G_n and in H_n. It follows that the first group is included in the second, hence is equal to it.

To see that the inclusion of the third group in the fourth can be proper, take G to be the group of cyclic permutations on $\{1,2,3\}$ and take H to be formed of the identity and the transposition (1,2) on the same set $\{1,2,3\}$. Then the intersection $G \cap H$, and so $(G \cap H)_2$ reduces to the identity. Yet G_2 and H_2, hence their intersection, are identical to the symmetric group on $\{1,2\}$. •

13.5.3 Generation of some dilated or contracted groups

(1) Let G be an m-ary group, and let $n \geq m$. The dilated group G^n is generated (in the usual sense, by composition) from the union of those n-ary groups whose contracted m-ary group is included in G.

Problem. Is the dilated group G^n the intersection of the n-ary groups whose m-ary contracted group includes G.

(2) Let G be an m-ary group and let $n < m$. The contracted group G_n is not necessarily generated by the union of those n-ary groups whose m-ary dilated group is included in G.

Indeed let G consist of the identity on $\{1,2,3,4,5\}$. Thus G_4 consists of the identity. Yet the union of those 4-ary groups whose dilated 5-ary group reduces to the identity contains the transposition (2,4): see 13.4.3.

Problem. Letting $n < m$, is the contracted group G_n the intersection of those groups whose dilated m-ary group includes G.

13.5.4 On G-compatible extensions

There exist a 5-ary group G and two chains on $\{1,2,3,4\}$ which are G_4-compatible, yet have no G-compatible extensions to $\{1,2,3,4,5\}$.

• Let A be the chain 1 2 3 4 and B the chain 3 2 1 4. Let G be the 5-ary group which preserves 4 and 5 and circularly permutes 1,2,3. The contracted group G_4 symmetrically permutes 1,2,3 hence takes A into B. Let A', B' be two extensions of A, B to $\{1, 2, 3, 4, 5\}$. For A' and B' to be G-compatible, it is necessary that 5 be after 1,2,3 (mod A') and (mod B'); but then we do not have G-compatibility. •

13.6 Indicative group, indicator (Frasnay)

A group of permutations is said to be **indicative** iff it is generated by a bichain with infinite base (which we can assume to be denumerable). The indicative group goes back to [83] FRASNAY 1963; rediscovered independently by [111] HIGMAN 1977.

13.6.1 Finite bichain generating an indicative group

For each integer m, there exists an $m^* \geq m$ such that every bichain on a base with cardinal $\geq m^*$ generates an m-ary indicative group.

• Suppose the contrary, and consider an ω-sequence of bichains B_i(i integer) with finite bases E_i where Card E_i is strictly increasing with i, such that each B_i generates the same non-indicative group G. There are only finitely many, say h permutations in G. So that for each integer i, there exists a subset F_i of E_i with cardinality $p \leq h.m$, such that the restriction B_i/F_i generates G.

We can suppose that the E_i are finite sets of integers, and that $F_i = \{1, ..., p\}$ for each i. Extract from the sequence of the B_i's a first infinite sequence giving a common restriction C_0 to $\{1, ..., p\}$; this C_0 generates the group G. From this first sequence we extract a second infinite sequence which gives a common restriction C_1 extending C_0 to $\{1, ..., p, p+1\}$. Again C_1 exactly generates G since it is a restriction of some B_i and an extension of C_0. Then we extract a third infinite sequence which gives a common restriction C_2 extending C_1 to $\{1, ..., p, p+1, p+2\}$; etc.

Finally the common extension of $C_0, C_1, C_2, ..$ to the set of integers is an infinite bichain which generates G; hence G is indicative: contradiction. •

13.6.2 Description of indicative groups; indicator

Given an infinite bichain, for each integer m it generates an m-ary indicative group G_m.

The function G which to each m associates G_m is said to be the **indicator** of the bichain (called "fiche" in FRASNAY's terminology).

Let us describe as follows all the indicators, hence all indicative groups.

The indicator S gives by definition, for each m, the m-ary **symmetric group** denoted by S_m. Take for instance the chain of non-negative integers, and another chain which for each integer m realizes all possible permutations of m consecutive integers.

The indicator I gives for each m the group I_m consisting only of the **identity** on m elements. Take a bichain formed of two identical infinite chains.

The indicator J gives for each m the group J_m formed of the identity and of the **reflection** $(1, m)(2, m-1)(3, m-2)\ldots$: take an infinite chain and its converse.

The indicator T gives for each m the group T_m of **translations**, generated (in the usual sense) by the cyclic permutation $(1, 2, 3, \ldots, m)$. Take for instance two infinite chains A, B with disjoint bases, then the two sums $A + B$ and $B + A$.

The indicator D gives the **dihedral group** D_m, which is generated (usual sense) by the union of T_m and J_m. Take for instance two infinite chains A, B with disjoint bases, then the sums $A + B$ and $A^- + B^-$ (sum of converses).

Given two integers p, q **the indicator** $I^{p,q}$ gives for each m the group $I_m^{p,q}$ formed of those permutations which preserve the initial interval $\{1, \ldots, p\}$ and the final interval $\{m - q + 1, \ldots, m\}$ and which reduce to the identity on $\{p + 1, p + 2, \ldots, m - q\}$, assuming that $m \geq p + q$. In the opposite case where $m < p + q$, then $I_m^{p,q}$ denotes the m-ary symmetric group.

Take for instance the chain $12 \ldots p$ followed by an infinite chain A followed by the chain $1'2' \ldots q'$; and on the other hand the chain $23 \ldots p1$ followed by A followed by $2'3' \ldots q'1'$.

Note that $I^{1,1}$ is identical with I.

Given an integer r, **the indicator** J^r gives the group J_m^r which is generated by the union of $I_m^{r,r}$ and J_m. Then the initial interval $(1, \ldots, r)$ is taken into the final interval $(m - r + 1, \ldots, m)$ and conversely; moreover the median interval $(r + 1, \ldots, m - r)$ is reversed, assuming that $m \geq 2r$. In the opposite case where $m < 2r$, then J_m^r denotes the m-ary symmetric group.

Take for instance the chain $12 \ldots r$ followed by an infinite chain A followed by $1'2' \ldots r'$; and on the other hand 2' 3' ... r' 1' followed by the converse A^- followed by $23 \ldots r1$.

Note that J^1 is identical with J.

13.6.3 Connections between indicators

(1) Let A, B be two chains with common base of cardinality $\geq m$. If A and B are D_m-compatible, then either A and B, or A and the converse B^- are T_m-compatible.

Let m, r be two integers with $2r \leq m$. If A and B (with common base) are J_m^r-compatible, then either A and B, or A and B^- are $I_m^{r,r}$-compatible.

(2) **The dilated groups of** T_3 **are the** $T_m (m \geq 3)$.

The dilated groups of D_4 **are the** $D_m (m \geq 4)$.

If $p + q \leq m$, then the dilated groups of $I_m^{p,q}$ are the $I_n^{p,q}$ for $n \geq m$.

If $m \geq 3$ and $2r \leq m$, then the dilated groups of J_m^r are the J_n^r for $n \geq m$.

Consequently, **two chains whose intersection of bases has cardinality** $m \geq 3$, **are** T_m**-compatible iff they are** T_3**-compatible**.

For $m \geq 4$, **they are** D_m**-compatible iff they are** D_4**-compatible**.

For $m \geq 2$ and $p + q \leq m$, if the intersection of bases has cardinality $n \geq m$, then the chains are $I_n^{p,q}$-compatible iff they are $I_m^{p,q}$-compatible.

For $m \geq 3$ and $2r \leq m$, if the intersection of bases has cardinality $n \geq m$, then the chains are J_n^r-compatible iff they are J_m^r-compatible.

In general, the dilated or the contracted group of one of the previously described indicative groups is simply obtained by keeping the indicator I, J, T, D, S and by modifying the arity m, keeping the indices p, q or r, where it is understood that, if $p + q > m$ or if $2r > m$, then the group $I_m^{p,q}$ or the group J_m^r is identical with the symmetric group S_m.

13.6.4 The number of m-ary indicative groups

For $m = 0, 1, 2, 3$, all m-ary permutation groups are indicative. For instance the ternary groups are S_3 (symmetric group), $I_3^{1,1}$ (identity group), $I_3^{2,1}$ (transposition (1,2)), $I_3^{1,2}$ (transposition (2,3)), J_3^1 (transposition (1,3)) and T_3 (cyclic group).

There exist exactly eleven quaternary indicative groups, namely S_4 (symmetric group) $I_4^{1,1}$ (identity group), $I_4^{1,2}$ (transposition (3,4)), $I_4^{2,1}$ (transposition (1,2)), $I_4^{2,2}$ (union of the two preceding groups), $I_4^{1,3}$ (symmetric group on the set 2,3,4), $I_4^{3,1}$ (symmetric group on 1,2,3), J_4^1 (reflection), J_4^2 (generated by the reflection and the two transpositions (1,2) and (3,4)), T_4 (cyclic group), D_4 (dihedral group). The famous Klein group is not indicative; yet it is the intersection of two indicative groups: J_4^2 and D_4.

We leave it to the reader to show that, for $m \geq 4$, the number of indicative m-ary groups is the maximum integer $\leq 3 + m^2/2$.

13.7 Q-bichain, Q-indicative group; five Q-indicative group theorem

13.7.1 Q-bichain, Q-indicative group

We shall call a **Q-bichain** any bi-relation both of whose components are chains each isomorphic with **Q**.

We say that a group is **Q-indicative** iff it is generated by a **Q**-bichain.

We see that the five following groups are **Q**-indicative, for each arity m: the identity I_m; the group J_m (identity and reflection); the group T_m of translations; the dihedral group D_m generated by the union of J_m and T_m; and finally the symmetric group S_m. We call these temporarily the **canonical groups**.

The group generated by the union of two canonical groups is canonical. Indeed the only case where two of these groups are non-inclusive, is the case of J_m and T_m whose union generates D_m.

Consequently **for each group G there exists a maximum canonical group included in G**.

13.7.2 The five Q-indicative group theorem

The five canonical groups are the only Q-indicative groups.

The theorem is proved by the eight following lemmas. They show that **if G is Q-indicative, then it is equal to the maximum canonical group included in G.** This is obvious for the symmetric group S_m; so we shall only consider the four cases of I_m, T_m, J_m, D_m.

Consider a **Q**-bichain with components A, B. We say that a pair of elements x, y of the base is **preserved** or **inverted** according to whether x and y are in the same order modulo A and modulo B, or in the opposite order.

13.7.3 First lemma

Let AB be a Q-bichain having at least one preserved pair and one inverted pair. Then either the group J_m is included in the group generated by AB, for every m, or T_m is included in the group, for every m.

• Let u and v denote two elements such that $u < v \pmod A$ and $v < u \pmod B$. Since A is isomorphic with **Q**, there exist infinitely many elements x between u and $v \pmod A$. For each such x, we have $v < x$ or $x < u \pmod B$, hence there exist infinitely many x satisfying, for example, $x < u \pmod B$. Using RAMSEY theorem, either there exist infinitely many of these x which form mutually inverted pairs, in which case the group J_m is included in the m-ary group generated by AB, for each m. Or there exist infinitely many of these x forming preserved pairs. Then since u is less than $(\bmod A)$ and greater than $(\bmod B)$ these elements, the group T_m of translations is included in the group generated by AB, for each m. •

13.7.4 Second lemma

Let G be the m-ary group generated by a given **Q**-bichain. **If the maximum canonical group included in G is I_m, then G is identical to I_m.**

• Either there only exist preserved pairs in the given bichain, in which case $G = I_m$. Or there only exist inverted pairs, in which case $G = J_m$. Or finally there exists at least one pair of each kind, in which case J_m or T_m is included in G by 13.7.3. Under our assumption only the first case is possible. •

13.7.5 Third lemma

Let AB be a **Q**-bichain with base E. **Suppose that there exists an infinite subsest U of E all of whose pairs are preserved, hence $A/U = B/U$. Let m be a positive integer; suppose that the group T_m is generated by the bichain. Then either $G = T_m$ or $G = S_m$** (the symmetric group).

• Suppose that $m \geq 2$, since the case where $m = 1$ is obvious. The set U is different from E, since T_m is included in G. For each element x in $E - U$, let x_A denote the cut defined on A/U by the initial interval of those elements of U less than $x \pmod A$, and the complementary final interval. Let x_B denote the cut analogously defined with B.

First suppose that for each x in $E-U$, we have $x_A = x_B$. By hypothesis, there exists an inverted pair, say $\{x,y\}$ which is thus included in $E-U$; this implies that $x_A = x_B = y_A = y_B$. If, before this cut, we have infinitely many elements of U, then G contains the transposition $(m-1,m)$, which together with T_m generates S_m. Similarly, if there exist infinitely many elements greater than this cut, then G contains the transposition (1,2), which together with T_m generates S_m.

From this point on, we are in the case where there exists an x in $E-U$ with $x_A \neq x_B$. Suppose that at least one of these cuts is neither the initial cut lying before A/U, nor the final cut. Then there exist at least two intervals bounded on U by x_A and x_B, and at least one of these intervals is infinite.

If it is the interval between x_A and x_B, and if the other non-empty interval lies before it, then G contains the permutation which preserves 1 and which is defined by the cycle $(2,3,...,m)$. By composition with $(1,2,...,m)$ we generate S_m. Similarly if the interval between x_A and x_B is infinite and the other non-empty interval lies after it. If the interval between x_A and x_B is finite and the interval lying before is infinite, then G contains the transposition $(m-1,m)$ which, together with the translation $(1,2,...,m)$ generates S_m. Same result in the case of an infinite interval lying after.

Suppose now that, if x_A and x_B are distinct, then they are extremal (i.e. one of them is the initial cut and the other is the final cut). We can require that any two elements x,y which give the same non-extremal cut $x_A = x_B = y_A = y_B$, form a preserved pair. Indeed otherwise, we obtain again the transposition (1,2) or $(m-1,m)$. Thus augment U by all these x corresponding to non-extremal cuts.

From this point on, every x in $E-U$ yields two extremal cuts. Either there exist x,y in $E-U$ with x_A and y_B initial cut and x_B and y_A final cut. Then G contains the transposition $(1,m)$ and hence again is identical with S_m.

Or we are in the case where, for instance, x_A is initial and x_B final for each x in $E-U$. Then either there exists an inverted pair among these; then G contains the permutation which transforms $1,2,3,...m-1,m$ into $3,4,...m,2,1$ and this together the translation generates S_m.

Or finally all the pairs in $E-U$ are preserved, in which case A is the sum $A/(E-U)+A/U$ and B is the sum $A/U+A/(E-U)$, hence $G=T_m$. •

13.7.6 Fourth lemma

Let G be generated by a **Q**-bichain. **If the maximum canonical group included in G is T_m, then $G = T_m$.**

• Because of RAMSEY's theorem, there exists an infinite subset U of E, all of whose pairs are preserved, or an infinite subset U all of whose pairs are inverted. In the second case, the group J_m is included in G, and since by hypothesis T_m is included in G as well, we have the dihedral group D_m included in G, contradicting the assumption that T_m is the maximum canonical group included in G. Thus by the preceding statement $G = T_m$ or $G = S_m$, this last case contradicting our assumptions. •

13.7.7 Fifth lemma

Let AB be a Q-bichain with base E. **Suppose that there exists an infinite subset U of E, all of whose pairs are preserved, and an infinite subset V all of whose pairs are inverted. Then for each m, the m-ary group generated by AB is the symmetric group S_m.**

• Note first that U and V have at most one element in common. Moreover, the existence of V shows that the group J_m of the reflection is generated. Using 13.7.5, it suffices to prove that the group T_m of translations is generated, since then the entire generated group cannot be reduced to the single group T_m and is then identical with S_m.

First suppose that for each integer h, there exists an x in V for which the cuts x_A and x_B defined by x on the chain $A/U = B/U$ are separated by at least h elements of U. In this case, the translation $(1, 2, ..., m)$ is obtained, hence our proposition holds.

Suppose now that there exists an integer h such that for each x in V, we have at most h elements in U between the cuts x_A and x_B. Take an ω-sequence of elements in V, which is for example decreasing $(\text{mod}\, A)$, hence decreasing $(\text{mod}\, B)$. Then from some point on, the cuts x_A become identical, as well as the x_B become identical. Thus there exist infinitely many elements in U which are either all greater than these cuts, or all less than these cuts.

This yields for instance, for each m and each $p \leq m$, the permutation which preserves $1, 2, ..., p$ and which interchanges $(p+1, m), (p+2, m-1)$, etc. This suffices to generate the symmetric group S_m. •

13.7.8 Sixth lemma

Let AB be a Q-bichain having at least one preserved pair and one inverted pair. Then for each m the group T_m is included in the group generated by AB.

• Consider two elements $u < v$ $(\text{mod}\, A)$ with $v < u$ $(\text{mod}\, B)$. Take infinitely many elements between u and v $(\text{mod}\, A)$. Then either there exist infinitely many of them which are $> u$ $(\text{mod}\, B)$. In this case, by RAMSEY's theorem, extract an infinite subset of these elements, all of whose pairs are preserved, or all of whose pairs are inverted. In the case where the pairs are preserved, by our choice of the element v, we obtain the group T_m. In the case where the pairs are inverted, we obtain J_m. By this reflection and by our choice of u, we again obtain T_m.

Or there exist infinitely many elements $< v$ $(\text{mod}\, B)$, and still between u and $v (\text{mod}\, A)$. Then the preceding argument still works.

Or finally there exist infinitely many elements between u and v $(\text{mod}\, A)$ and between v and u $(\text{mod}\, B)$. Then in the case of preserved pairs, we obtain the group T_m.

There remains the case of an infinite set of inverted pairs. We shall prove that if T_m is not already obtained, then we obtain also an infinite set of preserved pairs: hence by the preceding statement, we have the symmetric group S_m.

To this end, consider two new elements still denoted by u and v, such that $u < v(\bmod A$ and $\bmod B)$. Take again infinitely many elements between u and $v(\bmod A)$. The same argument as before proves that, if there exist infinitely many of these elements which are $< u \pmod B$, or infinitely many which are $> v \pmod B$, then we again obtain the group T_m.

Finally suppose that there exist infinitely many of these elements between u and $v, \bmod A$ and $\bmod B$ as well. In the case of an infinite set of inverted pairs, we obtain the group J_m of the reflection. Taking the images, under this reflection, of u, v and our infinite inverted set, we again obtain T_m. There remains the case of infinitely many preserved pairs. Together with the previously obtained infinite set of inverted pairs, they give the symmetric group S_m, in view of 13.7.7. •

13.7.9 Seventh lemma

Let G be an m-ary group generated by a Q-bichain. If the maximum canonical group included in G is J_m (reflection), then $G = J_m$.

• By hypothesis, there exists an inverted pair. If there exists as well a preserved pair, then by the preceding proposition the group T_m is included in G. By hypothesis J_m is also included in G. Thus the maximum canonical group included in G is at least D_m: contradiction which proves that all the pairs are inverted, so that $G = J_m$. •

13.7.10 Eighth lemma

Let G be an m-ary group generated by a Q-bichain. **If the maximum canonical group included in G is D_m (dihedral group), then $G = D_m$.**

• Let AB be the considered bichain, and E its base. By RAMSEY's theorem, there exists an infinite subset U of E, all of whose pairs are preserved, or all of whose pairs are inverted. If they are preserved, then since T_m is included in G, we have that $G = T_m$ or $G = S_m$ by 13.7.5, contradicting our assumption that D_m is the maximum canonical group included in G.

Hence we have an infinite set U with inverted pairs. Let B^- denote the converse chain of B, and let G' be the m-ary group generated by AB^-. We shall prove that G' is included in G. The group G' is generated by some permutations s, each obtained by taking m elements which we denote by $1, 2, ..., m$ in the ordering A and $s(1), s(2), ..., s(m)$ in the ordering B^-. For such an s, the composition $r_m \circ s$ of the reflection r_m of the integers $1, 2, ..., m$ with s, is an element of G. Since r_m belongs to G, then s belongs to G as well.

By the preceding, the bichain AB^- has preserved pairs. Then either all pairs are preserved, and so $A = B^-$, hence $G = J_m$, contradicting our assumptions. Or AB^- has also an inverted pair, and then T_m is included in G' by 13.7.8. Furthermore $G' \neq S_m$, since by hypothesis D_m is the maximum canonical group included in G. Hence by 13.7.5 we have $G' = T_m$. Then G, which is generated by the compositions $r_m \circ s$ where r_m is the m-ary reflection and s belongs to G', satisfies the inclusion $G \subseteq D_m$, and thus $G = D_m$. •

13.8 Set-transitive group theorem (Cameron)

13.8.1 Relational system freely interpretable in Q

Let R be a relational system which is freely interpretable in the chain $\mathbf{Q}$ of the rationals. **Then the group of automorphisms of R is either the symmetric group, or the group of increasing bijections, or the group of increasing and decreasing bijections (i.e. automorphisms of the betweenness relation), or the group of automorphisms of the cyclic relation associated with Q, or finally the group of bijections which preserve or which inverse the cyclic relation** (i.e. automorphisms of the dihedral relation of $\mathbf{Q}$) (see [197] POUZET 1979; a generalization, replacing $\mathbf{Q}$ by any infinite chain having neither a minimum nor a maximum, is studied by [87] FRASNAY 1984: see 13.12.4 below).

• Consider a positive integer m and an automorphism f of R. To each strictly increasing m-sequence $x_1 < x_2 < ... < x_m (\bmod \mathbf{Q})$ associate the m-ary permutation s such that $f(x_{s(1)}) < f(x_{s(2)}) < ... < f(x_{s(m)}) \ (\bmod \mathbf{Q})$.

Firstly fix m and f, and let G_m be the m-ary group generated by s when the sequence $(x_1, x_2, ..., x_m)$ varies. This G_m is identical to the Q-indicative group generated by the bichain $(\mathbf{Q}, \overline{f}(\mathbf{Q}))$. Indeed the inverse s^{-1} is nothing else but the reordering of the $x_1, ..., x_m$ by $\overline{f}(\mathbf{Q})$. From the five $\mathbf{Q}$-indicative group theorem 13.7.2 it follows that G_m is one of the five groups I_m, J_m, T_m, D_m, S_m. More precisely if f remains fixed yet m (and the m-sequences) vary, then G_m runs over one of the five indicators I, J, T, D, S.

Now if f runs on the entire set of automorphisms of R, then some groups can be replaced by their union, yet again G_m runs over one of the five previous indicators. Each of these five cases gives one of the five conclusions of our statement. •

Remark 1. Each of the five case actually occurs. Indeed $R = \mathbf{Q}$ leads to the indicator I, the betweeness relation of $\mathbf{Q}$ leads the indicator J, the ternary cyclic relation leads to T, the quaternary dihedral relation leads to D and a constant relation leads to S.

Remark 2. Let us change the proof by firstly fixing the m-sequence with f variable. Then using obvious automorphisms of $\mathbf{Q}$ we can suppose that the images fx are the x themselves, permuted by s; moreover:

(i) the group G_m generated by fixing a m-ary sequence (with f variable) does not depend upon the choice of this sequence;

(ii) **G_m is a subgroup of the group of automorphisms of the restriction** $R/\{x_1, ..., x_m\}$. Note that it can be a proper subgroup: for instance, take the relation R such that $R(x, y, z) = +$ iff $x < z$ and $y < z$ and $x \neq y$. In this example automorphisms of the restriction $R/\{x_1, ..., x_m\}$ give $I_m^{2,1}$ yet G_m reduces to the identity group I_m.

13.8.2 Set-transitive group theorem

The n-set-transitive group is defined in 12.3.3.

The 2-set-transitive group of finite arity was already used in 9.7.

Let G be a group of permutations of a denumerable set E. Then G is n-set-transitive for every integer n iff the closure of G under adherence is one of the five groups obtained as follows, starting from a chain $\mathbf{Q}$ on E, isomorphic with the chain of the rationals: symmetric group; group of automorphisms of $\mathbf{Q}$; group of automorphisms of betweenness (mod $\mathbf{Q}$); group of automorphisms of the cyclic ternary relation associated with $\mathbf{Q}$; group of automorphisms of the dihedral relation of $\mathbf{Q}$. Due to [24] CAMERON 1976.

• Each of the five mentioned groups is obviously n-set-transitive for each n, and similarly so for any group yielding one of these five groups under adherence.

Conversely, let G be a group of permutations of a denumerable set E, which is n-set-transitive for every n. By 12.3.5 proposition (2), there exists a homogeneous relational system R whose automorphism group is the closure G^* of G under adherence. Since G, and so G^* as well, is n-set-transitive for every n, the system R is n-monomorphic for every n; in other words R is monomorphic: see 12.3.6 proposition (2). It follows that R is chainable: see 9.6.2, which can be extended to relational systems which are w-sequences of relations with base E, by simply taking the limit. This holds in spite of the use of RAMSEY theorem in the beginning, which supposes for each integer n, a finite number of isomorphism types of multirelations of cardinality n.

Let A be a chain in which R is freely interpretable. Take the chain $\mathbf{Q}$ of the rationals, then the system S freely interpretable in $\mathbf{Q}$, which is obtained as follows: take each local isomorphism f from A into $\mathbf{Q}$ with a finite domain F , then take the image $\overline{f}(R/F)$, and finally the common extension of these images.

The systems R and S have the same age. It suffices to prove that S is homogeneous. Indeed by 12.1.3 proposition (2), which easily extends to relational systems, R will be isomorphic with S, hence R will be freely interpretable in an isomorphic copy of $\mathbf{Q}$. Thus our proposition will follow from the previous subsection.

Now let us prove the homogeneity of S. Let f be a local automorphism of S, with finite domain F and range F', and let u be an arbitrary element in the base. Take an isomorphism h from the finite chain $\mathbf{Q}/(F \cup \{u\})$ onto a restriction of A, and an isomorphism h' from $\mathbf{Q}/F'$ onto another restriction of A. Let $v = hu$. Then the bijective composition $h' \circ f \circ (h^{-1})$ is a local automorphism of R. Since R is homogeneous, by 12.1.2 there exists a local automorphism of R which extends the preceding local automorphism to the domain $\overline{h}(F)$ augmented by v. Let g denote this extension. Due to the fact that $\mathbf{Q}$ is dense and without endpoints, there exists a local isomorphism from A into $\mathbf{Q}$, which is an extension of h'^{-1} to the domain $\overline{h'}(F')$ augmented by gv. Let h" denote this extension. Finally the bijective composition $h" \circ g \circ h$ is a local automorphism of S and extends f to its domain F augmented by u. Then S is homogeneous by 12.1.2. •

13.8.3 A strengthening of the previous theorem

The following strengthening of the theorem on set-transitive groups is due to [24] CAMERON 1976, as is the theorem itself, and is more easily obtained by [201] POUZET 1981 p.323.

Let m be a positive integer and G be a permutation group on a denumerable set E. **If G is $m, m+1, \ldots, p(m)$-set-transitive but not m-transitive, then the closure of G under m-adherence is n-set-transitive for every positive integer n.**

Then the conclusion of the prevous theorem holds, except for the symmetric group, since G is not m-transitive. More precisely, there exists a chain $\mathbf{Q}$ based on E and isomorphic with the chain of the rationals, such that the closure of G under m-adherence is either the group of automorphisms of $\mathbf{Q}$, or the group of automorphisms of intermediacy $(\mathrm{mod}\, Q)$, or the group of automorphisms of the ternary cyclic relation associated with $\mathbf{Q}$, or finally the group of automorphisms of the dihedral quaternary relation associated with $\mathbf{Q}$.

These groups are at most 3-transitive. So it follows that for $m \geq 5$, every $m, m+1, \ldots, p(m)$-set-transitive group which is 4-transitive is also m-transitive. The case where $m = 4$ is interesting. Indeed there exists a group whose closure is none of the four groups listed above, and which is 4 and 5-set-transitive yet not 4-transitive ([27] CAMERON 1983). Thus $p(4) \neq 5$; since $p(4) = 5$ or 6 by [85] FRASNAY 1965, hence necessarily $p(4) = 6$.

13.9 Indicative group theorem, reduction theorem

13.9.1 Indicative group theorem

The only indicative groups are those of the five preceding families I, J, T, D, S ([85] FRASNAY 1965).

Sketch of the proof, which is little more complicated than for $\mathbf{Q}$-indicative groups.

We now call **canonical groups** all those groups I, J, T, D, S described in 13.6.2. This notion is more general than "canonical groups" in 13.7.1, where I reduced to the identity and J to identity plus reflection.

• For each pair of m-ary canonical groups, verify that the union group, more precisely the group generated by their union, is canonical (if it seems too long, the reader can assume this in a first reading). Consequently, for each group G, there exists a maximum canonical group included in G.

1 - Let G be an indicative m-ary group, i.e. a group generated by an infinite bichain AB. Suppose that there exists an infinite subset U of the base $|AB|$ for which $A/U = B/U$. Either G contains a permutation which, together with $(1, 2, \ldots, m)$, generates the symmetric group S_m; or G is included in T_m (group of translations, generated by $(1, 2, \ldots, m)$).

Conclude from the preceding that, if G is an indicative m-ary group, and if the maximum canonical group included in G is T_m, then $G = T_m$.

2 - Let G be an indicative m-ary group. If the maximum canonical group included in G is D_m, then $G = D_m$. Furthermore, if AB is a bichain generating D_m, then every subset U of the base for which $A/U = B/U$ is finite.

3 - Let G be an indicative m-ary group generated by an infinite bichain AB, and let U be an infinite set with $A/U = B/U$. Suppose that G does not contain the permutation $(1, 2, ..., m)$ and that there exist $p, q (p + q \leq m)$ such that G contains none of the following types of permutations: those which take $p + 1$ into an integer $\leq p$ and preserve $p + 2, p + 3, ..., m$; and those which take $m - q$ into an integer $< m - q$ and preserve $1, 2, ..., m - q - 1$. Then G is included in $I_m^{p,q}$.

4 - Let G be an indicative m-ary group. If the maximum canonical group included in G is $I_m^{p,q}$ (with $p + q \leq m$), then $G = I_m^{p,q}$.

5 - If the maximum canonical group included in G is J_m^r (with $2r \leq m$), then $G = J_m^r$. Moreover, if AB is a bichain generating J_m^r, then every U for which $A/U = B/U$ is finite. •

Corollaries.

(1) **The group generated by the union of any two indicative groups is indicative.**

(2) **A group is indicative iff it is generated by an arbitrary set of chains on the same denumerable base.** Because there are finitely many m-ary groups for each integer m.

(3) **If G is an m-ary indicative group and $n \geq m$, then G^n is indicative and $(G^n)_m = G$.**

Indeed, given an integer m, there exists an $m^* \geq m$ such that every set of chains on a base of cardinality $\geq m^*$ generates an indicative m-ary group (generalization of 13.6.1).

(4) **Given an m-ary group G, the dilated group G^n is indicative for** $n \geq m^*$ (given by the previous alinea).

13.9.2 Reduction theorem

Given an m-ary group G, there exists a maximum indicative m-ary group $H(G)$ included in G. Because the finiteness of G and because the union of two indicative subgroups of G is again an indicative subgroup of G.

Moreover for $n \geq m$ we have $H(G^n) = (H(G))^n$.

Taking $n \geq m$ such that G^n is indicative, then we have $(H(G))^n = G^n$ and $H(G) = (G^n)_m$.

13.10 Reduction, compatibility and monomorphy thresholds, G-chain

13.10.1 Reduction threshold

Given an m-ary group G of permutations, we define the **reduction threshold** of G, denoted by $s(G)$, as the least integer s such that the dilated group G^{m+s} is indicative: see 13.9.1.

In particular $s(G) = 0$ iff G is indicative.

Given an integer m, we define the m-**ary reduction threshold**, denoted by $s(m)$, as being the maximum of the $s(G)$ for all m-ary groups G.

These definitions are due to FRASNAY 1965, who obtains that $s(1) = s(2) = s(3) = 0$, $s(4) = 2$ and for $m \geq 5$, the inequalities $1 \leq s(m) \leq (3m-8)^2 - m + 1$ (Ibid. p.493-494). The upper bound is improved to $s(m) \leq m - 3$ (again for $m \geq 5$) by [117] HODGES, LACHLAN, SHELAH 1977; see 13.11 below.

Finally it is proved by FRASNAY 1984 that $s(m) = m - 3$ for $m \geq 5$. More precisely **the value** $s(G) = m - 3$ **is reached by taking** G **to be the group on** $\{1, 2, ..., m\}$ **which preserves the extremities 1 and** m.

• The first dilated group G^{m+1} preserves 1,2 and $m, m+1$; then G^{m+2} preserves 1,2,3 and $m, m+1, m+2$, and so on. Finally the dilated group G^{2m-3} preserves 1,2,...,$m-2$ and $m, m+1, ..., 2m-3$, hence it is the identity group, which is obviously indicative. •

13.10.2 Compatibility threshold

Given an m-ary group G, we define the **compatibility threshold** of G, denoted by $t(G)$, as the least integer t such that $n = m+t$ satisfies the group-compatibility theorem 13.3.2.

If G is not indicative, then $t(G) \leq s(G) + 1$ ([85] FRASNAY 1965 p.500).

However, for the indicative group J_4^2 on $\{1, 2, 3, 4\}$ generated by (1,4),(2,3) and the two transpositions (1,2) and (3,4) , we have $s = 0$ and $t = 2$ (Ibid. p.500).

Given an integer m, we define the m-ary **compatibility threshold**, denoted by $t(m)$, as being the maximum of the $t(G)$ for all m-ary groups G.

We have $t(1) = 0, t(2) = t(3) = 1, t(4) = 2$, and for $m \geq 5$ we have $1 \leq t(m) \leq s(m) + 1$ ([85] FRASNAY 1965 p.500 and [129] JULLIEN 1966). Hence $t(m) \leq m - 2$ for $m \geq 3$, in view of the preceding improved upper bound of $s(m)$. Finally, it is proved by [87] FRASNAY 1984 that $t(m) = s(m) + 1 = m - 2$ for $m \geq 5$, by taking again the group on $\{1, ..., m\}$ which preserves 1 and m.

Problem. For each group G, do we have that $s(G) \leq t(G)$.

13.10.3 (G, A)-chain, G-chain; connection with the reducton threshold

Consider a chain A with base E, and an m-ary group G. Following [86] FRASNAY 1973, inspired by [34] CLARK, KRAUSS 1970, we define the (G, A)-**chain** as being the m-ary relation R based on E and A-chainable, such that $R(x_1, ..., x_m) = +$ iff $x_1, ..., x_m$ are all distinct and there exists a permutation s belonging to G with $x_{s(1)} < x_{s(2)} < ... < x_{s(m)} \bmod A$.

In the particular case where G is the binary group which reduces to the identity, then we have again the usual chain A, more precisely the strict chain $<$ (mod A).

We say that R is a G-**chain** iff there exists a chain A such that R is the (G, A)-chain (G-rangement in FRASNAY's terminology).

Starting from the chain ω of the integers, consider the (G, ω)-chain R and denote by m the arity of G. Then we see that, for each $n \geq m$, the dilated group G^n is the group of automorphisms of any restriction of R to n elements (obviously we replace each element by its rank, which is an integer). **Consequently G^n becomes an indicative group for $n \geq s(G)$ = reduction threshold.**

More generally, consider an m-ary relation R which is ω-chainable. For every $n \geq m$, denote by $G(n)$ the n-ary group of automorphisms of any restriction of R to n elements. Then there exists an n such that, from this point on, the group $G(n)$ is indicative ([111] HIGMAN 1977).

Problem posed by [87] FRASNAY 1984. For $n \geq s(m)$, the group $G(n)$ is it always indicative.

Another non-trivial example of universal class (see 13.3.4): given a permutation group G, the class formed of all the G-chains is universal; moreover the maximum cardinality of bounds is $m + t(G)$ (FRASNAY in 1979).

13.10.4 G-chains and compatibility threshold

The compatibility threshold led [85] FRASNAY 1965 (p.517) to the following result (1), and then [201] POUZET 1981 (p.307) to the following (2) and (3).

Let G be an m-ary group. There always exists an m-ary relation R which is freely interpretable in the chain ω of the integers, such that G is the group of automorphisms of the restriction of R to an arbitrary m-element set, when ordered by increasing values: for instance we can take R to be the (G, ω)-chain. Then denoting again by $t(G)$ the compatibility threshold:

(1) **The maximum $b(R)$ of cardinalities of all bounds of R satisfies** $b(R) \leq m+t(G)$; the latter value is reached if R is the (G, ω)-chain ([86] FRASNAY 1973);

(2) **There exists a bound of R with cardinality m; hence $b(R) \geq m$;**

(3) **Either $b(R) = m + t(G)$ or $b(R) = m$ and $t(G) = 1$, hence $m + t(G) = m + 1$.**

• (1) Let $\mathcal{P}$ be a free operator which takes ω into R (see 9.3.1). Let U be an m-ary relation, bound of R with cardinality $> m + t(G)$. Thus restrictions of U with cardinality $< (m + t(G))$ are all embeddable in R . Thus each $(m + t(G))$-element subset X of the base $|U|$ is the base of a chain C_X satisfying to $U/X = \mathcal{P}(C_X)$. By our hypotheses, the C_X are mutually G-compatible. By the group-compatibility theorem 13.3.2, there exists a chain C with base $|U|$, which is G-compatible with each C_X. It follows that $U = \mathcal{P}(C)$, hence that U is embeddable in R, so U is not a bound: contradiction. •

• (2) Take the restriction H of R to the first m integers. Preserve the values of H for each sequence of m terms for which at least two terms are equal. However, change the value of H when the m terms are all distinct, either by taking always the value (+) (if it is not already the case for H), or by always taking (-). The relation H thus modified is a bound of R with cardinality m. •

• (3) It suffices to prove that $n = \mathrm{Max}(m+1, b(R))$ satisfies the inequality $n \geq m + t(G)$, hence satisfies the group-compatibility theorem.

Take a set E of cardinality $\geq n$, in which each n-element subset X is the base of a chain D_X, where these chains are assumed to be mutually G-compatible. Then the image relations $\mathcal{P}(D_X)$ are mutually compatible, i.e. they have a common restriction to the intersection of their bases. To see this, use the inequality $n \geq m+1$: we can always pass from an n-element set X to another Y by a finite number of intermediate n-element sets $X(i)$, each of which includes the intersection $X \cap Y$, and each of which has at least m common elements with the preceding one. Thus for these successive $X(i)$, the images $\mathcal{P}(D_{X(i)})$, then $\mathcal{P}(D_X)$ and $\mathcal{P}(D_Y)$ have a common restriction to the intersection $X \cap Y$.

Under these conditions, there exists a common extension S of these $\mathcal{P}(D_X)$. Moreover, each restriction of S to $\leq n$ elements is the image of a chain under $\mathcal{P}$, hence is embeddable in R. Additionally $n \geq b(R)$, hence S admits no embedding of any bound of R. It follows that S is itself embeddable in R. Thus there exists a chain D with base E, such that $S = \mathcal{P}(D)$. This D is G-compatible with each D_X, thus the group-compatibility theorem holds. •

Problem. Existence of a "teratological" ω-chainable m-ary relation R whose bounds have maximum cardinality $b(R) = m$ and yet with $t(G) = 1$, where G denotes the group of automorphisms of any restriction of R to an m-element set.

13.10.5 Bounds and compatibility threshold

Given the integer m, the maximum cardinality of the bounds of all m-ary chainable relations with infinite base is equal to the m-ary compatibility threshold in its complete form $m + t(m)$ ([86] FRASNAY 1973; this result is extended by [201] POUZET 1981 to chainable and almost chainable multirelations with maximum arity equal to m).

• In view of the previous subsection, it suffices to construct an m-ary relation R freely interpretable in ω, which admits at least one bound of cardinality $m+1$. Take the (G, ω)-chain where G is the identity; in other words, take the relation $x_1 < x_2 < \ldots < x_m (\mathrm{mod}\, \omega)$. Then consider the binary cycle C with cardinality $m+1$. On the base $|C|$, define the m-ary relation U which takes the value (+) iff $x_1, x_2, \ldots, x_m$ are consecutive modulo C: this U is a bound of R. •

13.10.6 Monomorphy threshold

If an m-ary relation R is $(m + t(m))$-monomorphic and its cardinality is either infinite or finite but sufficiently large, then R is chainable ([85] FRASNAY 1965 p.508 prop.12.1.1 and [201] POUZET 1981 p.311 prop.V.3.8).

• By the finitary form of RAMSEY's theorem, if the cardinality of the base is either infinite or finite but sufficiently large, then there exists a restriction of R which is chainable and of cardinality $m+t(m)$. By monomorphism, all restrictions with cardinality less than or equal to $m+t(m)$ are chainable. Starting with such a restriction, and extending the chain in which it is freely interpretable to the chain ω, we obtain an m-ary relation S which is freely interpretable in ω, and all of whose restrictions to at most $m+t(m)$ elements are embeddable in R.

We shall show that every finite restriction of R is embeddable in S, which will prove that these restrictions are chainable, and consequently that R is itself chainable. Suppose that every restriction of R to at most k elements is embeddable in S, and suppose $k \geq m+t(m)$. If there exists a restriction with cardinality $k+1$ which is not embeddable in S, then this restriction is a bound of S, contradicting the previous subsection. •

Moreover it is proved ([88] FRASNAY 1990) that $m+t(m)$ **is the lowest possible value satisfying the preceding statement**.

Let us define the **monomorphy threshold** $p(m)$ as being the lowest integer p such that any m-ary and p-monomorphic relation with a sufficiently large finite cardinality be necessarily chainable (this being also true for infinite cardinalities, by the coherence lemma). Then FRASNAY's result becomes that the monomorphy threshold is equal to the complete compatibility threshold: $p(m) = m+t(m)$. Hence $p(1) = 1$, $p(2) = 3$, $p(m) = 2m-2$ for $m \geq 3$: see below 13.12.1.

13.11 Adjacent elements modulo a relation; application to reduction threshold (Hodges, Lachlan, Shelah)

We follow [117] HODGES, LACHLAN, SHELAH 1977.

Given an m-ary relation R with base E, two elements i, j in E are said to be **adjacent modulo** R iff the bijection f with domain $E-\{i\}$ and range $E-\{j\}$, defined by $f(j) = i$ and $f(x) = x$ for each $x \in E-\{i,j\}$ is a local automorphism of R. The notion is obviously symmetric: exchange i and j and replace f by f^{-1}.

In the particular case where R is a chain (arity 2, base with ≥ 3 elements), we find again the trivial notion of two consecutive elements (without any other element between them).

If a relation S is freely interpretable in R, then any two elements are S-adjacent provided they be R-adjacent.

Given a m-ary relation R with base E and a subset $F \subset E$, any two adjacent elements modulo R are again adjacent modulo R/F.

Suppose the m-ary relation R is reflexive or antireflexive, i.e. $R(x_1, ..., x_m) = -$ if there are at least two equal values among $x_1, ..., x_m$. Then for any subset F of the base with Card $F \leq m$, any two elements of F are adjacent modulo R/F. In particular this is the case if R is a G-chain, where G is a m-ary group.

The noton is interesting when restrictions of R to any two singletons are isomorphic; for instance when R is reflexive or antireflexive. Then i **and** j **are adjacent modulo** R **iff the pair** $\{i,j\}$ **is an interval** (in the sense of 9.8) **modulo** R **concatenated with the poset** $(i<j)(i,j$ **incomparable with other elements in** $|R|)$.

13.11.1 Adjacence lemma

Consider the base $E=\{1,2,...,n\}$, an integer m $(1\leq m<n)$ called the **arity**. Let A be the chain $1<2<...<n$ and R an m-ary relation on E, assumed to be freely interpretable in A. Given two integers h,k with $1\leq h\leq k\leq m$, the cycle $(h,h+1,...,k)$ denotes this permutation of S_m which tranforms h into $h+1,...,k-1$ into k and k into h and reduces to identity on $\{1,2,...h-1,k+1,...,m\}$.

Let i,j be two integers, elements of E with $1\leq i<j\leq n$. Following [87] FRASNAY 1984 p.255, let us denote by $C^{i,j}_{m,n}$ any cycle $(h,h+1,...,k)$ where h,k satisfy inequalities $1\leq h\leq k\leq m$ and $0\leq i-h<j-k\leq n-m$; then the lemma of [117] p.213 reinterpreted in [87] p.255 writes:

Lemma. Given $x_1<x_2<...<x_m$ **(mod** A**), the group of automorphisms of the restriction** $R/\{x_1,...,x_m\}$ **includes all** $C^{i,j}_{m,n}$ **with** $\{i,j\}(i<j)$ **running on all couples of adjacent elements modulo** R.

• Given adjacent elements $i,j>i$, by definition the bijection f from $E-\{i\}$ onto $E-\{j\}$ which reduces to identity except that $f(j)=i$ is a local automorphism of R. From our hypotheses $h\leq i$ and $k-h<j-i$ and $m-k\leq n-j$; thus we can take in E an increasing m-sequence $x_1<...<x_m$ with $x_1<...<x_{h-1}$ strictly before i; then $x_h<x_{h+1}<...<x_{k-1}<x_k$ with $i<x_h$ and $x_k=j$; and finally $x_{k+1}<...<x_m$ strictly after j.

Firstly transform this m-sequence by the local automorphism f of R, restricted to values x: then each term is saved except $x_k=j$ which goes from j to i, so that: $x_1<...<x_{h-1}<f(x_k)<x_h<x_{h+1}<...<x_{k-1}<x_{k+1}<...<x_m$ (mod A).

Secondly apply the local automorphism g of A (therefore of R) which saves $x_1,...,x_{h-1}$, then changes $f(x_k)=i$ into $g(f(x_k))=g(i)=x_h$, then x_h into x_{h+1}, then x_{h+1} into x_{h+2}, ... , then x_{k-1} into $x_k=j$, and finally saves $x_{k+1},...,x_m$. The composition $g\circ f$ is again a local automorphism of R; moreover it is nothing else than the permutation of values x which saves x_1 to x_{h-1}, which transforms $x_h,...x_k$ according to the cycle $(h,h+1,...,k)$ of their indices, and finally saves $x_{k+1},...,x_m$. •

First example. $R=A$, then $m=2<n$. Now $i<j$ are adjacent modulo A iff $j=i+1$, so that our condition writes $0\leq i-h<i+1-k$; hence $k=h$ so that all our cycles reduce to the identity on $\{1,2\}$.

Second example. R is a constant relation. Now every $i,j>i$ are R-adjacent. To fix ideas, take $m=n-1$; then $0\leq i-h<j-k\leq 1$ hence $h=i$; let us take $h=i=1$ and $k=j-1$; consider successively $j=2,3,...,n$ hence $k=1,2,...,m=n-1$, we get the entire symmetric group S_m. We leave it to the reader the calculation in cases $m=1$ to $m=n-2$.

Third example. R is the ternary cycle ($m = 3$) associated with the chain A based on $\{1, ..., n\}(n \geq 4)$. Then $i < j$ are adjacent modulo R iff either $j = i+1$ or $i = 1, j = n$. Consider $j = i + 1$: we find again the identity on $\{1, 2, 3\}$. Consider $i = 1, j = n$: we get $0 \leq 1 - h < n - k \leq n - 3$ hence $h = 1$ and $3 \leq k \leq m = 3$ hence $k = 3$ which gives the unique cycle $(1, 2, 3)$ as desired.

Fourth example. R is the ternary betweeness relation on A. Then we see that, as for the chain A, the only couples of adjacent elements are of the form $i, j = i+1$ which only give the identity.

Fifth example. R is the quaternary dihedral relation on A; by using adjacent elements we only get the cyclic group (1,2,3,4).

13.11.2 Calculation of the reduction threshold: first and second case

The quoted authors use adjacence lemma in order to calculate the reduction theshold $s(G)$ of a m-ary group G of permutations, then the upper bound of this threshold for a given integer m: see 13.10.1.

To do that, we replace the previous notation R by G **which denotes the m-ary (G, A)-chain** where G is any m-ary group and A is the chain $1 < 2 < ... < n$. Recall that the m-ary relation G is freely interpretable in A and $G(x_1, ..., x_m) = +$ iff $x_{u(1)} < x_{u(2)} < ... < x_{u(m)}$ for $u \in G$. We already noticed in 13.10.3 that for every $n \geq m$ the n-ary group of automorphisms of any restriction of R to n elements is the dilated group G^n. Consequently it suffices to prove the following:

Theorem. If $n \geq$ Max$(2m - 3, m + 2)$, then the dilated group G^n is indicative.

1 - The first case considered by the quoted authors is $G = S_m$, which is obviously indicative and leads to indicative dilated groups S_n.

2 - The second case is defined by conditions $(1, 2, ..., m) \in G \neq S_m$. Assuming that G is not indicative (thus $m \geq 4$) and $n \geq m + 2$, then G^n is indicative.

3 - The third case is defined by $(1, 2, ..., m) \notin G$. Again assuming that G is not indicative (thus $m \geq 4$) and $n \geq$ Max$(2m - 3, m + 2)$, then G^n is indicative.

In order to prove the second case, we need the following:

Lemma leading to a cyclic or a dihedral relation. Let G be an m-ary non-indicative group (thus $m \geq 4$) with $(1, 2, ..., m) \in G \neq S_m$ and set $n > m$. Suppose that the couples of adjacent elements modulo the m-ary G-chain based on $\{1, 2, ..., n\}$ are exactly all couples $(i, j = i+1)$ plus $(1, n)$. Then the dilated group G^n is indicative. More precisely G^n is either the n-ary cycle or the n-ary dihedral relation on the chain $1 < 2 < ... < n$, according to whether reflection does not belong or belongs to G.

• The dilated group $H = G^n$ contains at least all chains deduced from $123...n$ by translation, and possibly by reflection. For any other chain B on $\{1, ..., n\}$ which would belong to H, there would necessarily exist two integers i, j which be consecutive modulo B (or modulo a chain translated from B), yet be neither consecutive in the trivial sense nor equal to $1, n$. These i, j would be adjacent modulo B, hence by free interpretability they would be adjacent modulo the G-chain based on $\{1, ..., n\}$: contradiction. •

Proof of the theorem, second case, with the assumptions $m \geq 4, n \geq m+2$. By hypothesis all couples $(i, j = i+1)(i = 1 \text{ to } n-1)$ plus $(1, n)$ are formed of adjacent elements. If there are no other cases of adjacence, then the previous lemma solves the problem.

• Suppose to the contrary that there exist at least two adjacent elements i, j with $j - i \geq 2$ and either $j < n$ or $i > 1$.

Either $j \leq m+1$, hence $j < n$ since $n \geq m+2$. Then take $h = i$ and successively $k = j-2$ and $k = j-1$; so that $1 \leq h \leq k \leq m$ and $0 = i-h < j-k \leq n-m$ since by hypothesis $n \geq m+2$. Then G contains cycles $(i, ..., j-2)$ and $(i, ..., j-1)$, hence also the transposition $(j-2, j-1)$. Since by hypothesis G contains the cycle $(1, 2, ..., m)$, then $G = S_m$ which contradicts our case assumption.

Or $i \geq m$ and $j \geq m+2$ (with $i \leq j-2$). Then take $h = m-1$ (recall that $m \geq 4$) and $k = m$, so that $1 \leq h < k = m$ and $0 \leq i-h < j-k = j-m \leq n-m$ since $j-i \geq 2$ and $k-h = 1$. Then G contains the transposition $(m-1, m)$ which contradicts our case assumption.

Or finally $i \leq m-1$ and $j \geq m+2$; taking into account the exception $i = 1, j = n$, we must distinguish two subcases.

Either $i \leq m-1$ and $m+2 \leq j < n$. Then take $h = i$ and successively $k = m-1$ and $k = m$, so that $1 \leq h = i \leq m-1 \leq k \leq m$ and $0 = i-h < j-k \leq n-m$ since $j < n$. Then G contains the transposition $(m-1, m)$ hence $G = S_m$ which contradicts our case assumption.

Or $2 \leq i \leq m-1$ and $m+2 \leq j = n$. Then take $k = m$ and successively $h = i-1$ and $h = i$, so that $1 \leq h \leq i < k = m$ and $0 \leq i-h < j-k = n-m$ since $j-k = n-m \geq 2$. Then G contains the transposition $(i-1, i)$ hence $G = S_m$ which contradicts our case assumption. •

Take for example $m = 5$ and suppose that $G = A_5$ is the 5-ary alternating group (example given by the quoted authors, p.214). Then for $n = 6$, not only $(i, i+1)$ and (1,6) but also for instance (1,4) and (2,5) and (3,6) are adjacent modulo the G-chain based on $\{1, ..., 6\}$. For instance $(i = 1, j = 4)$ lead to $h = 1, k = 3$ since $1 \leq h \leq k \leq m = 5$ and $0 = i-h < j-k = n-m = 1$. Indeed the cycle (1,2,3) belongs to the alternating group A_5.

The dilated group $(A_5)^6$ is formed of 36 chains, i.e 1 2 3 4 5 6 plus its 6 translated chains plus 6 other chains obtained by reflection: 6 5 4 3 2 1 and so on. Moreover $(A_5)^6$ contains 12 analogous chains obtained from 1 4 3 6 5 2 plus again 12 chains from 1 4 5 2 3 6.

Now let us go from $m = 5$ to $n = 7$. Then the only couples of adjacent elements modulo the 5-ary G-chain based on $\{1, ..., 7\}$ are $(i, i+1)(i = 1 \text{ to } 7)$ plus (1,7): the lemma works and the dilated group $(A_5)^7$ is the dihedral group formed of 14 chains, say 1 2 3 4 5 6 7 and all chains deduced by translation and reflection.

Note that the dihedral group D_5 is the maximum indicative group included in the alternating group A_5. After the first dilatation we get $(A_5)^6$ including, as its maximum indicative group, the dihedral group D_6. After the second dilatation we get D_7, as well from D_6 as from $(A_5)^6$.

13.11.3 Calculation of the reduction threshold: third case

We need the following:

Lemma leading to indicative groups I, J. Let G be an m-ary, non-indicative group (thus $m \geq 4$) with $(1, 2, ..., m) \notin G$; set $n > m$. Define p to be the greatest integer such that every permutation of $\{1, 2, ..., p\}$ is in G, and q to be the least integer ($q \leq m$) such that every permutation of $\{q, q+1, ..., m\}$ is in G. Note that $p < q$ in order to respect our case assumption. Then we consider two cases:

(1) **General case where $p \geq 2$ or $q \leq m-1$ or $n \neq 2m-3$. Suppose that couples (i, j) are adjacent (modulo the G-chain based on $\{1, 2, ..., n\}$) iff either $j = i+1$ or $i < j \leq p+1$ or $i < j$ with $i \geq n-m+q-1$. Then the dilated group G^n is indicative. More precisely either $G^n = I_n^{p, m-q+1}$ or $p+q = m+1$ and $G^n = J_n^p$, according to whether reflection does not belong or belongs to G.**

(2) **Exceptional case where $p = 1, q = m$ and $n = 2m-3$. Suppose that couples (i, j) are adjacent (modulo the G-chain based on $\{1, 2, ..., n\}$) iff $j = i+1$ and possibly $i = m-2, j = m$. Then the dilated group G^n is indicative. More precisely either $G^n = I_n^{1,1}$ (identity alone) or $G^n = J_n^1$ (identity plus reflection) according to whether reflection does not belong or belongs to G.**

• General case. The dilated group $H = G^n$ contains at least all chains which begin with an arbitrary initial interval based on $\{1, ..., p\}$ followed by $p+1, p+2, ..., q-1+n-m$, followed by an arbitrary final interval based on $\{q+n-m, ..., n\}$. Plus possibly chains deduced by reflection. For any other chain B based on $\{1, ..., n\}$ which would belong to H, there would necessarily exist two integers i, j which be consecutive modulo B yet are not adjacent modulo our G-chain based on $\{1, ..., n\}$: contradiction.

Analogous proof for the exceptional case, where, since $p = 1, q = m$, the dilated group G^n reduces to the chain $12...n$ plus eventually the reflected chain $n....21$. For any other chain B which would belong to H, there would necessarily exist two integers $i < j$ with $j \neq i+1$ **and moreover** $i \neq m-2$ or $j \neq m$, which would be consecutive modulo B yet are not adjacent modulo our G-chain based on $\{1, ..., n\}$: contradiction. •

Proof of the theorem, third case, with the assumptions $m \geq 4, n \geq$ Max$(2m-3, m+2)$. We take again notations p and q in the previous lemma, and we recall that $p < q \leq m$.

1 - **If $i \leq p < j$ and i, j adjacent, then $j = p+1$.**

• Suppose $i \leq p < j$ and i, j adjacent but $j \geq p+2$; note that $i < m$. First we have $j < n$. Indeed if $j = n$ then set $h = i$ and $k = m$, so that $1 \leq h = i \leq p < k = m$ and $0 = i - h < j - k = n - m$. Then G contains $(i, ..., m)$; since by hypothesis G also contains every permutation of $\{1, ..., p\}$, then G contains the cycle $(1, ..., m)$, contradicting our case assumption.

Either $j \leq m+1$. Then set $h = i$ and successively $k = j-2$ and $k = j-1$, so that $1 \leq h = i \leq j-2 \leq k \leq j-1 \leq m$ and $0 = i - h < j - k \leq n - m$

since $j - k = 1$ or 2 and $n \geq m + 2$ by hypothesis. Then G contains $(i, ..., j - 2)$ and $(i, ..., j - 1)$, then G contains every permutation of $\{i, ..., j - 1\}$. Since G also contains every permutation of $\{1, ..., p\}$, we infer that it contains every permutation of $\{1, ..., j - 1\}$. But $j - 1 > p$, contradicting the choice of p.

Or $m + 2 \leq j < n$. Then set $h = i$ and successively $k = m - 1$ and $k = m$, so that $1 \leq h = i \leq k \leq m$ and $0 = i - h < j - k \leq n - m$ since $j - k = j - m$ or $j - m + 1$, and since $j < n$. Then G contains $(i, ..., m - 1)$ and $(i, ..., m)$; hence G contains every permutation of $\{i, ..., m\}$. Since $i \leq p$, then $G = S_m$, contradicting our case assumption. •

2 - **If** $i < n - m + q \leq j$ **and** i, j **adjacent, then** $i = n - m + q - 1$. The proof is the mirror image of our previous 1.

3 - **For all** i, j, **if** $p < i < j < n - m + q$ **and** i, j **adjacent, then either** $j = i + 1$ **or** $i = m + 2$ **and** $j = m$ with $p = 1, q = m$ and $n = 2m - 3$.

Firstly **we show that neither** $i + m < n + p$ **nor** $q < j$ **can occur**.

• By our hypothesis $p < i < j < n - m + q$ and i, j adjacent. Moreover we assume $j \geq i + 2$.

Let us show that $i + m < n + p$ is impossible: we argue ad absurdum.

Either $j \geq i + m - p + 1$. Then set $k = m$ and successively $h = p$ and $h = p + 1$; so that $1 \leq h \leq k = m$ (since $p + 1 \leq m$). Also $0 \leq i - h \leq i - p < j - k = j - m \leq n - m$ since by assumption $i > p$ and $j > i + m - p$. Finally G contains the transposition $(p, p + 1)$, contradicting the choice of p.

Or $j \leq i + m - p$. Then set $k = j - i + p - 1$ and successively $h = p$ and $h = p + 1$; so that $1 \leq h \leq p + 1 \leq k = j - i + p - 1 < m$ (since $j - i \geq 2$) and by assumption $i + m - p \geq j$. Also $0 \leq i - h \leq i - p < j - k = i - p + 1 \leq n - m$ since $i + m < n + p$ (our hypothesis). Finally G contains the transposition $(p, p + 1)$, contradicting the choice of p.

Let us show that $q < j$ is impossible: again ad absurdum.

Either $j \leq i + q$, then set $h = q + 1 + i - j$ and successively $k = q$ and $k = q - 1$; so that $1 \leq h \leq k \leq m$ since $q + i - j \geq 0$ and $q \leq m$ and $q + 1 + i - j \leq q - 1$ (since $j - i \geq 2$). Also $0 \leq i - h < j - k \leq n - m$ since $j \geq q + 1$ (our hypothesis) and $j < n - m + q$. Then G contains the transposition $(q - 1, q)$, contradicting the choice of q.

Or $j \geq i + q + 1$, then set $h = 1$ and successively $k = q - 1$ and $k = q$; so that $1 = h \leq k \leq m$ since $2 \leq q \leq m$, and $0 \leq i - 1 < j - q < j - q + 1 \leq n - m$ since by assumption $j \geq i + q + 1$ and $j < n - m + q \leq n$. Then G contains the transposition $(q - 1, q)$, contradicting the choice of q. •

Secondly **we prove statement** (3).

• Start from:

(i) $(i + m - p) + (q - j) = m + (q - p) - (j - i) \leq m + (m - 1) - 2 = 2m - 3 \leq n$ (since $j - i \geq 2$ and $q - p \leq m - 1$).

Assume $(i + m - p) + (q - j) < n$; then either $i + m < n + p$ or $q < j$: both are false by our previous proof.

Consequently $(i + m - p) + (q - j) = n$; so that equalities hold throughout (i); hence $p = 1, q = m, q - p = m - 1, j - i = 2$ and $n = 2m - 3$; moreover:

(ii) $m = q \geq j = i+2 \geq n+p-m+2 = 2m-3+1-m+2 = m$ (since $i+m \geq n+p$ by our previous proof).

Hence equalities hold throughout (ii), so that $j = i+2 = m$. •

First example. Take $m = 5$ and G defined by the six chains $1 < a < b < c < 5$ where a, b, c are the three integers 2, 3, 4 in any order. We see that $p = 1$ and $q = m = 5$ (p, q defined in the lemma). The first dilated group G^6 is formed of two chains $1 < 2 < a < b < 5 < 6$ where a, b is either 3, 4 or 4, 3. The second dilated group G^7 reduces to identity on $\{1,2,3,4,5,6,7\}$. Since 7 = 2.5 - 3, we are in the above "exceptional case".

Note that $m-2 = 3$ and $m = 5$ are adjacent modulo the G-chain based on $\{1, ..., 7\}$. Indeed take the domain $\{1,2,4,5,6,7\}$, the range $\{1,2,3,4,6,7\}$ and the bijection which saves 1, 2, 4, 6, 7 and changes 5 into 3. If we reduce this bijection to a permutation on the base $\{1,2,3,4,5,6\}$ it gives the exchange of 3 and 4, which duly belongs to G^6.

Second example. Take $m = 4$ and $G = A_4$ (quaternary alternating group). Then $(1,2,3,4) \notin G$ yet the reflection belongs to G. We have $p = 1, q = m = 4$. The dilated group G^5 is formed of six chains: 1 2 3 4 5, 3 2 5 4 1, 5 2 1 4 3 plus three reflected chains (non trivial couples of adjacent inegers: (1,4) and (2,5)). The next dilated group G^6 is indicative and reduces to identity and reflection. Since $2m-3 = 5 \neq 6$, we are in the "general case".

Third example. Take $m = 4$ and $G =$ KLEIN group formed of the four chains 1 2 3 4, 3 4 1 2, 2 1 4 3, 4 3 2 1. Then $p = 1, q = m = 4$. The first dilated group G^5 is indicative: it reduces to identity and reflection 5 4 3 2 1. Here $n = 5 = 2m-3$: we are in the "exceptional case". Note that $m = 4$ and $m-2 = 2$ **are not adjacent**.

A refinement. In [87] FRASNAY 1984, it is proved that in the third case where $(1,2,...,n) \notin G$ ($m \geq 4$, G non-indicative), then **the dilated group G^n is indicative of the form I or J, provided that $n = \text{Max}(m+2, m+q-p-2)$.**

13.12 Exercises

13.12.1 Minoration of the monomorphy threshold

Due to [88] FRASNAY 1990.

Take an arity $m \geq 3$. Let H be the group of all permutations of $\{1,2,...,m\}$ which saves 1 and m. Let ω' be the chain deduced from the usual chain ω by exchanging $m-2$ and $m-1$. Let us denote $A = \{0,1,2,...,2m-3\}$ and $B = \{1,2,3,...,2m-2\}$. To each $(2m-3)$-element subset $|X|$ of the set of integers, we associate the chain $X = \omega/|X|$ if either 0 or 1 or ... or $m-1$ does not belong to $|X|$, and the restriction $\omega'/|X|$ if 0 and 1 and ... and $m-1$ all belong to $|X|$.

1 - Show that all chains X are mutually H-compatible. On the base of integers, define the m-ary relation R, common extension of all (H, X)-chains (defined in 13.10.3): then this R is $(2m-3)$-monomorphic.

2 - Show that the restrictions R/A and R/B are not isomorphic: so that R is not a $(2m-2)$-monomorphic relation. Consequently the monomorphy threshold (see 13.10.6) is $\geq 2m-2$, hence is exactly $2m-2$ (for $m \geq 3$).

Hint: R/A admits two restrictions of cardinal m respectively constructed from the chains $0 < 1 < 2 < ... < m-3 < m-1 < m-2$ (a restriction of ω') and $m-2 < m-1 < m < ... < 2m-3$ (a restriction of ω). Hence we have $R(0,1,2,...,m-3,m-1,m-2) = R(m-2,m-1,m,...,2m-3) = +$ with the common element $m-2$ in last, then in first position (joint element in FRASNAY's terminology). Yet R/B is chainable by ω/B; more precisely R/B is a $(H,\omega/B)$-chain with cardinality $2m-2$: then the structure of H does not allow the existence in B of two restrictions with cardinality m and with a joint element. Incidentally R/A is monomorphic yet not chainable (see 13.12.2 below).

3 - Show that R/A is a bound (of maximum cardinal) for the (H,ω)-chain based on non-negative integers.

13.12.2 Degree of a universal class

Given a universal class U, we define the **degree** of U as being the maximum cardinal of all bounds of U (see 5.10.1). If no bound exists, by convention the degree is 0.

For instance the universal class of all m-ary relations has degree 0. The class of all m-ary relations with cardinalities $\leq p$, has degree $p+1$.

Other examples. The universal class of unary constant relations has degree 2; the universal class of all chainable binary relations has degree 3; the universal class of all monomorphic binary relations has degree 4.

For $m \geq 3$, we shall show that the class of all m-ary chainable relations has degree $\geq 2m-2$ (a result of FRASNAY in 1996; we may ask if this degree is exactly $2m-2$).

Consider the group H, the relation R and the finite sets A, B in the preceding exercise: we know that R/A is not a H-chain and is a bound (having maximum cardinal $2m-2$) of the universal class of all H-chains.

1- Show that $R/(A \cup B)$ is a bound of the class of monomorphic m-ary relations; consequently the class of monomorphic m-ary relations has degree $\geq 2m-1$.

2- For each integer $h \in A$ (hence $0 \leq h \leq 2m-3$), let $A_h = A - \{h\}$; let $C_h = \omega/A_h$ for $0 \leq h \leq m-1$ and $C_h = \omega'/A_h$ for $m \leq h \leq 2m-3$ (same chains ω and ω' than in the preceding exercise). For each couple (i,j) with $0 \leq i,j \leq 2m-3$, let us denote $f_{i,j}$ the (unique) isomorphism which maps C_i onto C_j. Show that $f_{i,j}$ is the unique isomorphism which maps R/A_i onto R/A_j.

3- Show that each $f_{i,j}$ admits at least one fixed point. Yet given a finite chain $a_0 < a_1 < ... < a_n (n \geq 1)$, then the local isomorphism which maps $a_0 < ... < a_{n-1}$ on $a_1 < ... < a_n$ has no fixed point. So that R/A is not chainable.

Moreover every proper restriction of R/A is a H-chain; so that R/A is a bound for the class of m-ary chainable relations.

13.12.3 An alternative definition of indicators

Start from canonical groups as they are described in 13.6.2. An m-ary group H is said to be a **spring** iff H is canonical and there is no $(m-1)$-ary group whose first dilated would be H. Then we call **natural indicator** the sequence of all successive dilated groups starting from a spring.

For example the cyclical group T_3 which generates the indicator T, the dihedral group D_4 which generates the indicator D, are springs. Example for an arbitrary great arity: the (p+q)-ary group $I_{p+q}^{p,q}$ is a spring.

Prove the following lemma.

Given an arbitrary m-ary group G and the maximum m-ary canonical group $H(G)$ included in G, then:

(i) **for any $n > m$ the dilated group $H(G)^n$ is the maximum canonical group included in the dilated group G^n;**

(ii) **there exists an integer $n \geq m$ from which point on we have the equality $H(G)^n = G^n$.**

From the previous lemma, deduce the indicative group theorem: every indicative group is canonical.

13.12.4 Relational system freely interpretable in a chain having neither a minimum nor a maximum element

We follow [87] FRASNAY 1984, specially p.263.

(1) Let A be a non-empty chain having neither a minimum nor a maximum (for example the chain $\mathbf{Z}$ of positive and negative integers). We don't mind about the cardinal of A, provided it be infinite. Show that, given an arbitrary set $\mathcal{A}$ of chains on the base $|A|$ with $A \in \mathcal{A}$, then the ω-sequence G_m(m integer ≥ 1) generated by $\mathcal{A}$ is associated with one of the five indicators S, I, J, T, D. Two possibilities for the proof: either adapt the eight lemmas beginning from 13.7.3; or show that the presence of a chain having neither minimum nor maximum forbids indicators $I^{p,q}$ and J^r.

(2) Show that the statement 13.8.1 subsists with a relational system R freely interpretable in any non-empty chain A having neither a minimum nor a maximum. The only difference: one cannot always transform, by an automorphism of A, the m-sequence of values fx into the values x. Yet using the **local** automorphism of A which transforms the sequence of fx into the x, we find again G_m as a subgroup of automorphisms of $R/\{x_1, ..., x_m\}$.

Problem. The group G_m is, among the five $\mathbf{Q}$-indicative groups, the maximum which be included in the group of automorphisms of $R/\{x_1, ..., x_m\}$.

Appendix A

Partitions of countable homogeneous systems

The following Appendix is kindly communicated by Norbert SAUER.

In this text several increasingly severe notions of divisibility of homogeneous relational systems are introduced. Those divisibility properties are related to the type of amalgamation of the age of the homogeneous relational system. The divisibility of some of the more prominent homogeneous relations is investigated.

A **relational language** $\mathbf{L}$ is a set $\{R_i | i \in I\}$ of relation symbols. Each of the relation symbols R_i is associated with a non-negative number n_i, the **arity** of R_i. The set $\{n_i | i \in I\}$ is the **arity** of $\mathbf{L}$.

A **relational system** S in the language $\mathbf{L}$ is a set $|S|$, the **base** of S, together with a set $\{R_i^S | i \in I\}$ so that $R_i^S \subseteq |S|^{n_i}$ for every $i \in I$. If $(x_0, x_1, ..., x_{n-1}) \in |S|^{n_i}$ then we write $R_i^S(x_0, x_1, ..., x_{n-1}) = +$ if $(x_0, x_1, ..., x_{n-1}) \in R_i^S$; value (-) in the contrary case. The R_i^S are the **components** of S (see 1.7). A relational language $\mathbf{K}$ subset of $\mathbf{L}$ is a **reduct** of $\mathbf{L}$ and $\mathbf{L}$ is an **expansion** of $\mathbf{K}$.

This Appendix is to a certain degree a continuation of Chapter 12 with more emphasis on relational systems instead of single relations. In order to become familiar with the use of the notions, some of Chapter 12 is repeated. The relational system R is **denumerable** if the base $|R|$ is denumerable. All relational systems and languages considered in this Appendix will be denumerable or finite.

A.1 Some prominent homogeneous relations

The notions of **restriction, isomorphism, isomorphic copy, embedding, local isomorphism or automorphism** defined for single relations in chapters 1, 5, 9, immediately extend to relational systems, as already quickly seen in 12.3.

Two relational systems (12.3.1, see also multirelation 1.7) A and B of the same language are **compatible** (1.7.2) if the restriction of A to $|A| \cap |B|$ is equal to the restriction of B to $|A| \cap |B|$.

Let A and B be compatible and D a relational system of the same language. A function f from $|A| \cup |B|$ to $|D|$ is an **amalgamation** function of A and B if f restricted to $|A|$ is an embedding of A into D and if f restricted to $|B|$ is an embedding of B into D. The relational system D is an **amalgam** of the relational systems A and B.

A set $\mathcal{R}$ of relational systems of the same language is **amalgamable** if for any two compatible relational systems A and B belonging to $\mathcal{R}$ there exists an amalgam D in $\mathcal{R}$ (12.2).

The set $\mathcal{R}$ is **directed under embeddability** if any two elements S, T of $\mathcal{R}$ with $|S| \cap |T|$ empty have an amalgam in $\mathcal{R}$ (the set $\mathcal{R}$ is 0-amalgamable).

An **age** (10.2.1) is a set of finite relational systems of the same language closed under isomorphism, restriction and directed under embeddability. Let R be a relational system. The **age of** R is the set of relational systems which are isomorphic to a finite restriction of R. A **representative** of an age $\mathcal{R}$ is a relational system R whose age is $\mathcal{R}$. Every age has a representative (10.2).

The relational system S is **younger** than the system R, or equivalently R is **older** than S, if the age of S is a subset of the age of R (see 10.1.3).

The **skeleton** of a relational system R is the set of restrictions of R to a finite subset of the base of R. The age of R is the set of relational systems A which are isomorphic to an element of the skeleton of R.

We have the following **mapping extension property** of an amalgamable age:

Let $\mathcal{R}$ be an amalgable age. Let A and B be elements of $\mathcal{R}$. Let $u \in |A|$ and f an embedding of $A - u$ into B. Then there is an extension $D \in \mathcal{R}$ of B so that f has an extension to an embedding of A into D.

A.1.1 The mapping extension property

The relational system R has the **mapping extension property** if:

for every element A in the age of R and every element $u \in |A|$ and every embedding f from $A-u$ to R there exists an extension of f which is an embedding of A to R.

The relational system R has the mapping extension property iff for every relational system S younger than R and every embedding f from a finite restriction of S to R, there exists an embedding from S to R which is an extension of f.

• If for every relational system S younger than R and every embedding f from a restriction of S to a finite subset of $|S|$ to R there exists an embedding from S

to R which is an extension of f, then R has the mapping extension property.

Conversely let R have the mapping extension property. Enumerate the elements of $|S|$ setminus the domain of f into a sequence $s_0, s_1, s_2, \ldots$. Extend step by step the function embedding f to $|S| \cup \{s_0\}, |S| \cup \{s_0\} \cup \{s_1\}, |S| \cup \{s_0\} \cup \{s_1\} \cup \{s_2\}, \ldots$ using the mapping extension property. Take the limit of the preceding extensions. •

Let R, S be two relational systems with the same language and the same age, and both having the mapping extension property. Then every embedding g of a finite restriction of R into S can be extended to an isomorphism of R to S.

• Enumerate $|R|$ into the ω-sequence $r_0, r_1, \ldots$ and $|S|$ into the ω-sequence $s_0, s_1, \ldots$. Suppose we get an embedding f_n of a finite restriction of R to S, such that $\operatorname{Dom} f_n$ contains as elements all $r_i (i < n)$ and $\operatorname{Rng} f_n$ contains all $s_i (i < n)$. Then using the mapping extension property we get an extension f_{n+1} of f_n which satisfies the same properties for $i < n+1$. Finally we get the ω-sequence $f_0 \subset f_1 \subset f_2 \subset \ldots$: the union of these f_n is an isomorphism from R to S. •

A.1.2 Homogeneity

A relational system R is said to be **homogeneous** if every local automorphism of R has an extensiion to an automorphism of R (see 12.1).

The relational system R is homogeneous if and only if it has the mapping extension property.

• Suppose that R has the mapping extension property. Let F be a finite subset of $|R|$ let f be a local automorphism with domain F, let u be an element in $|R| - F$. The restriction of R to $F \cup \{u\}$ is a relational system, say A, in the age of R and f is an embedding from $A - u$ to R. Hence, using the mapping extension property, there is an embedding, which of course is a local automorphism, which extends f to the domain $F \cup \{u\}$. Hence by the preceding R is homogeneous.

Conversely suppose that R is homogeneous. Let A be a relational system in the age of R, the element u in the base of A and f an embedding from $A - u$ to R. Because A is an element in the age of R there is an embedding g from A to R. Let g_1 be the restriction of g to $|A| - \{u\}$, let h be the local automorphism given by $f \circ g_1^{-1}$, let h' be an extension of the local automorphism $f \circ g_1^{-1}$ to an automorphism of R. Then $h' \circ g$ is an embedding from A to R which extends f. •

A.1.3 The amalgamation theorem

The age of a homogeneous relational system is closed under isomorphism, restriction, directed under embeddability and it is amalgamable. An amalgamable age has a homogeneous representative (see 12.2.1).

• Let H be a homogeneous relational system. The age of H is closed under isomorphism, restriction and directed under embeddability. Let A, B be two compatible elements in the age of H and let C be the common restriction of A and B to the intersection $|A| \cap |B|$. Then C is in the age of H and hence there is an embedding f of C into H. The function f has an extension f_A to an embedding of A into H and an extension f_B to an embedding of B into H. The union $f_A \cup f_B$ is an amalgamation function.

Let $\mathcal{R}$ be an amalgable age with language $\mathbf{L}$, assumed to be denumerable. Hence the pairs (A, u) with $A \in \mathcal{R}$ and $u \in |A|$ are denumerable into an ω-sequence $(A_0, u_0), (A_1, u_1), \ldots$. Let $B_0 = A_0$ and B_n an element of the age so that for every $i < n$ and every embedding f of $A_i - u_i$ into B_{n-1} there is an extension of f to an embedding of A_i into D_n (such an element $D_n \in \mathcal{R}$ exists because of the mapping extension property of the age $\mathcal{R}$). Then B_{n+1} is an extension of B_n for every integer n. Let H be the limit. Then the age of H is $\mathcal{R}$ and H has the mapping extension property and is therefore homogeneous. •

Corollary. There is, up to isomorphism, exactly one countable homogeneous representative of an amalgamable age.

A.1.4 Graphs and hypergraphs

A **graph** G is a binary symmetric antireflexive relation. The base of G is denoted by $V(G)$ and $E(G)$ denotes the set of all two element subsets of $V(G)$ on which the relation G is positive (as usual in the theory of graphs the base of a graph G is denoted by $V(G)$ instead of $|G|$). The elements of $V(G)$ are the **vertices** of G and the elements of $E(G)$ are the **edges** of G. The vertices x and y of G are **adjacent**, $x \sim y$, if $\{x, y\}$ is an edge of G.

A graph G is **complete** if $E(G) = [V(G)]^2$. By K_n we denote the **complete graph** with $n \leq \omega + 1$ vertices.

A k-**uniform hypergraph** G is a relation of arity k so that $G(x_0, x_1, \ldots, x_{k-1}) = +$ implies $x_i \neq x_j$ for all $i \neq j$ in k and $G(x_{\pi(0)}, \ldots, x_{\pi(k-1)}) = +$ for all permutations π of k. The base of G is denoted by $V(G)$.

The set of **hyperedges** of G is the set $E(G) = \{\{x_0, \ldots, x_{k-1}\} | G(x_0, \ldots, x_{k-1}) = +\} \subseteq [V(G)]^k$.

The hypergraph G is **complete** if $\cup_{A \in E(G)} [A]^2 = [V(G)]^2$.

Let $\mathbf{L} = \{E_0, E_1, \ldots, E_{n-1}\}$ be a binary relational language. The relational system G with language $\mathbf{L}$ is a graph with n-**types of edges** if all of the relations $E_0^G, E_1^G, \ldots E_{n-1}^G$ are graph relations. We write $E_i(G)$ for the set of edges of type i, with $i \in n$. Analogously we define a k-uniform hypergraph with several types of edges.

Let the arities of the relation symbols E_i in $\mathbf{L}$ be arbitrary numbers. A relational system G with language $\mathbf{L}$ is a hypergraph with several types of edges if each of the relations E_i^G is a hypergraph relation. The hypergraph G with several types of edges is **complete** if $\cup_{A \in \cup_{i \in n} E_i(G)} [A]^2 = [V(G)]^2$; that is if for every two elements of $V(G)$ there is an hyperedge of G which contains both.

A.1.5 The rationals and the Rado graph

The age of the chain $\mathbf{Q}$ of rationals is the set of all finite chains. Clearly if f is an embedding of a finite chain A to $\mathbf{Q}$ and if B is a finite chain extension of A, then there is an extension of f to an embedding of B. Hence $\mathbf{Q}$ is a homogeneous relation by A.1.2.

Another important example of a homogeneous relation is the **Rado graph**. It is constructed similar to the rich relation in 10.3.1. The Rado graph is the union of finite graphs $A_0 \subset A_1 \subset A_2 \subset A_3 \subset \cdots$ where A_0 is the graph with exactly one vertex. For every subset $F \subseteq V(A_i)$ there is a vertex $a \in V(A_{i+1})$ which is adjacent to all vertices in F and not adjacent to the vertices in $V(A_i) - F$. It follows from the construction that the age of the Rado graph is the set of all finite graphs and that the Rado graph has the mapping extension property. Hence the Rado graph is a homogeneous relation.

The defining property of the Rado graph is:

The Rado graph is the graph in which for any two finite subsets $F_1 \subseteq F$ of vertices there exists a vertex which is not in F, adjacent to all vertices in F_1 and to no vertex in $F - F_1$.

• It follows from the construction that the Rado graph has this porperty. If a graph has this property then it has the mapping extension property. •

A.1.6 The K_n-free homogeneous graphs H_n

The graph G is K_n-**free** if there is no embedding of K_n into G. The subset $T \subseteq V(G)$ is K_n-free if there is no embedding of K_n to the restriction of G to T.

The **K_n-free homogeneous graph** H_n is constructed in a similar way as the Rado graph. It is the union of finite graphs $A_0 \subset A_1 \subset A_2 \subset A_3 \subset \cdots$ where A_0 is the graph with exactly one vertex. For every K_{n-1}-free subset $S \subseteq V(A_i)$ there is a vertex $a \in V(A_{i+1})$ which is adjacent to all vertices in S and not adjacent to the vertices in $V(A_i) - S$. It follows from the construction that the age of the graph H_n is the set of all finite K_n-free graphs and that H_n has the mapping extension property. Hence the graph H_n is a homogeneous relation.

The defining property of H_n is:

The K_n-free homogeneous graph H_n is the graph in which for every finite subset F of vertices and every K_{n-1}-free subset $F_1 \subseteq F$ there exists a vertex which is not in F, adjacent to all vertices in F_1 and to no vertex in $F - F_1$.

• It follows from the construction that the Graph H_n has the porperty above. If a graph has the property above then it has the mapping extension property. •

Let $\mathbf{L} = \{E_i | i \in n\}$ be a relational language and $\mathcal{T}$ a set of finite complete hypergraphs with language $\mathbf{L}$. A hypergraph G with several types of edges whose language is $\mathbf{L}$ is $\mathcal{T}$-**free** if there is no embedding of any of the hypergraphs of $\mathcal{T}$ into G.

The $\mathcal{T}$-free homogeneous hypergraph $H_{\mathcal{T}}$ is constructed in a similar way as the K_n-free graph H_n. The hypergraph A_0 is the hypergraph with exactly one vertex.

For every possible way of extending A_i, without creating one of the forbidden hypergraphs of $\mathcal{T}$, we add a single vertex to A_i. The hypergraph A_{i+1} consists of the extensions of A_i to all of those additional vertices.

A.1.7 Known characterizations of homogeneous relational systems

The complement of a graph G has the same set of vertices as G and two vertices are adjacent in the complement if and only if they are not adjacent in G (the complement of a graph is the negation of the graph viewed as a relation as defined in 1.7.4). Note that a graph G is homogeneous if and only if its complement is homogeneous. The complement of the Rado graph is again the Rado graph.

More generally let R be a relational system and S a set of components of R. The relational system R' derived from R by negating some of the components in S is called **essentially equivalent** to R. If R and R' are essentially equivalent then R is a homogeneous relational system iff R' is a homogeneous relational system (the automorphism group of R is the same as the automorphism group of R').

The relation which consists of denumerably many copies of K_n or of finitely many or denumerably many copies of K_ω is a homogeneous relation. The Rado graph, the K_n-free graphs the graphs which consist of copies of K_n or K_ω and the complements of those graphs are all the homogeneous graphs: [148] LACHLAN, WOODROW 1980.

The homogeneous tournaments are characterised in [146] LACHLAN 1984, the homogeneous partial orders are characterised in [223] SCHMERL 1979, the homogeneous directed graphs are characterised in [32] CHERLIN 1998, the countable stable structures which are homogeneous for relational systems with finitely many components are characterised in [147] LACHLAN 1984. As every finite relational system is stable also all finite homogeneous graphs together with other finite homogeneous systems are characterised in [147]. All the finite homogeneous graphs where exhibited earlier in [92] GARDINER 1976. The pentagon for example is a finite homogeneous graph.

A.2 Various types of amalgamation

Strong amalgamation has already been defined in 12.2. The definition given below is equivalent. Free amalgamation is also called independent amalgamation.

A.2.1 Strong amalgamation

Let $\mathcal{R}$ be a set of relational systems of the same language. The set $\mathcal{R}$ is said to be **strongly amalgamable** (12.2), if for any two compatible relational systems A, B belonging to $\mathcal{R}$ there exists an amalgam $D \in \mathcal{R}$ which is an extension of

A and B. That is if there is an amalgamation function which is the identity on $|A| \cup |B|$. Or equivalently if there is a one to one amalgamation function whose image is an element of $\mathcal{R}$. Let $K_\omega + K_\omega$ be the graph which consists of two disjoint copies of K_ω. The age of $K_\omega + K_\omega$ is strongly amalgamable. The set of all finite chains is strongly amalgamable with $\mathbf{Q}$ as representative. The set of all finite posets is strongly amalgamable (1.7.3). The **universal homogeneous poset** is the homogeneous representative of the set of finite posets. The homogeneous graph which consists of denumerably many disjoint triangles is not strongly amalgamable.

A.2.2 Free amalgamation

Let A and B be two compatible relations of the same language $\mathbf{L} = \{R_i | i \in I\}$. The **free amalgam** of A and B is the extension D of A and B, with base $|A| \cup |B|$ so that $R_i^D = R_i^A \cup R_i^B$ for all $i \in I$.

The set $\mathcal{R}$ is said to be **freely amalgamable** if the free amalgam of any two compatible relational systems A, B belonging to $\mathcal{R}$ is again an element of $\mathcal{R}$. The set of all finite graphs which do not embed the complete graph K_n is a freely amalgamable age. The representative of the set of all finite graphs which do not embed the complete graph K_n is the K_n-free homogeneous graph H_n. A freely amalgamable age is strongly amalgamable. The age of $K_\omega + K_\omega$ is strongly but not freely amalgamable. The set of all finite chains and the set of all finite posets are examples of strongly but not freely amalgamable ages.

Let $\mathcal{T}$ be a set of finite complete hypergraphs with several types of edges with relational language $\mathbf{L}$. Let $\mathcal{R}$ be the set of finite hypergraphs A of the language $\mathbf{L}$ so that there is no embedding of an element of $\mathcal{T}$ to A. Then $\mathcal{R}$ is a freely amalgamable age. The homogeneous hypergraph which is the homogeneous representative of $\mathcal{R}$ is denoted by $H_{\mathcal{T}}$.

A.2.3 Amalgamation over a particular relational system

The set $\mathcal{R}$ is **amalgamable over the relational system** C if any two compatible relational systems A and B belonging to $\mathcal{R}$ whose intersection is C have an amalgam in $\mathcal{R}$. The set $\mathcal{R}$ is n **amalgamable** if any two compatible relational systems A and B belonging to $\mathcal{R}$ whose intersection has n elements have an amalgam in $\mathcal{R}$. Hence a set $\mathcal{R}$ of relational systems is 0-amalgamable if and only if it is directed under embeddability (10.2). Of interest are amalgamable ages which are strongly 0-amalgamable, freely 0-amalgamable, strongly or freely 1-amalgamable and so on.

A.2.4 Bounds, complete relational system

The relational system C with language $\mathbf{L} = \{R_i | i \in I\}$ is **complete** if for any two elements $y, z \in |C|$ there is an $i \in I$ and an n_i-tuple $(x_0, \ldots, x_{n_i-1}) \in R_i^S$ so that $\{x, y\} \subseteq \{x_0, \ldots, x_{n_i-1}\}$.

Let $\mathcal{R}$ be a set of finite relational systems all having the same arity. A **bound** of $\mathcal{R}$ is a finite relational system C of that same arity which is not in $\mathcal{R}$ but every proper restriction of C is an element of $\mathcal{R}$. If $\mathcal{R}$ is closed under restriction and isomorphism it is completely determined by its bounds. That is, for a given set $\mathcal{C}$ of finite relational systems of the same arity there is one and only one set $\mathcal{R}$ of finite relational systems closed under restriction and isomorphism whose set of bounds is $\mathcal{C}$. The set $\mathcal{R}$ is the set of all finite relational systems of the given arity which do not embed any one of the elements of $\mathcal{C}$.

Let $\mathcal{C}$ be a set of finite relational systems all of the same arity which are pairwise incomparable under embedding. Let $\mathcal{R}$ be the set of relational systems whose bound is $\mathcal{C}$. The set $\mathcal{R}$ is the set of $\mathcal{C}$-**free** relational systems of the given arity denoted by Forb($\mathcal{C}$). If $\mathcal{C}$ is a finite set then the $\mathcal{C}$-free set $\mathcal{R}$ of relational systems is a universal class (see 5.10).

A.2.5 Characterization of freely amalgamable ages

Let $\mathcal{C}$ be a set of finite relational systems having the same language L $= \{R_i | i \in I\}$ of arity $\{n_i | i \in I\}$ so that no relational system in $\mathcal{C}$ can be embedded into another relational system of $\mathcal{C}$. The $\mathcal{C}$-free set Forb($\mathcal{C}$) of relational systems is a freely amalgamable age if and only if every element of $\mathcal{C}$ is complete.

• Let every element of $\mathcal{C}$ be complete. Clearly Forb($\mathcal{C}$) is closed under isomorphism and restriction. Let A and B be two compatible elements of Forb($\mathcal{C}$). Let D be the free amalgam of A and B. Assume for a contradiction that there is an embedding f from an element $C \in \mathcal{C}$ into D. As both A and B do not embed C there are two elements y and z belonging to $|C|$ so that $f(y) \in |A| - |B|$ and $f(z) \in |B| - |A|$. Assume that the relation R_C of arity $n \in \omega$ is a component of C with the property that there is an n-tupel $(x_1, x_2, \ldots, x_n)$ of elements in $|C|$ so that $R_C(x_1, x_2, \ldots, x_n) = +$ and $\{y, z\} \subseteq \{x_1, x_2, \ldots, x_n\}$. Let R_A and R_B be the corresponding components of A and B. Then $R_D(f(x_1), f(x_2), \ldots, f(x_n)) = +$ but $R_A(f(x_1), f(x_2), \ldots, f(x_n)) = -$ and $R_B(f(x_1), f(x_2), \ldots, f(x_n)) = -$ because the set $\{f(x_1), f(x_2), \ldots, f(x_n)\}$ contains the element $f(y)$ not in the base of B and the element $f(z)$ not in the base of A.

Assume Forb($\mathcal{C}$) is a freely amalgamable age and that there is a relational system $A \in \mathcal{C}$ and elements $y \in |A|$ and $z \in |A|$ so that for every $n \in \omega$ and n-tupel $(x_1, x_2, \ldots, x_n)$ with $\{y, z\} \subseteq \{x_1, x_2, \ldots, x_n\}$ and every n-ary component R of A the relation $R(x_1, x_2, \ldots, x_n)$ has value $-$. The relational systems $A - y$ and $A - z$ are compatible and elements of Forb($\mathcal{C}$). The relational system A is the free amalgam of $A - y$ and $A - z$. Hence $A \in$ Forb($\mathcal{C}$), a contradiction. •

A.2.6 More homogeneous relational systems with freely amalgamable age

The first theorem A.2.5 implies that there is large number of homogeneous structures. If $\mathcal{C}$ is a set of finite complete relational systems of the same arity so that no relational system in $\mathcal{C}$ can be embedded into another relational system of $\mathcal{C}$, the homogeneous representative of Forb($\mathcal{C}$) is denoted by $H_{\mathcal{C}}$. Let L be the binary relation consisting of a singleton base, say $\{x\}$, so that $L(x,x) = +$. Let R be the binary relation with base $\{x,y\}$ so that $R(x,x) = R(y,y) = R(y,x) = -$ and $R(x,y) = +$. Then the Rado graph is the homogeneous structure $H_{\{L,R\}}$. The K_n-free homogeneous graph, which in the context of graphs is just denoted by H_n, is then the homogeneous structure $H_{\{L,R,K_n\}}$.

A **graph G with two types of edges** is a relational system with two binary components each of which is symmetric and antireflexive. The base of G is denoted by $V(G)$ and the **sets of edges** of G are denoted by $E_1(G)$ and $E_2(G)$. Also $E_1(G) \cap E_2(G) = \emptyset$. A graph with n types of edges is defined accordingly. The set of finite graphs with two types of edges is an age; the homogeneous representative is the **universal graph with two types of edges**.

Let A_n be the graph with two types of edges and base n so that $E_1(A_n) := \{\{i,i+1\} \mid i \in n\}$ and $E_2(A_n) := [n]^2 - E_1(A_n)$. The graph A_n is complete and if $n \neq m$ then there is no embedding from A_n into A_m. Hence for every subset $\mathcal{C} \subseteq \{A_n \mid n \in \omega\}$ there exists the homogeneous graph with two types of edges $H_{\mathcal{C}}$. It follows that there are continuum many homogeneous graphs with two types of edges.

[108] HENSON 1972 has constructed an infinite set of finite tournaments not one of which can be embedded into another. Hence there are continuum many homogeneous directed graphs. Similarly there are continuum many 3-uniform hypergraphs and so on.

A.2.7 The cyclic order

From the complete characterisations obtained so far it seems that for every language there are the homogeneous relations of the type $H_{\mathcal{C}}$ with $\mathcal{C}$ a set of finite complete relations and then some special homogeneous relations whose age has a more complicated amalgamation. Here is just one example of such a special homogeneous relation.

The **cyclic order** has as base any countable dense subset of the unit circle with the property that no two points make an angle of $2\pi/3$ at the center. Two vertices will be **adjacent** if the acute angle they make at the center is less than $2\pi/3$.

A.2.8 Orbits

Let H be a homogeneous relational system and $\mathrm{Aut}(H)$ the automorphism group of H. If A is a finite subset of $|H|$ then $\mathrm{Aut}_A(H)$ is the subgroup of all elements $g \in \mathrm{Aut}(H)$ so that $g(a) = a$ for all $a \in A$; the stabilizer subgroup of $\mathrm{Aut}(H)$ stabilizing the set A.

The elements $x, y \in |H| - A$ are in the same **orbit** of A if there is a $g \in \mathrm{Aut}_A(H)$ so that $g(x) = y$. The relation of being in the same orbit of A is an equivalence relation on $|H| - A$. The corresponding equivalence classes are said to be the **orbits of** A. If X is an orbit of A then A is the **base** of X which we denote by $F(X)$.

An **orbit of** H is an orbit of A for some finite subset A of $|H|$. It follows that if H does not have any unary components then $|H|$ is the only orbit of the empty set.

Let T be an orbit of H and x and y two elements of T. Then, according to the definition of orbit, there is a local isomorphism f so that $f(x) = y$ and $f(a) = a$ for all elements $a \in F(T)$. This implies that the *type* of x over $F(T)$ is the same as the type of y over $F(T)$. If for example H is a graph then a vertex a in $F(T)$ is adjacent to x if and only if it is adjacent to y. Hence, given F, an orbit T of F is completely determined by the subset $F_1 \subseteq F$ of vertices in F which are adjacent to one and hence to all of the elements of T. If H is a 3-uniform hypergraph then two vertices a and b in $F(T)$ form a hyperedge with x if and only if they form a hyperedge with y.

Many notions for homogeneous relational structures are dependent only on the automorphism group. For example orbit. Given a group G of permutations of a set E which is closed under adherence (12.3.4) then there exists a homogeneous relational system whose group of automorphisms is G (12.3.5). This group theoretic aspect of homogeneous structures has been investigated in [24], [26], [28], [29] CAMERON 1976 to 1990 and [164] MACPHERSON 1985.

A.2.9 Universal homogeneous relations

Let T be a relational system with ω as base and language $\mathbf{L} = \{R_i | i \in I\}$ with corresponding arities $\{n_i | i \in I\}$. The **square** S of T has base $|S| = \{(x, y) | x \in \omega \wedge y \in \omega\} = |S| \times |S|$. The interpretation of the relation-symbols is

$$R_i^S = \{(x_0, y_0), ..., (x_{n_i-1}, y_{n_i-1})|$$
$$(\forall i, j \in n_i (x_i = x_j) \wedge R_i^T(y_0, ..., y_{n_i-1}))\vee$$
$$(x_i \neq x_j \text{ for all } i \neq j \wedge R_i^T(x_0, ..., x_{n_i-1}))\}.$$

A homogeneous relational system U is called **universal** if the age of the square of U is equal to the age of U. Examples of universal relations are the Rado graph, also called the universal graph, the universal tournament which is the homogeneous representative of the set of all finite tournaments, the universal directed graph which is the representative of the set of all finite directed graphs, the universal 3-uniform hypergaph which is the representative of the set of all 3-uniform hypergraphs, and so on.

A.3 Cutting finite pieces from homogeneous systems

A.3.1 Inexhaustible ages of homogeneous relational systems

Let H be a homogeneous relational system. If the age of H is inexhaustible (10.6.2) then there exists an inexhaustible representative of the age of H which is embeddable into H and hence for every finite subset $A \subseteq |H|$ the age of H restricted to $|H| - A$ is equal to the age of H. We will then say that H is **age-inexhaustible**. If H is age-inexhaustible then clearly the age of H is strongly 0-amalgamable. If the age of H is strongly 0-amalgamable then the age of H is inexhaustible according to 10.6.3.

A.3.2 Orbits of the empty set and inexhaustible ages

Let H be a homogeneous relational system. Every orbit in H of the empty set is infinite if and only if H is age-inexhaustible: [201] POUZET 1981.

• If $K \subseteq |H|$ is a finite orbit of the empty set then the restriction of H to $|H| - K$ cannot embed the restriction of H to K.

Assume that every orbit of the empty set is infinite. It suffices to prove that if A is a finite subset of $|H|$ then there are infinitely many pairwise disjoint subsets $A = A_0, A_1, A_2, \ldots$ of $|H|$ so that the restriction of H to A is isomorphic to the restriction of H to A_i for every $i \in \omega$. The assertion is certainly true if A is a singleton set $\{a\}$. We procede by induction on the number of elements in $|A|$.

Let $a \in A$ and $B = A - a$. Assume that $A = A_0, A_1, A_2, \ldots, A_{n-1}$ is a maximal sequence of pairwise disjoint subsets of $|H|$ so that the restriction of H to A is isomorphic to the restriction of H to A_i for every $i \in \omega$. Let $a = a_0, a_1, a_2, \ldots, a_{n-1}$ be the corresponding sequence of a's and $B = B_0, B_1, B_2, \ldots, B_{n-1}$ the corresponding sequence of B's.

Using induction, there are infinitely many pairwise disjoint subsets of $|H|$ so that the restriction of H to B is isomorphic to the restriction of H to each of those subsets. Only finitely many of those subsets have a non empty intersection with the set $\bigcup_{i \in n} B_i$. Hence there is an infinite extension of the sequence $B = B_0, B_1, B_2, \ldots, B_{n-1}$ to a sequence $B = B_0, B_1, B_2, \ldots, B_{n-1}, B_n, B_{n+1}, B_{n+2} \ldots$ of pairwise disjoint sets so that the restriction of H to B is isomorphic to the restriction of H to B_i for every $i \in \omega$.

Let f_i be an extension of the local isomorphism from B_0 to B_i to an automorphism of H. The automorphism f_i maps the set $\{a_0, a_1, \dots, a_{n-1}\}$ onto itself because of the maximality of the sequence $A = A_0, A_1, A_2, \dots, A_{n-1}$. Hence there is $j \in n$ so that $f_i(a) = a_j$ for infinitely many $i \in \omega$. That is a_j forms with infinitely many pairwise disjoint copies of B a copy of A.

Let g be an automorphism of H which maps a_j into $|H| - \{a_0, a_1, \dots, a_{n-1}\}$. Then $g(a_j)$ forms with infinitely many pairwise disjoint copies of B a copy of A. That is there are infinitely many pairwise disjoint sets $C_0, C_1, \dots$ so that H restricted to $\{g(a_j)\} \cup C_i$ forms a copy of A for every $i \in \omega$. This is a contradiction to the maximality of the sequence $A = A_0, A_1, A_2, \dots, A_{n-1}$ because only finitely many of the sets C_i intersect the set $\bigcup_{i \in n} B_i$. •

A.3.3 Inexhaustible homogeneous systems and inexhaustible ages

The proof of the next theorem is from [21] BÖRÖCKY, SAUER, ZHU 1993; the theorem follows directly from general results in [201] POUZET 1981.

A homogeneous relational system H with finitely many components which is age-inexhaustibe is inexhaustible.

• Assume that the base of H is ω and that I_n is the restriction of H to n. Let $A \subseteq |H|$ be finite. To prove that there is an embedding k of H into the restriction of H to $|H| - A$.

Let F_n be the set of local isomorphisms from I_n to $|H| - A$. Because the age of H is inexhaustible the set F_n is not empty. For f and g in F_n let $f \overset{n}{\sim} g$ iff $(f \circ g^{-1}) \cup \mathrm{id}_A$ is an isomorphism from the image of g union A to the image of f union A. The relation $\overset{n}{\sim}$ is an equivalence relation on F_n. The equivalence relation $\overset{n}{\sim}$ has finitely many classes because H has finitely many components (if H has finitely many components then there are only finitely many pairwise non isomorphic extensions of disjoint copies of A and H_n). Let P_n be the set of equivalence classes of $\overset{n}{\sim}$ and P the union of the P_n. Let F be the union of the F_n and $\sim = \bigcup_{n \in \omega} \overset{n}{\sim}$.

For f and g in F put $g \leq f$ if $(f \circ g^{-1}) \cup \mathrm{id}_A$ is an embedding from the image of g union A into the image of f union A. If $f' \sim f$ and $g' \sim g$ and $g \leq f$ then $g' \leq f'$. Hence $\leq$ induces a partial order on P. The partial order $(P; \leq)$ is denumerable, every element has a finite cover and it is well founded. Hence $(P; \leq)$ contains according to Königs Lemma an infinite chain $C = \{C_0 < C_1 < C_2 < \cdots\}$ which contains an element from every level of P.

Let $g_0 \in C_0$. Assume $g_0 \subseteq g_1 \subseteq g_2 \subseteq \cdots \subseteq g_{n-1}$ with $g_i \in C_i$ for all $i \in n$. If $f \in C_n$ then $g_{n-1} < f$ and hence $h = (f \circ g_{n-1}^{-1}) \cup \mathrm{id}_A$ is a local isomorphism which can be extended to an automorphism l of H. Then $l^{-1}f \sim f$ because $(l^{-1}f)f^{-1} = l^{-1}/\mathrm{im}(f)$ is an isomorphism and $l^{-1}/A = \mathrm{id}_A$. Because $g_{n-1} \subseteq l^{-1}f$ the embedding $g_n = l^{-1}f$ extends the sequence $g_0 \subseteq g_1 \subseteq g_2 \subseteq \cdots \subseteq g_{n-1}$ as required.

Hence there is an infinite sequence $g_0 \subseteq g_1 \subseteq g_2 \subseteq \cdots$ with $g_i \in F_i$ for all $i \in \omega$. Then $k = \bigcup_{n\in\omega} g_n$ is an embedding of H into the restriction of H to $|H| - A$. •

It follows that:

A homogeneous relational system with finite language is inexhaustible iff its age is strongly 0-amalgamable.

Also:

If the homogeneous relational system H has finite language L and none of the components of L is unary, then H is inexhaustible.

A.3.4 An example with infinitely many components

Example. The base of H is ω. The elements $x, y \in \omega$ are in the relation τ_n if and only if $|x - y| = n$. This is a homogeneous structure, the translations, the reflection $x \mapsto -x$ and their compositions are the only isomorphisms even between finite isomorphic substructures. The automorphism group of H acts transitively and there is no proper substructure of H isomorphic to H. Hence H is not inexhaustible. The restriction of H to the non-negative integers is an inexhaustible relational system.

Hence H is age-inexhaustible but not inexhaustible. The only orbit of the empty set is $|H|$, the automorphism group is transitive. The relational system H and its restrictions to subsets of $|H|$ are the only representatives of the age of H. Hence 10.6.2 (4) fails in this case. Nevertheless the age of H does have an inexhaustible representative.

A.3.5 Strongly inexhaustible homogeneous relational systems

The relational system R is **strongly inexhaustible** if for every finite $F \subseteq |R|$ the restriction of R to $|R| - F$ is isomorphic to R. The homogeneous relational system H has the **strong embedding property** if for every A in the age of H and $a \in A$ and embedding f from $A - a$ into H there are infinitely many embeddings from A into H which are an extension of f.

Let H be a homogeneous relational system. The age of H is strongly amalgamable if and only if H has the strong embedding property if and only if H is strongly inexhaustible: [49] EL-ZAHAR, SAUER 1991.

• Assume that the age of H is strongly amalgamable and there is an element A in the age of H with $a \in A$ and an embedding f from $A - a$ into H which has only finitely many extensions $f_0, f_1, \ldots, f_{n-1}$ to A which are embeddings. The number n is larger than 0 because H is homogeneous.

Let g be the function with domain $|A|$ so that g restricted to $|A| - a$ is f restricted to $|A| - a$ and $g(a) = a$. Let B be the relational structure with $|B| = g(|A| - a) \cup \{a\}$ so that g is an isomorphism from A to B. Let C be the restriction of H to $f(|A| - a) \cup \{f_0(a), f_1(a), \ldots, f_{n-1}(a)\}$. The relational systems B and C are compatible and in the age of H. Hence there exists a common extension of B and C to an element D in the age of H. The identity map on $|C|$ has an extension h to an embedding of D which maps a into H with $h(a) \notin \{f_0(a), f_1(a), \ldots, f_{n-1}(a)\}$. Put $f_n = h \circ g$. Hence H has the strong embedding property.

Assume that H has the strong embedding property and that F is a finite subset of $|H|$. The restriction of H to $|H| - F$ will be isomorphic to H if it has the embedding property for the age of H. Let A be an element of the age of H and $a \in |A|$ and f an embedding from $A - a$ into the restriction of H to $|H| - F$. Because H has the strong embedding property, there is an extension f^* of f so that $f^*(a) \in> |H| - F$. Hence H is strongly inexhaustible.

Assume that H is strongly inexhaustible. Let A and B be two compatible elements in the age of H. We may assume without loss of generality that A is a restriction of H. Let C be the common restriction of A and B to $|A| \cap |B|$. The identity map is an embedding from C into the restriction of H to $|H| - (|A| - |C|)$ which is isomorphic to H. Hence there is an extension f of the identiy map to an embedding from B into the restriction of H to $|H| - (|A| - |C|)$. Hence the age of H is strongly amalgamable. •

A.4 Partitions of homogeneous relational systems

The relational system R is **indivisible** if whenever $|R| = X \cup Y$ then R has an embedding into its restriction to X or an embedding into its restriction to Y (6.8).

The relational system R is **weakly indivisible** if whenever $|R| = X \cup Y$ and the age of R restricted to X is a proper subset of the age of R then R has an embedding into its restriction to Y.

The relational system is **age-indivisible** if whenever $|R| = X \cup Y$ then the age of the restriction of R to X or the age of the restriction of R to Y is equal to the age of R.

A set $\mathcal{R}$ of relational systems of the same arity is a **Ramsey class** if for every element A of $\mathcal{R}$ there is an element B of $\mathcal{R}$ so that if $|B| = X \cup Y$ then A embeds into the restriction of B to X or into the restriction of B to Y. The next theorem implies that R is age-indivisible if and only if the age of R is a Ramsey class.

A.4.1 Age-indivisible homogeneous relational systems

Let R be a relational system; then:

The system R has **property** $P(1, n)$ if for every partition of $|R|$ into n classes $C_0, \ldots, C_{n-1}$ the age of the restriction of R to one of the classes is equal to the age of R.

The system R has property $P(2, n)$ if for every partition of $|R|$ into n classes the union of the ages of the restrictions of R to those classes is equal to the age of R.

The system R has property $P(3, n)$ if for every A in the age of R and for every partition of the base $|B|$ of every B in the age of R into n classes $C_0^B, C_1^B, C_2^B, \ldots, C_{n-1}^B$

there is an element B in the age of R and an $i \in n$ so that there is an embedding from A into the restriction of B to C_i^B.

The system R has property $P(4,n)$ if for every A in the age of R there is a B in the age of R so that for every partition of $|B|$ into n classes there is an embedding of A into the restriction of B to one of the classes of the partition.

Let R be a relational system and $n \geq 2$; then $P(1,n)$ implies $P(2,n)$ implies $P(3,n)$ implies $P(4,n)$ implies $P(4,2)$ implies $P(4,n)$ implies $P(1,n)$: [49] EL-ZAHAR, SAUER 1991.

• Obviously $P(1,n)$ implies $P(2,n)$. Assume $P(2,n)$ and let for every B in the age of R a partition $(C_i^B; i \in n)$ of $|B|$ be given. Assume that the base of R is ω and let I_m be the restriction of R to $m \in \omega$. Let $\mathcal{U}$ be an ultrafilter on ω which contains all of the cofinal subsets of ω. Let $\prod_{\mathcal{U}} I_m$ be the ultraproduct of the I_m. The system R is isomorphic to the restriction of the untraproduct to the set of constant sequences. We may assume therefore that R is a restriction of the ultraproduct . Partition $|R|$ into classes $(C_i; i \in n)$ with $x \in C_i$ if and only if the set of indices m so that $x \in C_i^{I_m}$ is an element of $\mathcal{U}$. There is an $i \in n$ so that A has an embedding into the restriction of R to C_i. Hence there is an $m \in \omega$ so that A has an embedding into the restriction of I_m to $C_i^{I_m}$.

Let A be an element in the age of R and assume that not $P(4,n)$. Then for every B in age R there is a partition $(C_i^B; i \in n)$ of $|B|$ so that A can not be embedded into the restriction of B to any of the classes of the partition. Then not $P(3,n)$. Hence $P(3,n)$ implies $P(4,n)$ and clearly $P(4,n)$ implies $P(4,2)$.

In order to prove that $P(4,2)$ implies $P(4,n)$ it suffices to prove that $P(4,n)$ implies $P(4,2n)$. Assume $P(4,n)$ and let A be an element of the age of R. Let B be an element of the age of R so that for every partition of $|B|$ into n classes there is an embedding of A into the restriction of B to one of the classes. Let C be an element of the age of R so that for every partition of $|C|$ into two classes B has an embedding into the restriction of C into one of the classes. Then for every partitiono of $|C|$ into $2n$ classes there is an embedding of A into the restriction of C to one of the two classes.

Finally $P(4,n)$ implies $P(1,n)$. Let $(C_i; i \in n)$ be a partition of $|R|$ into n classes. Assume that for every $i \in n$ there is A_i in the age of R but not in the age of the restriction of R to C_i. Because each A_i has an embedding into R there is an element A in the age of R so that there is an embedding from A_i into A for every $i \in n$. Let B in the age of R be such that for every partition of $|B|$ into n classes A has an embedding into the restriction of R to one of the classes. There is an embedding f from B into R. The function f induces a partition of $|B|$ into n classes $(C_i^B; i \in n)$ via $x \in C_i^B$ if $f(x) \in C_i$. The system A has an embedding into the restriction of B to one of the classes C_i^B and hence an embedding into C_i. •

A.4.2 Divisibility and strong amalgamation

If the relational system R is indivisible then it is weakly indivisible and if it is weakly indivisible then it is age-indivisible. In the case that H is a homogeneous relational system there does not seem to be any relationship between those three properties of H and the property of strong amalgamation of its age. The age of the homogeneous graph $K_\omega + K_\omega$ which consists of two disjoint copies of K_ω has strong amalgamation but is not age-indivisible.

There is an indivisible homogeneous system H which does not have strong amalgamation. In [147] LACHLAN 1984 the following homogeneous relational system is describe. Let $|H| := [S]^2$ for some set S. The relational system H has a binary relation $B(\ ,\ )$ and a ternary relation $T(\ ,\ ,\ )$ with $B(a, b) = +$ if a and b have an element of S in common and $T(a, b, c) = +$ if a, b and c form a triangle. Then H is homogeneous. The amalgamation of the age of H is not strong because if two triangles have two edges in common they must have the third edge in common. If $A \cup B = [S]^2 = |H|$ then there is an infinite subset $T \subseteq S$ so that $[T]^2 \subseteq A$ or $[T]^2 \subseteq B$ by Ramsey's theorem. Hence H is indivisible.

A.4.3 Weak indivisibility and free 1-amalgamation

On the other hand it follows from the next theorem that if the age of the homogeneous relational system H is freely 1-amalgamable then H is weakly indivisible: [49] EL-ZAHAR, SAUER 1991.

Let H be a homogeneous relational system no component of which is unary and whose age is freely 1-amalgamable. Then H is weakly indivisible.

• Because the age of H is freely 1-amalgamable it is also freely 0-amalgamable.

Given an element A of the age of H so that $|A| \cap |H| = \emptyset$ denote by $A + H$ the free amalgam of A and H. That is the common expansion of A and H so that $R(x_1, x_2, \ldots, x_n) = -$ for every component R of H if $|A| \cap \{x_1, x_2, \ldots, x_n\} \neq \emptyset$ and $|H| \cap \{x_1, x_2, \ldots, x_n\} \neq \emptyset$. The age of H is freely amalgamable and hence it is equal to the age of $A + H$. Hence $A + H$ has an embedding into H. It follows that for every finite subset F of H there is a copy H^F of H in H so that for every finite $F' \subseteq H/H^F$ the relational system $H/(F \cup F')$ is a free amalgam of H/F and H/F'.

The homogeneous system H is **weakly j-indivisible** if for every partition of H into two sets X and Y for which there is no embedding from H into X the age of H/Y contains all elements of the age of H whose base consists of at most j elements. We proceed by induction and proof that if H is weakly j-indivisible then it is weakly $j+1$-indivisible. Let A be an element of the age of H whose base has $j + 1$ elements and $a \in |A|$.

Assume that $X \cup Y = |H|$ and $X \cap Y = \emptyset$ and that the base of H is ω. Denote by I_n the restriction of H to n. If there is no embedding from H into H/X then there is an $n \in \omega$ and an isomorphism g from I_n to H/X which does not have an extension h which is an embedding of I_{n+1} into H. Let F be the base of the image of g. The partition (X, Y) induces a partition $(X \cap |H^F|, Y \cap |H^F|)$ of the base of H^F. The homogeneous relational system H can not be embedded into the restriction of H to $X \cap |H^F|$ otherwise H would have an embedding into H/X. Hence, using induction, there is a subset L in $Y \cap H^F$ so that there is an isomorphism f of $A - a$ to H/L.

Let b be an element which is not in the base of H. Let A' be the relational system with base $L \cup \{b\}$ so that the extension f' of f with $f'(a) = b$ is an isomorphism of A. Let I'_n be the relational system with base $F \cup \{n\}$ so that the extension g' of g with $g'(n) = b$ is an isomorphism of I_{n+1}. Then, because no component of H is unary, the relational systems I'_n and A are compatible and hence can be freely amalgamated to the amalgam B (the age of H is freely 1-amalgamable). The identity map on $F \cup L$ must have an extension to an embedding h of B. If $h(b) \in X$ then $h \circ g'$ would be an extension of g to an embedding of I_{n+1} into H. Hence $h(b) \in Y$ This implies that $h \circ f'$ is an extension of f to an embedding of A into Y. •

The argument in this proof can easily be adapted to homogeneous relational systems whose age is not freely 1-amalgamable but which are essentially equivalent to a homogeneous relational systems whose age is freely 1-amalgamable. For example. Let H be the homogeneous relational system which has a binary component which is a graph and a ternary component which is a 3-uniform hypergraph. There is exactly one boundary element B, the 3-uniform hypergraph which consists of two hyperedges having exactly two elements in common while there are none of the binary graph edges in B. Negating the binary edges of the graph we obtain a freely amalgamable age. The corresponding homogeneous representative is weakly indivisible according to the previous theorem. Of course then the original homogeneous system, before negation of the edges, is also weakly indivisible.

A.4.4 Free amalgamation and age-indivisibility

A relational system R without unary components, not necessarily homogeneous, whose age is freely amalgamable is age-indivisible.

• Let H be the homogeneous representative of the age of R. Then H is weakly indivisible hence age-indivisible and hence the age of H, that is the age of R, is a Ramsey class (A.4.1). Using A.4.1 again, it follows that R is age-indivisible. •

A.4.5 Necessary condition for indivisibility

Let R and S be two relational systems of the same arity. The relational system R **precedes** the relational system S, $R \preceq S$, if there exists a partition of $|R|$ into finitely many sets $C_0, C_1, C_2, \dots, C_{n-1}$ so that there is an embedding of R/C_i into S for every $i \in n$. The set $\mathcal{R}$ of relational systems of the same arity satisfies the **chain condition** if for any two elements R and S of $\mathcal{R}$ at least one of $R \preceq S$ of $S \preceq R$ holds.

If the homogeneous relational system H is indivisible then the restrictions of H to its orbits satisfy the chain condition: [51] EL-ZAHAR,

SAUER 1994.

• Assume that the base of H is ω. Let T be an orbit of the finite subset $F \subseteq |H|$. For $x \in \omega = |H|$ and $F' \subseteq \omega$ we write $x \geq_T F'$ to mean that x is a larger number than any of the numbers in F' and that there is a local isomorphism f from F' to F so that one and hence every extension of f to an automorphism of H maps x into T. For each number x let $\mathrm{dom}_T(x)$ be the set F' of finite subsets of $|H|$ so that $x \geq_T F'$.

Assume that H is indivisible and let T be an orbit of the finite subset F of $|H|$ and S an orbit of the finite subset D of $|H|$. We wish to prove that either $T \preceq S$ or $S \preceq T$. If S is finite then $S \preceq T$ and we are done. Assume that S and T are infinite. The finite subsets of $|H| = \omega$ are totally ordered by the lexicographic ordering (the subset A is larger than the subset B in this lexicographic ordering if the smallest element which is not in the intersection of A and B belongs to A).

Let $X_{T,S}$ be the set of all elements x of $|H|$ so that the lexicographically largest set in $\mathrm{dom}_T(x)$ is lexicographically larger than the lexicographically largest set in $\mathrm{dom}_S(x)$ and neither $\mathrm{dom}_T(x)$ nor $\mathrm{dom}_S(x)$ is empty. Let $X_{S,T}$ be the set of all elements x of $|H|$ so that the lexicographically largest set in $\mathrm{dom}_S(x)$ is lexicographically larger than the lexicographically largest set in $\mathrm{dom}_T(x)$ and neither $\mathrm{dom}_T(x)$ nor $\mathrm{dom}_S(x)$ is empty. Let X_T be the set of all elements x of $|H|$ so that $\mathrm{dom}_S(x)$ is empty. Let X_S be the set of all elements x of $|H|$ so that $\mathrm{dom}_T(x)$ is empty. Clearly $X_T \cup X_S \cup X_{T,S} \cup X_{S,T} = |H|$.

Because $|H|$ contains elements x for which $\mathrm{dom}_T(x) \neq \emptyset$ and elements x for which $\mathrm{dom}_S(x) \neq \emptyset$ every copy of H in H contains elements x for which $\mathrm{dom}_T(x) \neq \emptyset$ and elements x for which $\mathrm{dom}_S(x) \neq \emptyset$. Hence there is no embedding from H into H/X_T or into H/X_S. Therefore we may assume without loss of generality that there is a copy H^* of H with $|H^*| \subseteq X_{T,S}$. Let $F^* \subseteq |H^*|$ so that there is an isomorphism f from $H|F$ to $H|F^*$. Extend f to an embedding g from $H|(T \cup F)$ into H^*. Denote the image of T under g by T^*.

There are only finitely many elements of T^* which are not larger than all of the elements of F^*. If x is in T^* and larger than all of the elements of F^* then $x \geq_T F^*$. Because such an x is also an element of $X_{T,S}$ it follows that $x \geq_S D'$ for some copy D' of D. and all of those copies of D are lexicographically smaller than F^*. Hence there are only finitely many of those copies of D. Label x with the smallest such copy of D. This labeling partitions the elements of T^* which are larger than all of the elements of F^* into finitely many classes so that the restriction of H to any one of those classes has an embedding into S. Hence $T^* \preceq S$ and therefore $T \preceq S$. •

A.4.6 Orbits of homogeneous systems whose age is freely amalgamable

Let H be a homogeneous relational system, let F be a finite subset of $|H|$ and let T be an orbit of F. There exists a homogeneous relational system H^T with base T so that every component of H/T is a component of H^T and the automorphism group of H^T is G_F. If the age of H is

freely amalgamable then the age of H^T is freely amalgamable. If every component of H has arity at most two then H/T is equal to H^T.

• Let f be a function from a subset M of F into n. Assume that M has m elements. The n-tupel $(x_1, x_2, \dots, x_n)$ agrees with f if $x_{f(y)} = y$ for all $y \in M$. If $(x_1, x_2, \dots, x_n)$ agrees with f then the $n - m$-tupel $(x_1, x_2, \dots, x_n)_f$ is obtained from $(x_1, x_2, \dots, x_n)$ by removing all of the elements of the form $x_{f(a)}$ for $a \in M$ and retaining the order of the remaining elements.

For all $m < n \in \omega$, for every n-ary component R of H and for every function f from a subset of F of size m into n let R_f be the $n - m$-ary relation with base T so that $R_f(a_0, a_1, \dots, a_{n-m-1})$ is positive if and only if there is an n-tuple $(x_0, x_1, \dots, x_{n-1})$ which agrees with f and $(x_0, x_1, \dots, x_{n-1})_f = (a_0, a_1, \dots, a_{n-m-1})$ and $R(x_0, x_1, \dots, x_{n-1}) = +$. Note that if the arity of R_f is one and R_f is positive on some element of T then R_f is positive on all elements of T because G_F, the stabiliser group of F, is transitive on T.

Let H^T be the relational structure with base T whose components are of this form R_f and not unary. Every component of H/T is a component of H^T. (The relation R does not change if f is the empty function.) If every component of H has arity at most two then every component in H^T which is not a component of H/T has arity at most one. Hence in this case H/T is equal to H^T except for unary relations which are positive on every element of T. We disregard such unary relations.

It follows from the construction of H^T that the stabilizer group G_F is the automorphism group of H^T and hence H^T is a homogeneous relational system.

Assume that the age of H is freely amalgamable. Let A and B be two compatible elements in the age of H^T and let C be the common restriction of A and B to the intersection of their bases. Let f be an embedding from A into H^T and A^f the copy of A in H^T which is the image of f. Let g be an embedding of B into H^T and B^g the copy of B in H^T which is the image of g (note that f and g are embeddings which preserve the relations of H together with the additional relations in H^T). Let C^f be the copy of C in H^T which is the image of C under f and C^g the copy of C in H^T which is the image of C under g. The function $f \circ g^{-1}$ is a local H^T isomorphism of C^g to C^f. That is it has an extension to a local isomorphism of H which fixes all of the elements of F. Hence it can be extended to an automorphism h of H^T, that is an automorphism of H which fixes the elements of F (the function h is an element of G_F). Let B' be the image of B^g under h. Let A_F be the restriction of H to $|A^f| \cup F$. Let B_F be the restriction of H to $|B'| \cup F$. Then A_F and B_F are compatible. Hence they have a free amalgam D. The restriction of D to $|A^f| \cup |B'|$ is a free amalgam of A^f and B'. •

Example: Let B be the complete three-uniform hypergraph with vertices $V(B) = \{a, b, 0, 1, 2\}$ and edges $E(B) = \{\{a, b, 0\}, \{a, b, 1\}, \{a, b, 2\}, \{a, 0, 1\}, \{0, 1, 2\}\}$. Let H_B be the homogeneous B-free three-uniform hypergraph. The age of H_B is freely amalgamable by A.2.5. Let a', b' be two vertices of H_B and T the orbit of vertices x in $V(H_B)$ so that $\{x, a', b'\}$ is a hyperedge of H_B. The set T is then an orbit of the set $F = \{a', b'\}$.

The homogeneous relational system H_B^T has three components G/T, G_a and

G_b. The component G_a is a graph so that $x, y \in T$ are adjacent in G_a if $\{x, y, a\}$ is a hyperedge of H_B. The component G_b is a graph so that $x, y \in T$ are adjacent in G_b if $\{x, y, b\}$ is a hyperedge of H_B. It follows from the previous argument that H_B^T is a homogeneous relational system whose age is freely amalgamable.

Let R be the relational system with components $(H_B/T)'$, G'_a and G'_b. The component $(H_B/T)'$ is a three uniform hypergraph and the components G'_a and G'_b are graphs. The base of R consists of three elements, say $\{r, s, t\}$. The set $\{r, s, t\}$ is the ony hyperedge of the three-uniform component and $E(G'_a) = \{r, s\}$ and $E(G'_b) = \emptyset$. Then $\{R, B\}$ is the boundary of H_B^T.

Extend the example as follows. Let A be the 3-uniform hypergraph with $V(A) = V(B) = \{a, b, 0, 1, 2\}$ and $E(A) = E(B) \cup \{\{a, 1, 2\}\}$. Let C be the 3-uniform hypergraph with $V(C) = V(B)$ and $E(C) = E(B) \cup \{\{a, 1, 2\}, \{a, 2, 3\}\}$. Let $\mathcal{D} = \{A, B, C\}$ and $H_\mathcal{D}$ the $\mathcal{D}$-free homogeneous directed graph. Let a', b' be two vertices of $H_\mathcal{D}$ and T the orbit of vertices x in $V(H_\mathcal{D})$ so that $\{x, a', b'\}$ is a hyperedge of $H_\mathcal{D}$. The set T is an orbit of the set $F = \{a', b'\}$. While as before the system $H_\mathcal{D}^T$ is homogeneous with freely amalgamable age, the restriction of $H_\mathcal{D}$ to T does not have the mapping extension property and hence is not homogeneous.

A.4.7 Determining the orbits from the boundary of the age

Let H be a homogeneous relational system whose age is freely amalgamable and the arity of each of the components of H is two. Let F be a finite subset of H and T an orbit of F. Then according to A.4.6 the restriction of H to T is a homogeneous relational system whose age is freely amalgamable. According to A.2.5 the boundary of the age of H/T consists then of complete relational systems.

Let A be an element of the boundary of the age of H/T. Because every component of H and hence of H/T has arity two the set $|A|$ must have at least two elements. Assume without loss that $|A| \subseteq T$ (of course H restricted to $|A|$ must then be different from A; just the elements of A are elements of T). Let B be the relational system of the same arity as H with base $A \cup F(T)$ so that A is a restriction of B and which has the property that for every $a \in |A|$ the restriction of B to $F(T) \cup \{a\}$ is equal to the restriction of H to $F(T) \cup \{a\}$. No component of B is positive on any argument unless required to be positive by those conditions. Because A is not in the age of H/T the relational system B is not in the age of H. Let $C \subseteq F(T)$ have a minimal number of elements so that the restriction of B to $|A| \cup C$ is not in the age of H. It follows that $|A| \cup C$ is in the boundary of the age of H and hence is complete. Let D be the restriction of B to $|A| \cup C$.

The base of D is partitioned into $|A|$ and C. For any two elements a and b in $|A|$ there is a local isomorphism of D which fixes every element of C and maps a to b. That is the type of a over C eqals the type of b over C. If D is a relational system and the base of D is partitioned into the sets A and C so that for every pair of vertices a and b of A there is a local isomorphism of D which fixes every element of C and maps a to b we say that the partition (A, C) is *type preserving.*

Let (A, C) be a type preserving partition of the boundary element D of the age of H so that $|A| \geq 2$. Then D/C is in the age of H and hence there is a finite subset $F \subseteq |H|$ so that there is an isomorphism f from D/C to H/F. Let $a \in A$ and T be the orbit of all elements x of H so that f has an extension to an isomorphism from $D/(\{a\} \cup C)$ which maps a to x. It follows that the restriction of D to A is in the boundary of H/T.

Assume that $D_0, D_1, D_2, \ldots$ is a finite or infinite sequence of boundary elements of H each D_i having the type preserving partition (A_i, C_i) with $|A_i| \geq 2$. Let a_i be an element of A_i for every i and C_i' the restricion of D_i to $\{a_i\} \cup |C_i|$. Every D_i/C_i' is in the age of H. Assume without loss that $a_i = a_j$ for all i and j. Then the C_i' are pairwise compatible and hence there is a free amalgam of the D_i/C_i' which is in the age of H. We procede as before and obtain an orbit T of H whose set of boundaries is the set $\{D_i/A_i \mid i\}$ (well, not quite: there might be D_i/A_i which have an embedding into another D_j/A_j; in such a case just remove the relational system D_j/A_j).

It is now easy to see how the type preserving partitions of the boundary of the age of H determine the sets of boundaries of the orbits of H and hence how the orbits of H are related under the relation younger. If T and S are two orbits of H then H/S will be younger than H/T if every element in the boundary of H/T is an element in the boundary of H/S. For every element A in the boundary of T there is an element D in the boundary of H and a type partition $(|A|, C)$ of D so that A is $D/|A|$. If H/S is younger than H/T there must be an element B in the boundaty of H/S so that B has an embedding into A. Hence there must be an element D' in the boundary of H which has a type preserving partition of the form $(|B|, C')$.

Let H be a homogeneous relational system all of whose components have arity two and whose age is freely amalgamable. Then the orbits of H satisfy the chain condition if and only if for any two sequences $D_0, D_1, D_2, \ldots$ and $D_0', D_1', D_2', \ldots$ of elements in the boundary of H which have type preserving partitions $(A_0, C_0), (A_1, C_1), (A_2, C_2), \ldots$ and $(A_0', C_0'), (A_1', C_1'), (A_2', C_2'), \ldots$ respectively there either is for every one of the i's a j so that D_i/A_i has an embedding into D_j'/A_j' or for every one of the j's there is an i sot that D_j'/A_j' has an embedding into D_i/A_i.

In the case that the age of H is a universal class, that is the boundary has finitely many elements, this theorem provides an effective checking if the orbits of H satisfy the chain condition. Every element of the boundary is finite and hence has only finitely many type preserving partitions.

Another consequence of the previous discussionis that it is not difficult to construct homogeneous structures whose restrictions to their orbits form any given denumerable partial order under embedding.

In particular it is possible to construct for every denumerable or finite partial order P a set $\mathcal{T}$ of tournaments so that the ages of the orbits of the $\mathcal{T}$-free homogeneous directed graph $H_{\mathcal{T}}$ form a partial order under $\subseteq$ which is isomorphic to P.

A.4.8 Indivisibility and the ages of the orbits

If H is a homogeneous relational system whose age is freely amalgamable and if T is an orbit of H then the H restricted to T is age-indivisible.

• The homogeneous relational system H^T is weakly indivisible hence age-indivisible. Every component of H/T is a component of H^T and H^T and H/T have the same basis T. Hence H/T is age-indivisible. •.

Let H be an indivisible homogeneous relational system whose age is freely amalgamable. Then the restrictions of H to the orbits of H form a total quasiorder under the relation of younger.

• Let S and T be two orbits of H. Because H is freely amalgamable both S and T are age indivisible. Because H is indivisible we may assume without loss that there is a partition of S into finitely many classes $(C_0, C_1, C_2, \dots, C_{n-1})$ so that H/C_i has an embedding into H/T for every $i \in n$. Because H/S is age-indivisible there is an $i \in n$ so that the age of S is equal to the age of S/C_i. Hence the age of H/S is a subset of the age of H/T. •

Problem: Is there an indivisible homogeneous relational system H which is indivisible but one of its orbits is not indivisible? How about this if the age of H is freely amalgamable? If the age of H is freely amalgamable and every component of H has arity at most two?

Problem: Is there a homogeneous relational system whose restriction to its orbits satisfies the chain condition and which is not indivisible? If the age is freely amalgamable? If the age of H is freely amalgamable and every component has arity at most two?

A.4.9 Examples, restrictions to all orbits are isomorphic to each other

The age of the homogeneous K_n-free graph H_n is freely amalgamable. Because the arity of H_n is two the restriction of H_n to any orbit is again a homogeneous graph which is freely amalgamable, hence again one of the homogeneous graphs of the form H_m with $m \leq n$. For every $m \leq n$ there is an orbit of H_n which is isomorphic to H_m. It follows that the homogeneous graphs H_n form a chain of type ω under embedding and the orbits of each H_n form a subchain of this chain. Certainly then the orbits of H_n satisfy the chain condition.

Within the class of graphs with two types of edges, E_1 and E_2, let K_3^1 be the graph with base $3 = \{0, 1, 2\}$, $E_1(K_3^1) = [3]^2$ and $E_2(K_3^1) = \emptyset$. Let K_3^2 be the graph with base 3, $E_2(K_3^2) = [3]^2$ and $E_1(K_3^2) = \emptyset$. Let $\mathcal{R} := \{K_3^1, K_3^2\}$ and H the $\mathcal{R}$-free homogeneous relational system. The elements of R are complete hence the age of H is freely amalgamable.

Let x be an element of $|H|$. The set $\{x\}$ has three orbits. The set T_0 of vertices which are not adjacent to x, the set T_1 of vertices which are via an edge of E_1 adjacent to x and the set T_2 of vertices which are via an edge of E_2 adjacent to x. Note that $E_1(H) \cap [T_1]^2 = \emptyset$ and that $E_2(H) \cap [T_2]^2 = \emptyset$. The age of H/T_1 is the set of all finite graphs which have edges only of type E_2. Hence because H/T_1 is homogeneous it is the Rado graph with edges entirely from E_2. Similarly H/T_2 is the Rado graph with edges entirely from E_1. Hence the ages of H/T_1 and H/T_2 are not comparable by set inclusion. It follows that H is not indivisible.

If the restriction of the homogeneous relational system H to all of its orbits is isomorphic to H, that is H is universal, then the orbits of H satisfy the chain condition. Examples of such homogeneous relational systems are the Rado graph, the universal tournament, the universal directed graph, the universal three uniform hypergraph, the universal n-uniform hypergraph for any $n \in \omega$, the rational numbers as an order structure, the universal partial order etc.

If the restriction of the homogeneous relational system H to any of its orbits is isomorphic to H then H is indivisible.

• Let ω be the base of H and I_n the restriction of H to n. Let (A, B) be a partition of ω. Assume that H can not be embedded into the restriction of H to A. Then for some $n \in \omega$ there is an embedding f of I_n into the restriction of H to A which does not have an extension to an embedding of I_{n+1} into the restriction of H to A. Let X be the set of all elements $g(n)$ so that g is an extension of f to I_{n+1}. Then X is an orbit of H and $X \subseteq B$. Hence there is an embedding of H into the restriction of H to B. •

A.4.10 Indivisibility of universal relational systems

The proof of the next theorem is similar the the argument in 10.3.3.

If H is a universal relational system then H is indivisible.

If H is the Rado graph or the universal three uniform hypergraph or the universal tournament or the universal directed graph etc. then the age of the square of H is equal to the age of H. Hence all such homogeneous relational systems are indivisible.

The k-uniform hypergraph G is **l-covering** if every element of $[|G|]^l$ is a subset of some hyperedge of G. Every l-covering hypergraph with $l \geq 2$ is complete. Hence if $\mathcal{T}$ is a set of finite l-covering k-uniform hypergraphs with $l \geq 3$ then Forb($\mathcal{T}$) is freely amalgamable and we obtain the $\mathcal{T}$-free homogeneous hypergraph $H_\mathcal{T}$. Because the age of the square of $H_\mathcal{T}$ is equal to the age of $H_\mathcal{T}$ the homogeneous hypergraph $H_\mathcal{T}$ is indivisible; see [51] EL-ZAHAR, SAUER 1994.

A.4.11 Indivisibility of the homogeneous K_n-free graph

It is proved in [138] KOMJÁTH, RÖDL 1986, that the triangle-free homogeneous graph is indivisible and in [50] EL-ZAHAR, SAUER 1993, that if $\mathcal{T}$ is a finite set of tournaments then the $\mathcal{T}$-free homogeneous graph $H_\mathcal{T}$ is indivisible if and only if it satisfies the chain condition. The orbits of the homogeneous K_n-free graph H_n satisfy the chain condition and indeed:

The K_n-free graph H_n is indivisible for every $n \in \omega$ ([48] EL-ZAHAR, SAUER 1989)

A.5 Coloring copies of relational systems

The previous section dealt with partitions of the set of elements of homogeneous relational systems. Now we are going to discuss partitions of the set of edges of homogeneous graphs and digraphs. In principle we are interested in partitions of sets of other restrictions of homogeneous relational systems, like complete subgraphs etc, but outside of exactly formulating the problem very little is known. We do not have to consider the case where finitely many finite restrictions are removed from a homogeneous relational system, as then only finitely many elements are removed. A case we have already studied.

A.5.1 Sets of copies and the Ramsey arrow

A **copy** of the relational system S in the relational system R is a restriction of R which is isomorphic to S. The set of copies of S in R is denoted by

$$\binom{R}{S}$$

The relational system R is **S-indivisible** if for every partition (A, B) of $\binom{R}{S}$ there is a copy R^* of R in R such that

$$\binom{R^*}{S} \subset A \quad \text{or} \quad \binom{R^*}{S} \subset B.$$

The fact that R is S-indivisible is also expressed by writing $R \longrightarrow (R)^S_2$.

In particular, if S reduces to an edge, we get the notion of an **edge-indivisible** relational system. The complete graph K_ω is edge-indivisible according to Ramsey's theorem.

A.5.2 An obstruction to edge-indivisibility of graphs

Let G be a graph with ω as set of vertices. Associate with every vertex $n \in \omega$ the 0-1-chain $\sigma(n) = (\sigma(n)_0, \sigma(n)_1, \sigma(n)_2, \ldots, \sigma(n)_{n-1})$ where $\sigma(n)_i = 1$ if and only if n and i are adjacent. For $n \neq m$ let $\sigma(n) < \sigma(m)$ if $\sigma(n)_i = 0$ for the smallest i with $\sigma(n)_i \neq \sigma(m)_i$ or if $n < m$ and $\sigma(n)_i = \sigma(m)_i$ for all $i \in n$. (Lexicographic ordering.)

Let A be the set of all edges $\{n, m\}$ of G so that if $n < m$ then $\sigma(n) < \sigma(m)$. Let B be the set of all the other edges of G. The edges in A are called **up edges** and the edges in B are called **down edges**. This partition of the edges of G into up and down edges is called **lexicographic partition** of the edges of G.

Let $K_{\omega,\omega}$ be the complete bipartite graph, that is the graph on ω as set of vertices in which two vertices are adjacent if and only if they have different parity. It was noticed in [53] ERDÖS, HAJNAL, PÓSA 1975 that:

If $K_{\omega,\omega}$ has an embedding into the graph G then G is not edge-indivisible.

• Assume that ω is the set of vertices of G and (A, B) is the lexicographic partition of the edges of G with A the set of up edges and B the set of down edges. Assume that there is an embedding of $K_{\omega,\omega}$ into G.

It suffices to prove that there is no copy $K^*_{\omega,\omega}$ of $K_{\omega,\omega}$ in G such that all of the edges of $K^*_{\omega,\omega}$ are in A or all of the edges of $K^*_{\omega,\omega}$ are in B.

Assume for a contradiction that there is a copy $K^*_{\omega,\omega}$ of $K_{\omega,\omega}$ all of whose edges are in A. Let $L \cup R = V(K^*_{\omega,\omega})$ with $L \cap R = \emptyset$ and every edge of $K^*_{\omega,\omega}$ contains a vertex of L and a vertex of R. Let n be the smallest number in R. For $x \in R$ denote by $\gamma(x)$ the restriction of the sequence $\sigma(x)$ to n. ($\gamma(x)$ consists of the first n elements of $\sigma(x)$.) Let $x \in R$ have maximal $\gamma(x)$ for all elements in S. Let $r \in L$ with $r > x$. Then $\gamma(x) \leq \gamma(r)$. Let $y \in R$ be larger than r. Then $\gamma(r) \leq \gamma(y)$ and hence $\gamma(x) = \gamma(r) = \gamma(y)$. But r is adjacent to n and y is not. Hence $\sigma(r) > \sigma(y)$ in contradiction that all of the edges of $K^*_{\omega,\omega}$ are elements of A.

Assume next that there is a copy $K^*_{\omega,\omega}$ of $K_{\omega,\omega}$ all of whose edges are in B. Let $L \cup R = V(K^*_{\omega,\omega})$ with $L \cap R = \emptyset$ and every edge of $K^*_{\omega,\omega}$ contains a vertex of L and a vertex of R. Let n be the smallest number in R. For $x \in L$ denote by $\gamma(x)$ the restriction of the sequence $\sigma(x)$ to n. Let $x \in S$ have minimal $\gamma(x)$ for all elements in L. Let $r \in R$ with $r > x$. Then $\gamma(x) \geq \gamma(r)$. Let $y \in L$ be larger than r. Then $\gamma(r) \geq \gamma(y)$ and hence $\gamma(x) = \gamma(r) = \gamma(y)$. But r is not adjacent to n and y is. Hence $\sigma(r) < \sigma(y)$ in contradiction that all of the edges of $K^*_{\omega,\omega}$ are elements of B. •

Let $\mathbf{Q}$ be the rationals as order structure. There is a partition (A, B) of $[\mathbf{Q}]^2$ so that for every copy $\mathbf{Q}^*$ of $\mathbf{Q}$ in $\mathbf{Q}$ the set $[\mathbf{Q}^*]^2$ has a non empty intersection with $[A]^2$ and with $[B]^2$. That is

$$\neg(\mathbf{Q} \to (\mathbf{Q})_2^{\{a,b\}}).$$

• Enumerate $\mathbf{Q}$ into the ω-sequence $r_0, r_1, r_2, \ldots$. let $A := \{\{r_i, r_j\} \mid \text{if} \, r_i < r_j \, \text{then} \, i < j\}$ and $B = [\mathbf{Q}]^2 - A$. If there would be a copy $\mathbf{Q}^*$ of $\mathbf{Q}$ in $\mathbf{Q}$ so that $[\mathbf{Q}^*]^2 \subseteq A$ then $\mathbf{Q}$ would be order isomorphic to ω. Similarly there is no copy $\mathbf{Q}^*$ of $\mathbf{Q}$ in $\mathbf{Q}$ so that $[\mathbf{Q}^*]^2 \subseteq [\mathbf{Q}]^2 - A$. •

A.5.3 Weak indivisibility, canonical partitions

It follows that neither the Rado graph nor any of the graphs H_n are edge-indivisible and the rationals are not pair-indivisible.

Let

$$R \to (R)^S_{r\setminus l}$$

mean that for every partition of $\binom{R}{S}$ into r parts there is a copy R^* of R in R so that $\binom{R^*}{S}$ has a non empty intersection with at most l parts of the partition.

The relational system R is **weakly S-indivisible** if for every partition (X, Y) of $\binom{R}{S}$, either for every element A of the age of R there is a copy A^* in R with $\binom{A^*}{S} \subseteq X$ or there is a copy R^* of R in R with $\binom{R^*}{S} \subseteq Y$.

Let $A \subseteq \binom{R}{S}$. Let R' be the $|S|$-uniform hypergraph whose hyperedges are the base sets of the copies of S in A. Let R_A be the relational system whose components are the components of R together with R'. Let S' be the $|S|$-uniform hypergraph with the single hyperedge $|S|$. Let S_A be the relational system whose components are the components of S together with S'. The partition $(P_0, P_1, \dots, P_{n-1})$ of the set $\binom{R}{S}$ is **canonical** if:

(i) for every copy R^* of R in R the set $\binom{R^*}{S}$ has a non empty intersection with every part of the partition and

(ii) $R_{A_i} \to (R_{A_i})_2^{S_{A_i}}$ for every $i \in n$.

It follows that if $\binom{R}{S}$ has a canonical partition into l parts then $R \to (R)^S_{r\backslash l}$ for every $r \in \omega$ and the number l is minimal under the condition that $R \to (R)^S_{r\backslash l}$ for every $r \in \omega$. If $\binom{R}{S}$ has a canonical partition into two parts (X, Y) and for every element A in the age of R there are copies A^* and A^{**} of A in R so that $\binom{A^*}{S} \subseteq X$ and $\binom{A^{**}}{S} \subseteq Y$ then R is weakly S-indivisible. For a more detailed discussion see [221] SAUER.

A.5.4 Partitioning pairs of rationals

The following is an unpublished result of Fred GALVIN. It follows also from a much more general result in [41] DEVLIN 1979 which has a proof of almost hundred pages.

Let $r_0, r_1, r_2, \dots$ be an enumeration of the rationals $\mathbf{Q}$ and $P \subseteq [\mathbf{Q}]^2$ be those elements $\{r_i, r_j\}$ of $[\mathbf{Q}]^2$ so that if $r_i < r_j$ then $i < j$. Then $(P, [\mathbf{Q}]^2 - P)$ is a canonical partition of $[\mathbf{Q}]^2$.

We will prove that if $L \subseteq P$ and $\bar{L} := P - L$ then there is an embedding f from $\mathbf{Q}$ into $\mathbf{Q}$ so that if $f(r_i) = r_{i'}$ and $f(r_j) = r_{j'}$ and $i < j$ then $i' < j'$ and for which either $[f(\mathbf{Q})]^2 \subseteq L$ or $[f(\mathbf{Q})]^2 \subseteq \bar{L}$. This will suffice to establish the theorem. We have already shown that for every copy $\mathbf{Q}^*$ of $\mathbf{Q}$ in $\mathbf{Q}$ the set $[\mathbf{Q}^*]^2$ has a non emtpy intersection with P and with $[\mathbf{Q}]^2 - P$. The rest follows because $\mathbf{Q}$ is isomorphic under reversing the order.

• For $a \in \mathbf{Q}$ let $\Gamma_L(a) := \{b \in \mathbf{Q} \mid \{a, b\} \in L\}$. The set $D \subseteq \mathbf{Q}$ is **large** if there is an open interval of $\mathbf{Q}$ in which D is dense. Denote by large(A) the set of large subsets of $A \subseteq \mathbf{Q}$. If A and B are subsets of $\mathbf{Q}$ then $A < B$ means that every element of A is smaller than every element of B. Let ψ_L be the following formula:

$$\psi_L := \exists U \in \text{large}(\mathbf{Q})\, \forall V \in \text{large}(U) \exists A, B \in \text{large}(V)\, (A < B)$$

$$\forall A' \in \text{large}(A)\, \forall B' \in \text{large}(B)$$
$$\{a \in A' \colon \Gamma_L(a) \cap B' \in \text{large}(B')\} \in \text{large}(A').$$

The last two lines of the formula ψ_L are denoted by $\gamma_L(A, B)$, that is

$$\gamma_L(A, B) := \forall A' \in \text{large}(A)\, \forall B' \in \text{large}(B)$$
$$\{a \in A' \colon \Gamma_L(a) \cap B' \in \text{large}(B')\} \in \text{large}(A').$$

Then
$\psi_L = \exists U \in \text{large}(\mathbf{Q}) \forall V \in \text{large}(U) \exists A,B \in \text{large}(V)\,(A < B)(\gamma_L(A,B))$.

Note that

$$\gamma_L(A,B) \text{ and } A' \in \text{large}(A) \text{ and } B' \in \text{large}(B) \text{ implies } \gamma_L(A',B'). \qquad (1)$$

Because $\neg\psi_L$ implies $\psi_{\bar{L}}$ we may assume ψ_L. The large set U given by ψ_L contains a copy of $\mathbf{Q}$ and hence we may assume that $U = \mathbf{Q}$. It follows that every large subset V of $\mathbf{Q}$ contains two large subsets $A < B$ so that $\gamma_L(A,B)$. Hence, because of (1), every large subset of $\mathbf{Q}$ contains three large subsets $A < C < B$ so that $\gamma_L(A,B)$.

Let I_n be the set $r_0, r_1, r_2, \ldots, r_{n-1}$. The subset $X = \{x_0, x_1, x_2, \ldots, x_{n-1}\}$ of $\mathbf{Q}$ is an **initial segment** if the function f which maps r_i to x_i is a local isomorphism of $\mathbf{Q}$. For every initial segment $X = \{x_0, x_1, x_2, \ldots, x_{n-1}\}$ let π be the permutation of n so that $x_{\pi(0)} < x_{\pi(1)} < x_{\pi(2)} < \cdots < x_{\pi(n-1)}$.

The initial segment $X = \{x_0, x_1, x_2, \ldots, x_{n-1}\}$ is **well chosen** if there is a large set V_i^n for every $i \in n+1$ so that for all $i,k \in n$, $j \in n+1$ with $i < j$:
(i) $V_i^n < \{x_{\pi(i)}\} < V_{i+1}^n$.
(ii) $\gamma(V_i^n, V_j^n)$.
(iii) $\{x_{\pi(i)}\} \times V_j^n \subseteq L$.
(iv) If $\{x_{\pi(i)}, x_{\pi(k)}\} \in R$ then $\{x_{\pi(i)}, x_{\pi(k)}\} \in L$.
(v) If $x_i = r_{i'}$ and $x_j = r_{j'}$ and $i' < j'$ then $i < j$.

The empty initial segment is well chosen. Hence if there is a well chosen extension for every well chosen initial segment we are done. Let $X = \{x_0, x_1, x_2, \ldots, x_{n-1}\}$ be a well chosen initial segment. Let $v \in n$ be the smallest number so that $r_n < r_{\pi(v)}$ if such a number exists and otherwise let $v = n$.

There are large subsets $A_0 < C < B_0$ of V_v^n so that $\gamma_L(A_0, B_0)$. Because $\gamma_L(C,B)$ there is a large subset $C_0 \subseteq C$ so that $\Gamma_L(a) \cap B$ is a large set for all $a \in C_0$. Because $\gamma_L(C_0, V_{v+1}^n)$ there is a large subset $C_1 \subseteq C_0$ so that $\Gamma_L(a) \cap V_{v+1}^n$ is a large set for all $a \in C_1$. There is a large subset $C_2 \subseteq C_1$ so that $\Gamma_L(a) \cap V_{v+2}^n$ is a large set for all $a \in C_2$. Finally there is a large subset $C_n \subseteq C$ so that for all $a \in C_n$ and all $v < i \leq n$ the intersection of $\Gamma_L(a)$ with V_i^n is large. Let $x_n \in C_n$ to obtain the extension Y of the initial segment X (because there are infinitely many choices for x_n we can make sure that condition (v) is satisfied). Because of condition (iii), condition (iv) is satisfied for Y.

For $i < v$ let $V_i^{n+1} = V_i^n$, let $V_v^{n+1} = A$, let $V_{v+1}^{n+1} = B \cap \Gamma_L(a)$ and let $V_{v+1+i}^{n+1} = V_{v+i}^n$ for all $i \in n - v$. It is now easy to check that Y is a well chosen initial segment. •

A.5.5 Some additional partitioning results

Assume that ω is the set of vertices of the Rado graph R. Let (A,B) be the lexicographic partition of the edges of the Rado graph. Because there is an embedding of $K_{\omega,\omega}$ into R there is no copy R^* of R in R so that all of the edges of R^* are in A or all of the edges of R^* are in B. It is proved in [208] POUZET, SAUER 1996 that for every partition (X, Y) of A there is an order preserving embedding f from R into R so that all of the up edges of $f(R)$ are in X or all of the up edges of $f(R)$ are in Y. And the same for any partition of B. The proof of this is a more involved version of the proof in A.5.4:

The Rado graph is weakly edge-indivisible

• We have to prove that for every finite graph G there is a copy G^* of G in R so that all edges of G^* are in the set A of up edges and that there is a copy G^{**} of G in R so that all of the edges of G^{**} are in the set D of down edges.

Every finite graph has a linear ordering of its vertices so that all of the edges are up edges. This can be seen by induction. Put an arbitrary vertex lowest. Then use induction to order the vertices not adjacent to this lowest vertex and put the ordering of the vertices adjacent to the lowest vertex above.

Every graph H whose vertices can be ordered inot a sequence $v_0, v_1, v_2, \dots$ so that all of its edges are up edges has an order preserving embedding f into the Rado graph R so that every edge of H is mapped to an up edge of R. Assume that embdding f has already been defined for v_i with $i \in n \in \omega$. We wish to extend it by finding an apropriate image of $f(v_n)$. Let F be the numbers less than or equal to $f(n-1)$ and $F_1 := \{f(v_i) \mid i \in n \,\text{and}\, v_n \,\text{is adjacent to}\, v_i\}$. Let T be the orbit in R with $F(T) = F$ and $F_1(T) = F_1$. This orbit has infinitely many elements and hence an element $f(v_n)$ larger than $f(n-1)$.

A similar argument shows that every finite graph has an embedding into the Rado graph which maps all of its edges to down edges. •

A very similar situation exists for the triangle free homogeneous graph H_3. Let ω be the set of vertices of H_3. It is proved in [222] SAUER 1998 that for every partition (X, Y) of A, the set of up edges of H_3, there is an order preserving embedding f from H_3 into H_3 so that all of the up edges of $f(H_3)$ are in X or all of the up edges of $f(H_3)$ are in Y. And the same for any partition of B. The proof is again along the lines of the proof in A.5.4 but contains some considerable technical difficulties. It follows as above that:

The triangle-free homogeneous graph H_3 is weakly edge-indivisible.

The problem of partitioning $[\mathbf{Q}]^n$, the n-element subsets of the rationals $\mathbf{Q}$ is completely solved: [41] DEVLIN 1979. Let $\phi(n)$ be given by $\phi(1) = 1$ and

$$\phi(n) = \sum_{l=1}^{n-1} \binom{2n-2}{2l-1} \phi(l)\phi(n-l).$$

We write

$$\eta \to (\mathbf{Q})^n_{<\omega/\phi(n)}$$

to mean that for every partition of the n-element subsets of $\mathbf{Q}$ into $k \in \omega$ parts there is an order embedding from $\mathbf{Q}$ into $\mathbf{Q}$ which contains n-element subsets of only $\phi(n)$ parts of the partition. Then by [41]:

$$\mathbf{Q} \to (\mathbf{Q})^n_{<\omega/\phi(n)} \quad \text{and} \quad \mathbf{Q} \not\to (\mathbf{Q})^n_{<\omega/(\phi(n)-1)}.$$

This theorem followed by finding a canonical partition of the n-element subsets of

Q into $\phi(n)$ parts.

The function $\phi(n)$ is quite interesting. Let $E(n)$ be the sequence of Entringer or Euler numbers, not Eulerian numbers. Their exponential generating function is

$$\mathcal{E}(x) = \sum_{n=0}^{\infty} E_n \frac{x^n}{n!} = \frac{1}{\cos(x)} + \tan(x).$$

The number of the increasing complete binary trees on $2n+1$ elements is $E_{2n+1} = \phi(n+1)$.

Bibliography

[1] H. L. ABBOTT, D. HANSON 1972 - A problem of Schur and its generalizations. Acta Arithm. vol.20 p.175-187.

[2] Uri ABRAHAM, Robert BONNET 1999 - Hausdorff's theorem for posets that satisfy the finite antichain property. Fundamenta Math. vol.159 p.51-69.

[3] Marc Roland ASSOUS 1974 - Caractérisation du type d'ordre des barrières de Nash-Williams. Public. Dep. Math. Univ. Claude Bernard (Lyon) vol.11 p.89-106.

[4] Heinz BACHMANN 1955 - Transfinite Zahlen; 2nd ed. 1967, Springer Verlag (Berlin), 228 p.

[5] Leonard D. BAUMERT, S.W. GOLOMB 1965 - Backtrack Programming. Journal Assoc. Comput. Machinery vol.12 p.516-524.

[6] J. BAUMGARTNER 1973 - All $\aleph_1$-dense sets of reals can be isomorphic. Fundamenta Math. vol.79 p.101-106.

[7] 1982 - Order types of real numbers and other uncountable orderings. Ordered Sets (Rival ed.) vol.83 p.239-277.

[8] J.L. BELL, A.B. SLOMSON 1969 - Models and ultraproducts: an introduction, 322p. North-Holland (Amsterdam).

[9] Jean-Pierre BENEJAM 1970 - Une remarque sur des théorèmes de Rado et Fraïssé. Comptes rendus vol.270(A) p.1656-1658.

[10] R. D. BERCOV, C. R. HOBBY 1970 - Permutation groups on unordered sets. Math. Zeitschrift vol.115 p.165-168.

[11] Claude BERGE 1970 - Graphes et hypergraphes. Dunod (Paris), 502 p.

[12] Paul BERNAYS 1958 - Axiomatic set theory (Introduction by Abraham Fraenkel), 2nd ed. 1968, 228 p. North-Holland (Amsterdam).

[13] F. BERNSTEIN 1908 - Leipzig Ber. vol.60 p. 329.

[14] Garrett BIRKHOFF 1948 - Lattice Theory.

[15] Andreas BLASS 1977 - A model without ultrafilters. Bulletin Acad. Polon. Sci. vol.25 p.329-331.

[16] J. Adrian BONDY 1991 - A graph reconstructor's manual. London Math. Soc. Lecture Note Ser. 166, p. 221-252. Cambridge University Press.

[17] Robert BONNET 1975 - On the cardinality of the set of initial intervals of a partially ordered set. Infinite sets, Keszthely 1973 vol.3 p.189-198. North-Holland (Amsterdam).

[18] Robert BONNET, Maurice POUZET 1969 - Extension et stratification d'ensembles dispersés. Comptes-rendus vol.268(A) p.1512-1515.

[19] 1982 - Linear extensions of ordered sets, p. 125-170. Ivan Rival ed. Ordered Sets, N.A.T.O. Advanced Study Institutes Series vol.83, Proceedings A.S.I. Banff, Canada 1981 (966 p.).

[20] Robert BONNET, Saharon SHELAH 1985. Narrow Boolean Algebras. Annals of Pure and Applied Logic vol.28 p.1-12.

[21] K. BÖRÖCKY Jr., Norbert W. SAUER, X. ZHU 1993 - Inexhaustible homogeneous structures. Discrete Maths. vol.115 p.57-63. North-Holland (Amsterdam).

[22] Y. BOUDABBOUS, Gérard LOPEZ 1995 - La relation différence et l'anti-isomorphie. Math. Logic Quart. vol.41 p.268-280.

[23] Jean-Pierre CALAIS 1967 - Relation et multirelation pseudo-homogènes. Comptes rendus vol.265(A) p.2-4.

[24] Peter J. CAMERON 1976 - Transitivity of permutation groups on unordered sets. Math. Zeitschrift vol.148 p.127-139.

[25] 1981 - Finite permutation groups and finite simple groups. Bulletin London Math. Soc. vol.13 p.1-22.

[26] 1981 - Orbits of permutation groups on unordered sets, II. Journal London Math. Soc. (2) vol.23 p.249-264.

[27] 1983 - Orbits of permutation groups on unordered sets, III: Imprimitive groups; IV: Homogeneity and transitivity. Journal London Math. Soc. vol.27 p.229-247.

[28] 1987 - Some treelike objects. Quart. Journal Math. Oxford (2) vol.38 p.155-183.

[29] 1990 - Oligomorphic permutation groups. London Math. Soc. Lecture Notes, C.U.P. (Cambridge).

[30] Chen-Chung CHANG, Jerome KEISLER 1977 - Model Theory. Studies in Logic vol.73, 554 p. North-Holland (Amsterdam).

[31] Claude CHAUNIER, Nik LYGEROS 1992 - The number of orders with thirteen elements. Order vol.9 p.203-204.

[32] G. CHERLIN 1998 - The classification of countable homogeneous directed graphs and countable homogeneous n-tournaments. Mem. Amer. Math. Soc. 131, no.621, xiv + 161p.

[33] Fan Rong CHUNG 1973 - On the Ramsey numbers N(3,3,...,3;2). Discrete Math. vol.5 p.317-321.

[34] David CLARK, Peter KRAUSS 1970 - Universally complete universal theories III. Notices Amer. Math. Soc. vol.17 p.1019.

[35] Paul J. COHEN 1966 - Set theory and the continuum hypothesis, 154 p. Benjamin (New-York).

[36] Ernest COROMINAS 1984 - On better quasi ordering countable trees. Discrete Math., Actes Conférence Ensembles Ordonnés et applications, L'Arbresle 1982.

[37] Martin DAVIS 1965 - The Undecidable, 440 p. Raven Press Hewlett (New-York).

[38] Richard DEDEKIND 1888 - Was sind und was sollen die Zahlen? Braunschweig.

[39] D. H. J. DE JONGH, R. PARIKH 1977 - Well partial orderings and Hierarchies. Indag. Math. vol.39 p.195-206.

[40] Peter DEMBOWSKI 1968 - Finite Geometries. Ergebnisse der Math. und ihrer Dengebiete vol.44, 375 p. Springer Verlag (Berlin, New-York).

[41] Denis DEVLIN 1979 - Some partition theorems and ultrafilters on ω. Ph. D. Thesis, Dartmouth College, mimeog.

[42] Keith J. DEVLIN, Havard JOHNSBRĂTEN 1974 - The Suslin Problem. Lecture Notes in Math. vol.405, Springer Verlag (Berlin).

[43] R.P. DILWORTH 1950 - A decomposition theorem for partially ordered sets. Annals of Math. vol.51 p.161-166.

[44] R.P. DILWORTH, A.M. GLEASON 1962 - A generalized Cantor theorem. Proceedings Amer. Math. Soc. vol.13 p.704-705.

[45] Ben DUSHNIK, E.W. MILLER 1940 - Concerning similarity transformations of linearly ordered sets. Bulletin Amer. Math. Soc. vol.46 p.322-326.

[46] 1941 - Partially ordered sets. Amer. Journal Math. vol.63 p.600-610.

[47] Erik ELLENTUCK 1974 - A new proof that analytic sets are Ramsey. Journal Symb. Logic vol.39 p.163-165.

[48] M. EL-ZAHAR, Norbert W. SAUER 1989 - The indivisibility of the homogeneous K_n-free graphs. Journal of combinatorial Theory (B), vol.47 no.2.

[49] 1991 - Ramsey-type properties of relational structures. Discrete Math. vol.115 p.57-63. North-Holland (Amsterdam).

[50] 1993 - On the divisibility of homogeneous directed graphs. Canadian Journal Math. vol.45(2) p.284-294.

[51] 1994 - On the divisibility of homogeneous hypergraphs. Combinatorica vol.14 (2) p.1-7.

[52] Paul ERDÖS 1942 - Some set-theoretical properties of graphs. Revista Univ. Nacion. Tucuman (A) vol.3 p.363-367.

[53] Paul ERDÖS, A. HAJNAL, L. PÓSA 1975 - Strong embeddings of graphs into colored graphs. Colloquia Mathematica Societatis János Bolyai vol.10, Infinite and finite sets, p.585-595, edited by A. Hajnal, Richard Rado, Vera T. Sós.

[54] Paul ERDÖS, Richard RADO 1952 - Combinatorial theorems on classifications of subsets of a given set. Proceedings London Math. Soc. vol.3 p.417-439.

[55] 1953 - A problem on ordered sets. Journal London Math. Soc. vol.28 p.426-438.

[56] 1956 - A partition calculus in Set Theory. Bulletin Amer. Math. Soc. vol.62 p.427-489.

[57] 1959 - Partition relations connected with the chromatic number of graphs. Journal London Math. Soc. vol.34 p.63-72.

[58] 1960 - A construction of graphs without triangles bearing pre-assigned order and chromatic number. Journal London Math. Soc. vol.35 p.445-448.

[59] Robert ERRA, Nik LYGERŌS, Nigel STEWART - On minimal strings containing by decimation the elements of the symmetric group. The Electronic Journal of Combinatorics.

[60] G. EXOO 1989 - A lower bound for R(5,5). Journal Graph Theory vol.13 p.97-98.

[61] 1994 - A lower bound for Schur numbers and multicolor Ramsey numbers of K_3. The Electronic Journal of Combinatorics, http://www.combinatorics.org/, ♯ R 8,1 (3 pages).

[62] Solomon FEFERMAN 1965 - Some applications of the notions of Forcing and Generic Sets. Fundamenta Math. vol.56 p.325-345.

[63] Ulrich FELGNER, J.K. TRUSS 1999 - The independence of the prime ideal theorem from the order-extension principle. Journal of Symb. Logic vol.64 p.199-215.

[64] Istvan FOLDES 1973 - Relations denses et dispersées: extension d'un théorème de Hausdorff. Comptes rendus vol.277(A) p.269-271

[65] J. FOLKMAN 1974 - Note on the Ramsey number N(3,3,3,3). Journal Combin. Theory (A) vol.16 p.371-379.

[66] W.L. FOUCHÉ, P.H. POTGIETER 1998 - Kolmogorov complexity and symmetric relational structures. Journal of Symbolic Logic vol.63 number 3, p.1083-1094.

[67] Abraham FRAENKEL 1925 - Untersuchungen über die Grundlagen der Mengenlehre. Math. Zeitschrift vol.22 p.250-273.

[68] 1953 - Abstract set theory; 3rd ed. 1966, 297 p. North-Holland (Amsterdam).

[69] Roland FRAÏSSÉ 1954 - Sur l'extension aux relations de quelques propriétés des ordres. Annales Scientifiques de l'Ecole Normale Supérieure, vol.71, p.363-388.

[70] 1954 - Sur quelques classifications des systèmes de relations (Doctoral Dissert. Paris 1953), Alger-Math. vol.1 p.35-182.

[71] 1956 - Etude de certains opérateurs dans les classes de relations, définis à partir d'isomorphismes restreints. Zeitschrift für Mathematische Logik und Grundlagen der Mathematik, vol.2 p.59-75; followed by: Application des γ-opérateurs au calcul logique du premier échelon. Ibid. p.76-92.

[72] 1958 - Un modèle définissant une Théorie aberrante des ensembles où sont niés les axiomes du choix et d'extensionalité. Alger-Math. vol.5 p.17-98.

[73] 1966 - Une généralisation de l'ultraproduit. Journal Sympolic Logic vol.31 p.235-244.

[74] 1971 to 1975 - Cours de Logique mathématique, vol.1 (1971) Relation et formule logique, 197 p.; vol.2 (1972) Théorie des modèles, 177 p.; vol.3 (1975) Récursivité et constructibilité, 137 p.; Gauthier-Villars (Paris). English translation: Course of Mathematical Logic, vol.1 (1973) 186 p.; vol.2 (1974) 192 p. (Synthese Library vol.54 and 69); Reidel (Dordrecht).

[75] 1982 - La zérologie, une recherche aux fronfières de la logique et de l'art; application à la logique des relations de base vide. International Logic Review, vol.26 p.67-79 (Bologna).

[76] 1984 - L'intervalle en théorie des relations; ses généralisations; filtre intervallaire et clôture d'une relation. Orders, Descriptions and Roles (ed. Pouzet, Richard) p.313-342. North-Holland (Amsterdam).

[77] 1986 - Theory of Relations, also denoted ToR-86. Studies in Logic vol.118, 397 p. North-Holland (Amsterdam).

[78] Roland FRAÏSSÉ, Walied HAZIM SHARIF 1993 - L'extension respectueuse à hauteur constante entre posets finis. Comptes rendus Série I vol.316 p.637-642.

[79] Roland FRAÏSSÉ, Gérard LOPEZ 1991 - La reconstruction d'une relation dans l'hypothèse forte: isomorphie des restrictions à chaque partie stricte de la base. Séminaire de Mathématiques Supérieures vol.109, 139 p. Les Presses de l'Université de Montréal.

[80] Roland FRAÏSSÉ, Nik LYGERŌS 1991 - Petits posets: dénombrement, représentabilité par cercles et "compenseurs". Comptes rendus Série I vol.313 p.417-420.

[81] Roland FRAÏSSÉ, Maurice POUZET 1971 - Interprétabilité d'une relation par une chaîne. Comptes rendus vol.272(A) p.1624-1627.

[82] 1971 - Sur une classe de relations n'ayant qu'un nombre fini de bornes. Ibid. vol.273(A) p.275-278.

[83] Claude FRASNAY 1963 - Groupes de permutations finies et familles d'ordres totaux; application à la théorie des relations. Comptes rendus vol. 257(A) p.2944-2947.

[84] 1964 - Groupes de compatibilité de deux ordres totaux; application aux n-morphismes d'une relation n-aire. Comptes rendus vol.259(A) p.3910-3913.

[85] 1965 - Quelques problèmes combinatoires concernant les ordres totaux et les relations monomorphes. Doctoral Dissert. Paris. Annales Institut Fourier Grenoble vol.15 p.415-524.

[86] 1973 - Interprétation relationniste du seuil de recollement d'un groupe naturel. Comptes rendus vol.277(A) p.865-868.

[87] 1984 - Relations enchaînables, rangements et pseudo-rangements. Orders, Descriptions and Roles, Pouzet, Richard ed., p.235-268. North-Holland (Amsterdam).

[88] 1990 - Détermination du degré optimal d_m de monomorphie pour les structures relationnelles au plus m-aires. Math. Rep. Acad. Sci. Canada vol.12 (4) p.141-146.

[89] H. FREDRICKSON 1979 - Schur numbers and the Ramsey numbers N(3,3,...,3;2). Journal Combin. Theory (A) vol.27 p.376-377.

[90] Fred GALVIN 1968 - A generalization of Ramsey's theorem. Notices Amer. Math. Soc. vol.15 p.548. Abstract 68 T 368.

[91] Fred GALVIN, Karel PRIKRY 1973 - Borel sets and Ramsey's theorem. Journal Symb. Logic vol.38 p.193-198.

[92] A. GARDINER 1976 - Homogeneous graphs. Journal combin. Theory(B) vol.20 p.94-102.

[93] Guy Roger GIRAUD 1969 - Majoration du nombre de Ramsey ternaire bicolore en (4,4). Comptes rendus vol.269(A) p.620-622.

[94] Michèle GIRAUDET, Dugald MACPHERSON, Norbert SAUER 1994 - Set-Homogeneous Graphs. Journal Combin. Theory (B), vol.62 p.63-95.

[95] André GLEYZAL 1940 - Order types and structure of orders. Transactions Amer. Math. Soc. vol.48, p.451-466.

[96] A.M. GLEASON, R.E. GREENWOOD 1955 - Combinatorial relations and monochromatic graphs. Canadian Journal Math. vol.7 p.1-7.

[97] Kurt GŐDEL 1931 - Uber formal unentscheidbare Sätze der Principia Mathematica und verwandter Systeme. Monats. Math. Phys. vol.38 p.173-198.

[98] 1940 - The consistency of the axiom of choice and of the generalized continuum hyopothesis with the axioms of set theory, 69 p. Princeton Univ. Press.

[99] R. J. GRAHAM, B. L. ROTHSCHILD, J. H. SPENCER 1990 - Ramsey Theory. John Wiley and sons (New-York).

[100] J.E. GRAVER, James YACKEL 1968 - Some graph theoretic results associated with Ramsey's theorem. Journal Combin. Theory vol.4 p.125-175.

[101] C.M. GRINSTEAD, S.M. ROBERTS 1982 - On the Ramsey numbers R(3,8) and R(3,9). Journal Combin. Theory (B) vol.33 p.27-51.

[102] Jean Guillaume HAGENDORF 1971 - Deux problèmes de représentation de l'abritement entre relations. Comptes rendus vol.273(A) p.71-72.

[103] 1972 - Extensions immédiates de chaînes et de relations. Comptes rendus vol.274(A) p.607-609.

[104] 1977 - Extensions de chaînes. Doctoral Diss. Univ. Provence (Marseille), mimeog.

[105] James D. HALPERN, Azriel LEVY 1971 - The boolean prime ideal theorem does not imply the axiom of choice. Axiomatic Set Theory vol.1 p.83-134. Proceedings Symposium Pure Math. Amer. Math. Soc., Los Angeles 1967.

[106] F. HARTOGS 1915 - Uber das Problem der Wohlordnung. Math. Annalen vol.76 p.436-443.

[107] Felix HAUSDORFF 1914 - Grundzüge der Mengenlehre, Leipzig; 2nd printing 1955, Chelsea (New-York).

[108] C. Ward HENSON 1972 - Countable homogeneous relational structures and $\aleph_0$-categorical theories. Journal Symbolic Logic vol.37 p.494-500.

[109] G. HESSENBERG 1906 - Grundbegriffe der Mengenlehre. Abhandlungen der Friesschen Schule N.S. vol.1.4 n.75, Göttingen; citation in BACHMANN 1967 p.107-112.

[110] Graham HIGMAN 1952 - Ordering by divisibility in abstract algebras. Proceedings London Math. Soc. vol.3 p.326-336.

[111] 1977 - Homogeneous relations. Quarterly Journal Math. vol.28 p.31-39.

[112] R. HILL, R. W. IRVING 1982 - On Group Partitions Associated with Lower Bounds for Symmetric Ramsey Numbers. European Journal of Combin. vol.3 p.35-50.

[113] T. HIRAGUCHI 1951 - On the dimension of partially ordered sets. Science Reports Kanazawa Univ. vol.1 p.77-94.

[114] 1955 - On the dimension of orders. Ibid. vol.4 p.1-20.

[115] Joram HIRSCHFELD, William H. WHEELER 1975 - Forcing, arithmetic, division rings. Lecture Notes in Math. vol.454. Springer Verlag (Berlin).

[116] Wilfrid HODGES 1993 - Model theory. Cambridge University Press. See also: 1997 - A shorter Model Theory

[117] Wilfrid HODGES, A. H. LACHLAN, Saharon SHELAH 1977 - Possible orderings of an indiscernible sequence. Bulletin London Math. Soc. vol.9 p.212-215.

[118] Pierre ILLE 1990 - Clôture intervallaire et extension logique d'une relation. Zeitschrift Math. Logik vol.36 p.217-227.

[119] 1991 - L'ensemble des intervalles d'une multirelation binaire et réflexive. Ibid. vol.37 p.227-256.

[120] 1993 - Recognition problem in reconstruction for decomposable relations. Finite and Infinite Combinatorics in Sets and Logic (ed. Sauer et al.) p.189-198.

[121] Pierre ILLE, Xavier RAMPON 1998. Reconstruction of posets with the same comparability graph. Journal Comb. Theory (B) vol.74 p.368-377.

[122] J.R. ISBELL 1969 - $N(4,4;3) \geq 13$. Journal Combin. Theory vol.6 p.210.

[123] Michel JEAN 1967 - Sur deux classes universelles de relations. Comptes rendus vol.264 p.591-593.

[124] 1969 - Line-symmetric tournaments. Recent Progress in Combinatorics, p. 265-271. Academic Press (New-York).

[125] Thomas J. JECH 1973 - The axiom of choice, 202 p. Studies in Logic. North-Holland (Amsterdam).

[126] 1978 - Set theory, 621 p. Academic Press (New-York).

[127] Bjarni JONSSON 1965 - Extensions of relational structures. The Theory of Models, Proceedings Symposium Berkeley 1963 p.146-157. Second printing 1970.

[128] Camille JORDAN 1872-73 - Sur la limite de transitivité des groupes non alternés. Bulletin Soc. Math. France vol.1 p.40-71. Oeuvres Complètes 1961 vol.1 p.365-396, Gauthier-Villars (Paris).

[129] Pierre JULLIEN 1966 - Sur quelques problèmes de la théorie des chaînes permutées. Comptes rendus vol.262(A) p.205-208.

[130] 1969 - Contribution à l'étude des types d'ordres dispersés. Doctoral Diss. Univ. Provence, Marseille. Mimeog. Univ. Saint-Martin d'Hères (Grenoble).

[131] J.G. KALBFLEISCH 1967 - Upper bounds for some Ramsey numbers. Journal Combin. Theory vol.2 p.35-42.

[132] William M. KANTOR 1972 - On incidence matrices of finite projective and affine spaces. Math. Zeitschrift vol.124 p.315-318.

[133] G. KERY 1964 - Ramsey egy grafelmeléti tététeröl. Mat. Lapok vol.15 p.204-224.

[134] Dieter KLAUA 1959 - Transfinite reelle Zahlenräume. Wissensch. Humboldt Univ. Berlin Math. Nat. Reihe vol.9 p.169.

[135] 1960 - Zur Struktur der reellen Ordnungszahlen. Zeitschrift Math. Logik vol.6.

[136] Bronislaw KNASTER 1928 - Un théorème sur les fonctions d'ensembles. Ann. Soc. Polon. Math. vol. 70 p. 507-520.

[137] Simon KOCHEN 1958 - Filtration systems. Notices Amer. Math. Soc. vol.5 p.605.

[138] P. KOMJÁTH, V. RÖDL 1986 - Coloring of universal graphs. Graphs and Combinatorics vol.2, p.55-60.

[139] Tat-Yung KONG, Paulo RIBENBOIM 1994 - Chaining partially ordered sets. Comptes rendus vol. 319 p. 533-537.

[140] Peter KRAUSS 1971 - Universally complete universal theories. Zeitschrift Math. Logik vol.17 p.351-370.

[141] D. L. KREHER, Wei LI, Stanislaw P. RADZISZOWSKI 1988 - Lower Bounds for Multi-Colored Ramsey Numbers From Group Orbits. Journal Combin. Math. and Combin. Computing vol.4 p.87-95.

[142] Joseph B. KRUSKAL 1960 - Well quasi-ordering, the tree theorem and Vazsonyi's conjecture. Transactions Amer. Math. Soc. vol.95 p.210-225.

[143] Kazimierz KURATOWSKI 1966 - Topology vol. I , 550 p. ed. PWN (Warszawa) and Academic Press (New-York).

[144] Djuro KUREPA 1935 - Ensembles ordonnés et ramifiés. Doctoral Dissert. Paris. Publications Math. Univ. Beograd vol.4 p.1-138.

[145] 1952 - On a characteristic property of finite sets. Pacific Journal Math. vol.2 p.323-326.

[146] A.H. LACHLAN 1984 - Countable homogeneous tournaments. Transactions Amer. Math. Soc. vol.284 p.431-461.

[147] 1984 - On countable stable structures which are homogeneous for a finite relational language. Israël Journal Math. vol.49 p.69-153.

[148] A.H. LACHLAN, Robert WOODROW 1980. Countable ultrahomogeneous undirected graphs. Transactions Amer. Math. Soc. vol.262 p.51-94.

[149] Jean A. LARSON 1978 - A solution for scattered order types of a problem of Hagendorf. Pacific Journal Math. vol.74 p.373-379.

[150] Hans LÄUCHLI 1964 - The independence of the ordering principle from a restricted axiom of choice. Fundamenta Math. vol.54 p.31-43.

[151] Richard LAVER 1968 - Fraïssé's conjecture. Notices Amer. Math. Soc. vol.15 p.934-935.

[152] 1971 - On Fraïssé's order type conjecture. Doctoral Diss. Univ. of California, Berkeley 1969. Annals of Math. vol.93 p.89-111.

[153] 1973 - An order type decomposition theorem. Annals of Mathematics vol.98 p.96-119.

[154] 1978 - Better quasi-orderings and a class of trees. Studies in Foundat. and Combin. Advances in Math. vol.1 p.31-48. Academic Press (New-York).

[155] J.H. van LINT, R.M. WILSON 1992 - A course in combinatorics, 530 p. Cambridge University Press.

[156] D. LIVINGSTONE, A. WAGNER 1965 - Transitivity of finite permutation groups on unordered sets. Math. Zeitschrift vol.90 p.393-403.

[157] Gérard LOPEZ 1969 - Une propriété des automorphismes d'une relation. Comptes rendus vol.269(A) p.7-8.

[158] 1973 - Problème d'extension respectueuse. Comptes rendus vol.277(A) p.567-569.

[159] 1977 - La p-extensivité d'un tableau bivalent pour $p \leq 5$. Comptes rendus vol.284(A) p.1245-1248.

[160] 1978 - L'indéformabilité des relations et multirelations binaires. Zeitschrift Math. Logik vol.24 p.303-317.

[161] 1983 - Reconstruction d'une S-expansion. Zeitschrift Math. Logik vol.29 p. 11-24.

[162] 1983 - Une nouvelle preuve du théorème de Galvin utilisant les ensembles-Ramsey. Comptes rendus vol.297(ser. I) p.275-277.

[163] Gérard LOPEZ, Claire RAUZY 1992 - Reconstruction of binary relations from their restrictions of cardinality 2, 3, 4 and $(n-1)$. Zeitschrift Math. Logik vol.38 p.157-168.

[164] H.D. MACPHERSON 1985 - Orbits of infinite permutation groups. Proceedings London Math. Soc.(3) vol.51 p.246-284.

[165] H.D. MACPHERSON, Maurice POUZET, Robert E. WOODROW 1992 - Countable structures of given age. Journal Symb. Logic vol.57, p.992-1010.

[166] Jerome MALITZ 1967 - On extending multirelations. Notices Amer. Math. Soc. vol.14 p.828.

[167] A.R.D. MATHIAS 1974 - The order extension principle. Axiomatic Set Theory vol.2, Proceedings Symposium Pure Math. Amer. Math. Soc. Los-Angeles 1967 p.179-183.

[168] Brendan D. Mc KAY, Stanislaw P. RADZISZOWSKI 1991 - The first classical Ramsey number for Hypergraphs is computed. Proceedings of the Second Annual Symposium on Discrete Algorithms, San Francisco, p.304-308.

[169] 1992 - A new upper bound for the Ramsey number R(5,5). Australasian Journal Comb. vol.5 p.13-20.

[170] 1995 - R(4,5) = 25 . Journal Graph Theory vol. 19 p.309-322.

[171] Eric C. MILNER 1981 - On the decomposition of partially ordered sets into directed sets. Proceedings Intern. Math. Conference Singapore. Math. Studies vol.74 p.85-90. North-Holland (Amsterdam).

[172] 1985 - Basic wqo and bqo theory. Graphs and Orders, Rival ed., p.487-502. D. Reidel Publishing Co (Dordrecht).

[173] Eric C. MILNER, Maurice POUZET 1982 - The cofinality of partially ordered sets in ordered sets. I. Rival ed., A.S.I. series vol.83, p.279-298. Reidel (Dordrecht).

[174] MISLOVE 1984 - Orders: description and roles. Pouzet, Richard, ed. Annals of Discrete Maths. vol.23, p.61-80.

[175] J. Donald MONK (with the cooperation of Robert BONNET) 1989 - Handbook of Boolean Algebras, vol.3. North-Holland (Amsterdam).

[176] Andrzej MOSTOWSKI 1948 - On the principle of dependent choices. Fundamenta Math. vol.35 p.127-130.

[177] C. St. NASH-WILLIAMS 1965 - On well quasi-ordering transfinite sequences. Proceedings Cambridge Phil. Soc. vol.61 p.33-39.

[178] 1965 - On well quasi-ordering infinite trees. Ibid. p.697-720.

[179] 1968 - On better quasi-ordering transfinite sequences. Ibid. vol.64 p.273-290.

[180] John von NEUMANN 1929 - Uber eine Widerspruchsfreiheitsfrage in der axiomatischen Mengenlehre. Journal reine und angew. Math. vol.160 p.227-241.

[181] Jean-François PABION 1972 - Relations pré-homogènes. Comptes rendus vol.274(A) p.529-531

[182] Giuseppe PEANO 1894-1908 - Formulaire de mathématiques; Introduction and 5 volumes. Torino.

[183] PELCZYNSKI, SEMADENI 1959 - Spaces of continuous functions. Studia Math. vol.18, p. 211-222.

[184] Micha A. PERLES 1963 - A proof of Dilworth's decomposition theorem for partially ordered sets. Israël Journal Math. vol.1 p.105-107.

[185] 1963 - On Dilworth's theorem in the infinite case. Ibid. p.108-109.

[186] David PINCUS 1972 - Zermelo-Fraenkel consistency results by Fraenkel-Mostowski methods. Jounal Symb. Logic vol.37 p.721-743.

[187] K. PIWAKOWSKI, S. RADZISZOWSKI 1998 - Journal of Comb. Mathematics and Comb. Computing vol.27 p. 135-141.

[188] Maurice POUZET 1969 - Généralisation d'une construction de Ben Dushnik - E.W. Miller. Comptes rendus vol.269(A) p.877-879.

[189] 1970 - Sur des conjectures de Fraïssé et les prémeilleurs ordres. Comptes rendus vol.270(A) p.1-3.

[190] 1970 - Algèbre ordinale prémeilleur ordonnée. Ibid. p.300-303.

[191] 1972 - Modèle universel d'une théorie n-complète. Comptes rendus vol.274(A) p.433-436; modèle uniformément préhomogène, p.695-698; modèle préhomogène, p.813-816.

[192] 1972 - Sur les prémeilleurordres. Annales Inst. Fourier (Grenoble) vol.22 p.1-20.

[193] 1972 - Le belordre d'abritement et ses rapports avec les bornes d'une multirelation. Comptes rendus vol.274(A) p.1677-1680.

[194] 1973 - Une remarque sur la libre-interprétabilité d'une relation par une autre. Algebra Universalis vol.3 p.67-71.

[195] 1976 - Application d'une propriété combinatoire des parties d'un ensemble aux groupes et aux relations. Math. Zeitschrift vol.150 p.117-134.

[196] 1978 - Note sur certains tournois reconstructibles; application à leurs groupes d'automorphismes. Discrete Math. vol.24 p.225-229.

[197] 1979 - Relation minimale pour son âge. Zeitschrift Math. Logik vol.25 p.315-344.

[198] 1979 - Relations non reconstructibles par leurs restrictions. Journal Combin. Theory (B) vol.26 p.22-34.

[199] 1979 - Chaînes d'un ensemble partiellement bien ordonné. Pub. Dept. Math. Lyon vol.16 p.21-26.

[200] 1981 - Relation impartible. Dissertationes vol.103 p.1-48.

[201] 1981 - Application de la notion de relation presque-enchaînable au dénombrement des restrictions finies d'une relation. Zeitschrift Math. Logik vol.27 p.289-332.

[202] 1981 - Ordered sets (Rival ed.), Problem sessions, Lattices. Nato A.S.I. series, Reidel, Dordrecht.

[203] 1985 - Applications of well quasi-ordering and better quasi-ordering in Graphs and Orders. I. Rival, ed. A.S.I. series vol.147 p.503-519. Reidel (Dordrecht).

[204] 1993 - Graphs and Posets with no Infinite Independent Set. Finite and Infinite Combinatorics in Sets and Logic, p.313-335. Kluver Academic Publishers.

[205] Maurice POUZET, Ivo ROSENBERG 1986 - Sperner properties for groups and relations. European Journal of Combinatorics vol.7 p.349-370.

[206] Maurice POUZET, B. ROUX 1996 - Ubiquity in category for metric spaces and transition systems. European Journal of Combinatorics vol.17 p.291-307.

[207] Maurice POUZET, N. ZAGUIA 1985 - Dimension de Krull des ensembles ordonnés. Discrete Math vol.53 p.173-192.

[208] Maurice POUZET, Norbert W. SAUER 1996 - Edge partitions of the Rado graph. Combinatorica vol.16 p.1-16.

[209] William C. POWELL 1975 - Extending Gődel's negative interpretation to Z.F. Journal Symbolic Logic vol. 40 p. 221-229.

[210] Richard RADO 1949 - Axiomatic treatment of rank in infinite sets. Canadian Journal of Math. vol.1 p. 337-343.

[211] 1954 - Partial well-ordering of sets of vectors. Mathematika vol.1 p.89-95.

[212] Stanislaw P. RADZISZOWSKI 1994 - Small Ramsey Numbers. Technical Report Rochester Institute of Technology (1993). The Electronic Journal of Combinatorics; 28 p.

[213] Frank P. RAMSEY 1930 - On a problem of formal logic. Proceedings London Math. Soc. (2) vol.30 p.264-286.

[214] Claire RAUZY 1986 - Sur l'extensivité des tableaux bivalents à deux colonnes. Comptes rendus vol.303 (Série I , n.15) p.721-724.

[215] Abraham ROBINSON 1963 - Introduction to model theory and to metamathematics of algebra; 284 p. Studies in Logic, North-Holland (Amsterdam).

[216] B. ROSE, Robert WOODROW 1981. Ultrahomogeneous structures. Zeitschrift Math. Logik vol.27 p. 23-30.

[217] Joseph G. ROSENSTEIN 1982 - Linear orderings, 487 p. Academic Press (New-York).

[218] Herman RUBIN, Jean RUBIN 1963 - Equivalents of the axiom of choice. Studies in Logic, North-Holland (Amsterdam).

[219] W. RUDIN 1957 - Continuous functions on compact spaces without perfect subsets. Proceedings Amer. Math. Soc. vol.8, p. 39-42.

[220] Adolfo SANCHEZ-FLORES 1995 - An improved upper bound for Ramsey number N(3,3,3,3;2). Discrete Math. vol.140 p.281-286.

[221] Norbert W. SAUER Coloring finite substructures of countable structures. Bolyai Society Mathematical Studies, vol.11.

[222] 1998 - Edge partitions of the countable triangle free homogeneous graph. Discrete Math. vol.185 p.137-181.

[223] J. SCHMERL 1979 - Countable homogeneous partially ordered sets. Algebra Universalis vol.9 p.317-321.

[224] J. SCHUR 1916 - Uber die Kongruenz $xm+ym = zm(\mathrm{mod}\,p)$. Jahr. Deutsch. Math. Verein. vol.25 p.114-117.

[225] Dana SCOTT 1955 - Definition by abstraction in axiomatic set theory. Bulletin Amer. Math. Soc. vol.61 p. 442 (abstract).

[226] J.C. SHEPHERDSON 1952 - Inner models for set theory. Journal Symb. Logic vol.17 p.225-237.

[227] Waclaw SIERPINSKI 1933 - Sur un problème de la théorie des relations. Annali Scuola Norm. Sup. Pisa vol.2 p.285-287.

[228] 1947 - Sur les types d'ordre de puissance du continu. Revista de Ciencias Univ. Nac. Mayor San Marcos, Lima (Peru) n.456 p.305-307.

[229] 1950 - Sur les types ordinaux des ensembles linéaires. Atti Accad. Naz. Lincei vol.8 p.427-428.

[230] 1958 - Cardinal and ordinal numbers. Monog. Mat. n.34, Polska Akad. Nauk, Warszawa, 487 p., second ed. 1965, 491 p.

[231] Roman SIKORSKI 1948 - On an ordered algebraic field. Soc. Sci. Lett. Varsovie C.R. Class III, Sci. Math. Phys. vol.41 p.69-96.

[232] Jack SILVER 1970 - Every analytic set is Ramsey. Journal Symb. Logic vol.35 p.60-64.

[233] Harold SIMMONS 1976 - Large and small existentially closed structures. Journal Symb. Logic vol.41 p.379-390.

[234] Ernst SPECKER 1956 - Teilmengen von Mengen mit Relationen. Comment. Math. Helvet. vol.31 p.302-314.

[235] Emanuel SPERNER 1928 - Ein Satz über Untermengen einer endlichen Menge. Math. Zeitschrift vol.27 p.544-548.

[236] Dennis STANTON 1990 - Unimodality and Young's lattice. Journal Combin. Theory (A) vol.54 p.41-45.

[237] P.K. STOCKMEYER 1977 - The falsity of the reconstruction conjecture for tournaments Journal Graph Theory vol.1 p.19-25.

[238] Patrick SUPPES 1960 - Axiomatic set theory, 265 p. Van Nostrand (Princeton and Toronto).

[239] Michail SUSLIN 1920 - Problème 3. Fundamenta Math. vol.1 p.223.

[240] E. SZPILRAJN 1930 - Sur l'extension de l'ordre partiel. Fundamenta Math. vol.16 p.386-389.

[241] Alfred TARSKI 1924 - Sur les ensembles finis. Fundamenta Math. vol.6 p. 45-95

[242] 1938 - Unerreichbare Kardinalzahlen. Ibid. vol.30 p.68-89.

[243] 1954 - Theorems on the existence of successors of cardinals, and the axiom of choice. Proceedings Konin. Nederl. Akad. Wetenschappen vol. A 57, and Indag. Math. vol.16 p.26-32 (Amsterdam).

[244] 1955 - A lattice-theoretical fixpoint theorem and its applications. Pacific Journal of Math. vol.5 p. 285-309.

[245] Stéphane THOMASSÉ 1993 - Belordre des séries parallèles dénombrables. Comptes rendus Série I vol.317 p.909-912.

[246] 1997 - Indivisibility and aleph-morphisms. European Journal Comb. vol.18 p.445-454.

[247] Stanislas M. ULAM 1960 - A collection of mathematical problems. Interscience Tracts in pure and appl. Math. vol.8, Interscience Publishers (Groningen), 150 p.

[248] Robert L. VAUGHT 1953 - Remarks on universal arithmetical classes. Bulletin Amer. Math. Soc. vol.59 p.391.

[249] 1961 - Denumerable models of complete theories. Infinistic Methods p.303-321. Pergamon Press (Oxford).

[250] Robert WOODROW 1979 - There are four countable ultrahomogeneous graphs without triangles. Journal Comb. Theory (B) vol.27 p. 168-179.

[251] Ernst ZERMELO 1904 - Beweis dass jede Menge wohlgeordnet werden kann. Math. Annalen vol.59 p.514-516.

[252] 1908 - Untersuchungen über die Grundlagen der Mengenlehre. Math. Annalen vol. 65 p.261-281.

[253] 1930 - Uber Grenzzahlen und Mengen bereiche. Fundamenta Math. vol.16 p.29-47.

Index

D

E

H

I

N

O

P

S

Y

Z